# PATHOBIOLOGY OF CELL MEMBRANES

VOLUME II

## CONTRIBUTORS

John U. Balis
M. Bessis
Janet V. Collins
A. Demsey
P. M. Grimley
Hal K. Hawkins
T. Katsuyama
M. Locke
Reginald G. Mason
Stephen R. Max
N. Mohandas
G. William Moore
Norman B. Ratliff
Urs N. Riede
David H. Rifenberick
Walter A. Sandritter
J. A. V. Simson
S. S. Spicer
Kyuichi Tanikawa
Kenneth R. Wagner
R. I. Weed

# PATHOBIOLOGY OF CELL MEMBRANES

## *Volume II*

*Edited by*

*BENJAMIN F. TRUMP*, M.D.

*Department of Pathology*
*University of Maryland*
*School of Medicine*
*Baltimore, Maryland*

*ANTTI U. ARSTILA*, M.D.

*Department of Cell Biology*
*University of Jyväskylä*
*Jyväskylä, Finland*
*and*
*Department of Pathology*
*University of Maryland*
*School of Medicine*
*Baltimore, Maryland*

1980

ACADEMIC PRESS
*Subsidiary of Harcourt Brace Jovanovich, Publishers*

New York London Toronto Sydney San Francisco

ACADEMIC PRESS, INC.
111 Fifth Avenue, New York, New York 10003

*United Kingdom Edition published by*
ACADEMIC PRESS, INC. (LONDON) LTD.
24/28 Oval Road, London NW1 7DX

**Library of Congress Cataloging in Publication Data**
Main entry under title:

Pathobiology of cell membranes.

Includes bibliographies and indexes.
1. Pathology, Cellular. 2. Cell membranes.
I. Trump, Benjamin F. II. Arstila, Antti U.
RB25.P39 611'.0181 74-27793
ISBN 0-12-701502-7 (v. 2)

PRINTED IN THE UNITED STATES OF AMERICA

80 81 82 83 9 8 7 6 5 4 3 2 1

# CONTENTS

# LIST OF CONTRIBUTORS

Numbers in parentheses indicate the pages on which the authors' contributions begin.

JOHN U. BALIS (425), Department of Pathology, College of Medicine, University of South Florida, Tampa, Florida 33612

M. BESSIS (41), Faculty of Medicine, Institut de Pathologie Cellulaire, INSERM, Hôspital de Bicêtre, Paris, France

JANET V. COLLINS (223), Department of Biology, Dalhousie University, Halifax, Nova Scotia, Canada

A. DEMSEY (93), Laboratory of Viral Carcinogenesis, National Institutes of Health, Bethesda, Maryland 20205

P. M. GRIMLEY (93), Division of Laboratories and Research, New York State Department of Health, Albany, New York 12201

HAL K. HAWKINS* (251), Department of Pathology, Duke University Medical Center, Durham, North Carolina 27710

T. KATSUYAMA† (1), Department of Pathology, Medical University of South Carolina, Charleston, South Carolina 29403

M. LOCKE (223), Department of Zoology, University of Western Ontario, London, Ontario N6A 5B7, Canada

*Present address: Department of Pathology, Emory University, Atlanta, Georgia 30322.

†Present address: Department of Pathology, Shinshu University, Faculty of Medicine, Matsumoto 390, Japan.

REGINALD G. MASON (425), Department of Pathology, College of Medicine, University of South Florida, Tampa, Florida 33612

STEPHEN R. MAX (291), Departments of Neurology and Pediatrics, University of Maryland, School of Medicine, Baltimore, Maryland 21201

N. MOHANDAS (41), Cancer Research Institute, University of California, San Francisco, California 94143

G. WILLIAM MOORE (173), Department of Pathology, The Johns Hopkins Hospital, Baltimore, Maryland 21205

NORMAN B. RATLIFF* (345), Department of Pathology, Duke University Medical Center, Durham, North Carolina 22710

URS N. RIEDE (173), Klinikum der Albert-Ludwigs-Universität, Pathologisches Institut, 78 Freiburg, Den, Germany

DAVID H. RIFENBERICK (291), Departments of Neurology and Pediatrics, University of Maryland, School of Medicine, Baltimore, Maryland 21201

WALTER A. SANDRITTER (173), Department of Pathology, University of Freiburg i. Br., 78 Freiburg, Germany

J. A. V. SIMSON (1), Departments of Anatomy and Pathology, Schools of Medicine and Dentistry, Medical University of South Carolina, Charleston, South Carolina 29403

S. S. SPICER (1), Department of Pathology, Medical University of South Carolina, Charleston, South Carolina 29403

KYUICHI TANIKAWA (381), The Second Department of Medicine, Kuruma University School of Medicine, Kuruma, Japan

KENNETH R. WAGNER (291), Departments of Neurology and Pediatrics, University of Maryland, School of Medicine, Baltimore, Maryland 21201

R. I. WEED† (41), Hematology Unit, Department of Medicine and Dentistry, Rochester School of Medicine, Rochester, New York 14642

*Present address: Department of Laboratory Medicine and Pathology, Medical School, University of Minnesota, Minneapolis, Minnesota 55455.

†Deceased.

# PREFACE

The main purpose of this multivolume treatise on pathobiological aspects of cell membranes is to give the reader an overview of the recent developments concerning the role of altered cell membranes in various pathological processes. In human disease the membranes themselves, as well as various types of tissue barriers such as the alveolar wall of the lung and the renal glomerular capillary wall, play an extremely important role which has not been fully understood or appreciated. The knowledge that is rapidly being gained in the normal biology of cell membranes needs to be applied to disease processes.

It is important to consider cell membranes from the broadest possible point of view. The title of this treatise does not imply that we are concerned only with the cell membrane. It is, to be sure, included, but the aim is to consider the totality of membranous systems in the cell, including their interactions with cytoplasmic components such as microtubules and microfilaments. For many years our understanding of the basic pathobiology of human disease rested mainly on knowledge concerning disturbances of macromolecular information and replication systems such as disturbances in DNA or RNA metabolism or in the chromosomes. However, it has become clear that in addition to self-replicating informational macromolecules, one prerequisite of life is a strict and selective compartmentalization of the various biological functions to small compartments surrounded by membranes.

In a philosophy similar to that adopted in the first volume we have attempted in this second volume not only to include information on pathobiological aspects of cell membranes as studied at the molecular and subcellular level but also important new advances in the role of membranes in human diseases such as multiple sclerosis, shock lung, muscle dystrophies, and hematological disorders.

It is appropriate that this volume be dedicated to Dr. Robert Weed whose tragic and untimely death interrupted a brilliant career in membrane research as related to human disease. His work, much of it in collaboration with Professor Marcel Bessis of Paris, has yielded an entirely new set of concepts regarding the red cell membrane, its normal and abnormal metabolism, and its changes in relation to hemolytic disease. It is on the shoulders of researchers such as these that the future of biomedical research will be based.

BENJAMIN F. TRUMP, M.D.
ANTTI U. ARSTILA, M.D.

# CONTENTS OF VOLUME I

## **Robert Inslee Weed** (1928-1976)

Dr. Robert Inslee Weed, Professor of Medicine and of Radiation Biology and Biophysics and Chief of the Hematology Unit in the Department of Medicine at the University of Rochester School of Medicine, died August 18, 1976.

Bob Weed had a special interest in the structure, the shape, the metabolism, and the mechanisms of destruction of the red cell, and particularly in the role of the red cell membrane in these matters. He wrote many articles in the fields of hematology, membrane physiology, and biophysics. Through his studies, he made many contributions to our understanding of the red cell membrane, its susceptibility to injury, and the effects of these injuries on red cell survival. He provided a unifying concept for the relationship of the type of red cell injury to the shape of the red cell in the circulation, a concept that gives insight to the diagnostician as well as the physiologist. His studies of red cell membrane damage were performed within a sound frame. He understood the laws of physics and chemistry that dictated the boundaries within which the normal and abnormal were likely to be explained; yet his imaginative mind was never constrained by dogma, permitting the introduction of bold ideas. The concepts that nonlethal loss of membrane occurred from circulating red cells, that contractile proteins may be present in or near the red cell membrane, and that calcium accumulation may be a mechanism of red cell membrane injury antedated recent interest in these topics and were under investigation in his laboratory nearly 15 years ago.

One cannot do justice to the ideas that his keen mind generated. Perhaps their greatest impact will be through his students whose careers they fostered. His enthusiasm, leadership, and intellect led him to make many contributions to the affairs of his school, to academic medicine, and to the education of physician-scientist. It is appropriate that this volume be dedicated to Bob Weed because he fostered the conviction that membrane injury is important in the pathogenesis of cell pathology and because he was devoted to the pursuit of knowledge.

Marshall A. Lichtman
University of Rochester
Rochester, New York

# CHAPTER I

# CELL MEMBRANE CATION LOCALIZATION BY PYROANTIMONATE METHODS: CORRELATION WITH CELL FUNCTION

J. A. V. Simson, S. S. Spicer, and T. Katsuyama



## I. Introduction

### A. *Cations and Cellular Membranes*

Among the important functions of the plasma membrane (plasmalemma or "cell membrane") is the activity of segregating cations of the extracellular



PATHOBIOLOGY OF CELL MEMBRANES, VOL. II

ISBN 0-12-701502-7

environment from the internal cellular milieu. The necessity of this capability was perhaps an accident of changing environmental conditions during early cellular evolution. It has been suggested that during some period of biological evolution, the external environment changed from a relatively high potassium and magnesium concentration to a high sodium and calcium concentration (Rasmussen, 1972). The problem of preserving the internal cationic composition and excluding external cations may have been solved by those cells destined to evolve toward the animal kingdom with the elaboration of a pumping mechanism coupled to a selectively permeable, slightly leaky membrane. Once developed, the ion-segregating potential possessed by the plasmalemma was harnessed to serve several specialized cellular activities, including excitability, motility, secretion, and transcellular transport, which have become essential for the survival of multicellular animals. Those cells responding to the demand for cell/environment segregation by developing a highly impermeable cell coat, characteristic of the plant kingdom, achieved a more efficient solution, perhaps, but one which precluded rapid motion and impulse propagation characteristic of the animal kingdom.

*1. Transport of Cations*

The prototype enzyme for plasma membrane cation transport is the $Na^+,K^+$-activated ATPase described originally in crab neurons (Skou, 1957, 1965). In mammals this enzyme was initially characterized in erythrocyte membranes (Post *et al.*, 1960; Dunham and Glynn, 1961). Adenosinetriphosphatases with similar properties, i.e., activation by $Na^+$ and $K^+$, and inhibition by cardiac glycosides, have since been obtained from a wide variety of tissues (see Bonting, 1970), including mammalian brain and kidney and avian salt gland. Interestingly, plant cells appear to possess little membrane-associated $Na^+,K^+$-activated ATPase of the classical type (see Rothstein, 1968).

Two other cation-dependent ATPases localized in cell membrane fractions are activated by $Ca^{2+}$ and/or $Mg^{2+}$. All membrane-bound ATPases have a requirement for $Mg^{2+}$, apparently for binding ATP in the first step of the reaction; the specific $Mg^{2+}$-activated enzyme requires only this cation. An understanding of calcium transport and the influence of calcium on other transport mechanisms is complicated by the fact that calcium exerts an inhibitory effect on the $Na^+,K^+$-activated enzyme (Epstein and Whittam, 1966; Davis and Vincenzi, 1971). Although $Ca^{2+}$ binds to phospholipids, presumably stabilizing membranes, and inhibits "leakage" of $Na^+$ and $K^+$ down their chemical gradients (Shanes, 1958; Manery, 1966), it also inhibits the $Na^+,K^+$-activated ATPase, and thus may permit accumulation of intracellular $Na^+$. The key role of the calcium ion in a large number of cellular processes has been emphasized repeatedly (Rasmussen, 1970; Rubin, 1974).

Calcium-binding proteins isolated from tissues responsive to ambient calcium levels (Wasserman and Taylor, 1968; Oldham *et al.*, 1974) probably play a role in calcium regulation of cellular function. Recent evidence suggests that the acetylcholine receptor may be composed, in part, of a calcium ionophore (Shamoo and Eldefrawi, 1975).

### 2. *Stimulus and Response*

The translocation of cations across cell membranes is known to play an important role in such diverse phenomena as electrical conduction in neurons (Shanes, 1958; Hodgkin, 1958; Eccles, 1966), muscle contraction (Huxley, 1969), and secretion (Douglas, 1968; Selinger and Naim, 1970). All of these cellular events involve alterations in transmembrane potential mediated by cation shifts. The precise locations of extra- and intracellular reservoirs of loosely bound, stimulus-responsive cations generally have been conjectural, with the exception of the known binding sites for $Ca^{2+}$ in the sarcoplasmic reticulum of striated muscle (see Section II, C, 2).

### 3. *Control of Enzyme Function*

The importance of precise regulation of the intracellular cation composition is readily understood when one considers the number of cellular enzymes that require or are regulated by cations (Westerfeld, 1961; Malmström, 1961; Pestka, 1971; Busch, 1971; Vallee, 1971). Many of the enzymes involved in nucleic acid transformations require $Mg^{2+}$; most oxidative enzymes require either di- or trivalent iron. And, of course, enzyme activity is very sensitive to hydrogen ion concentration.

### 4. *Can Membrane-Associated Cations Be Demonstrated Ultrastructurally?*

A major difficulty in determining the localization of subcellular cation pools that influence various biological processes arises from the fact that cellular fractionation procedures permit extensive redistribution of all but the most tightly bound cations (Clemente and Meldolesi, 1975). This difficulty applies especially to those cation-binding sites which are responsible for triggering stimulated cellular responses since these reactions in all probability involve loosely bound cations. Other approaches than cell fractionation seem requisite for determining the location of biologically active electrolytes. The most promising current methods for localizing cellular cations are ultracryotomy coupled with x-ray microanalysis (Trump *et al.*, 1976) and ion-capture cytochemistry, including pyroantimonate (antimonate) precipitation of cations, again in conjunction with x-ray microanalysis (Chandler, 1978; Simson *et al.*, 1979).

### *B. The Antimonate Method for Cation Localization*

#### *1. Background*

A method for the localization of cations at the fine-structural level, using pyroantimonate as a cation capture agent, was devised originally by Komnick (1962) for the purpose of localizing sodium in the avian salt gland. This method consisted of fixation in a solution containing 2% potassium pyroantimonate in unbuffered osmium tetroxide. A slight modification of this method (using a more nearly saturated concentration of potassium pyroantimonate in unbuffered osmium tetroxide) has been utilized in several studies performed in this laboratory (Grand and Spicer, 1967; Hardin and Spicer, 1970a,b; Hardin *et al.*, 1969, 1970; Spicer *et al.*, 1968, 1969; Simson and Spicer, 1974).

Since its introduction into the repertory of electron microscopic cytochemical techniques, the antimonate method has undergone a great many modifications in the course of its utilization in several laboratories (see Simson and Spicer, 1975; Simson *et al.*, 1979). Most modifications have involved including a buffer in the solution, or altering the concentration of pyroantimonate, and/or altering the fixative (e.g., glutaraldehyde versus osmium tetroxide). Since different methods frequently can be expected to yield disparate results, it should not be surprising that differences in localization and density of antimonate precipitates in certain tissue sites have been reported in studies using different methods. Variants of the antimonate method, used in only a few laboratories, have involved multiple-step processes. These multiple-step methods fall into two broad categories: (a) fixation with an antimonate-free solution prior to treatment with an antimonate containing solution, which we have termed *prefixation* procedures (Simson and Spicer, 1975), and (b) treatment with an antimonate-containing solution *prior* to immersion in any recognized fixative. This has been termed *pyroantimonate fixation* (Tandler *et al.*, 1970). Prefixation procedures have been utilized chiefly in studies by Ackerman and co-workers (Ackerman, 1972; Ackerman and Clark, 1972; Clark and Ackerman, 1971a,b) on peripheral blood and bone marrow cells in suspension. The limitations of this approach are that it permits cellular redistribution both of loosely bound ions and of ions released from their binding sites by fixation, and it allows ions to enter the cell from the extracellular fluid or the fixative buffer (often a sodium buffer) prior to capture by antimonate. The principal virtue that prefixation may have, however, is that it probably demonstrates *potential* cation-binding sites, particularly of the acidic glycoprotein variety. This is suggested by the observation that cations of prefixed tissue tend to accumulate in sites of high acidic glycoprotein concentration, e.g., on the outer surface of the plasmalemma and in the Golgi apparatus (Simson and Spicer, 1975). The "antimonate fixation" procedure utilized largely by Tandler and co-workers (Tandler *et al.*, 1970; Tandler and Kierzenbaum, 1971; Kierzenbaum *et al.*, 1971) has been repeatedly attempted in this laboratory but has yielded poor morphology and scanty intracellular deposits (Simson and Spicer, 1975). Antimonate apparently does not readily penetrate the plasmalemma of unfixed cells, and is not, itself, a good "fixative," at least for tissue possessing hydrolytic enzymes such as parotid and pancreas. When the cell membrane has become sufficiently permeable to permit penetration of antimonate with intracellular precipitation of cation, the resultant precipitate pattern probably represents an agonal distribution of cations.

Initially, the antimonate-containing fixative methods were considered to localize primarily sodium in tissues, but it soon became obvious that several other cations (notably calcium and magnesium), as well as potassium and organic amines, and perhaps even hydrogen ions, were precipitated by the pyroantimonate anion (Bulger, 1969; Legato and Langer, 1969; Shiina *et al.*, 1970; Torack and LaValle, 1970; Spicer and Swanson, 1972). Because of these observations, and because tissues fixed with pyroantimonate-containing fixatives often exhibited poor retention of sodium (Hartmann, 1966; Spicer and Swanson, 1972; Garfield *et al.*, 1972), some disenchantment with the technique has arisen and its cytochemical usefulness has been questioned.

### 2. *Selectivity of Antimonate Methods*

Recently, several methodological variants of the antimonate procedure have been tested by analytic methods *in vitro* (Simson and Spicer, 1975; Simson *et al.*, 1979). These data have provided evidence that certain methodological modifications may provide a degree of selectivity for different cations. The concentrations of cations precipitable *in vitro* by the methodological variants of particular interest are listed in Table I. As may be seen, several cations are precipitated from test solutions by a high concentration, unbuffered antimonate–osmium method. Buffering an antimonate-containing fixative with phosphate or collidine diminishes precipitation of the pyroantimonate anion with $K^+$, $-NH_3^+$, and $H^+$ in the test tube and presumably also in intracellular sites. With phosphate or ethylene glycol tetraacetate (EGTA, a calcium chelator) in the fixative, calcium *and* sodium precipitates of antimonate are decreased considerably *in vitro*, and this correlates with a diminution in tissue precipitation in sites known or suspected to possess sodium and/or calcium. With room-temperature fixation, precipitates tend to be diminished, particularly along the inner plasmalemma compared with fixation at ice-bath temperatures. Thus, it seems reasonable to make some deductions concerning the subcellular localization of cations of certain

TABLE I

*Selectivity of Antimonate Methods for Certain Cations*[a]

| *Methods* | *Cations precipitated (concentration at or above which precipitation occurs in vitro)* |
|---|---|
| High concentration unbuffered antimonate–osmium (2% antimonate or above) | $Na^+$ (10 m*M*) |
| | $Ca^{2+}$ ( 1 m*M*) |
| | $Mg^{2+}$ ( 1 m*M*) |
| | $K^+$,$NH_3^+$ (100 m*M*, pH 7.2 or below) |
| Low concentration antimonate–osmium (1.5% antimonate or below) | $Na^+$ (10 m*M*) |
| | $Ca^{2+}$ ( 1 m*M*) |
| | *$Mg^{2+}$* (10 m*M*) |
| | *$K^+$*, $NH_3^+$ (100 m*M*, pH 6.5 or below) |
| Collidine-buffered antimonate–osmium | $Na^+$ (10 m*M*) |
| | $Ca^{2+}$ ( 1 m*M*) |
| | $Mg^{2+}$ ( 1 m*M*) |
| | *$K^+$*, $NH_3^+$ (100 m*M*, pH 7.0 or below) |
| Phosphate-buffered antimoante–osmium | *$Na^+$* (100 m*M*) |
| | *$Ca^{2+}$* ( 10 m*M*) |
| | $Mg^{2+}$ ( 1 m*M*) |
| Collidine-buffered antimonate–glutaraldehyde | $Na^+$ ( 10 m*M*) |
| | $Ca^{2+}$ ( 1 m*M*) |
| | $Mg^{2+}$ ( 1 m*M*) |
| | *$K^+$* (100 m*M*, pH 7.0 or below), no $NH_3^+$ |

[a] Data from Simson and Spicer (1975). The ions set in italics show decreased precipitability relative to the high concentration unbuffered osmium-antimonate method.

species by observing the presence or absence of cellular precipitates in tissues fixed with selected modifications of the antimonate method.

## II. Membrane-Associated Antimonate-Precipitable Cations in Different Cell Types

### A. *General Considerations*

Cell membranes comprise a highly complex, often interconnected system of glycoprotein, protein, and lipid hydrophobic domains which interface with the aqueous cellular milieu and separate it into functional compartments. Cellular membranes can be conveniently divided into two major types: *outer membrane* (plasma membrane, plasmalemma, sometimes termed simply cell membrane) and *internal membranes* (organelle-associated membranes). The well-characterized organelle-associated membrane systems include: nuclear membranes, rough and smooth endoplasmic reticulum, Golgi membranes, and mitochondrial membranes. Most other cellular membranes, such as those surrounding secretory granules and endocytic vacuoles, are largely derived from one or another of the above-mentioned systems. Each membrane system has its own particular set of characteristics, e.g., cross-sectional diameter (Sjöstrand, 1963; Yamamoto, 1963); carbohydrate, lipid, and protein composition (Bendetti and Emmelot, 1968; Meldolesi, 1971); and enzymes (Siekevitz and Palade, 1958; Van Lancker and Holtzer, 1959; de Thé, 1968; Pratten *et al.*, 1978). Moreover, the various membrane systems possess selective and characteristic affinities for certain ions (Carvalho *et al.*, 1963; Dallner and Nilsson, 1966), and the ion binding to membranes correlates with their functional specialization, e.g., permeability, metalloenzyme composition, and pumping capacity. For example, muscle microsomes bind and transport $Ca^{2+}$ (Hasselbach, 1964; Weber *et al.*, 1966) as does a smooth microsomal fraction of parotid glands (Selinger *et al.*, 1970). Moreover, inorganic cations appear to be essential for optimal binding of at least some membrane proteins to the membrane lipid bilayer (Reynolds, 1972).

Figure 1 illustrates the differences in pyroantimonate-precipitable cation concentration associated with several cell organelles: nuclear envelope, granular reticulum, Golgi membranes, and secretory granules of the rat parotid acinar cell. In this cell the outer nuclear membrane is lined with antimonate-precipitable cations, except in the region of nuclear pores. The rough endoplasmic reticulum, which can be considered an extension of the outer nuclear membrane, possesses similar deposits along the ribosomal side of its membranes. Golgi membranes and secretory-granule membranes of parotid acinar cells are typically devoid of precipitates; the plasma membrane, on the other hand, usually possesses prominent deposits primarily on the cytoplasmic face (Fig. 2). The tissues illustrated in Figs. 1–11 and 15–17 were fixed with a high standard concentration, unbuffered, antimonate–osmium tetroxide solution and rinsed briefly (5 min) prior to dehydration and embedding (Simson and Spicer, 1975).

The cellular membranes involved in mediating exchanges between various cellular compartments, and between the cell and its environment, display considerable heterogeneity and variability in the concentration and distribution of antimonate-precipitable cation, depending on cell type and functional state of the membranes examined. The outer nuclear membrane, for example, exhibits abundant deposits in parotid and pancreatic acinar cells, in which there is active synthesis of exportable protein, but often lacks precipitates in lymphocytes and other nonsecretory cells. Mitochondria also display variability in the quantity and localization of antimonate precipitates depending upon cell type and function. Mitochondria of some cell types possess primarily matrix deposits whereas mitochondria of other cell types exhibit deposits in intracristal spaces (Grand and Spicer, 1967; Parmley *et al.*, 1976). Golgi cisternae vary from lacking antimonate deposits as in the parotid acinus (Simson and Spicer, 1974) to containing moderate precipitates as in neutrophil myelocytes (Hardin *et al.*, 1969). The prevalence of cation-accumulating sites on the plasma membrane of any given cell is at least in part a function of the cell type, since the plasma membranes of adjoining

---

**Fig. 1.** A region of the cytoplasm and part of the nucleus (N) of a parotid acinar cell in which antimonate precipitates may be seen along some cellular membranes but not others. The nucleus possesses deposits along the inner and outer membranes except in sites of nuclear pores (arrows). The membranes of rough endoplasmic reticulum (RER) possess deposits, whereas Golgi (G) membranes and membranes surrounding secretory granules (SG) do not. Lead citrate stained to enhance membrane structure. ×37,500.

**Fig. 2.** Two juxtaposed plasma membranes of parotid acinar cells. Note the double line of medium-sized antimonate deposits, largely on the inner faces of the membranes. Unstained. ×30,000.

**Fig. 3.** Basal infoldings of cells of the salt gland from a salt-adapted duck. Antimonate deposits are present (arrows) but infrequent on the plasma membrane (compare with Fig. 2). Mitochondria (M) also contain fine deposits. Unstained. ×40,000. (Courtesy of Dr. A. J. Garvin.)

**Fig. 4.** Lateral infoldings of striated duct cells of rat parotid. The membranes have very few deposits but intercellular spaces contain occasional large precipitates (arrows). Mitochondria (M) often have heavy grains in the matrix. Unstained. ×28,000.

**Fig. 5.** Human sweat coil showing a portion of both light (L) and dark (D) cells. The light cell has no deposits along its plasmalemma, and few deposits in mitochondria except for the large mitochondrial dense bodies. The dark cell possesses numerous deposits along its plasmalemma except in sites of junctions (arrow). Mitochondria of the dark cell have numerous deposits in the intracristal space. Unstained. ×20,000.

**Fig. 6.** Apical region of three adjacent rat pancreatic acinar cells showing heavy deposits in intercellular spaces (arrows), except at the apical junctions. Secretory granules do not normally possess appreciable precipitate, but the one illustrated (SG) has apparently just fused with the apical membrane and shows a fine stippling of precipitate, similar to that present in the acinar lumen (L). Unstained. ×37,000.

**Fig. 7.** Lateral plasma membranes of rat parotid acinar cells in which the membrane-associated precipitates are very fine (arrows). In one site (opposing arrows) the deposit appears to be double and to be on both the outer and inner faces of the membrane. Fine mitochondrial (M) deposits may also be seen. Lead citrate stained. ×60,000.

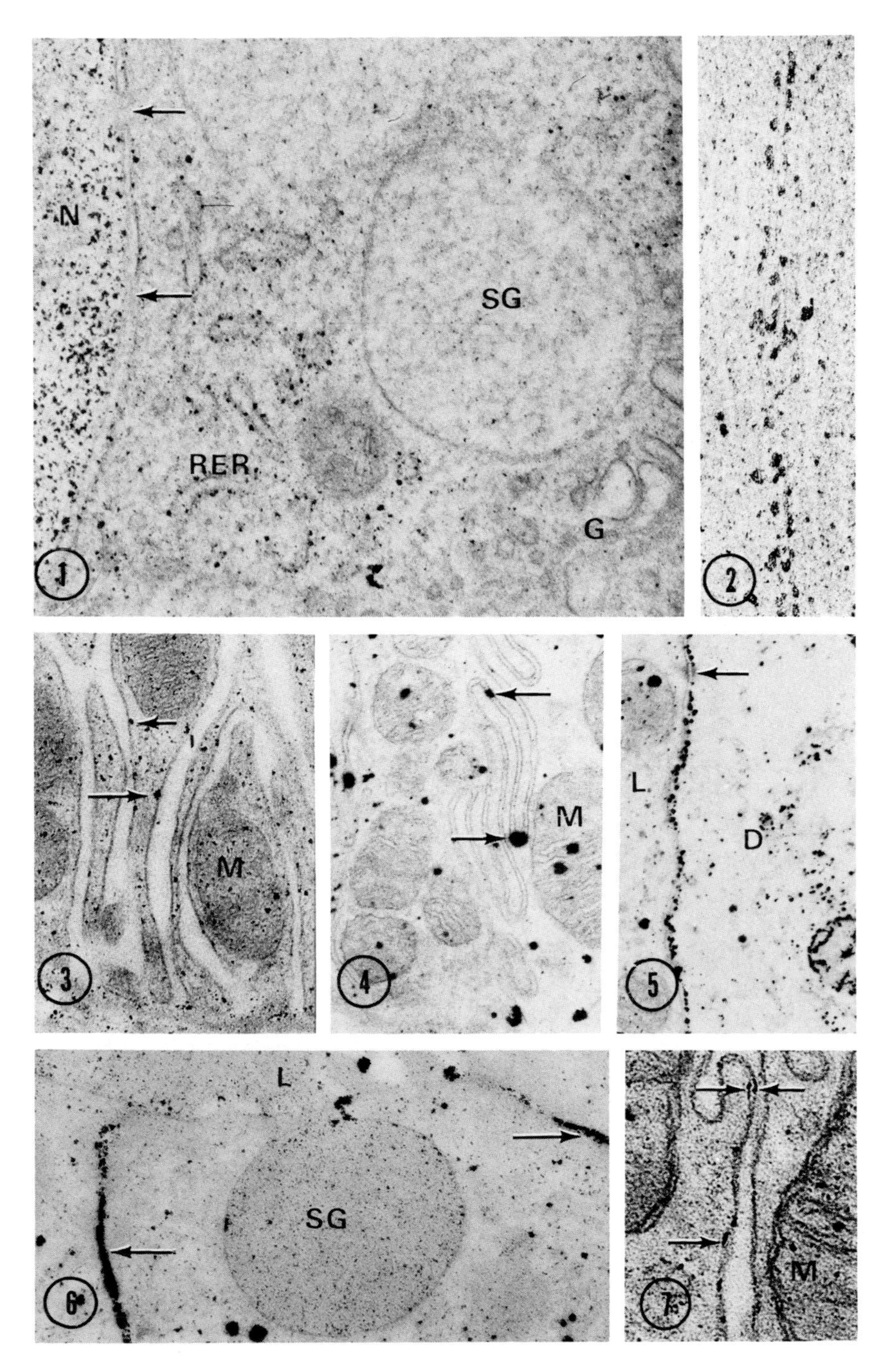
N
SG
RER
G
1
2
3
M
4
M
5
L
D
L
SG
6
7
M

cells of different histological type often differ markedly in antimonate-precipitate pattern with any given antimonate-containing fixation procedure.

With the above general considerations in mind, we can begin to correlate the distribution of antimonate-precipitable cation in several cell types with cellular function, membrane composition, and specialized biochemical activity. Antimonate deposits in various sites can also be compared with ion content determined by other methods to evaluate the cytochemical specificity of the antimonate technique. In this laboratory, the antimonate-precipitable cation distribution has been examined predominantly in two general cell types: blood-forming elements (Hardin and Spicer, 1970a; Spicer *et al.*, 1968) and secretory epithelia such as parotid (Simson and Spicer, 1974, 1975), pancreas (Katsuyama and Spicer, 1977a, b), and sweat glands (Grand and Spicer, 1967).

### *B. Transporting and Secreting Epithelia*

Komnick's (1962) pyroantimonate–osmium method and several variants have been employed to localize sites of cation binding indicative of transport, particularly sodium transport. The original studies utilizing the pyroantimonate method were performed on herring gull salt gland, a salt-transporting epithelium (Komnick, 1962; Komnick and Komnick, 1963). Several other transporting epithelia studied with these methods have included: kidney tubules (Nolte, 1966; Bulger, 1969; Tischer *et al.*, 1969, 1972; Tandler and Kierzenbaum, 1971), choroid plexus (Torack and LaValle, 1970), mussel gill (Satir and Gilula, 1970), chloride cells of the eel (Shirai, 1972), corneal endothelium (Kaye *et al.*, 1965), gallbladder (Kaye *et al.*, 1966), sweat gland (Grand and Spicer, 1967; Ochi, 1968), and small intestine (Yamada, 1967).

#### *1. Sodium Transport*

In most epithelia known to transport sodium actively, e.g., salt water adapted avian salt gland (Komnick and Komnick, 1963), kidney tubules (Nolte, 1966; Bulger, 1969), and sweat gland ducts (Grand and Spicer, 1967), antimonate precipitates along the plasmalemma have been disappointingly sparse. When present, deposits are generally localized along the inner or cytoplasmic face of the plasmalemma (Tisher *et al.*, 1969; Ochi, 1968) and in the intercellular spaces between interdigitated cell processes. Examples of the sparse deposition in epithelia functioning in active sodium transport are presented in Figs. 3 and 4, illustrating secretory cells of salt-adapted duck salt gland (Fig. 3) and striated duct cells of rat parotid gland (Fig. 4).

#### *2. Transport of Other Ions and Molecules*

By contrast, transporting epithelia not selectively specialized for sodium transport often show prominent deposits along the inner face of the plasmalemma. This localization of antimonate deposition has been observed in corneal endothelial cells (Kaye *et al.*, 1965), rat small intestine (Yamada, 1967), and mussel gill (Satir and Gilula, 1970), and is similar in distribution to presumed calcium-binding sites seen with the Oschman and Wall (1972) technique.

### 3. *Secretory Epithelia*

Several secretory cell types also exhibit plasmalemmal deposits when antimonate-containing fixatives are used (Figs. 2 and 5). These include dark cells but *not* light cells of human sweat coil (Grand and Spicer, 1967); the secretory portion of rat sweat gland (Ochi, 1968); pancreatic islet cells (Herman *et al.*, 1973); and parotid acinar cells (Simson and Spicer, 1974). Pancreatic acinar cells also possess heavy antimonate precipitates associated with their plasmalemma (Fig. 6), but unlike those of the aforementioned epithelia, the precipitates are largely on the extracellular (outer) surface of the plasmalemma (Clemente and Meldolesi, 1975; Katsuyama and Spicer, 1977a). Most membranes possessing cation-accumulating sites fail to show antimonate precipitates at sites of cell-to-cell junctions (Bulger, 1969; Simson and Spicer, 1974) (Fig. 5).

More recent studies of pyroantimonate-precipitable cation distribution in endocrine secretory cells (Stoekel *et al.*, 1975; Boquist and Lundgren, 1975; Schechter, 1976; Cramer *et al.*, 1978) have suggested that calcium is an important pyroantimonate-precipitable cation. These studies have indicated that redistribution in membrane- and granule-associated cations may accompany the secretory process. Another pyroantimonate-precipitable cation present in at least some secretory epithelia is zinc (Chandler and Battersby, 1976; Chandler *et al.*, 1977).

Figure 2 illustrates pyroantimonate precipitates along the plasmalemma of rat parotid acinar cells with the high concentration, unbuffered antimonate–osmium method which is the method most commonly utilized in this laboratory. The plasmalemma of most parotid acinar cells can be seen to possess precipitates with this method, either on the inner surface or spanning the membrane. The deposits may be coarse and frequent (Fig. 2) or finer and less frequent (Fig. 7). With collidine or cacodylate-buffered antimonate–osmium tetroxide solutions, the membrane deposits tend to be prominent, in part because of decreased cytoplasmic background precipitates. On the other hand, with phosphate buffered, antimonate-containing solutions, the membrane deposits are absent altogether (Simson and Spicer, 1975), as are most cytoplasmic deposits. Likewise, with EGTA in the antimonate-containing fixative, plasmalemmal deposits are seldom seen, despite heavy cytoplasmic deposits. The absence of deposits in this site with phosphate- and EGTA-containing solutions suggests that these are probably sites of calcium or sodium deposition (see Table I). The ability of both phosphate and EGTA to diminish the antimonate precipitability of sodium as well as calcium *in vitro* was a somewhat surprising but consistent observation in our methodological studies (Simson and Spicer, 1975). Intercalated duct cells in the rat parotid gland normally possess a few cytoplasmic precipitates but seldom exhibit membrane-associated antimonate even when adjacent acinar cells have abundant deposits along the membrane. Likewise, the membrane of nerve terminals abutting on acinar cells seldom possesses antimonate deposits whereas adjacent acinar plasma membrane often exhibits heavy precipitates (Fig. 8).

It has been postulated that disturbances in membrane permeability—either primarily *or* as a result of decreased $Na^+,K^+$ transport activity—result

---

**Fig. 8.** Site of contact between rat parotid acinar cell (A) and its secretory innervation (N). Most of the membrane-associated antimonate deposits line the acinar plasma membrane, although some deposits lie between the membranes. Deposits are also present in vesicles (arrows) of the nerve terminus. Unstained. ×37,500.

**Fig. 9.** Peripheral nerve fiber bundle from rat parotid gland. Membrane-associated antimonate deposits are not heavy in these peripheral nerves, either myelinated (M) or unmyelinated (U). Note antimonate deposits in mitochondria of the Schwann cell (arrow) but not in the nerve fiber itself. The fibroblast (F) of the perineurium has antimonate deposits along the inner and outer faces of its plasma membrane. Unstained. ×19,000.

**Fig. 10.** Rat cardiac muscle showing the lateral sacs of the sarcoplasmic reticulum (arrows) containing abundant antimonate precipitates and wrapping around the central T tubule (T). The T tubule on the left contains a large antimonate deposit. Lead citrate stained. ×25,000.

**Fig. 11.** Rat skeletal muscle, fixed 5 min after an intraperitoneal injection of 5 mg isoproterenol. Note the heavy deposits in the A–I junction on either side of the Z line (Z). Fine precipitates may be seen in the sarcoplasmic reticulum (arrows) surrounding the fiber in this oblique section. Unstained. ×19,000.

**Fig. 12.** Fibroblast from the connective tissue of rat parotid gland, fixed with the standard high concentration unbuffered antimonate–osmium followed by a 30 min distilled water rinse. Insoluble antimonate precipitates persist along the inner plasma membranes, on vesicle membranes, and in mitochondria. Unstained. ×19,000.

**Fig. 13.** Fibroblast from rat tracheal cartilage fixed with phosphate-buffered antimonate–osmium tetroxide. Deposits on the inner surface of the plasmalemma are virtually absent, but large deposits (arrows) on the external face of the plasmalemma and associated with extracellular collagen, are especially heavy with this fixative. Unstained. ×17,000.

**Fig. 14.** Fibroblast from rat parotid gland fixed for routine morphology with cacodylate-buffered 3% glutaraldehyde. Blebs on the cell surface and similar configurations embedded in adjacent collagen bundles (arrows) appear to be the morphological concomitants of the large (> 500 Å) antimonate deposits. These may be anionic materials which selectively bind and sequester extracellular cations. Uranyl acetate, lead citrate stained. ×25,000.

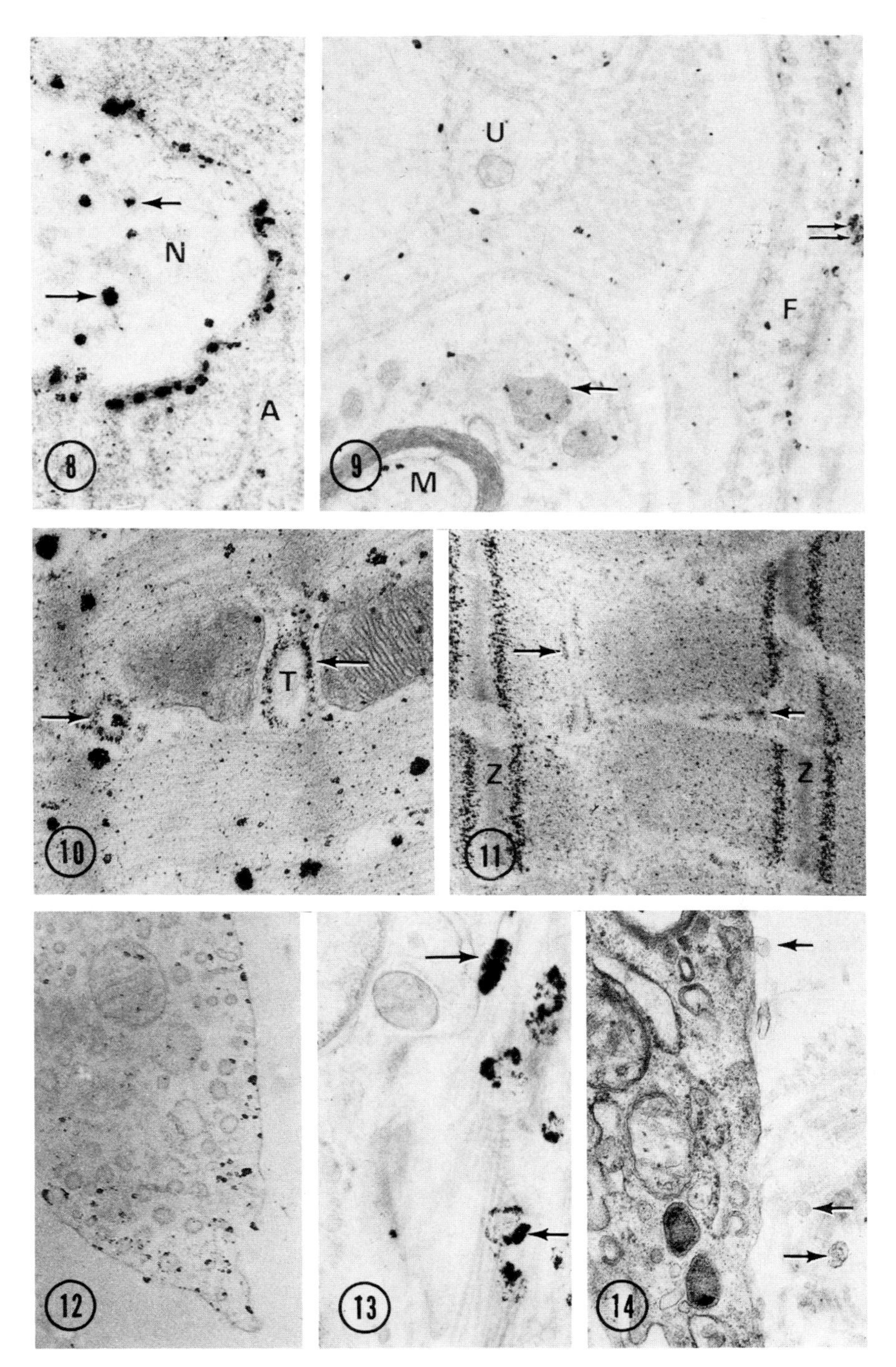
8
N
A
9
U
F
M
10
T
11
Z
Z
12
13
14

in accumulation of $Ca^{2+}$ along membrane sites. This $Ca^{2+}$ accumulation is considered important in the functional control of secretory epithelia (Case and Clausen, 1973) and of transport epithelia not selectively specialized for sodium (Elbrink and Bihler, 1975). The deposits demonstrated along the plasmalemma of parotid acinar cells by several antimonate variants mentioned above might therefore result, at least in part, from precipitation with $Ca^{2+}$. However, the increased abundance of deposits along the inner plasma membrane in epithelia of ouabain-treated animals (Kaye *et al.*, 1965) suggests the involvement of $Na^{+}$ in the formation of these precipitates. It is postulated, therefore, that the plasmalemmal precipitates in secretory epithelia are either calcium or sodium salts (or possibly both) since these precipitates are observed with most antimonate-containing fixatives (except those with EGTA or phosphate), and the precipitates are increased under conditions in which the sodium pump is inhibited.

### C. *Excitable Cells*

#### 1. *Nerve*

Efforts to localize antimonate-precipitable cations in nervous tissues have been hampered by the low permeability of this tissue to most fixatives and fixation additives. In addition, most studies on nervous tissue fixation with antimonate-containing solutions have utilized aldehyde fixatives, further limiting penetrability into the specimen. The frequent use of antimonate-free solutions for prefixation of this tissue prior to the capture step with pyroantimonate in osmium tetroxide may have resulted in extensive redistribution of cation in several studies. Hartmann (1966) employed an unbuffered osmium tetroxide prefixation procedure, followed by an antimonate–osmium tetroxide solution for capturing cations, and observed heavy antimonate deposits in cortical astrocytes. These results were interpreted to indicate a high $Na^{+}$ content in astrocytes comparable to that of extracellular spaces in other tissues. Few deposits were observed in association with nerve cell membranes. Siegesmund (1969), examining cat cerebellum prefixed with glutaraldehyde–acrolein and then treated with antimonate, found appreciable intracellular precipitates in mitochondria, Golgi membranes and endoplasmic reticulum as well as in synaptic vesicles. However, the extent to which cation redistribution occurred during the ½-hr prefixation period is unknown. Sumi (1971) and Sumi and Swanson (1971) have investigated antimonate reactivity in rat cerebral cortex and guinea pig cortical slices *in vitro*, employing several combinations of aldehyde fixatives and phosphate buffers, both of which decrease antimonate precipitability of several cations (see Table I; Simson and Spicer, 1975). Scanty and irregular precipitates with these methods led the authors to question the usefulness of the antimonate method for sodium localization in the brain. Likewise, Torack (1969, 1971) and Torack and LaValle (1970), studying brain stem, area postrema, and choroid plexus with various antimonate-containing combinations of phosphate buffer and aldehyde fixatives, observed only extracellular precipitates in tissue unless the sodium concentration was abnormally high. Tani *et al.* (1969a,b), using antimonate-containing collidine-buffered glutaraldehyde, observed precipitates on the inner surface of the membrane of dendritic terminal boutons, but they found essentially no precipitates in the region of synaptic clefts, thus confirming Hartmann's (1966) observations. Fine precipitates were also observed in dilatations between myelin lamellae (Tani *et al.*, 1969a; Siegesmund, 1969). Such spaces, and the antimonate precipitate contained therein, were exaggerated by triethyltin intoxication (Tani *et al.*, 1969b; Torack, 1969). Villegas (1968), utilizing phosphate-buffered glutaraldehyde as the antimonate-containing vehicle, reported antimonate precipitates indicative of high sodium concentration in the Schwann cell sheath of squid axons. Hardin and co-workers (Hardin *et al.*, 1970; Hardin and Spicer, 1970a), studying rat trigeminal neurons by means of the high concentration antimonate–osmium tetroxide method, observed heavy deposits in nuclei and nucleoli and fine precipitates along neurofilaments but found few deposits along the nerve cell membrane.

In our material fixed by various antimonate methods, peripheral nerves, usually unmyelinated, have been encountered in a variety of tissues. Deposits were sparse along the cytoplasmic face of the axolemma (Fig. 9), and were essentially absent from the membranes of nerve terminals (Fig. 8). Vesicles of nerve terminals often contained precipitates, probably reflecting the presence of $Ca^{2+}$ in these vesicles (Politoff *et al.*, 1974). Peripheral myelinated neurons rarely possessed antimonate deposits in the myelin sheaths, even when these latter appeared disrupted. In the spinal cord, however, the sheaths of myelinated axons often contained numerous deposits, intercalated into spaces, possibly artifactual, in the myelin sheaths. This distribution indicates that the spaces between leaves of myelin sheath are at least accessible to cations. Whether cations are normally present in these sites in undischarged neurons remains to be determined. Neurofilaments often possessed antimonate deposits distributed longitudinally along the axon. The characteristically small mitochondria of neurons seldom possessed antimonate deposits in our material (Fig. 9), although Schwann cell mitochondria revealed moderate precipitates.

### 2. *Muscle*

Komnick's original fixation procedure, as well as several modifications, have been utilized in studies of cation distribution in striated muscle cells

(Zadunaisky, 1966; Legato and Langer, 1969; Shiina *et al.*, 1970; Thureson-Klein and Klein, 1971; Klein *et al.*, 1970, 1972; Yeh, 1973; Yarom and Meiri, 1971, 1972, 1973; Yarom *et al.*, 1974; Davis *et al.*, 1974; Saetersdal *et al.*, 1974). In the first of these studies (Zadunaisky, 1966), antimonate deposits were restricted to the T tubule of frog sartorius muscle fibers following fixation with a phosphate buffered antimonate-glutaraldehyde solution. The assumption can be made that sites of precipitation reported in that study were probably loci of high $Na^+$ or $Ca^{2+}$ concentration. Subsequently, Legato and Langer (1969), Klein *et al.* (1972), Yeh (1973), and Saetersdal *et al.* (1974), examining cardiac muscle, and Shiina *et al.* (1970), Yarom and coworkers (1972, 1974), and Davis *et al.* (1974), examining skeletal muscle, have reported pyroantimonate deposits in the sarcoplasmic reticulum as well as in certain portions of the A or I bands (especially M and N lines). In all of these studies, fixative solutions very similar or identical to Komnick's classical unbuffered method were utilized at least for a portion of the experiments. Chelation studies with ethylenediaminetetraacetic acid (EDTA) and EGTA (Legato and Langer, 1969; Klein *et al.*, 1972; Yeh, 1973; Saetersdal *et al.*, 1974; Davis *et al.*, 1974) led these authors to the conclusion that precipitates in sarcoplasmic reticulum and in the I band were largely antimonate salts of $Ca^{2+}$. This interpretation was supported by x-ray microanalysis data (Yarom and Chandler, 1974). Using the high concentration antimonate-osmium technique, we have also observed numerous, fine antimonate deposits at these sites in both cardiac and skeletal muscle (Figs. 10, 11), but the concentration of deposits in the I-band region was variable. In agreement with the observations of Legato and Langer (1969), we have observed coarse deposits extracellularly and, occasionally, in sites of T tubules, particularly where these structures approach the sarcolemma. Finer deposits were consistently present in the lateral sacs of the sarcoplasmic reticulum. Striated muscle, obtained fortuitously along with parotid tissue 5 min following an injection of isoproterenol (Fig. 11), exhibited heavy deposits of antimonate in the zone of the I band that overlaps the A band. The interpretation that precipitates in T tubules represent $Na^+$ is consistent with the known continuity of these structures with the extracellular space (Peachey, 1965). The interpretation of $Ca^{2+}$ within the sarcoplasmic reticulum (SR) accords with the known $Ca^{2+}$-accumulating ability of this structure (Hasselbach, 1964; Weber *et al.*, 1966; Azzone *et al.*, 1966). Prior to antimonate studies, $Ca^{2+}$ had been localized in sarcoplasmic reticulum by oxalate binding (Pease *et al.*, 1965; Constantin *et al.*, 1965). Anionic sites, presumably for $Ca^{2+}$ binding, are present in the sarcoplasmic reticulum, as evidenced by Thorotrast staining of thin sections (Philpott and Goldstein, 1967), or by ruthenium red staining *en bloc* (Luft, 1966). The ATPase demonstrated cytochemically in the sarcoplasmic reticulum presumably functions in $Ca^{2+}$ transport to this site (Gordon *et al.*,

1967). In cardiac muscle, a "nonspecific" alkaline phosphatase is also present in the lateral sacs of the sarcoplasmic reticulum (Gordon *et al.*, 1967).

### *D. Connective Tissue and Circulatory Cells*

Fibroblasts generally exhibit antimonate deposits on the inner surface of the plasmalemma (Figs. 9 and 12), as well as occasional large precipitates on the outer plasmalemmal surface, with all the antimonate-containing fixatives except the EGTA–antimonate or phosphate–antimonate fixatives. Immature cartilage cells also display heavy antimonate deposits on the plasma membrane, as well as in the intracristal space of mitochondria (Brighton and Hunt, 1974). With the phosphate–antimonate method, the larger external particles are still generally present, although plasmalemmal deposits are rare (Fig. 13). An interpretation that the numerous, large extracellular aggregates of antimonate precipitate represent sodium is supported by the fact that in cartilage, where such large, extracellular deposits are most prominent, the sodium concentration is much higher than in the extracellular fluid of most tissues (Manery, 1954). However, antimonate deposits of similar size and morphology have been attributed to calcium in epiphyseal plates in the zone of calcifying cartilage by Brighton and Hunt (1974) and were presumed by these authors to represent initial sites of calcification. When routine, aldehyde-fixed preparations were examined for the morphological concomitant of the large extracellular antimonate precipitates, whisps of unidentified material have been observed attached to fibroblast membranes and interspersed among collagen fibrils (Fig. 14). It is possible that these are foci of acid mucopolysaccharide accumulation, which could selectively bind extracellular cations. Alternatively, these structures could be comparable to the calcium-accumulating matrix vesicles of calcifying tissues (Anderson, 1969).

Localization of antimonate deposits along the basement membrane of several epithelial cell types has been a common finding (Bulger, 1969; Tani *et al.*, 1969a; Torack, 1971; Tandler and Kierzenbaum, 1971; Thureson-Klein and Klein, 1971; Shirai, 1972; Simson and Spicer, 1974; Katsuyama and Spicer, 1977a). The presence of antimonate deposits in a nonrandom distribution in the extracellular matrix, particularly around collagen fibrils, has also been noted by several investigators (Komnick and Komnick, 1963; Bulger, 1969; Torack, 1971; Kierzenbaum *et al.*, 1971; Das Gupta *et al.*, 1971; Simson and Spicer, 1975). The antimonate precipitates associated with collagen are of two types: (a) fine precipitates measuring somewhat more than 100 Å in diameter, located at regular intervals of about 600 Å along the collagen fibrils in register with their periodicity (Simson and Spicer, 1975), and (b) coarse precipitates larger than 500 Å which often exist as rings, concentric circles or irregular aggregates (Fig. 15) usually near the edge of collagen bundles and similar to the large precipitates associated with fibroblasts. With phosphate buffer in the fixative, only the larger (>500 Å) precipitates were present extracellularly, possibly reflecting high extracellular sodium content.

Most capillaries in organ stroma possess only a few deposits of antimonate on their external membranes. The microendocytic vesicles of the endothelium, however, frequently contain antimonate precipitates. Formed elements of the bone marrow and serous (e.g., peritoneal) cavities seldom possess antimonate precipitates on the plasma membrane with techniques using simultaneous fixation and cation-capture. Buffy coat cells sometimes possess antimonate deposits on the inner plasmalemma (Hardin and Spicer, 1970b), and deposits are often present in vesicles and on the outer leaflet of the plasmalemma of mast cells (Parmley *et al.*, 1975a). Hardin and coworkers (Hardin *et al.*, 1969; Hardin and Spicer, 1970b; Spicer *et al.*, 1969) have studied the antimonate-precipitable cation localization in blood-formed elements, with particular emphasis on changes in granule antimonate with cell maturation. In these studies, a positive correlation was found between antimonate deposits and acid mucopolysaccharide content in granules of heterophils, eosinophils, and platelets. Antimonate precipitates were absent from Golgi lamellae during genesis of primary granules early in the development of heterophil leukocytes, but were present during the period of

---

**Fig. 15.** Connective tissue from rat parotid gland, illustrating the periodic precipitates of antimonate along collagen fibrils and the large deposits (arrow) often associated with collagen bundles. Lead citrate stained. ×30,000.

**Fig. 16.** Parotid acinar cell, 5 min following an injection of 5 mg isoproterenol, showing antimonate deposits along the membrane of a discharging granule (arrows). Large deposits are also present in the lumen (L). ×19,000.

**Fig. 17.** Parotid acinar cell from tissue that had been deliberately crushed. Note the accumulation of precipitates around granule membranes (arrows) and in the intracristal spaces of mitochondria (M). ×14,500.

**Fig. 18.** Dermal fibroblast from a patient with myositis ossificans. Note the heavy deposits on the inner plasma membrane. This tissue had been rinsed for 30 min after the standard fixation procedure in order to remove all but the highly insoluble antimonate precipitates. Lead citrate stained. ×19,000. (Material courtesy of Dr. A. Maxwell.)

**Fig. 19.** Rat pancreas fixed with silver acetate-containing unbuffered osmium tetroxide for anion localization, particularly chloride, bicarbonate, and phosphate. Note the abundant deposits in the apex of the centroacinar cell (C), particularly along its luminal surface (arrows), in contrast to the sparse apical deposits in the acinar cells (A), and their absence along the luminal surface of this cell type. Unstained. ×30,000.

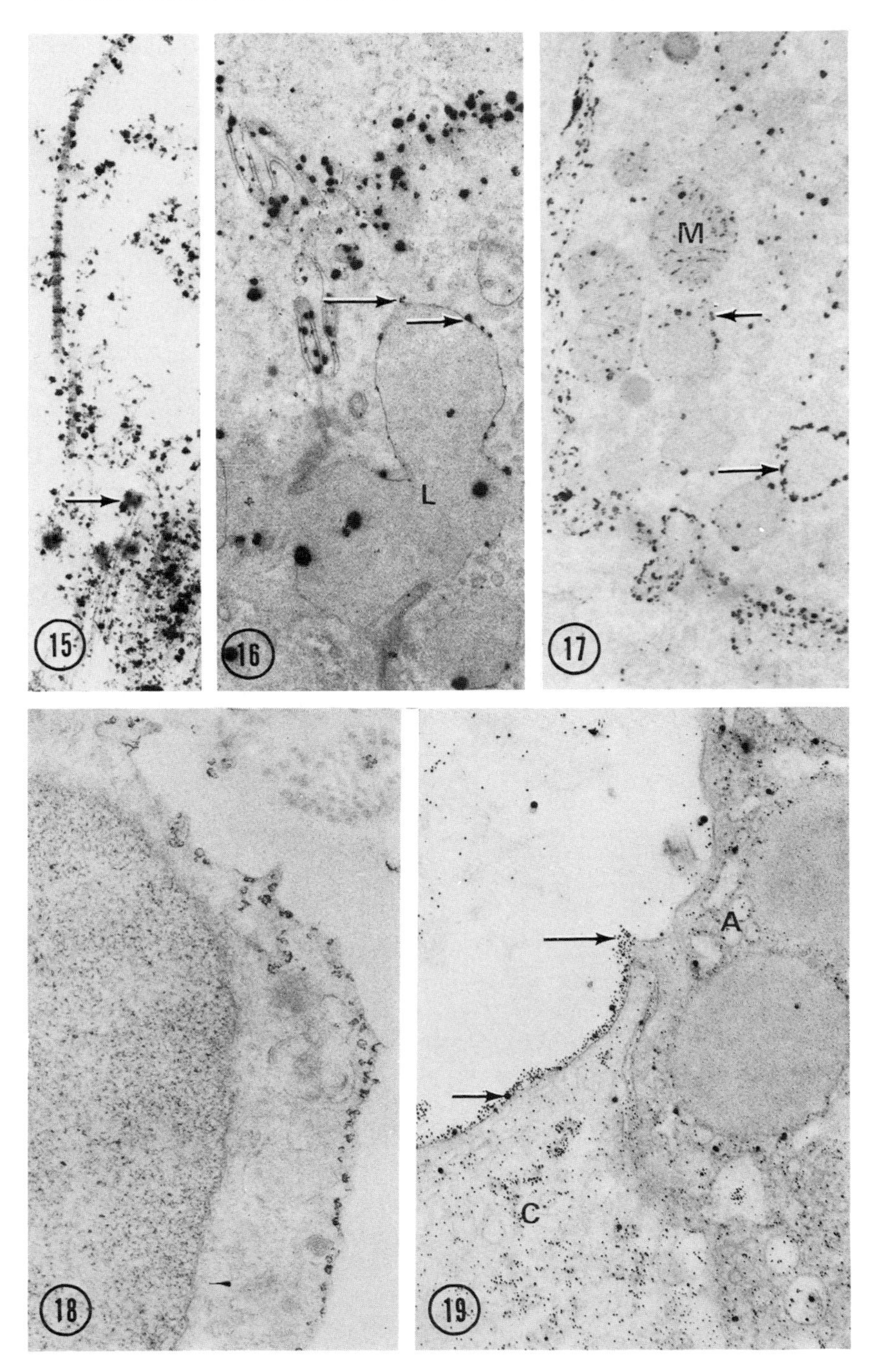
15
16
L
17
M
18
19
A
C

secondary granule production at the metamyelocyte stage. These observations provide additional evidence that an organelle in a given cell type may have altered cation composition and binding properties in different functional states. In this instance the cation localized in the Golgi region may migrate to the neutrophil secondary granule, probably accounting for antimonate precipitates in these granules. The failure of Golgi cisternae to possess antimonate deposits during genesis of primary granules remains puzzling, however, since these granules also contain antimonate-precipitable cations, at least in early stages of their maturation.

Ackerman and co-workers (Clark and Ackerman, 1971a,b; Ackerman, 1972; Ackerman and Clark, 1972) have also examined maturational stages of blood-formed elements, particularly erythrocytes and monocytes, using a prefixation procedure. They observed antimonate deposits on the *outer* leaflet of the plasmalemma of mature erythroid cells, which they suggest is correlated with sialic acid-rich glycoprotein present in this site. In these studies, monocytes and macrophages showed no surface antimonate precipitability.

## III. Alteration in Membrane Cations in Response to Physiological and Pathological Stimuli

### A. *Secretion*

#### 1. *Parotid*

One of the responses of parotid acinar cells to the secretory stimulus of isoproterenol, a sympathomimetic amine, is an increase of antimonate-precipitable cation along the inner face of the plasmalemma (Simson and Spicer, 1974). Following isoproterenol administration, antimonate-precipitable cations also accumulate on secretory granule membranes which are normally devoid of deposits. Those granules which either are ready to fuse or have just fused with the plasmalemma possess deposits along the cytoplasmic face of the granule membrane (Fig. 16), resulting in abundant deposits lining the luminal membrane (Simson and Spicer, 1974; Bogart, 1975). It is possible that the accumulation of cations along the granule membrane serves to neutralize internal negative charges on the granule membranes, permitting fusion of granule membranes with the plasma membrane and with one another, as occurs during merocrine secretion (Ichikawa, 1965; Simson, 1969).

The membrane accumulation of antimonate-precipitable cation accompanying the secretory stimulus may represent calcium and/or sodium which leaks into the cell as a consequence of charge alterations on the membrane accompanying the secretory stimulus. Pilocarpine, a cholinergic secretory stimulus, does not produce the same heavy plasmalemmal accumulation of precipitates (J. A. V. Simson, unpublished). The plasmalemma adjacent to nerve terminals in the parotid gland often possesses heavy deposits of antimonate (Fig. 8). It is tempting to correlate these deposits with the $Ca^{2+}$-dependent ionophore that is a component of the acetylcholine receptor (Shamoo and Eldefrawi, 1975).

### 2. *Pancreas*

In rat pancreas, pancreozymin, which stimulates enzyme secretion from acinar cells, also causes a decrease of antimonate precipitation along the outer face of the plasma membrane and an increase in deposits in the region of rough endoplasmic reticulum (Katsuyama and Spicer, 1977a). This response contrasts with the redistribution of cation to the plasmalemma and away from granular reticulum following the isoproterenol secretory stimulus in the parotid gland. On the other hand, after administration of secretin, which is postulated to stimulate only water and electrolyte secretion, fine cytoplasmic deposits in the centroacinar cell are considerably increased in amount and a dense band of precipitation appears beneath the luminal plasma membrane. Owing to the similarity of precipitate density in the apical portion of the centroacinar cell and the widened lumen after secretin administration, it is difficult to recognize luminal plasma membrane of the centroacinar cell. The acinar cell apical surface is generally devoid of antimonate deposits, and no apparent change of distribution pattern is observed in the acinar cell following secretin administration (Katsuyama and Spicer, 1977a).

### 3. *Sweat Gland*

Preliminary observations also suggest that pilocarpine may cause increased plasmalemmal deposits in the epithelial cells of the rat sweat gland, and even more marked accumulation on the plasmalemma of myoepithelial cells (T. Katsuyama, unpublished).

## *B. Drugs*

The effects of drugs on antimonate-precipitable cation distribution in target cells have not been extensively investigated, but this would seem to be a fruitful area for future investigation. Most of the studies on record have already been mentioned under the appropriate categories above. The cardiac glycoside, ouabain, which inhibits sodium–potassium transport, was shown to cause increased plasmalemmal accumulations of antimonate precipitates on corneal endothelium (Kaye *et al.*, 1965). Sumi and Swanson (1971) examined antimonate-reactive cations in cerebral cortex slices incubated with or without ouabain and found little difference in the distribution of antimonate deposits in the presence or absence of the drug. The fixation for the latter experiments employed phosphate-buffered glutaraldehyde with antimonate which does not permit rapid penetration of antimonate into the tissue and is likely to precipitate sodium only when present in very high concentrations. The administration of triethyltin, a neuronal poison, resulted in increased accumulation of antimonate deposits in myelin sheaths (Tani *et al.*, 1969b; Torack, 1969).

Isoproterenol, a β-adrenergic secretagogue which stimulates cell replication in parotid acinar cells, also causes some cellular damage at the dose levels used to stimulate cell replication (Simson, 1969, 1972). Several of the cation-distribution patterns of isoproterenol stimulation resemble those seen in damaged cells (Section III, C), particularly the increased plasmalemmal deposits and aggregates of precipitates in the euchromatin region of the nucleus (Simson and Spicer, 1974). High doses of isoproterenol also damage cardiac muscle and result in a rather complicated sequence of antimonate-precipitable cation shifts, with a brief, transient increase in intracellular precipitates, followed by decreased deposits at 2 hr, then reaccumulation to above normal levels at later time intervals (Yarom *et al.*, 1972, 1974). Hartmann (1966) administered diphenylhydantoin and deoxycorticosterone to rats in an effort to decrease intracellular sodium. He reported a decrease in the concentration of antimonate precipitates in the neuronal endoplasmic reticulum and in small astrocytic processes in the vicinity of synapses. However, the fact that this tissue was fixed for a half hour prior to immersion in antimonate-containing fixatives calls for caution in the evaluation of this study.

### C. *Cell Damage*

There is considerable evidence that ion redistribution occurs in cellular injury (Trump and Arstila, 1971; Trump *et al.*, 1971; Trump *et al.*, 1979). Damaged membranes become more permeable to the predominantly extracellular cations sodium and calcium, leading to an influx of these cations into the cell. This results in rising intracellular osmotic pressure, which, in turn, causes swelling and possible cell lysis. Under conditions of prolonged or chronic injury, mitochondria also generally show evidence of osmotic disturbance (Christie and Judah, 1968) as well as accumulation of numerous large inclusions probably containing sequestered calcium (Heggtveit *et al.*, 1964; Reynolds, 1965; Carafoli *et al.*, 1971). Trump and co-workers (Trump and Arstila, 1971; Trump *et al.*, 1971) have presented a schema outlining the sequence of subcellular events occurring in response to a variety of injurious agents. Most of these cellular distortions involve shifts of ions and water into or out of different compartments of the cells, probably as a consequence of increased membrane permeability, decreased pumping ability, depressed energy (ATP) availability, or some combination thereof. Subcellular phenomena occurring during injury include an initial swelling, frequently simultaneous with vesiculation of the endoplasmic reticulum (hydropic degeneration), followed by loss of density of the background cytoplasm. This latter is often accompanied by condensation of the mitochondrial matrix compartment, reminiscent of the morphology of deenergized, state 5 mitochondria (Hackenbrock, 1972). The mitochondria subsequently undergo high-amplitude swelling followed by degenerative accumulation of calcium-containing crystals in the matrix compartment. Calcium accumulation may also occur on the inner face of the inner mitochondrial membrane (Vasington and Greenawalt, 1968). Similar subcellular events have been reported by others in tissue exposed to a variety of damaging agents, including endotoxin (Boler and Bibighaus, 1967), 9α-fluorocortisol (D'Agostino, 1964), hypoxia (Sulkin and Sulkin, 1965), aflotoxin (Theron, 1965), ethionine (Meldolesi *et al.*, 1967), and carbon tetrachloride (Reynolds, 1965).

The type and actual subcellular localization of cations involved in osmotic cell injury have been difficult to ascertain. Tisher *et al.* (1969, 1972) have used pyroantimonate-containing solutions in an effort to localize subcellular cations in injured flounder kidney. $^{22}$Na autoradiography correlated well with antimonate precipitates along the plasma membranes (Tisher *et al.*, 1969). Electron probe microanalysis confirmed that pyroantimonate-containing precipitates along the plasma membrane contained sodium as well as calcium and other cations (Tisher *et al.*, 1972). As mentioned previously, triethyltin-induced injury of nerve tissue resulted in increased pyroantimonate-precipitable cation in myelin sheaths (Tani *et al.*, 1969b). In aldehyde–antimonate fixed material, intracellular deposits were observed only in tissue that had been rendered anoxic (Torack, 1969).

If parotid gland tissue is deliberately crushed or minced in air prior to fixation in an antimonate-containing fixative, the distribution of antimonate precipitates differs from that in carefully handled tissue (J. A. V. Simson, unpublished). Precipitates in such mechanically injured cells are heavy along the plasmalemma, diminished along the membranes of endoplasmic reticulum (although sometimes present in cisternae), and often abundant in the intracristal space of mitochondria. This distribution is similar in many ways to the cation redistribution observed at early times after isoproterenol administration (Simson and Spicer, 1974), except that with isoproterenol, mitochondrial antimonate distribution is not altered as it is in damaged cells. If ouabain is administered to an animal and the parotid acinar cells are fixed with antimonate-containing solutions, a similar distribution of antimonate precipitates is observed, notably accumulation along the inner face of the plasmalemma and in mitochondria (J. A. V. Simson, unpublished). This could be interpreted to indicate that the damaged distribution may result nonspecifically from intracellular sodium and calcium accumulation consequent upon increased leakiness of membranes or decreased pump activity. Precipitates may also be present in damaged cells on the outer face of secretory granule membranes (Fig. 17), as occurs in response to a secretory stimulus. Clemente and Meldolesi (1975) have reported redistribution of cations, particularly calcium, to membranes of pancreatic zymogen granules during tissue homogenization. This indicates that these membranes possess cation-binding sites, many or most of which are probably not occupied by calcium in the intact, resting gland.

Increased intracellular antimonate precipitates have been reported in cardiac muscle, particularly in the lateral sacs of the sarcoplasmic reticulum, following administration of scorpion venom (Yarom and Braun, 1971). These increased cellular deposits may reflect intracellular accumulation of calcium and possibly sodium, resulting from membrane leakage.

### *D. Cell Replication*

One of the most interesting and puzzling findings with the antimonate procedure has been the fluctuating concentration of plasmalemmal cation precipitates during the isoproterenol-induced cell cycle in rat salivary gland acinar cells (Simson and Spicer, 1974; Bogart, 1975). During the early phase of the induced cell cycle, (4 hr after isoproterenol administration), as well as during periods of the DNA synthetic phase and during mitosis, antimonate deposits were essentially absent from the cell membrane. This was observed with both buffered and unbuffered antimonate-containing fixatives.

### *E. Disease*

Very few studies have been performed in which cation localization by the antimonate method has been attempted in diseased tissue. In this laboratory, some studies have been performed on skin biopsies from patients with cystic fibrosis of the pancreas and myositis ossificans. One of the symptoms of cystic fibrosis, also known as mucoviscidosis, is diminished sodium reabsorption from sweat ducts so that patients lose excessive amounts of sodium during hot weather. We have not, however, been able to establish consistent differences in the antimonate distribution in sweat glands—either in the duct or the coil—between normal subjects and cystic fibrosis patients.

In other studies, skin from a patient manifesting myositis ossificans, in which ectopic calcification is common at sites of trauma, has been fixed with antimonate–osmium tetroxide and rinsed for 30 min to remove all but the least soluble antimonate salts. Heavy precipitates lined the inner surface of the plasmalemma of dermal fibroblasts in this specimen (Fig. 18). Although the evidence is preliminary, precipitates appeared heavier and more consistent in these specimens than in the fibroblasts of normal human skin.

## IV. Interpretations and Conclusions

### A. *Pump versus Leak*

Widespread utilization of antimonate-containing fixatives has raised questions concerning the chemical identity of antimonate-reactive cations on the plasmalemma and the functional significance of these deposits. Data obtained from x-ray microanalysis, from test tube precipitation analysis, and from modifications of the antimonate cytochemical method indicate that precipitates on the plasmalemma are probably the more insoluble antimonate salts of sodium or calcium, or perhaps a coprecipitate of both which has accumulated at selected sites. Do these precipitates reveal the sodium-binding sites of the $Na^+$,$K^+$-ATPase which should be present on the inner surface of the membrane? This would seem to be a reasonable conclusion in view of the fact that they are augmented by ouabain, at least in certain tissue sites. However, disappointingly few deposits are present on the inner leaflet of the plasma membrane of most cell types specializing in active transport of sodium, e.g., kidney tubules, salivary gland striated ducts, and the nasal gland of ducks adapted to salt water. Increased antimonate deposits are present, however, on the plasma membrane of many cell types under conditions in which the "pump" is presumably operating inefficiently, so that the membrane has become "leaky" to extracellular cations which could then accumulate at membrane-associated cation-binding sites. Such an effect could explain the heavy plasmalemma deposits on cells fixed in the cold, following a secretory stimulus, after ouabain administration, or after nonspecific damage, all of which are known to result in increased accumulation of intracellular sodium. Thus, it may be that with a more efficient and intact pump, fewer deposits will be present along the inner plasma membrane. The deposits may indicate the spatial distribution and frequency of pump sites or of the cation-binding molecule associated with the transport mechanism. Evidence that the precipitates on the inner plasma membrane may also represent the localization of calcium is provided by recent work of

Debbas *et al.* (1975) in which smooth muscle cells incubated in $Na^+$-depleted, $Ca^{2+}$-containing medium exhibited a marked increase in precipitates in this site. Whether the cation localized normally on the inner plasmalemma is sodium or calcium (which usually leaks into the cell along with sodium), or both, is not yet determined.

The antimonate precipitates observed on rough endoplasmic reticulum apparently represent a labile (that is, relatively loosely bound) cation pool, readily available for cation exchanges under altered cellular conditions. The microsomal fraction of the cell is capable of binding both monovalent and divalent cations, and this binding of metal cations is very dependent on pH (Carvalho *et al.*, 1963; Dallner and Nilsson, 1966).

### *B. Binding of Cations to Macromolecules*

Considerable controversy continues to exist concerning the extent to which intracellular ions are bound to (or "associated" with) intracellular macromolecules and the extent to which selective ion binding or association is responsible for the unequal distribution of potassium and sodium in the inner and outer cellular compartments (see Ling *et al.*, 1973; Damadian, 1973; Edzes and Berendsen, 1975). Using equilibrium dialysis and ion-sensitive microelectrodes, evidence is accumulating to support the interpretation that an appreciable proportion of intracellular cations is bound, to some extent, to cellular macromolecules (Damadian, 1973; Hinke *et al.*, 1973).

The degree of association differs for different cations. Although a case has been made for extensive complexing of potassium to intracellular macromolecules (Ling and Cope, 1969; Cope and Damadian, 1970), it is probably in the "bound" state less than other cations (perhaps only 20%). Other cations, particularly intracellular calcium and magnesium and probably sodium, exist largely (60 to 80%) complexed to cellular macromolecules, probably via phosphate and carboxyl groups (Cope, 1967; Hinke *et al.*, 1973; Edzes and Berendsen, 1975). Although the latter authors conclude that intracellular $K^+$ and $Na^+$ behave as if they were freely diffusable, they fail to explain the invisibility of 40–60% of cellular potassium and sodium by nuclear magnetic resonance methods.* The degree of association of metal cations with cellular macromolecules and with isolated membrane fractions is highly sensitive to pH, temperature, osmolarity, and the presence of other species of cations (Carvalho *et al.*, 1963; Damadian, 1973; Hinke *et al.*, 1973; Williams, 1975).

*Moreover, it should be noted that, although some potassium is lost from RBCs incubated in the presence of rubidium for 72 hr at 5°C, followed by 20 hr at 37°C, over 60% of the original $K^+$ is retained (Kimsey and Burns, 1973).

Since the pyroantimonate technique most readily demonstrates sodium, calcium, and magnesium (Simson and Spicer, 1975), those ions most extensively complexed with cellular macromolecules, the technique with its modifications is a potentially powerful tool for localizing sites of important intracellular cations and for demonstrating shifts in these cations under conditions of physiological and pathological alterations. Indeed, the regular and reproducible localization of intracellular pyroantimonate deposits (Hardin *et al.*, 1970; Klein *et al.*, 1970; Yarom *et al.*, 1974; Simson and Spicer, 1975) and the alterations of pyroantimonate-precipitable cation in a reproducible fashion by physiological manipulations (Simson and Spicer, 1974; Sato *et al.*, 1975) as well as by varying temperature, pH, and osmolarity of the fixative solution (Simson and Spicer, 1975) tend to confirm conclusions drawn by the proponents of cation association hypotheses. Moreover, these observations provide information about the subcellular sites of cation-binding macromolecules.

A change in location of completely diffusable ions prior to or during fixation does not interfere appreciably with cytochemical observations because their precise localization within the cell is arbitrary and largely a statistical matter. Those cations that are loosely bound may shift location during the initial tissue handling and the fixation process. Such loosely bound ions comprise a very interesting category of cellular ions, however, since they may be precisely those cations involved in the regulatory mechanisms for such events as muscle contraction and secretion. The means exist for evaluating the extent to which such ions have shifted in response to physiological and pathological stimuli, as well as during the tissue handling (Simson and Spicer, 1975; J. A. V. Simson, unpublished). Very tightly bound cations may not react at all with antimonate if binding constants of the cation–macromolecule complex are such that equilibrium favors it over precipitation with the antimonate anion. If fixation with either $OsO_4$ or glutaraldehyde does not alter a cation-binding macromolecule in such a way that it releases its cations, or if fixation "traps" the cation in an inaccessible matrix, then the ion clearly will not be available for antimonate precipitation. Such is probably the case for the calcium which is present in high concentrations in the secretory granules of the parotid gland (Wallach and Schramm, 1971) and pancreas (Case and Clausen, 1973), and which does not yield antimonate precipitate in either site in undischarged granules (Simson and Spicer, 1974; Clemente and Meldolesi, 1975). Nonprecipitable $Ca^{2+}$ accounts also for much of the calcium in platelets in which only a portion of the "granule-associated" calcium (about 30–40%) is precipitated by antimonate (Sato *et al.*, 1975). The antimonate-precipitable calcium in platelets is apparently the most rapidly reactive pool, however, and the remainder is either so tightly bound that it is inaccessible to antimonate or is lost during fixation and processing. Yarom and co-workers (Yarom and Chandler, 1974; Yarom *et al.*, 1974) omitted dehydration by proceeding directly from fixation to embedding with a glutaraldehyde–urea polymerization procedure and found improved retention of dehydration-labile cation (which could be lipid bound) in muscle. Some evidence exists suggesting that certain cation-binding molecules may be phospholipids (Susat and Vanatta, 1965).

## C. *Correlation with Anions*

Anion localization using a silver acetate technique was also originally described by Komnick (1962). This method has been used in our laboratories to analyze anion distribution in secretory epithelia (Katsuyama and Spicer, 1977a,b). Results obtained with this technique indicate that it is possible to obtain reproducible anion localization and that silver-precipitable anion localization patterns are quite different from, and probably independent of, antimonate-precipitable cation distribution (Fig. 19). On the other hand, antimonate-precipitable cation localization correlates well with macromolecular anions (largely acidic glycoprotein or mucopolysaccharide) in many sites, notably in leukocyte granules (Hardin *et al.*, 1969; Spicer *et al.*, 1969), on erythrocyte membranes (Ackerman and Clark, 1972), in mitochondria (Parmley *et al.*, 1975b, 1976), and in connective tissue (Figs. 13 and 15).

Our current overall appraisal of cation localization patterns in "typical" epithelial cells is illustrated in Fig. 20. This composite picture has been derived largely from data obtained using several antimonate methods on the parotid gland (Simson and Spicer, 1975) as well as on several other tissues. It is understood, of course, that the relative amount of cation localized at any given site (plasma membrane, Golgi apparatus, mitochondria) depends greatly on the cell type and its functional state.

## D. *Methodological Considerations*

To what extent do antimonate precipitates observed fine-structurally reflect *in vivo* cation localization? This is a question which must still be answered with caution. The following conditions were previously suggested as those under which antimonate localization can be considered most nearly to parallel cation localization *in vivo* (Simson and Spicer, 1975): (a) the cell should be viable and undamaged before cation immobilization by an antimonate-containing fixative, (b) the fixative and antimonate should penetrate into the cell simultaneously, and (c) when cation-binding molecules are "fixed," their cations should be immediately released to be "captured" by antimonate. Certainly, any technique in which there is an appreciable fixation period prior to exposure to the antimonate captor may allow redistribution of

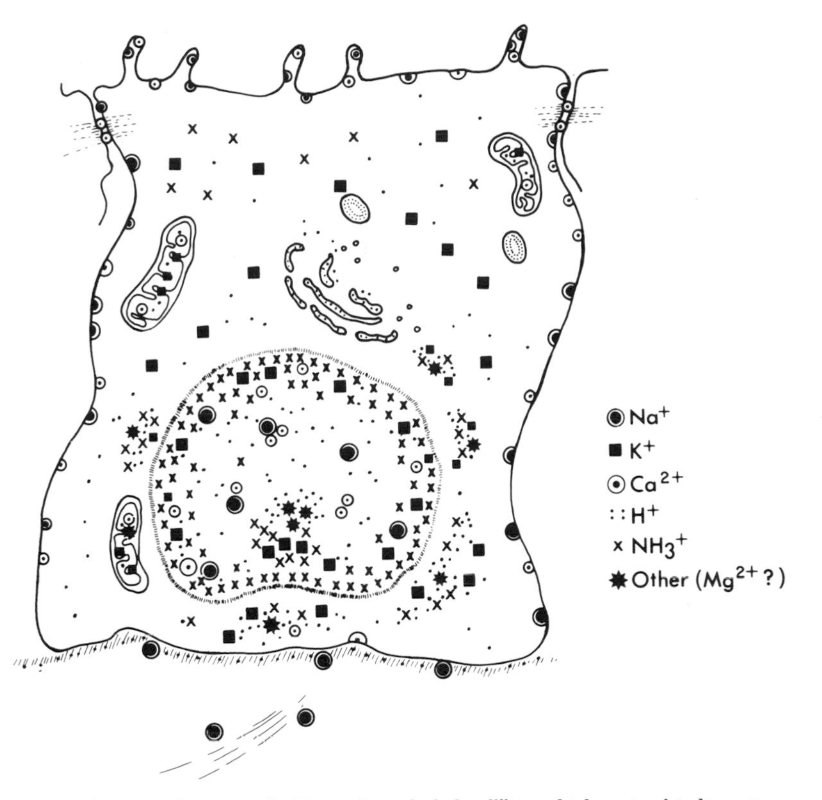


**Fig. 20.** Schematic drawing of a "typical epithelial cell" in which cation-binding sites are indicated by symbols. This is a composite in which data from *in vitro* precipitation and selective fixation and processing of several tissues have been combined. Any given cell will have more or less precipitate (and presumably cation) at any of these sites, depending upon cell type and functional state.

cations. This is probably also true to a lesser extent if glutaraldehyde is utilized as the primary antimonate-containing fixative instead of osmium, since it penetrates tissue less rapidly and may preserve cell membranes so well that they are poorly permeable to antimonate. Several studies in which osmium tetroxide and glutaraldehyde have been compared as fixative vehicles for antimonate have indicated that with glutaraldehyde there is little or no intracellular cation localization (Sumi, 1971; Yarom *et al.*, 1972; Simson and Spicer, 1975; Sato *et al.*, 1975).

Those interested in subcellular cation localization are not in complete accord concerning the significance of subcellular patterns of antimonate precipitation. However, our experience and the work of others (Simson and Spicer, 1975; Sato *et al.*, 1975; Yarom *et al.*, 1974) indicates that the antimonate method, if utilized and controlled appropriately, can localize biologically important cations and can indicate shifts of cations during physiological and pathological processes such as secretion, contraction, drug reactions, and cellular responses to injury. A large proportion of cytoplasmic antimonate-precipitable cation is associated with membranes (in contrast to nuclear and extracellular antimonate-precipitable cation). Since the distribution of these cations is dependent upon cell type and physiological state, pyroantimonate-containing fixatives hold considerable promise as cytochemical tools in investigations of the role of membranes in the binding of cellular cations and the role of cations in the functioning of cellular membranes.

## Acknowledgment

Much of the work performed by the authors during the course of studies reported herein was supported by NIH Grants AM-10956 and AM-11028 to S. S. Spicer and NIH GRS Grant RR-05420 to the Medical University of South Carolina. Technical assistance by Zandra Waters and Rosa Lampkin and artistic assistance of Betty Goodwin are gratefully acknowledged.

## References

Ackerman, G. A. (1972). *Z. Zellforsch. Mikrosk. Anat.* **134,** 153.
Ackerman, G. A., and Clark, M. A. (1972). *J. Histochem. Cytochem.* **20,** 880.
Anderson, H. C. (1969). *J. Cell Biol.* **41,** 59.
Azzone, G. F., Azzi, A., Rossi, C., and Millic, G. (1966). *Biochem. Z.* **345,** 322.
Benedetti, E. L., and Emmelot, P. (1968). *In* "The Membranes" (A. J. Dalton and F. Haguenau, eds.), pp. 33–120. Academic Press, New York.
Bogart, B. I. (1975). *J. Ultrastruct. Res.* **57,** 139.
Boler, R. K., and Bibighaus, A. J., III (1967). *Lab. Invest.* **17,** 537.
Bonting, S. L. (1970). *In* "Membranes and Ion Transport" (E. E. Bittar, ed.), Vol. I, pp. 257–363. Wiley (Interscience), New York.
Boquist, L., and Lundgren, E. (1975). *Lab. Invest.* **33,** 638.
Brighton, C. T., and Hunt, R. M. (1974). *Clin. Orthop. Rel. Res.* **100,** 406.
Bulger, R. E. (1969). *J. Cell Biol.* **40,** 79.
Busch, D. H. (1971). *Science* **171,** 241.
Carafoli, E., Tiozzo, R., Pasquali-Ronchetti, I., and Laschi, R. (1971). *Lab. Invest.* **25,** 516.
Carvalho, A. P., Sanui, H., and Pace, N. (1963). *J. Cell. Comp. Physiol.* **62,** 311.
Case, R. M., and Clausen, T. (1973). *J. Physiol.* **235,** 75.
Chandler, J. A. (1978). *In* "Electron Probe Microanalysis in Biology" (D. A. Erasmus, ed.), pp. 37–93. Chapman and Hall, London.
Chandler, J. A., and Battersby, S. (1976). *J. Histochem. Cytochem.* **24,** 740.
Chandler, J. A., Sinowatz, F., Timms, B. G., and Pierrepoint, C. G. (1977). *Cell Tissue Res.* **185,** 89.
Christie, G. S., and Judah, J. D. (1968). *Lab. Invest.* **18,** 108.

Clark, M. A., and Ackerman, G. A. (1971a). *J. Histochem. Cytochem.* **19**, 727.
Clark, M. A., and Ackerman, G. A. (1971b). *J. Histochem. Cytochem.* **19**, 388.
Clemente, F., and Meldolesi, J. (1975). *J. Cell Biol.* **65**, 88.
Costantin, L. L., Franzini-Armstrong, C., and Podolsky, R. J. (1965). *Science* **147**, 158.
Cope, F. W. (1967). *J. Gen. Physiol.* **50**, 1353.
Cope, F. W., and Damadian, R. (1970). *Nature (London)* **228**, 76.
Cramer, E. B., Cardases, C., Periera, G., Milks, L., and Ford, D. (1978). *Neuroendocrinology* **26**, 72.
D'Agostino, A. N. (1964). *Am. J. Pathol.* **45**, 633.
Dallner, G., and Nilsson, R. (1966). *J. Cell Biol.* **31**, 181.
Damadian, R. (1973). *Ann. N. Y. Acad. Sci.* **204**, 211.
Das Gupta, T. K., Moss, G. S., and Newson, B. (1971). *Acta Anat.* **80**, 426.
Davis, P. W., and Vincenzi, F. F. (1971). *Life Sci.* **10**, 401.
Davis, W. L., Matthews, J. L., and Martin, J. H. (1974). *Calcif. Tissue Res.* **14**, 139.
Debbas, G., Hoffman, L., Landon, E. J., and Hurwitz, L. (1975). *Anat. Rec.* **182**, 447.
De Thé, G. (1968). *In* "The Membranes" (A. J. Dalton and F. Haguenau, eds.), pp. 121–150. Academic Press, New York.
Douglas, W. W. (1968). *Br. J. Pharmacol.* **34**, 451.
Dunham, E. T., and Glynn, I. M. (1961). *J. Physiol. (London)* **156**, 274.
Eccles, J. C. (1966). *Ann. N. Y. Acad. Sci.* **137**, 473.
Edzes, H. T., and Berendsen, H. J. C. (1975). *Annu. Rev. Biophys. Bioeng.* **4**, 265.
Elbrink, J., and Bihler, I. (1975). *Science* **188**, 1177.
Epstein, F. H., and Whittam, R. (1966). *Biochem. J.* **99**, 232.
Garfield, R. E., Henderson, R. M., and Daniel, E. E. (1972). *Tissue & Cell* **4**, 575.
Gordon, G. B., Price, H. M., and Blumberg, J. M. (1967). *Lab. Invest.* **16**, 422.
Grand, R. J., and Spicer, S. S. (1967). *Bibl. Paediatr.* **86**, 100.
Hackenbrock, C. R. (1972). *Ann. N. Y. Acad. Sci.* **195**, 492.
Hardin, J. H., and Spicer, S. S. (1970a). *J. Ultrastruct. Res.* **31**, 16.
Hardin, J. H., and Spicer, S. S. (1970b). *Am. J. Anat.* **128**, 283.
Hardin, J. H., Spicer, S. S., and Greene, W. B. (1969). *Lab. Invest.* **21**, 214.
Hardin, J. H., Spicer, S. S., and Malanos, G. E. (1970). *J. Ultrastruct. Res.* **32**, 274.
Hartmann, J. F. (1966). *Arch. Neurol. (Chicago)* **15**, 633.
Hasselbach, W. (1964). *Fed. Proc. Fed. Am. Soc. Exp. Biol.* **23**, 909.
Heggtveit, H. A., Herman, L., and Mishra, R. K. (1964). *Am. J. Pathol.* **45**, 757.
Herman, L., Sato, T., and Hales, C. N. (1973). *J. Ultrastruct. Res.* **42**, 298.
Hinke, J. A. M., Caille, J. P., and Gayton, D. C. (1973). *Ann. N. Y. Acad. Sci.* **204**, 274.
Hodgkin, A. L. (1958). *Proc. R. Soc. Lond.* **148**, 1.
Huxley, H. E. (1969). *Science* **164**, 1356.
Ichikawa, A. (1965). *J. Cell Biol.* **24**, 269.
Katsuyama, T., and Spicer, S. S. (1977a). *Histochem. J.* **9**, 467.
Katsuyama, T., and Spicer, S. S. (1977b). *Am. J. Anat.* **148**, 535.
Kaye, G. I., Cole, J. D., and Donn, A. (1965). *Science* **150**, 1167.
Kaye, G. I., Wheeler, H. O., Whitlock, R. T., and Lane, N. (1966). *J. Cell Biol.* **30**, 237.
Kierszenbaum, A. L., Libanati, C. M., and Tandler, C. J. (1971). *J. Cell Biol.* **48**, 314.
Kimsey, S. L., and Burns, U. C. (1973). *Ann. N. Y. Acad. Sci.* **204**, 486.
Klein, R. L., Horton, C. R., and Thureson-Klein, A. (1970). *Am. J. Cardiol.* **25**, 300.
Klein, R. L., Yen, S. S., and Thureson-Klein, A. (1972). *J. Histochem. Cytochem.* **20**, 65.
Komnick, H. (1962). *Protoplasma* **55**, 414.
Komnick, H., and Komnick, U. (1963). *Z. Zellforsch. Mikrosk. Anat.* **60**, 163.
Legato, M. J., and Langer, G. A. (1969). *J. Cell Biol.* **41**, 401.

Ling, G. N., and Cope, F. W. (1969). *Science* **163**, 1335.
Ling, G. N., Miller, C., and Ochsenfeld, M. M. (1973). *Ann. N. Y. Acad. Sci.* **204**, 6.
Luft, J. H. (1966). *Anat. Rec.* **154**, 379.
Malström, B. G. (1961). *Fed. Proc. Fed. Am. Soc. Exp. Biol.* (Suppl.) **10**, 60.
Manery, J. F. (1954). *Physiol. Rev.* **34**, 334.
Manery, J. F. (1966). *Fed. Proc. Fed. Am. Soc. Exp. Biol.* **25**, 1804.
Meldolesi, J. (1971). *Adv. Cytopharmacol.* **1**, 145.
Meldolesi, J., Clemente, F., Chiesara, E., Conti, F., and Fanti, A. (1967). *Lab. Invest.* **17**, 265.
Nolte, A. (1966). *Z. Zellforsch. Mikrosk. Anat.* **72**, 562.
Ochi, J. (1968). *Histochemie* **14**, 300.
Oldham, S. B., Fischer, J. A., Shen, L. H., and Arnaud, C. D. (1974). *Biochemistry* **13**, 4790.
Oschman, J. L., and Wall, B. J. (1972). *J. Cell Biol.* **55**, 58.
Parmley, R. T., Simson, J. A. V., and Spicer, S. S. (1975a). *Exp. Mol. Pathol.* **22**, 252.
Parmley, R. T., Poon, K. C., Spicer, S. S., and Simson, J. A. V. (1975b). *J. Cell Biol.* **67**, 325a.
Parmley, R. T., Spicer, S. S., Poon, K., and Wright, J. (1976). *J. Histochem. Cytochem.* **24**, 1159.
Peachey, L. D. (1965). *J. Cell Biol.* (Suppl.) **25**, 209.
Pease, D. C., Jenden, D. J., and Howell, J. N. (1965). *J. Cell. Comp. Physiol.* **65**, 141.
Pestka, S. (1971). *In* "Membranes and Ion Transport" (E. E. Bittar, ed.), Vol. 3, pp. 279–296. Wiley (Interscience), New York.
Philpott, C. W., and Godlstein, M. A. (1967). *Science* **155**, 1019.
Politoff, A. L., Rose, S., and Pappas, G. D. (1974). *J. Cell Biol.* **61**, 818.
Post, R. L., Merritt, C. R., Kinsolving, C. R., and Albright, C. D. (1960). *J. Biol. Chem.* **235**, 1796.
Pratten, M. K., Williams, M. A., and Cope, G. H. (1978). *In* "Histochemistry of Secretory Processes" (J. R. Garrett, J. P. Harrison, and P. J. Stoward, eds.), pp. 109–139. Chapman and Hall, London.
Rasmussen, H. (1970). *Science* **170**, 404.
Rasmussen, H. (1972). *In* "Clinics in Endocrinology and Medicine" (I. MacIntyre, ed.), Vol. I, pp. 3–20. Saunders, Philadelphia, Pennsylvania.
Reynolds, E. S. (1965). *J. Cell Biol.* (Suppl.) **25**, 53.
Reynolds, J. A. (1972). *Ann. N. Y. Acad. Sci.* **195**, 75.
Rothstein, A. (1968). *Annu. Rev. Physiol.* **30**, 15.
Rubin, R. P. (1974). "Calcium and the Secretory Process." Plenum, New York.
Saetersdal, T. S., Myklebust, R., Berg Justesen, N. P., and Olsen, W. C. (1974). *Cell Tissue Res.* **155**, 57.
Satir, P., and Gilula, N. B. (1970). *J. Cell Biol.* **47**, 468.
Sato, T., Herman, L., Chandler, J. A., Stracher, A., and Detwiler, T. C. (1975). *J. Histochem. Cytochem.* **23**, 103.
Schechter, J. E. (1976). *Am. J. Anat.* **146**, 189.
Selinger, Z., and Naim, E. (1970). *Biochim. Biophys. Acta* **203**, 335.
Selinger, Z., Naim, E., and Lasser, M. (1970). *Biochim. Biophys. Acta* **203**, 326.
Shamoo, Q. E., and Eldefrawi, M. E. (1975). *J. Membr. Biol.* **25**, 47.
Shanes, A. M. (1958). *Pharmacol. Rev.* **10**, 59.
Shiina, S., Mizuhira, V., Amakawa, G., and Futaesaku, Y. (1970). *J. Histochem. Cytochem.* **18**, 644.
Shirai, N. (1972). *J. Faculty Sci.* (Univ. Tokyo) **12**, 385.
Siegesmund, K. A. (1969). *J. Anat.* **105**, 403.
Siekevitz, P., and Palade, G. E. (1958). *J. Biophys. Biochem. Cytol.* **4**, 203.
Simson, J. A. V. (1969). *Z. Zellforsch. Mikrosk. Anat.* **101**, 175.

Simson, J. A. V. (1972). *Anat. Rec.* **173,** 437.
Simson, J. A. V., and Spicer, S. S. (1974). *Anat. Rec.* **178,** 145.
Simson, J. A. V., and Spicer, S. S. (1975). *J. Histochem. Cytochem.* **23,** 575.
Simson, J. A. V., Bank, H. L., and Spicer, S. S. (1979). *In* "Scanning Electron Microscopy," Vol. II, pp. 779–792. SEM Incorporated, AMF O'Hare, Illinois.
Sjöstrand, F. S. (1963). *J. Ultrastruct. Res.* **9,** 561.
Skou, J. C. (1957). *Biochim. Biophys. Acta* **23,** 394.
Skou, J. C. (1965). *Physiol. Rev.* **45,** 596.
Spicer, S. S., and Swanson, A. A. (1972). *J. Histochem. Cytochem.* **20,** 518.
Spicer, S. S., Hardin, J. H., and Greene, W. B. (1968). *J. Cell Biol.* **39,** 216.
Spicer, S. S., Greene, W. B., and Hardin, J. H. (1969). *J. Histochem. Cytochem.* **17,** 781.
Stoekel, M. E., Hindelang-Gertner, C., Dellmann, H.-D., Porte, A., and Stutinsky, F. (1975). *Cell Tissue Res.* **157,** 307.
Sulkin, N. M., and Sulkin, D. F. (1965). *Lab. Invest.* **14,** 1523.
Sumi, S. M. (1971). *J. Histochem. Cytochem.* **19,** 591.
Sumi, S. M., and Swanson, P. D. (1971). *J. Histochem. Cytochem.* **19,** 605.
Susat, R. R., and Vanatta, J. C. (1965). *Proc. Soc. Exp. Biol. Med.* **119,** 534.
Tandler, C. J., and Kierszenbaum, A. L. (1971). *J. Cell Biol.* **50,** 830.
Tandler, C. J., Libanati, C. M., and Sanchis, C. A. (1970). *J. Cell Biol.* **45,** 355.
Tani, E., Ametani, T., and Handa, H. (1969a). *Acta Neuropathol.* **14,** 137.
Tani, E., Ametani, T., and Handa, H. (1969b). *Acta Neuropathol.* **14,** 151.
Theron, J. J. (1965). *Lab. Invest.* **14,** 1586.
Thureson-Klein, A., and Klein, R. L. (1971). *J. Mol. Cell. Cardiol.* **2,** 31.
Tisher, C. C., Cirksena, W. J., Arstila, A. U., and Trump, B. F. (1969). *Am. J. Pathol.* **57,** 231.
Tisher, C. C., Weavers, B. A., and Cirksena, W. J. (1972). *Am. J. Pathol.* **69,** 255.
Torack, R. M. (1969). *Acta Neuropathol.* **12,** 173.
Torack, R. M. (1971). *Z. Zellforsch. Mikrosk. Anat.* **113,** 1.
Torack, R. M., and LaValle, M. (1970). *J. Histochem. Cytochem.* **18,** 635.
Trump, B. F., and Arstila, A. U. (1971). *In* "Principles of Pathology" (M. F. La Via and R. Hill, Jr., eds.), pp. 9–95. Oxford Univ. Press, London and New York.
Trump, B. F., Croker, B. P., and Mergner, W. J. (1971). *In* "Cell Membranes: Biological and Pathological Aspects" (G. W. Richter and D. G. Scarpelli, eds.), pp. 84–128. Williams & Wilkins, Baltimore, Maryland.
Trump, B. F., Berezesky, I. K., Chang, S. H., and Bulger, R. E. (1976). *Virchows Arch. B.* **22,** 111.
Trump, B. F., Berezesky, I. K., Chang, S. H., Pendergrass, R. E., and Mergner, W. J. (1979). *In* "Scanning Electron Microscopy," Vol. III, pp. 1–11. SEM Incorporated, AMF O'Hare, Illinois.
Van Lancker, J. L., and Holtzer, R. L. (1959). *J. Biol. Chem.* **234,** 2359.
Vallee, B. L. (1971). *Fed. Proc. Fed. Am. Soc. Exp. Biol.* (Suppl.) **10,** 71.
Vasington, F. D., and Greenawalt, J. W. (1968). *J. Cell Biol.* **39,** 661.
Villegas, J. (1968). *J. Gen. Physiol.* (Suppl.) **51,** 61s.
Wallach, D., and Schramm, M. (1971). *Eur. J. Biochem.* **21,** 433.
Wasserman, R. H., and Taylor, A. N. (1968). *J. Biol. Chem.* **243,** 3987.
Weber, A., Herz, R., and Reiss, I. (1966). *Biochim. Zeitschr.* **345,** 329.
Westerfeld, W. W. (1961). *Fed. Proc. Fed. Am. Soc. Exp. Biol.* (Suppl.) **10,** 158.
Williams, R. J. P. (1975). *In* "Biological Membranes. Twelve Essays on Their Organization, Properties and Functions" (D. S. Parsons, ed.), pp. 106–121. Oxford Univ. Press (Clarendon), London and New York.
Yamada, E. (1967). *Arch. Histol. Jap.* **28,** 419.

Yamamoto, T. (1963). *J. Cell Biol.* **17**, 413.
Yarom, R., and Braun, K. (1971). *J. Mol. Cell. Cardiol.* **2**, 177.
Yarom, R., and Chandler, J. A. (1974). *J. Histochem. Cytochem.* **22**, 147.
Yarom, R., and Meiri, U. (1971). *Nature (London), New Biol.* **234**, 254.
Yarom, R., and Meiri, U. (1972). *J. Ultrastruct. Res.* **39**, 430.
Yarom, R., and Meiri, U. (1973). *J. Histochem. Cytochem.* **21**, 146.
Yarom, R., Ben-Ishay, D., and Zinder, O. (1972). *J. Mol. Cell. Cardiol.* **4**, 559.
Yarom, R., Peters, P. D., and Hall, T. A. (1974). *J. Ultrastruct. Res.* **49**, 405.
Yeh, B. K. (1973). *J. Mol. Cell. Cardiol.* **5**, 351.
Zadunaisky, J. A. (1966). *J. Cell Biol.* **31**, C11.

# EDITORS' SUMMARY TO CHAPTER I

This chapter deals with a very important and yet poorly known aspect of cellular metabolism; namely, the compartmentalization of intracellular cations such as sodium, calcium, and magnesium. The distribution and regulation of these cations is of primary importance in many physiological and pathological processes including secretion, contraction, fertilization, cell division, cell-to-cell communication, electric conduction, protein synthesis, etc. Furthermore, abnormal ion shifts within the cell and between the cell interior and exterior are of key importance in many pathological processes such as cell injury, cardiac failure and, possibly, neoplasia.

It is becoming quite clear that ionic mechanisms, recently shown to regulate cellular activity at the beginning of life including events at fertilization and early differentiation, are also important in later stages of the organism's existence. Both nerve and muscle are well-characterized examples of cells whose function is intimately related to controlled regulation of sodium, potassium, and calcium. More recently there are important indications that small ions may be important participants in the more complex processes of cell division and differentiation. Calcium in particular seems to be very important in this regard and will be covered in later volumes of this treatise.

In the localization of these intracellular cations there are many problems which have so far hampered exact knowledge of these distribution patterns. First of all there seems to be great variation in the binding of the cations to intracellular macromolecules. A small percentage is probably freely diffusible or tightly bound to macromolecules, e.g., calmodulin whereas most of the cations are probably loosely bound or completely unbound to cellular structures. It is these last mentioned cations which are probably most important in terms of intracellular regulation. Because most of the cations are either freely diffusible or loosely bound, it has not been possible to use conventional biochemical techniques such as cell fractionation techniques for their localization. Therefore, new approaches are needed in order to solve this problem. Among these methods is the so-called precipitation method such as the pyroantimonate method discussed in this chapter. In recent years other methods have also been developed and used, such as autoradiography and direct elemental analysis of cations using x-ray analytical instruments. In all these methods the extreme rapidity by which the cations may diffuse to other compartments makes careful tissue preparation and handling extremely important but extremely difficult. In x-ray analytical techniques this is usually done by cryofixation, combined with freeze-drying (Fig. 21) or direct analysis of frozen hydrated sections. The energy or wavelength dispersive x-ray spectrometers when attached to either transmission or scanning electron microscopes make it possible to obtain information on the elemental distribution of small areas about the size of the organelles (Figs. 22 and 23). As discussed in this chapter one special problem with the pyroantimonate method is to know exactly which cation has precipitated with pyroantimonate. In theory the new energy dispersive x-ray spectrometers attached to an electron microscope make it possible to combine the histochemical pyroan-

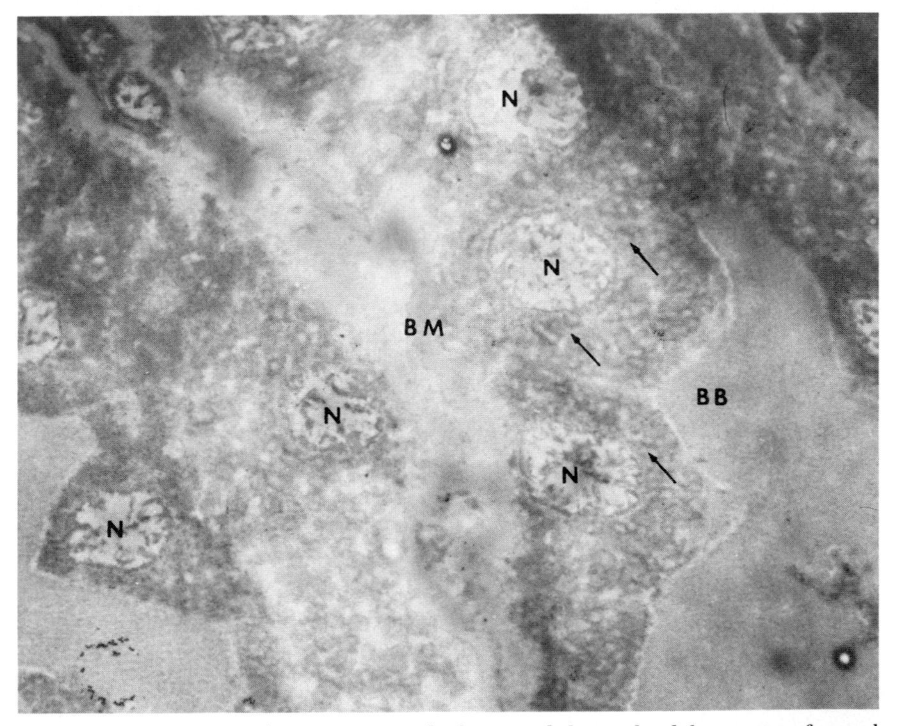


**Fig. 21.** Transmission electron micrograph of unstained, frozen-dried thin section of control kidney cortex. Portions of proximal convoluted tubules are seen with cells projecting into the lumens. Nuclei (N), mitochondria (arrows), brush border (BB), and basement membrane (BM) are identifiable. (Trump *et al.* 1978.)

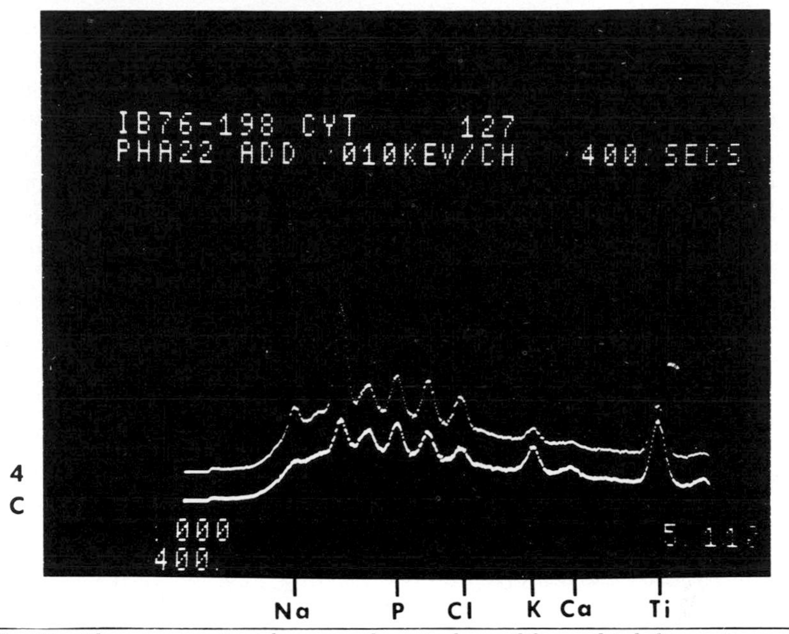


**Fig. 22.** The x-ray spectra taken over the cytoplasm of freeze-dried thin sections of rat kidney cortex after 4 hr ischemia *in vivo*. Control (C). (Berezesky *et al.*, 1977.)

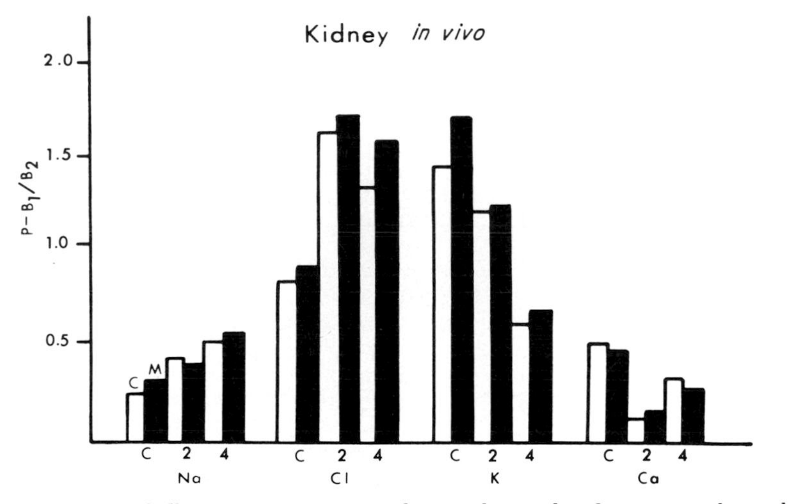


**Fig. 23.** Bar graph illustrating P–$B_1/B_2$ ratios for Na, Cl, K, and Ca from x-ray analyses taken over freeze-dried thin sections of rat kidney cortex after 2 and 4 hr ischemia *in vivo*. Cytoplasm (C), mitochondria (M). (Trump *et al.*, 1978.)

timonate technique with x-ray elemental analysis and by this means increase the accuracy of these methods.

In the future, increased use of these techniques may become of routine importance in diagnostic pathology. One of the classic problems of diagnosing human disease at autopsy has been the localization of early myocardial infarcts. Although these can be localized, at least roughly by using formazan techniques by 6 hr, infarcts less than 6-hr-old cannot be localized at autopsy because of the difficulties of superimposed autolytic change. Therefore, routine electron microscopy is of little or no value. In the future, further developments of methods, including x-ray analysis, possibly using markers

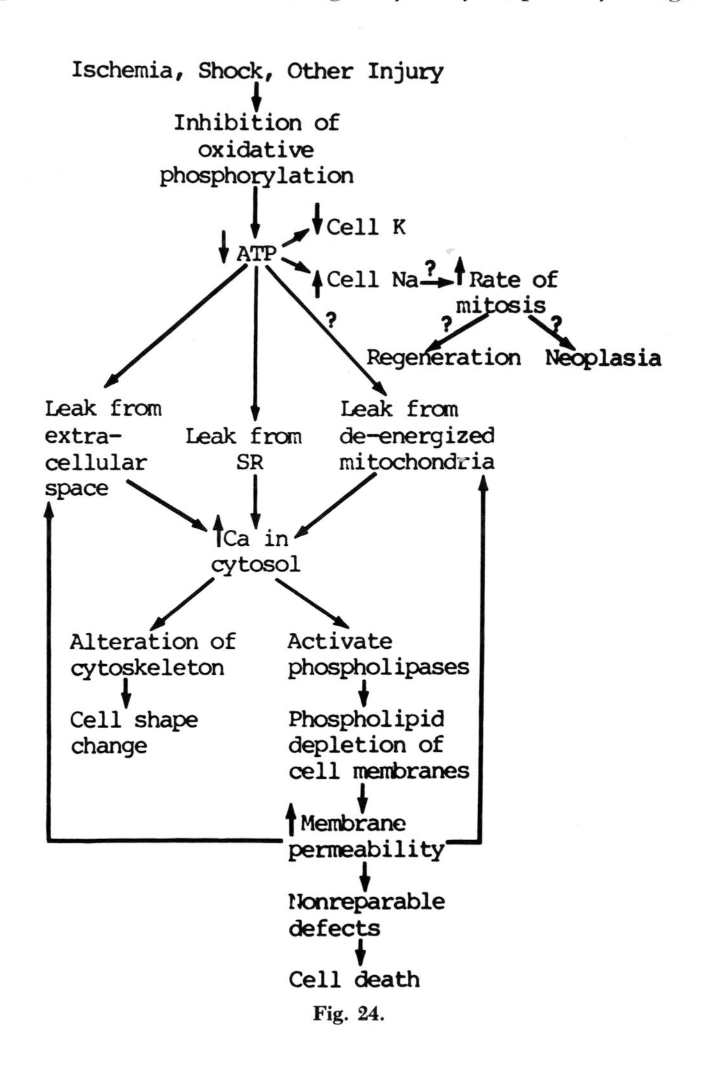


Fig. 24.

such as the pyroantimonate and related methods, may acquire routine use in this very important diagnosis. The same concept can be applied to many other organs including early recognition of strokes, studies of acute renal failure, and distribution of organ involvement in shock. Some ions, such as K which is normally low in the extracellular space, will still maintain their new distribution concentrations indefinitely after somatic death since further movement is not possible once the circulation stops. Therefore, such methods could even yield significant information at the autopsy table or surgical bench, and be of extreme value in the diagnosis of human disease.

Recent results from our laboratory indicate an important, if not primary, role for calcium in the pathogenesis of irreversible cell injury. Shifts of ionized calcium are probably often correlated with those of sodium and potassium and may trigger a series of events that lead to irreversible changes (Trump *et al.*, 1979) (Fig. 24). In the future we badly need methods to accurately measure ionized calcium in various organelle compartments.

## References

Berezesky, I. K., Chang, S. H., Pendergrass, R. E., Bulger, R. E., Mergner, W. J., and Trump, B. F. (1977). 35th Ann. Proc. Electr. Microsc. Soc. Amer. (G. W. Bailey, ed.), pp. 524–525. Boston.

Trump, B. F., Berezesky, I. K., Pendergrass, R. E., Chang, S. H., Bulger, R. E., and Mergner, W. J. (1978). *Scanning Electron Microsc.* **2**, 1027.

Trump, B. F., Berezesky, I. K., Chang, S. H., Pendergrass, R. E., and Mergner, W. J. (1979). *Scanning Electron Microsc.* **3**, 1.

# CHAPTER II

# RED CELL MEMBRANE PATHOLOGY

**N. Mohandas, R. I. Weed,* and M. Bessis**



## I. Introduction

Over the past 10 years, there have been several reviews (Jandl, 1965; Weed and Reed, 1966; Firkin and Wiley, 1966; Neerhout, 1968a; Weed, 1968, 1970a,b; Shohet and Lux, 1974) written concerning abnormalities of the red cell membrane, particularly in relation to hemolytic disease. In a way, the recent history of publications about the red cell membrane reflects the evolution of techniques for characterization of membranes in general, such as advances in electron microscopy, biochemical analysis, new immunological techniques, and most importantly, combinations of

*Because of Dr. Weed's untimely death during the preparation of this manuscript, Dr. Mohandas updated and edited the work.



PATHOBIOLOGY OF CELL MEMBRANES, VOL. II

ISBN 0-12-701502-7

these approaches with studies of membrane dynamics visualized by phase microscopy. In the context of membrane pathology, we shall attempt in this chapter to focus on abnormalities of structure, function, or chemical composition of the human erythrocyte membrane. Some of these are unique to red cells, others are found in other blood cells or other cells in the body.

We have used the word membrane abnormality rather than membrane pathology so as to include mention of any alterations which represent a departure from what is considered to be the normal membrane structure, conformation, and function. This definition includes reversible changes in membrane conformation that can provide insight into membrane dynamics. These changes, themselves, become irreversible and, therefore, pathologic. In addition, by analogy to certain intracellular enzyme deficiencies [e.g., catalase deficiency (Takahara and Miyamoto, 1948)] which are unassociated with any shortening in red cell life span, we intend to touch on functional abnormalities of the membrane which are not of consequence to the function or survival of the red cell. Pathological studies will be discussed in terms of physical properties, chemical abnormalities, and ultrastructural alterations.

## II. Ultrastructure and Chemical Composition of Normal Membranes

It is important to emphasize that no single definition of "the membrane" of the red cell is acceptable to everyone studying its properties. For physiologists, the membrane is that which separates the cell interior from its external environment, and mediates molecular traffic moving from one compartment to the other. For biochemists, the membrane is the red cell ghost, since hemoglobin in the cell interior makes up 97% of the dry weight of an erythrocyte, and would thus be a major contaminant of the membrane protein during analysis. As Ponder (1948) emphasized a long time ago, the composition of a red cell ghost (and inferences drawn from it about the composition of the membrane) depends entirely upon the manner by which the ghost was prepared. Thus, the pH at which the ghosts are prepared (Ponder, 1948; Dödge *et al.*, 1963), the metabolic state of the cell (Weed *et al.*, 1969), or even the composition of the hemolyzing solution (Rega *et al.*, 1967) may determine the quantity of protein that remains associated with the membrane preparation.

Similarly, because the electron opacity of hemoglobin makes it difficult to study details of membrane structure, in the past transmission electron microscopic studies of the membrane have been carried out on hemoglobin-free ghosts. Recent evidence (McMillan and Luftig, 1973), in fact, suggests that osmium treatment during preparation for electron microscopy, either alone or as "postfixation" after glutaraldehyde, is associated with the loss of

70–90% of the red cell membrane protein. The ultrastructural concomitant of this is the difference between a thicker (approximately 160 Å), less distinct membrane fixed with glutaraldehyde alone or the classic "railroad track" appearance of a 90–100 Å thick membrane after osmium treatment. It is obvious that membrane "structure" can only be interpreted or compared in the light of the fixative employed. Introduction of the scanning microscope has permitted us to appreciate the topology of the membrane in the intact cell, and the freeze-etch technique with intact cells has opened up possibilities for characterization of membrane structure. However, the problem remains to correlate these observations with those made on hemoglobin-free ghosts.

Current evidence suggests the existence of at least three layers of organization within the membrane. There is an external layer which contains glycoprotein. This is probably a discontinuous layer with the glycoprotein traversing the bilayer. It extends out from the surface in branchlike form, interspersed between lipid molecules (Tillack and Marchesi, 1970). The ABO blood group antigens exist primarily as glycoproteins in human erythrocytes (Whittemore *et al.*, 1969). These glycoproteins may represent some of the intramembrane particles which have been demonstrated in freeze-etch preparations of red cell membranes by many workers (Branton, 1966; Weinstein and Bullivant, 1967; Weinstein, 1969; Weinstein and McNutt, 1970). The Rh antigenic sites are located at or near the surface and number approximately 10,000/cell (Nicolson *et al.*, 1971). Rh antigenic activity is dependent upon lipoprotein integrity for its activity (Green, 1968a,b), and it is inactivated by sulfhydryl inhibitors and has a pH maximum. These findings suggest that the Rh antigen is associated with a structural lipoprotein that undergoes conformational rearrangement in response to changes in the environment. Finally, in addition to intrinsic membrane components found at the cell surface, varying amounts of plasma proteins may be adsorbed on the outer surface of the membrane. These adsorbed proteins may include albumin as suggested by Furchgott and Ponder (1940) and perhaps even a specific erythrophilic globulin (Fidalgo *et al.*, 1967). Less than 35 molecules of IgG can be demonstrated on the surface of saline-washed normal red cells (Gilliland *et al.*, 1970).

Beneath the layer of adsorbed protein and protruding glycoproteins, there is a double layer which is readily seen with conventional transmission electron microscopy, although as mentioned above (McMillan and Luftig, 1973), this appearance may result from extraction of protein by osmium. This bilayer structure is present in both ghosts and intact cells and because virtually all of the lipid present in intact erythrocytes can be recovered from hemoglobin-free ghosts (Reed *et al.*, 1960; Weed *et al.*, 1963; Dodge *et al.*, 1963), the lipid can be presumed to be present within this membrane structure. Although the classic model of Davson and Danielli (1943) requires modification today because of more recent experimental observations, the basic model remains the most likely configuration of the lipid molecules based on evidence from x-ray diffraction (Engelman, 1970; Wilkins *et al.*, 1971) as well as electron spin resonance studies (Tourtellote *et al.*, 1970). Red cell membranes contain $3 \times 10^{-16}$ moles per cell of free cholesterol, $4 \times 10^{-16}$ moles per cell of phospholipid and $0.5 \times 10^{-16}$ moles per cell of glycolipid. The glycolipids constitute about 10% by weight of the red cell lipid, but they contain virtually none of the sialic acid (Weed and Reed, 1966). The association between phospholipid and cholesterol molecules (Finean, 1962; Vandenheuval, 1963) plays a major role in orienting the phospholipid molecules within the membrane (Hsia and Boggs, 1972) and leads to close packing of the various lipid components. This may be of considerable significance in the determination of permeability to nonelectrolytes and to ions (Kroes and Ostwald, 1971).

There is also recent evidence for asymmetrical distribution of different phospholipids across the membrane bilayer. Bretscher (1972) and Gordesky and Marinetti (1973), examining the differences in the reactivity of phosphatidylethanolamine (PE) and phosphatidylserine (PS) to nonpenetrating alkylating agents in intact cells and ghosts, concluded that PE and PS are preferentially localized on the inner half of the bilayer, while phosphotidyl choline and sphingomyelin are localized on the outer half of the bilayer. These conclusions about the asymmetrical distribution of phospholipids are also supported by the interaction of pure phospholipases with intact red cells and ghosts (Zwaal *et al.*, 1973). The subject of asymmetrical distribution of proteins and lipids in biological membranes has been recently reviewed by Rothman and Lenard (1977).

An interesting general model for the organization of cell membranes has been proposed by Singer and Nicolson (1972). These authors have proposed a fluid-mosaic model which consists of a lipid bilayer matrix with globular proteins randomly dispersed within this lipid matrix. Maddy and Malcolm (1965) presented evidence that human erythrocyte proteins exist predominantly as randomly coiled globular molecules, but with approximately 17% in the form of $\alpha$ helices. Figure 1a and b is a diagrammatic model of the erythrocyte membrane which suggests the existence of both $\alpha$-helical and randomly coiled globular proteins within the bilayer. These proteins are probably bound to lipids by nonionic binding since organic solvents such as butanol are essential for the separation of lipid and protein components (Maddy, 1964; Maddy and Malcolm, 1965; Rega *et al.*, 1967).

A layer of proteins is attached to or extends from the double membrane into the interior of the cell. With the exception of major glycoproteins that are present on both sides of the membrane and also span the bilayer, the majority of these proteins are localized only on the cytoplasmic side. Two

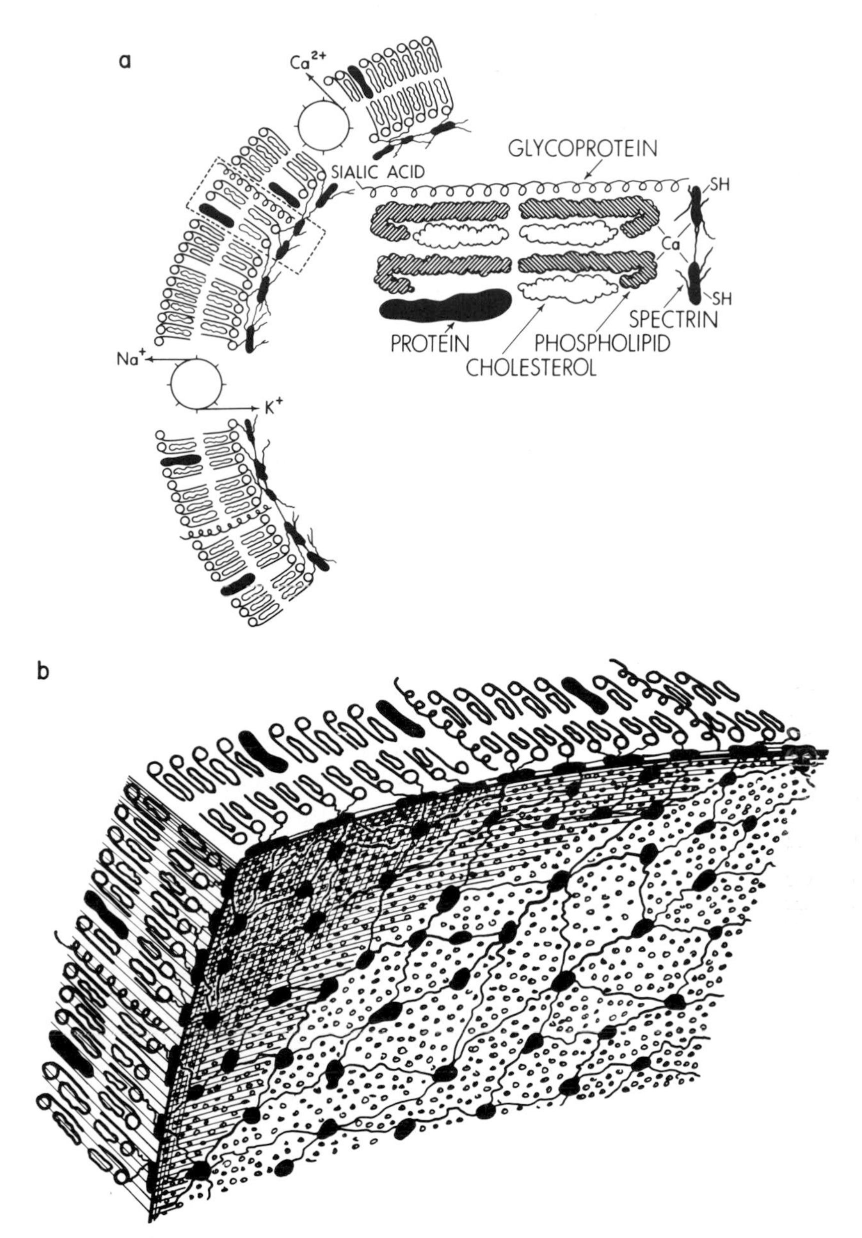


**Fig. 1.** (a) Model of erythrocyte membrane structure in cross-section. The insert is intended to suggest the detailed relationships between the various components. (b) Three-dimensional diagram of inner surface of the red cell membrane to indicate protein–protein, protein–lipid, and protein–Ca–lipid cross-linkages. (Reproduced from Greenwalt, 1973, with permission of the publisher.)

low ionic strength extractable proteins, spectrin and actin, have been increasingly implicated in the regulation of red cell shape and deformability (Marchesi and Palade, 1967; Marchesi and Steers, 1968; Marchesi *et al.*, 1969; Kirkpatrick, 1976; Wins and Schoeffeniels, 1966; Palek and Liu, 1979; Rosenthal *et al.*, 1970; Lux, *et al.*, 1976; Mohandas *et al.*, 1976; Lux, 1979). Sheetz *et al.* (1976), have recently demonstrated that the spectrin portion of the actin–spectrin complex may indeed be involved in contractile activities of the erythocyte membrane. It has also been postulated that calcium (as suggested in Fig. 1a and b) may play a critical role in determining the interaction of spectrin with the rest of the membrane and also, perhaps in determining the characteristics of the protein itself (Kirkpatrick *et al.*, 1975; Carraway *et al.*, 1975). In spite of significant indirect evidence of the important role played by spectrin and actin in the regulation of red cell shape and deformability, direct evidence for its topological distribution in the cell and its aberration in pathology are not available.

Figure 1 differs in one important respect from the Singer–Nicolson model in that it indicates cross-linkage between the globular proteins on the cytoplasmic side of the membrane. Although the erythrocyte membrane provides little or no resistance to bending forces perpendicular to the plane of the membrane, it is very resistant to stretch within the plane of the membrane, although there is a measurable elastic component (Hochmuth *et al.*, 1973). The fact that the biconcave red cell can transform to an osmotically swollen sphere indicates the membrane must have properties which dissipate tensional and compressional stress in the plane of the membrane while offering little resistance to tangential tension or pressure (Fung and Tong, 1968). Bull (1973) has proposed a model with cross-linkage between hexagonal components, although the latter have translational mobility. Evans (1973a,b) has proposed a two-dimensional elastomer model. The development of persistent myelin form extensions from the membranes of cells exposed to shear (Hochmuth and Mohandas, 1972) also suggests existence of intramembrane cross-links which, if ruptured, would result in plastic deformation. Evans and Hochmuth (1977), have recently proposed a model very similar to the one shown in Fig. 1 based on mechanical properties of the red cell membrane. They demonstrated that the red cell can behave as either viscoelastic or viscoplastic material under appropriate conditions, and have obtained the intrinsic material constants that characterize these properties. The model represented in Fig. 1a and b permits both translational mobility of protein particles within the lipid bilayer and interactions between proteins on the inner membrane surface that would provide the cross-linkage that appears necessary to explain the physical characteristics of red cell membranes that can undergo plastic deformation when stretched beyond a limit.

## III. Reversible Alterations

### A. *Discocyte–Echinocyte Transformation*

The biconcave shape of normal red cells results from a delicate equilibrium of forces. Among the various shape changes which red cells may undergo, the isovolumic "disk-sphere" shape change was clearly identified by Ponder in the early 1940s (Ponder, 1948). The fundamental membrane mechanism responsible for this transformation is becoming comprehensible only now. It is now apparent that the discocyte–spherocyte shape change, in fact, occurs via two routes, the discocyte–echinocyte and the discocyte–stomatocyte transformations (Bessis, 1972). The discocyte–echinocyte transformation can be seen by washing red cells with isotonic sodium chloride and examining them between a glass slide and coverslip. They change in shape from biconcave disks to spheres covered with crenations or spicules (echinocytes, from the Greek word for sea urchin). If these cells are reintroduced into fresh plasma, they will reassume the discoid shape even between glass slide and coverslip. This rapidly reversible echinocytic transformation, which does not imply any change in viability, is the phenomenon that Ponder described as the "disk-sphere" transformation. As he pointed out, there are several recognizable stages between the extremes of a discocyte on the one hand and a prelytic sphere on the other. Four stages can be recognized: echinocyte I, an irregularly contoured disk; echinocyte II, a discoid cell with spicules over its surface; echinocyte III, an ovoid or spherical cell with 30 to 50 spicules evenly distributed over its surface; and spheroechinocytes, distinctly spherical cells whose spicules have become fine needlelike projections (Fig. 2), often not visible with the light microscope.

The echinocyte transformation can be produced (a) by washing fresh red cells free of plasma and examining them between a slide and coverslip of glass, (b) by extrinsic factors, and (c) by intrinsic factors.

#### 1. *Washing and the "Glass Effect"*

As was pointed out by Ponder (1948) and Ponder and Ponder (1962), the shape changes seen in saline are not attributable to the latter per se but rather to the fact that a protective layer of plasma protein is removed from the red cell surface by washing. The evidence for this is that three washes are required to produce crenation of 100% of the cells, since after two washes only approximately 10% of the cells will undergo the shape change. In addition to washing, however, as Furchgott and Ponder (1940) pointed out, this type of shape change occurs only if the cells are examined between slides

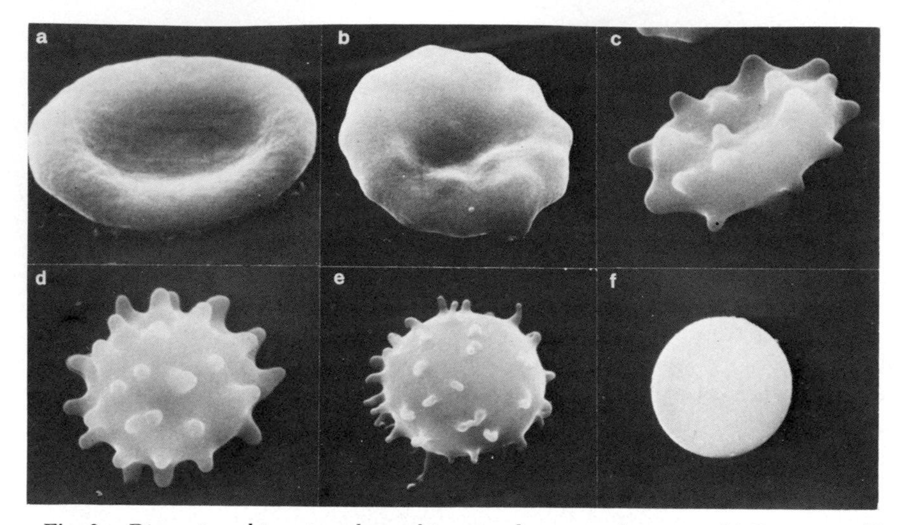


**Fig. 2.** Discocyte–echinocyte–spheroechinocyte shape transformation. (a) Discocyte. (b) Echinocyte I. (c) Echinocyte II. (d) Echinocyte III. (e) Spheroechinocyte I. (f) Spheroechinocyte II. Stages through echinocyte III are reversible, while spheroechinocytes I and II usually have lost membrane material and, therefore, cannot regain the biconcave shape.

and coverslips made of glass. The shape change does not take place if red cells that have been washed three times in saline are observed between two plastic coverslips, or between glass coverslips treated with silicone. It appears that close contact with glass or a high ratio of glass surface area to red cell surface area is necessary to produce the shape change, since exposure to the glass in a test tube is insufficient. This can be verified by observing red cells through the wall of a glass tube, using an objective with a long working distance. Alternatively, the shape of red cells in the original suspension itself can be evaluated simply by gentle agitation of a very dilute suspension, as Ponder pointed out (1948, p. 40). If the cells remain in their biconcave shape, the latter, through light diffraction, imparts a shimmering, pearllike appearance to the suspension; but if the cells have changed their shape to become crenated, this effect disappears. This phenomenon of light scattering by discocytes can be used to quantitate the disk-sphere change (Kuroda *et al.*, 1959; Hoffman, 1972; Oster and Zalusky, 1974).

Extensive washing (five to 12 times) will result in the crenation of some cells between plastic coverslips, although even these cells will revert to discocytes on reintroduction into normal fresh plasma. Thus, the difference between plastic and glass, although striking, is probably relative rather than absolute.

### 2. *Extrinsic Factors*

A variety of chemical agents will induce the echinocyte transformation. Some examples include fatty acids, lysolecithin, bile acids, barbiturates, dipyridamole, phenylbutazone, and salicylate. Although at first glance it is extremely difficult to appreciate any relationship between these various agents, Deuticke (1968) has suggested that echinocytogenic agents are lipid soluble nonpolar or anionic amphiphilic substances. Recently Mohandas and Feo (1975) quantitated the uptake of anionic phenothiazines by red cells and correlated the cellular concentrations of these drugs with the morphological changes induced. For a wide variety of agents they showed that red cell morphology depends only on the concentration of the substances in the red cells and not on the concentration of the agents in the suspending medium.

It is of interest and also of practical importance that normal plasma itself becomes echinocytogenic after incubation at 37°C for 24 hr (Bessis and Brecher, 1971). The echinocytogenic factor in plasma can be quantitated by dilution with normal plasma, and its evolution is via an enzymatic mechanism. Fresh plasma heated to 56°C for 30 min does not become echinocytogenic on subsequent incubation. Heating the plasma after it has become echinocytogenic does not destroy the echinocytogenic factor (Bessis and Brecher, 1971; Feo, 1972). It has been demonstrated that the echinocytogenic factor in incubated plasma is lysolecithin (Bessis and Brecher, 1971; Feo, 1972; Lichtman and Marinetti, 1972) and that the enzyme involved is lecithin–cholesterol acyltransferase (whose action results in the production of lysolecithin). The fatty acid content of plasma is not significantly changed by incubation at 37°C.

It is of interest that red cells are capable of recovering their discocyte shape after a period of incubation in an echinocytogenic environment (Lichtman and Marinetti, 1972), even though the environment remains echinocytogenic for fresh red cells introduced subsequently. This observation suggests that the red cell is able to compensate for the presence of echinocytogenic factors in the medium.

### 3. *Intrinsic Factors*

Nakao *et al.* (1960, 1961, 1962) demonstrated that the disk–echinocyte change also occurred when red cells were incubated in a glucose-free medium so as to deplete them of their ATP and that this change was reversible by reincubation in adenosine or adenine plus inosine to regenerate the cellular ATP. These observations were confirmed by Weed *et al.* (1969), who observed progressive loss of cellular deformability in conjunction with the shape changes and falling ATP. The latter authors also observed that the depleted cells tended to accumulate calcium. Although the addition of calcium to the medium has no immediate effect on red cell shape (Weed and Chailley, 1973), incorporation of calcium into the cell during hemolysis produces a ghost which is echinocytic. Simultaneous incorporation of sufficient ATP or ethylenediaminetetraacetic acid (EDTA) will prevent this shape change, suggesting that one function of ATP may be to maintain the free intracellular calcium at extremely low levels in the discocyte.

Echinocytes may also occur *in vivo*. "Burr" cells have been reported in uremia, bleeding peptic ulcer, carcinoma of the stomach, and heart disease,

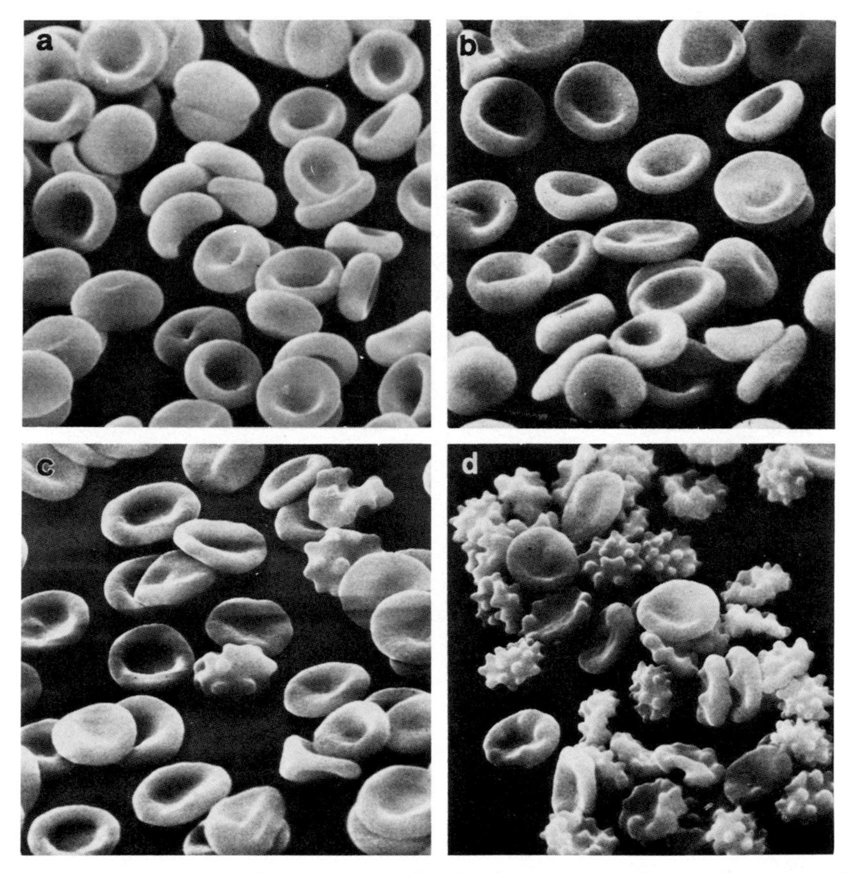


**Fig. 3.** (a–d). Scanning electron micrographs of erythrocytes washed and fixed at various pH values; (a) pH 6.0; (b) pH 7.4; (c) pH 9.2; (d) pH 10.1. (e–h) Erythrocytes washed and fixed at various pH values in the presence of $5 \times 10^{-3}$ *M* calcium; (e) pH 6.0; (f) pH 7.4; (g) pH 9.1; (h) pH 10.1. (Reproduced from *Nouvelle Revue française d'Hématologie*, 1972, with permission of the authors and publisher.)

and "spur" cells have been reported in certain anemias (Schwartz and Motto, 1949; Cooper, 1969). The term "burr" cell has been used to designate a nonreversible acanthocyte as well as an echinocyte. Such terms as "burr" cell and "spur" cell are used because they are often associated with specific disease states, implying that one can recognize the underlying disorder from the red cell morphology. This is not the case and the simpler designation, echinocyte, is preferable to indicate a cell with regularly arranged spicules or crenations that are reversible while the designation acanthocyte should be used for irregularly spiculated cells that do not revert to normal in fresh plasma or after washing.

Because of the possibility that changes *in vitro* may occur rapidly in the plasma of a patient thought to have echinocytes *in vivo*, fresh blood should be examined immediately between plastic coverslips to be sure that the crenated cells actually exist within the circulation. Examination of stained

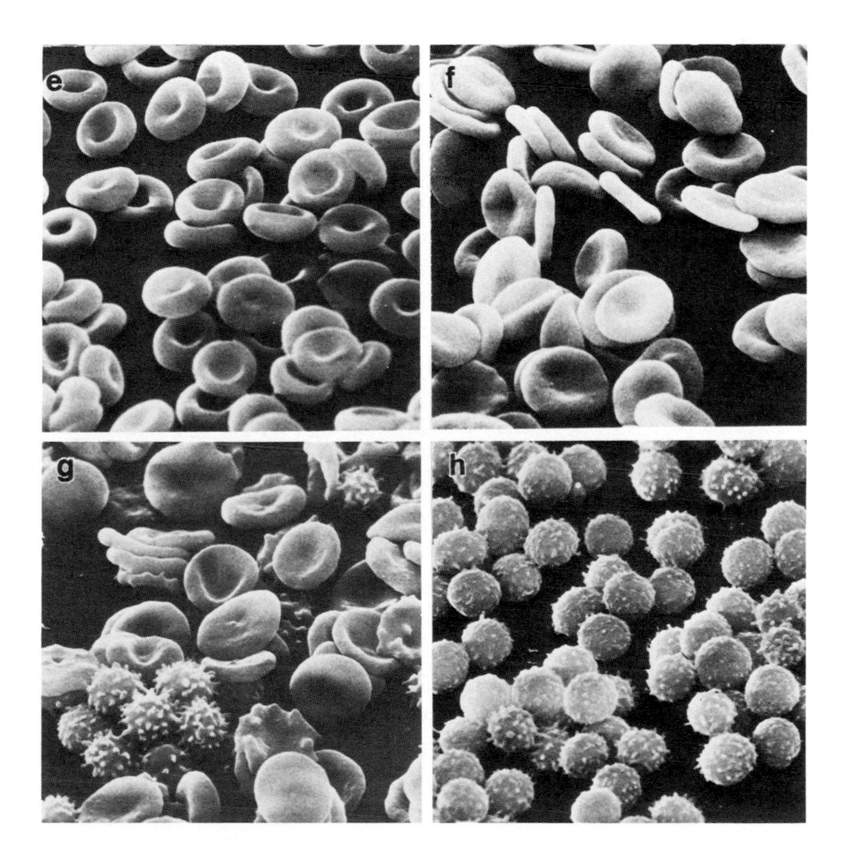

smears is less reliable even when they are prepared from free-flowing blood. When examined immediately, in the living state, between plastic coverslips, echinocytes that are truly present, for example, in pyruvate kinase deficiency (Tanaka *et al.*, 1962; Nathan *et al.*, 1966) will be evident.

#### 4. *Effect of pH*

Elevation of the pH above 8 produces the echinocytic shape change as illustrated in Fig. 3a. Although there is distinct heterogeneity within the population, at pH 10 irreversible changes (see below) take place and cell fusion may even be seen, particularly in the presence of calcium. Above pH 8, the presence of calcium enhances the echinocytic shape change produced by elevation of the pH (Weed and Chailley, 1973). It has been shown by Weed and Chailley (1973) that the charge on the glass surface significantly raises the pH, accounting for the "glass effect."

### B. *Discocyte–Stomatocyte Transformation*

This transformation is less well known than the discocyte–echinocyte equilibrium, although the end result in both cases is a spherocytic cell. Certain chemical agents that can be classified either as lipid-soluble cationic amphiphilic compounds or nonpenetrating anions (Deuticke, 1968) will produce the sequence of shape changes illustrated in Fig. 4, although the concentration required for each chemical varies considerably. Mohandas and Feo (1975) quantitated the relationship between cellular concentrations of cationic phenothiazines and the stomatocytic transformation induced. Progressive lowering of the pH below pH 6 will also induce the discocyte–stomatocyte transformation. These cup-shaped red cells are generally referred to as stomatocytes, because if these cells are examined on a dry blood smear, they will appear to have a stoma or mouth. As with the echinocytic agents, at low concentrations the morphological changes are completely reversible.

### C. *Mechanisms*

Various hypotheses to explain the mechanism or mechanisms that underly these dramatic and reversible changes in red cell shape exist at present. Hoffman (1972) has proposed that the echinocytic shape change induced by agents such as rose bengal depends upon a transmembrane concentration gradient of the responsible chemical agent, and that when the latter disappears the shape reverts to a disk. Deuticke (1968), on the other hand, has suggested that the stomatocytogenic effect of nonpenetrating anions relates to their ability to change the transmembrane pH gradient. Because of the

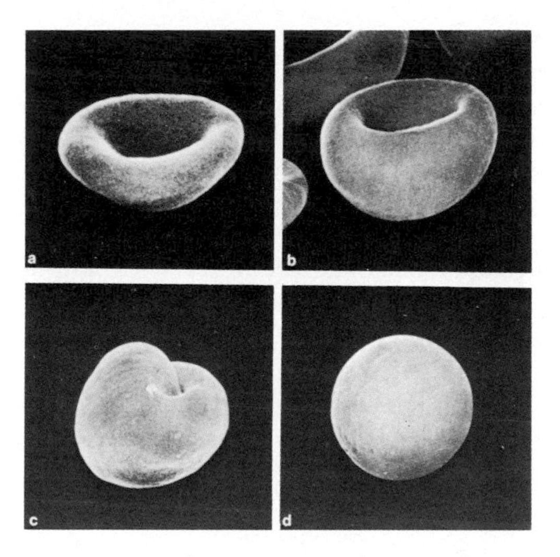


**Fig. 4.** Discocyte–stomatocyte–spherostomatocyte shape transformation. (a) Stomatocyte I. (b) Stomatocyte II. (c) Stomatocyte III. (d) Spherostomatocyte. Stages through stomatocyte III are reversible while spherostomatocytes have lost membrane into the cell interior by endovesiculation and cannot regain their biconcave shape.

augmentation of the echinocytic shape changes induced by drugs in the presence of calcium, and the appearance of the echinocytic shape change when red cells accumulate calcium during ATP depletion, it has been suggested (Weed and Chailley, 1973) that the echinocytic shape change is associated with a generalized increase in permeability to calcium. In such a model, the arrangement of a finite number of calcium-pump sites over the membrane surface might lead to heterogeneity in local calcium concentrations and account for the spicules in the echinocyte. Low pH and stomatocytic agents such as chlorpromazine may inhibit the calcium pump causing uniform accumulation of calcium or redistribution of calcium or both within the membrane itself (Weed and Chailley, 1973). Because of the extremely small quantity of calcium which is normally present in the red cell (Lichtman and Weed, 1973), the evidence is indirect but of interest, in view of the suggestion that there are proteins within the membrane which may undergo conformational changes in response to alterations in calcium–magnesium–ATP balance. Sheetz and Singer (1974) proposed an alternative hypothesis based on the differences in the protein and lipid compositions of the two halves of the bilayer. They proposed that the anionic compounds preferentially bind to the outer half of the bilayer, expand the outer surface area relative to the inner surface area of the membrane and cause the cell to crenate. The cationic compounds on the other hand preferentially bind to the inner half of the membrane, increase the inner surface area relative to the outer and cause the formation of cup-shaped cells. They attributed the difference between cationic and anionic drugs to electrostatic interaction between the drugs and the negatively charged lipid, phosphatidylserine, localized at the inner half of the bilayer.

## IV. Progression of Reversible to Irreversible Alterations

The reversible changes in erythrocyte shape induced by chemical agents are generally not accompanied by changes in red cell volume. Elevation of pH is accompanied by a decrease in volume and diminution in pH by an increase in volume, but the changes in volume can be clearly separated from the alterations in shape (Hoffman, 1972; Weed and Chailley, 1973). However, with both the echinocytic and stomatocytic shape changes, at high concentrations of the appropriate chemical agents the red cells will become spherical, usually with a decrease in volume. The change from a disk to a sphere is accompanied by a decrease in surface area resulting in irreversible changes in the membrane structure, properties, and function.

### A. *Microspherulation and Loss of Myelin Forms*

Erythrocytes which have undergone the discocyte–echinocyte transformation produced by extrinsic influences such as anionic or noncharged detergents or elevation of pH, or by an increase in intracellular calcium to ATP ratios ultimately lose membrane and become spherical as the process becomes more marked (Chailley *et al.*, 1973). As the cells become more crenated, the spicules become finer (Fig. 2), and ultimately the terminal portions of the spicules are lost from the cell as fragments, either in the form of long thin tubules or myelin forms, or in the form of microspherules (Fig. 5). These membrane fragments usually contain very small amounts of hemoglobin as can be demonstrated by the sickling of myelin forms from hemoglobin SS-containing cells (Bessis, 1973c). Such fragments contain all of the membrane lipids (Weed and Reed, 1966), and probably membrane protein as well. Figure 5 illustrates the loss of microspherules from a cell stored in plasma, a process which can lead to the loss of as much as 30% of the cell membrane from cells stored under blood bank conditions for 6 to 8 weeks (Haradin *et al.*, 1969). The phenomenon of microspherulation, budding or loss of entire portions of membrane to the exterior corresponds to exotropic vesiculation as discussed by Trump and Arstila (1980). Microspherulation is one type of fragmentation loss of membrane. Membranes vary in size and a useful definition is that "fragments" contain all of the components present in

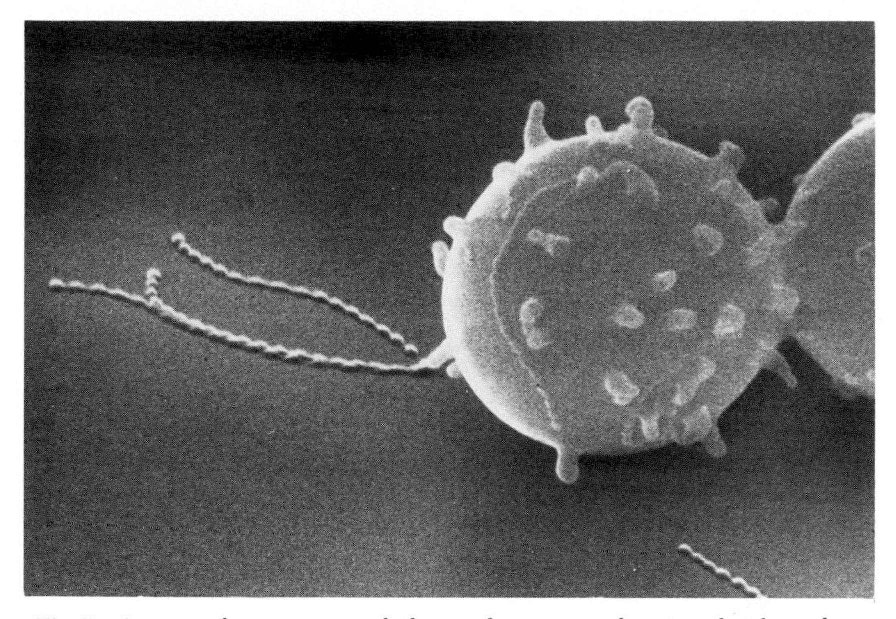

**Fig. 5.** Scanning electron micrograph showing fragmentation by microspherulation due to aging of blood. (Reproduced from *Nouvelle Revue française d'Hématologie*, 1972, with permission of the authors and publisher.)

the membrane of the intact cell and in the same proportion (Weed and Reed, 1966). The fact that such membrane fragments are dense enough to be centrifuged down indicates that they have a significant protein to lipid ratio (Weed and Reed, 1966; Fig. 6).

The ultimate consequence of loss of membrane material from the cell is a decrease in the surface area to volume ratio, a change which dictates that at the same volume, the cell must become spherical (Fig. 4). The increased osmotic fragility of these spherocytes clearly demonstrates the decreased surface area. This change, with the resultant decrease in cellular deformability, will markedly predispose stored cells that have lost membrane to rapid removal from the circulation after transfusion, e.g., in the spleen. Although with the light microscope, it may be difficult to distinguish the sphere which has arisen through the echinocytic transformation from one which has evolved through the stomatocytic transformation (see below), remnants of spicules may be seen in spheroechinocytes with the aid of the scanning electron microscope. Although loss of membrane material, and therefore surface area, as is seen in erythrocytes stored under blood-bank conditions for long periods of time, appears to preclude complete reversal of the spherical shape to a normal biconcave discocyte configuration, cells from stored

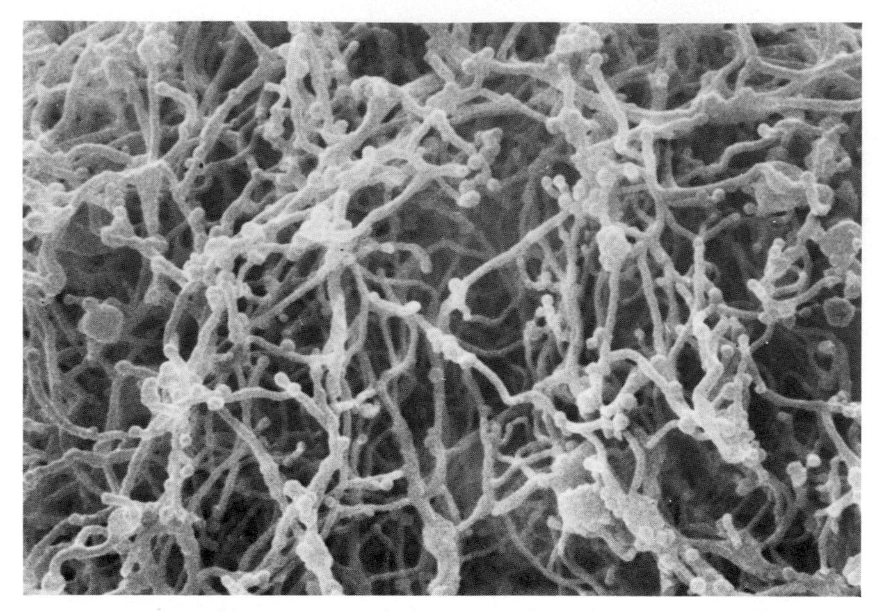

**Fig. 6.** Scanning electron micrograph showing membrane fragments from cells like those seen in Fig. 5, recovered from plasma. (Reproduced from *Nouvelle Revue française d'Hématologie*, 1972, with permission of the author and publisher.)

blood which have undergone the echinocytic transformation, almost to the point of sphering, revert to cup-shaped cells when their ATP is restored, but never to a biconcave disk. However, when spheroechinocytes are produced by drugs, they seem to be able to return to the biconcave discoid forms (presumably because of significantly less membrane loss).

### *B. Endovesiculation*

A comparable progression toward irreversible damage to the membrane occurs under the influence of environmental conditions or chemical agents which produce the stomatocytic change (Chailley *et al.*, 1973; Weed and Bessis, 1973; Fig. 7). Among the agents capable of inducing this type of shape change are the nonpenetrating anions, cationic amphiphilic agents as originally described by Deuticke (1968), primaquine (Ginn *et al.*, 1969), and vinblastine (Jacob *et al.*, 1972). As the cell becomes progressively more hemispheric, endovesiculation or esotropic vesiculation (Trump and Arstila, (1980) occurs in the portion of the rim adjacent to the remnant of the concavity. Penniston and Green (1968) and Ben-Bassat *et al.* (1972) have suggested that this process of endovesicle formation is dependent on intracellular levels

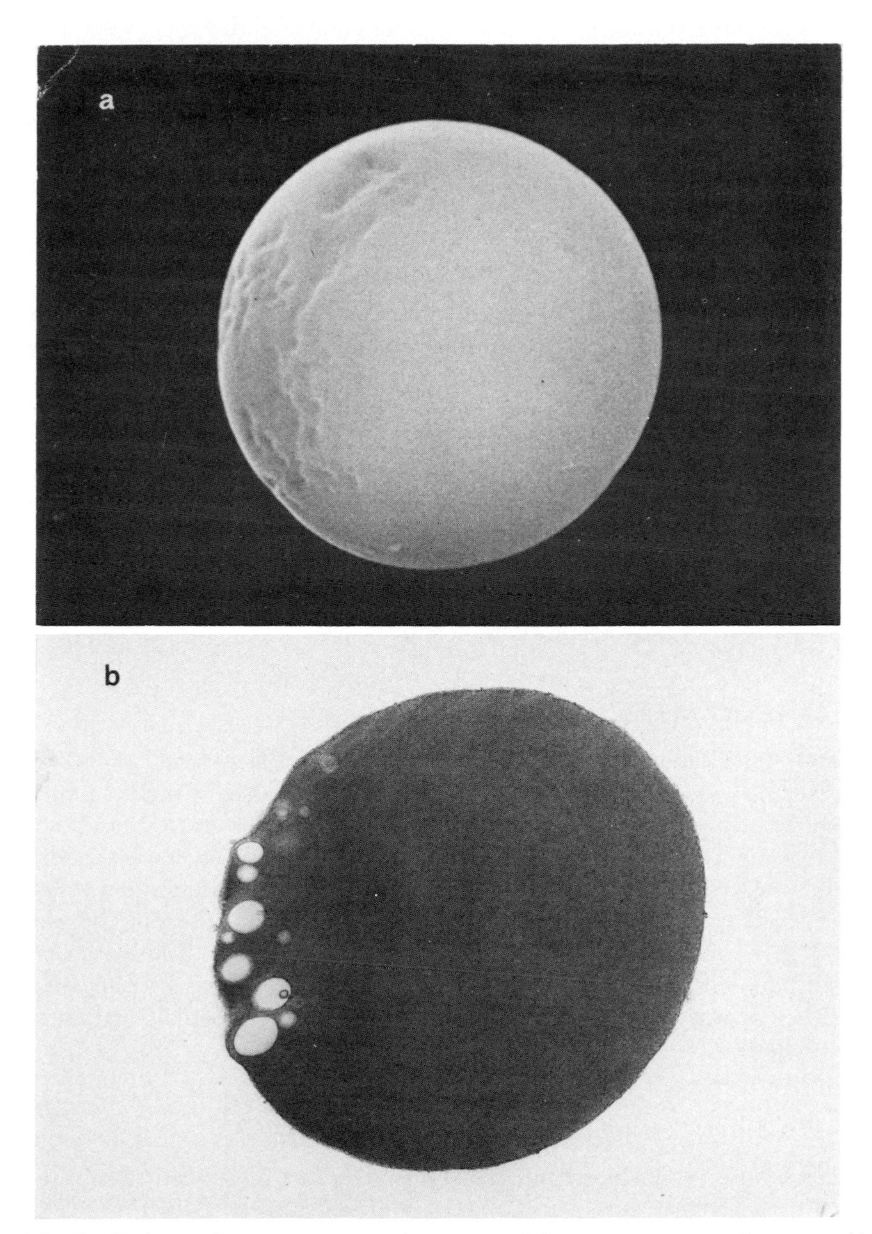


**Fig. 7.** Endovesiculation occurring in late stages of the stomatocytic transformation. (a) Scanning electron micrograph showing surface depressions at sites of vesicles. (b) Transmission electron micrograph showing vesicles in cross-section. (Reproduced from *Nouvelle Revue française d'Hématologie,* 1973, with permission of the authors and publisher.)

of ATP, although others have reported production of endovesiculation in red cells (Feo and Mohandas, 1977) and in red cell ghosts in the absence of ATP (Katsumata and Asai, 1972).

The morphological transformation of a biconcave erythrocyte to a spherostomatocyte, like that to a spheroechinocyte, is associated with loss of membrane material (in this case to the interior of the cell rather than to the outside) and this determines the formation of a nonreversible spherocyte. Apart from pH-induced shape changes which are accompanied by volume changes, it is important to emphasize that the end-stage spheres, both those produced by stomatocytic agents and those produced by echinocytic agents are in size and volume either equal to or smaller (10–15%) than the biconcave disks from which they were derived. The morphological similarity between cells in hereditary stomatocytosis, cells in hereditary spherocytosis (Leblond *et al.*, 1973), and those obtained experimentallfy from normal cells, suggests that this type of shape change, like the echinocytic shape change, may represent a final common pathway by which erythrocytes can respond to a variety of perturbations.

## V. Pathological States

### A. *General Comments*

The classification of the pathological states of the red cell membrane which follows has as its primary purpose the identification of various types of membrane pathology or pathophysiology. Second, in many cases such classification emphasizes how this pathology may lead to functional aberrations which result in shortened red cell life span. The classification reflects the current views of the authors and it is recognized that others might prefer alternative classification. This section on pathological states deals with structural, chemical, and functional abnormalities, primary or secondary, whether or not they lead to a shortened life span of the cells within the circulation.

### B. *Altered Ion Permeability and Transport*

The human erythrocyte has long been an excellent model cell for the study of the passive permeation and transport of $Na^+$ and $K^+$ ions through cell membranes. Reviews written over the past eight years dealing with the physiology of sodium and potassium transport through red cell membranes (Hoffman, 1966; Whittam and Wheeler, 1970; Dunham and Gunn, 1972) and reviews dealing with permeability and transport in pathological cells (Jandl,

1965; Nathan and Shohet, 1970; Parker and Welt, 1972) have traced the evolution of our understanding of these properties of erythrocyte membranes and the mechanisms by which they operate. In fact, abnormalities of monovalent cation permeability or transport or both have been found in a variety of pathological erythrocytes, as summarized in Table I. These observations have generated considerable interest in the possibility that such cells might suffer from faulty volume regulation and thereby be predisposed to hemolysis because of colloid osmotic swelling (Robinson, 1980). With the exception of the erythrocytes found in one type of stomatocytosis (Zarkowsky *et al.*, 1968), however, no pathological erythrocytes are, in fact, swollen. This suggests that alterations of permeability are generally balanced by compensating increases in active, outward transport of sodium. It is perhaps relevant to point out that dog erythrocytes circulate for approximately the same period of time (120 days) as human cells, yet dog cells are virtually lacking in an active transport mechanism for maintaining high intracellular potassium and low intracellular sodium levels.

In hereditary spherocytosis, which has been the focus of considerable study, there is a negative correlation between the increased sodium flux and shortened survival, i.e., the "leakiest" cells appear to have the best survival

**TABLE I**

*Abnormalities of $Na^+/K^+$ Permeability and/or Transport in Pathological Erythrocytes*

| *Disorder* | *Abnormality* | *References* |
|---|---|---|
| Hereditary spherocytosis | Increased $Na^+$ flux and transport | Harris and Prankerd (1953), Bertles (1957), Jacob and Jandl (1964), Wiley (1972) |
| Hereditary elliptocytosis | Increased $Na^+$ flux | Peters *et al.* (1966) |
| Stomatocytosis | Increased $Na^+$ and decreased $K^+$ content with increased osmotic fragility | Zarkowsky *et al.* (1968) |
| | Increased $Na^+$, slight decrease in $K^+$; decreased osmotic fragility | Miller *et al.* (1971b) |
| Sickle (SS) cell disease | $K^+$ loss in excess of $Na^+$ gain at low $pO_2$ | Tosteson *et al.* (1952), Tosteson *et al.* (1955) |
| Thalassemia | Increased $K^+$ permeability | Nathan *et al.* (1969), Gunn *et al.* (1972) |
| Malaria | Increased RBC $Na^+$ | Overman *et al.* (1949), Dunn (1969) |
| Amphotericin B therapy | Increased $Na^+$ and $K^+$ leak | Butler *et al.* (1965), Blum *et al.* (1969) |
| Pyruvate kinase deficiency | Increased $K^+$ leak and $H_2O$ loss | Nathan and Shohet (1970) |

(Wiley, 1970, 1972). In certain disorders, the propensity to lose $K^+$ appears to explain changes observed in the osmotic fragility pattern after incubation *in vitro* (Miller *et al.*, 1971b; Gunn *et al.*, 1972).

In point of fact, many of the studies summarized in Table I have been based on observations made with $Na^+$ and $K^+$ radioisotopes. As Sachs (1972) has pointed out, any studies of potassium movement unsupported by concomitant observations on chemical estimations of cation movement are subject to potential misinterpretation because of the phenomena of ouabain sensitive potassium–potassium or sodium–sodium exchanges which are evident when flux measurements are made using radioisotopes.

One disorder deserves special mention because it has been suggested (Nathan and Shohet, 1970) that loss of $K^+$ and water from pyruvate kinase-deficient cells may be critical in producing "dessicocytosis" or cellular rigidity secondary to $K^+$ and water loss from the cells. The ability of EDTA to block the effects of cyanide suggests that $Ca^{2+}$ accumulation may play a role in this disorder. The demonstration (Weed and Chailley, 1973) that $Ca^{2+}$ alone in high concentrations, or as a modifier of echinocytic agents, appears to augment the disk–echinocyte shape transformation, with changes in physical properties suggests that changes in $Ca^{2+}$ permeability or transport or both in pathological cells may prove to be fully as important as the $Na^+$ or $K^+$ abnormalities reported to date. The data in the literature suggests that the ion-transport abnormalities which lead to cellular dehydration (desiccytosis) such as those observed in pyruvate kinase deficiency and in irreversibly sickled cells may play a central role in the pathophysiology of these diseases but it appears that the tendency to accumulate sodium and water in red cells is much less threatening to the life span of the red cell. In fact, these transport abnormalities may actually be epiphenomena which accompany much more important, if more subtle, deviations of metabolic membrane integrity.

### C. *Altered Membrane Shape and Deformability*

The ability of the human erythrocyte to circulate for 120 days implies both considerable deformability and resistance to wear and tear. The normal human erythrocyte has a larger diameter of 8 $\mu$m, yet spends considerable time passing through capillaries in the general microcirculation which range from 3 to 12 $\mu$m in diameter, and occasionally must negotiate specialized regions of the microcirculation such as holes in the basement membrane, separating cords from sinuses in both bone marrow and spleen. These latter holes or potential openings range from 0.5 to 5.0 $\mu$m in diameter, obviously requiring considerable deformation of the cells in order to pass (Weed,

1970a,b; Weiss, 1974; deBoisfleury and Mohandas, 1977; Nightingale *et al.*, 1972). The excess of surface area to volume, which makes possible the biconcave disk shape of normal red cells, provides the geometric basis for normal deformability, while the fluid state of both the membrane and normal contents (primarily hemoglobin) are also critical. While the cell may change its shape in a reversible fashion without significant change in cellular deformability (Figs. 2 and 4), if the changes are accompanied by the loss of membrane material as described in Section IV, the decrease in surface area to volume ratio puts a geometric limitation on the cells' ability to pass into or through the smallest vessels. Assuming no prelytic loss of cations or changes in intrinsic membrane resistance to stretch, the osmotic fragility test is useful clinically to detect the presence of subtle spherocytosis, since it is a way of measuring red cell deformability as determined by the geometry of surface area to volume. Other techniques available for the measurement of red cell deformability are (a) viscometric techniques (Murphy, 1968), (b) filtration techniques (Messer and Harris, 1970), (c) the ektacytometer—a laser diffraction technique (Bessis and Mohandas, 1975), and (d) resistive pulse spectroscopy (Mel and Yee, 1975). By combining these different techniques it is possible to assess which of the three following factors contribute to reduced deformability in pathological states (a) intrinsic rigidity of the red cell membrane, (b) the internal viscosity of hemoglobin, and (c) the relationship of surface area to volume. Measurement of deformability is useful in understanding the pathological state since reduced red cell deformability invariably leads to reduced red cell survival and would also give clues to understand the primary abnormality (LaCelle, 1970; Weed, 1970a,b; Allard *et al.*, 1977).

### *1. Primary Membrane Disorders*

These are disorders in which the basic pathology is presumed to reside within the structure of the membrane itself.

*a. Hereditary Spherocytosis (HS).* This disorder is generally classified by most authors as a primary membrane defect. Membrane abnormalities which have been reported include abnormally low $Mg^{2+}$-ATPase activity (Nakao *et al.*, 1967), the absence of an electrophoretically identified protein component (Limber *et al.*, 1970; Gomperts *et al.*, 1972), although most investigators have failed to confirm the electrophoretic findings (Boivin and Galand, 1974). A functionally abnormal protein (Jacob *et al.*, 1971) has also been reported. As mentioned above, the hereditary spherocyte is characterized by increased passive sodium influx and extrusion, but this propensity is negatively correlated with the severity of the disorder (Wiley, 1970, 1972), and in fact, as pointed out by Selwyn and Dacie (1954) the HS cell has a decreased rather than an increased volume *in vivo* and *in vitro*, prior to hemolysis (Jandl, 1965; Weed and Bowdler, 1966).

Central to the question of whether the shape of the HS cell and shortened life span are related to an absent or abnormal structural membrane protein, is the important observation that HS marrow precursor cells are quite normal in appearance and physical characteristics (Leblond *et al.*, 1971a, 1971b, 1973). In fact it is well recognized that many of the cells seen in the peripheral blood from patients with HS appear to be biconcave disks on a stained blood smear. The underlying abnormality appears to predispose the cells to undergo the stomatocytic–spherocytic transformation, once released from the marrow (Leblond *et al.*, 1973). It has been suggested (LaCelle and Weed, 1969) that the shape and permeability abnormalities in HS cells are partially reversible by elevation of ATP levels, perhaps by ATP competing as a chelator for calcium with membrane-binding sites (Chau-Wong and Seeman, 1971). Recent evidence has been presented (Feig and Guidotti, 1974) that extends the observation of Nakao (1967), to suggest that the critical deficit in HS cells is a deficiency in the calcium-activated ATPase activity, which is linked to the calcium pump (Schatzmann and Rossi, 1971).

In regard to the possible role of increased membrane accumulation of calcium in hereditary spherocytes, it should be mentioned that Leblond *et al.* (1973) have observed that in the peripheral blood of patients with hereditary spherocytosis, only approximately 5% of the cells are true spherocytes while the majority of cells that are identifiably different from biconcave disks are stomatocytes (Fig. 8). This particularly is interesting since Schatzmann (1969) has demonstrated that chlorpromazine is an inhibitor of calcium-ATPase and transport in erythrocytes, while Deuticke (1968) and Weed and Chailley (1973) have pointed out that chlorpromazine is one of a group of chemical agents that will induce the stomatocytic shape change. The latter authors have also pointed out that preincubation in calcium enhances the stomatocytic shape change.

Under conditions of hypoxia and increased pH, such as may exist in the spleen, increased intracellular binding of ATP to hemoglobin (Udkow *et al.*, 1973) may make the ATP less available to bind intracellular calcium, thereby permitting more interaction of calcium with the membrane. This, in turn, may contribute to changes in membrane conformation and physical properties (Weed *et al.*, 1969; Leblond, 1972). Furthermore, as the shape change progresses in the environment of the spleen, perhaps at low pH and low oxygen tension, there is loss of membrane lipid (Reed and Swisher, 1966; Weed *et al.*, 1969; Bessis and Mandon, 1972; Weed and LaCelle, 1973). Such lipid loss contributes to the "conditioning" which occurs in the spleen, giving rise to cells with marked decreases in membrane lipid content

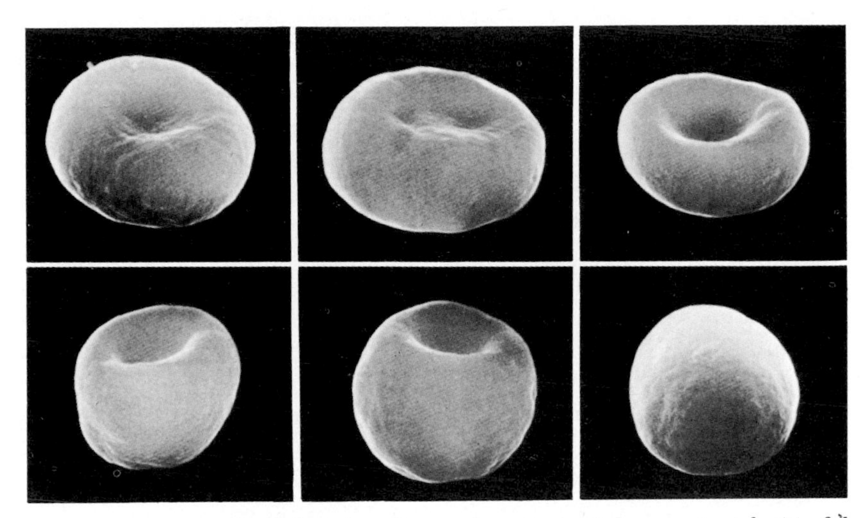

**Fig. 8.** Appearance of cells from hereditary spherocytosis. Only approximately 5% of the cells are truly small, smooth spherocytes as illustrated in the lower right photograph while the remainder represent varying steps in the discocyte–stomatocyte transformation as illustrated by the other cells in the sequence. (Reproduced from *Nouvelle Revue française d'Hématologie*, 1974, with permission of the authors and publisher.)

and, therefore, decreased surface area to volume ratios. With such loss of membrane material, changes in cell shape become irreversible (Chailley *et al.*, 1973). Observations by Greenquist and Shohet (1976) suggest another explanation for the membrane defect in hereditary spherocytosis. They observed decreased phosphorylation of spectrin in 22 patients with hereditary spherocytosis. This suggests that the primary membrane defect may reside in the kinase system and that the observed morphological and transport defects may result secondarily from failure of one of the unphosphorylated protein substrates to assume a required conformation.

In summary, the altered monovalent cation permeability and transport in HS appears to be secondary to the primary abnormality which may be either an altered membrane protein that predisposes to the stomatocytic–spherocytic shape change or a primary defect in the calcium-pump mechanism with the shape changes being secondary to calcium accumulation by the cells. Irreversibility in shape and deformability occur when membrane lipid loss takes place, possibly enhanced by the environment within the spleen.

*b. Elliptocytosis.* In this disorder, as with hereditary spherocytosis, the morphological abnormality is inherited as an autosomal dominant trait. Up to

90% of the erythrocytes may be elliptocytic or ovalocytic, or in the extreme, cigar-shaped. However, only 10% of the patients with this disorder actually have a hemolyltic anemia. In the latter patients, the laboratory features, clinical features and response to splenectomy are quite similar to the situation in hereditary spherocytosis, although to date no specific abnormalities of red cell energy metabolism, lipid, or protein composition of the membrane have been demonstrated (Cutting *et al.*, 1965). The morphological abnormality of this disorder is more or less reminiscent of the shape which a red cell assumes in slow passage through a narrow capillary. Experimentally, deliberate deformation of normal erythrocytes within a micropipette for prolonged periods of time, may produce a deformation which has a resemblance to the elliptocyte, suggesting the possibility that the membranes of elliptocytes may be more plastic, i.e., when forced to assume new configurations, they may tend to keep the new conformation rather than returning, as if elastic, to the original biconcave discocytic shape.

*c. Stomatocytosis.* As mentioned in Section III,B, stomatocytes may actually represent the morphological concomitants of several different pathophysiological disorders (Fig. 9). In one type of stomatocytosis (Zarkowsky *et al.*, 1968), the accumulation of $Na^+$ and cell water may be important in the pathogenesis of the hemolytic state. In a second variant (Miller *et*

**Fig. 9.** Stomatocytes from a case of hereditary stomatocytosis.

*al.*, 1971b), the cells actually have a decreased osmotic fragility, but no demonstrable hemoglobin or glycolytic defect, no abnormality of membrane lipid composition or abnormalities of membrane protein by electrophoresis. Thus, up to the present time, only the physiological abnormalities which accompany stomatocytosis have been characterized.

In addition to the variants of hereditary stomatocytosis which have been described, it is of interest to note that Rh null erythrocytes are morphologically stomatocytic and have a shortened life span (Levine *et al.*, 1973). This observation, plus the fact that most hereditary spherocytes are, in reality, stomatocytes, provides a strong stimulus to identify the structural abnormality which underlies this shape change.

### 2. *Abnormalities of Membrane Shape and Deformability Secondary to Intracellular or Extracellular Factors*

*a. Abnormalities of Hemoglobin.* Disorders of hemoglobin synthesis provide superb examples of disease states in which the entire pathophysiology can be traced from the gene to the alterations in physical properties of the erythrocyte which lead to red cell destruction. In recent years, evolution of increasingly sophisticated techniques for the study of hemoglobin synthesis and structure have contributed to our appreciation that a wide range of pathophysiological alterations may result from genetically determined abnormalities in hemoglobin synthesis. Over one-hundred structural variants of hemoglobin have been recognized as being attributable to specific qualitative amino-acid substitutions in the globin chains. Many of these, however, result in functional abnormalities of the hemoglobin without necessarily affecting cell membrane shape or deformability. However, two major types of disorders in hemoglobin synthesis are relevant to this consideration of erythrocyte pathology: aggregating hemoglobins, including hemoglobin S and C, in which intracellular polymerization or crystallization of the hemoglobin produces distortion of the membrane along with cellular and membrane rigidity; and second, an increasingly long list of abnormalities classified as unstable hemoglobins, involving amino acid substitutions in either the $\alpha$ or $\beta$ chains that are accompanied by physical instability of the hemoglobin, resulting in intracellular precipitation of globin chains. These intracellular precipitates, if extensive enough, may induce marked secondary deformation, tearing, or both of the erythrocyte membrane (see Section V,C,2,a,iii).

*i. Sickle hemoglobin.* The molecular defect and the various hypotheses regarding the structural abnormality and mechanism involved in the sickling process have been discussed extensively elsewhere by others. Although the physical properties of the membrane of an oxygenated sickle cell do not significantly differ from those of normal erythrocytes (except for the irreversibly sickled cell), when a cell sickles, it loses the characteristic flicker movement of a normal erythrocyte. The first slight deformation of the cell membrane becomes apparent as the hemoglobin rearranges itself within the cell and then the characteristic spicules of hemoglobin S appear and the cell becomes completely rigid (Bessis *et al.*, 1970). If the cell was a biconcave disk prior to sickling, the hemoglobin appears to polymerize within the plane of the disk, producing spicules at the ends of the cells or, alternatively, on one side giving rise to a "holly leaf" form (Fig. 10a). The parts of the cell that are most strikingly thinned out may not be visible under the light microscope and the membrane may even appear to be torn. However, with a phase microscope or scanning electron microscope, it is apparent that the membrane is not torn, but merely thinned by rearrangement of the cell contents. As mentioned above, the sickled cell manifests striking alterations in its monovalent cation permeability (Tosteson *et al.*, 1952; Tosteson *et al.*, 1955), although this phenomenon does not appear to be important in the pathogenesis of hemolysis of sickle cells. Potentially much more significant, however, is the suggestion that sickled cells also have increased permeability to divalent cations, specifically calcium. Eaton *et al.* (1973) and Palek (1973) have presented evidence that Hb SS cells contain increased amounts of calcium, although Statius Van Eps *et al.* (1971) found lower than normal calcium content in red cells from two patients. If intramembrane calcium is accumulated when the cell is sickled, the calcium may indeed play a role in the development of the irreversible sickle cell. Jensen *et al.* (1973) have presented evidence implicating ATP depletion in the presence of calcium in the generation of irreversibly sickled cells. Since ATP is the energy substrate for the calcium-extrusion pump in human erythrocytes, low ATP would also predispose to calcium accumulation.

The demonstration of the inability of sickled cells to pass through micropore filters (Jandl *et al.*, 1961) focused attention on the role of decreased sickle cell deformability in the pathogenesis of hemolysis and the microvascular occlusive phenomena of the disease (LaCelle and Weed, 1971; Usami *et al.*, 1975; Bessis and Mohandas, 1977).

In addition to the altered membrane cation permeability and the polymerization of the hemoglobin which induces membrane rigidity in sickle cells, additional membrane damage may occur. Fragmentation and loss of pieces of the sickle cell membrane may occur as a result of trauma to sickled cells within the microcirculation (Jensen *et al.*, 1967; Jensen and Lessin, 1970). Such traumatic separation of a portion of the cell may lead to osmotic swelling and lysis (Jensen *et al.*, 1967) and a decreased surface area to volume ratio with resultant loss of deformability (even when the cell is not sickled), or it may even contribute to the development of irreversibly sickled cells (Jensen, 1969).

**Fig. 10.** Drepanocytes (sickle cells). (a) Drepanodiscocyte in which sickling has taken place in a discocyte. (b) Drepanoechinocyte—this cell develops when a preformed echinocyte containing SS hemoglobin is then exposed to low oxygen tension. (c) Drepanostomatocyte—this cell results when a preformed stomatocyte containing SS hemoglobin is then deoxygenated.

The molecular interaction of the sickled hemoglobin with the membrane appears to depend upon the generation of helical rodlike structures (White and Krivit, 1967; Döbler and Bertles, 1968) which appear like microtubules that can be dissolved by cold or vinblastine (White and Krivit, 1967). These helical rods become arranged into parallel bundles of fibers in the long axis of the sickle cell spicules. They remain fixed to the internal surface of the membrane after freeze-fracture of sickle cells (Jensen and Lessin, 1970; Lessin, 1972). Lessin (1972) has also presented evidence that "microbodies" attached to the inner membrane surface of sickled cells play a role in microspherulation loss of membrane. Although as illustrated in Fig. 10a, sickling of a discocyte results in the classic deformation associated with the sickle cell, it is extremely interesting to note that if the oxygenated cell is exposed to an agent which induces the echinocytic shape transformation and the cell is then deoxygenated, the resulting morphological abnormality is the formation of a drepano-echinocyte (Bessis *et al.*, 1970) which has the three-dimensional regular distribution of spicules seen in an echinocyte formed from a normal erythrocyte. Some spicules may be as long as 3 $\mu$m, resembling the projections seen in sickled cells made from discocytes. However, the spicules usually terminate in a truncated cone with an average diameter of 0.2 $\mu$m at the tip (Fig. 10b). Similarly, if oxygenated sickled cells are treated with stomatocytic agents, such as chlorpromazine (Section III,B), and then deoxygenated, a drepano-stomatocyte is formed (Fig. 10c) (Weed and Bessis, 1973). These observations of cellular morphology, when sickling is induced after prior echinocytic or stomatocytic shape change, indicates that it is the conformation of the membrane which dictates the direction of polymerization of the sickled hemoglobin, and not the converse.

*ii. Hemoglobin C disease.* In this disorder, the molecular aberrations of one polar group in an external region of the $\beta$ chain predisposes to molecular aggregation that is favored somewhat by deoxygenation of the cell and by the relatively decreased solubility of hemoglobin C (Charache *et al.*, 1967). Increased organization of the hemoglobin C molecules even to the point of formation of intraerythrocytic crystals of hemoglobin C imparts a rigidity to the membrane which favors fragmentation and hemolysis of the cells, particularly within the spleen. Lessin *et al.* (1969) have demonstrated precrystalline linear alignment of hemoglobin molecules adjacent to the cell membrane. The red-cell fragmentation leads to the formation of microspherocytes which are seen in the circulation of homozygous hemoglobin C individuals. These cells in particular have decreased filterability and increased viscosity (Murphy, 1968; Allard *et al.*, 1977).

*iii. Heinz body disorders.* A Heinz body may be defined as an intracellular precipitate arising from a defect in the synthesis or denaturation of globin chains or hemoglobin itself. Included in this category are the unstable hemoglobin syndromes, the thalassemic disorders in which the abnormality in the relative rates of $\alpha$- and $\beta$-chain synthesis result in precipitation of the normal globin chain produced in excess and defects of the hexose monophosphate shunt which predispose to oxidative precipitation of hemoglobin–glutathione mixed disulfides. Table II summarizes the disease states in which Heinz bodies are encountered. The reader is referred to several well-referenced reviews (Jacob, 1970; Beutler, 1971; LaCelle, 1971; Miller *et al.*, 1971a; Miller, 1972) which discuss the biochemical basis for the formation of Heinz bodies in these disorders.

Through animal studies using phenylhydrazine, as well as studies of hemolytic anemia in workers exposed to coal tar, Heinz (1890) observed the anemia to be accompanied by the presence of spherical free-floating inclusions within the erythrocytes.

**TABLE II**

*Heinz Body Disorders*

|  |
|---|
| *Abnormalities of hexose monophosphate shunt predisposing to disulfide formation by oxidant drugs* |
| Glucose-6-phosphate dehydrogenase deficiency |
| Glutathione reductase deficiency |
| Glutathione synthetase deficiency |
| Glutathione peroxidase deficiency |
| *Unstable hemoglobins* |
| $\alpha$ Chain abnormalities, e.g., Hemoglobin Sinai |
| $\beta$ Chain abnormalities, e.g., Hemoglobins Köln, Zurich, Hammersmith |
| *Unbalanced synthesis of globin chains* |
| Thalassemia ($\alpha$ and $\beta$) |
| Hemoglobin H |
| Hemoglobin Lepore |

If the pathological erythrocytes contain a sufficient number of rigid Heinz bodies, the entire cell will be sequestered in the spleen and removed from the circulation (Weed and Weiss, 1966; Weiss and Tavassoli, 1970; Wennberg and Weiss, 1969). However, from the point of view of membrane pathology, interaction of individual Heinz bodies with the erythrocyte membrane may lead to recognizable alterations which are not necessarily lethal to the cell, but certainly predispose it to early removal from the circulation. It is clear that individual Heinz bodies may be "pitted" from the cell (Crosby, 1959; Wennberg and Weiss, 1969). This phenomenon presumably can occur in any restricted region of the microcirculation, but a particularly suitable location is the point at which Heinz body containing cells pass through the basement membrane separating splenic cords from sinuses (Koyama *et al.*, 1964; Leblond, 1973). As the Heinz bodies are removed from the cells during splenic passage, there is a loss of membrane material surrounding the Heinz bodies which leads to a decrease in surface area/volume ratio such that many cells which resume a symmetrical contour will become spherical (Miller *et al.*, 1971a).

In the thalassemias, prior to actual precipitation of a Heinz body, there may be partial insolubility or gelation of the globin chain which is present in excess. Whatever the pathogenesis, after pitting has occurred, the cell from which the Heinz body has been removed may retain a tear drop (dacryocytic) configuration (as illustrated in Fig. 11) possibly related to the hemoglobin adjacent to, or associated with the membrane. Following splenectomy, as might be anticipated, the fragmentation removal of Heinz bodies decreases markedly so that the Heinz bodies are readily evident within the circulating cells. There is a distinct decrease in dacryocytes and an increase in mean cell volume and mean cell hemoglobin because the cells are no longer being pitted (Slater *et al.*, 1968).

As indicated in Table I (Section V,B) thalassemic cells manifest altered cation permeability, although it is not clear that these abnormalities play any role in the pathogenesis of the hemolytic process itself. Jacob (1970) has presented evidence that Heinz bodies in the unstable hemoglobin syndromes may become associated with the membrane by formation of a mixed disulfide bonds that may, in turn, predispose the membrane to the development of pathological permeability properties. The key question is whether this process in unstable hemoglobin-containing cells and the association of Heinz bodies with the membrane in thalassemic cells is directly responsible for the altered permeability, or alternatively, whether pitting of even one Heinz body from some cells in the population may not be the proximate mechanism of membrane damage responsible for altered cation permeability.

Finally, apart from the formation of Heinz bodies, the use of some oxidant

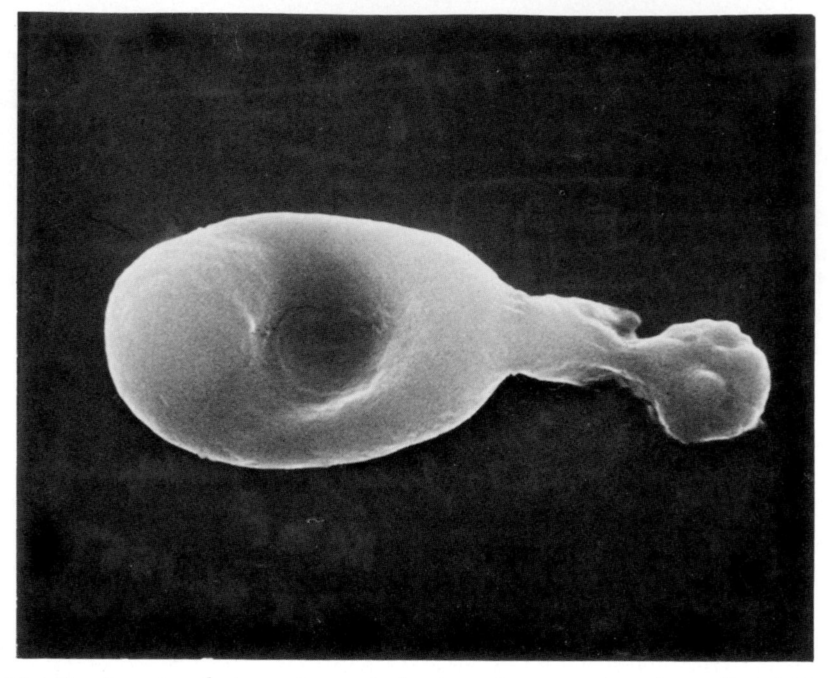

**Fig. 11.** Scanning electron micrograph showing a dacryocyte (tear-drop cell) with Heinz body being torn from the cell (G6PD-deficient blood).

drugs, such as primaquine (Ginn *et al.*, 1969) may induce endocytic vesicles in erythrocytes, but because of the concentration required by comparison to that which produces hemolysis *in vivo*, it seems unlikely that this is an important mechanism in the hemolytic process. In this regard, primaquine behaves like all other stomatocytogenic agents (see Section III,B).

*b. Immune Injury.* A wide spectrum of membrane pathology is induced by the interaction of various antibodies with erythrocytes, ranging from immediate breakdown of the membrane with escape of the hemoglobin, to sensitization without any evidence of altered membrane or cellular function. Table III summarizes the three general types of immune injury which can be recognized.

Direct hemolysis mediated by complement can be produced *in vitro* by some isoantibodies, such as anti-A or anti-B. Fortunately, direct release of hemoglobin from damaged cells is rarely seen *in vivo*, except in paroxysmal nocturnal hemoglobinuria (see Section V,D) or in rare cases of major transfusion mismatches. The physiological mechanism of complement (C)-mediated

TABLE III

*Types of Immune Injury to Erythrocytes*

| |
|---|
| Direct hemolysis (complement dependent) |
| Agglutination (formation of myelin forms) |
| Phagocytosis |

hemolysis is the production of defects in the membrane which may either be smaller than 32 Å in effective radius, resulting in colloid osmotic hemolysis, or larger than 32.5 Å in effective radius, permitting direct escape of hemoglobin (Sears *et al.*, 1963). The chemical and anatomical details of the action of complement in this type of immune hemolysis is a controversial and fascinating story. Observations by Humphrey and Dourmashkin (1969) suggested the production of distinct holes in the membrane. However, the fact that saponin produces similar appearing lesions in phospholipid–cholesterol mixtures (Lucy and Glauert, 1964) raises questions as to whether the "holes" might simply represent a rearrangement of membrane lipids. This whole subject has been reviewed in depth by Rosse and Lauf (1970). More recent evidence (Polley *et al.*, 1971) has suggested that attachment of the fifth component of complement (C5) alters membrane appearance to give the 80–100 Å lesions seen in immune lysis, but without hemolysis, if the C sequence is interrupted at that step. Freeze-etch studies by Seeman and Iles (1972) have identified the lesions as ringlike structures with a 90 Å central depression. These ringlike structures persist on the cell surface after lysis and resealing. Thus, the physiological defect is transient and apparently independent of the rings. In fact, the "rings" bear a striking resemblance to the ringlike proteins which Harris (1969) has suggested represent intrinsic structural constituents of the membrane. It is possible that the rings represent preexistent structures uncovered by the complement reaction.

A second type of membrane pathology arises secondary to agglutination. Although agglutination does not necessarily represent direct intercellular bridging by antibodies but rather some type of membrane–membrane adhesion or fusion once the membranes of the agglutinating cells have been altered by the antibody (Bessis, 1973b), forceful separation of agglutinated cells, either *in vitro* or as one may presume would occur *in vivo,* within the circulation, will cause myelin forms to break off (Fig. 12). Myelin forms contain all membrane constituents and small amounts of hemoglobin, as well. Their loss will contribute to a decrease in the surface area/volume ratio of the affected cells, i.e., the cells will be slightly more spherocytic after loss of myelin forms from their surface.

The third type of pathology resulting from immune injury is that which

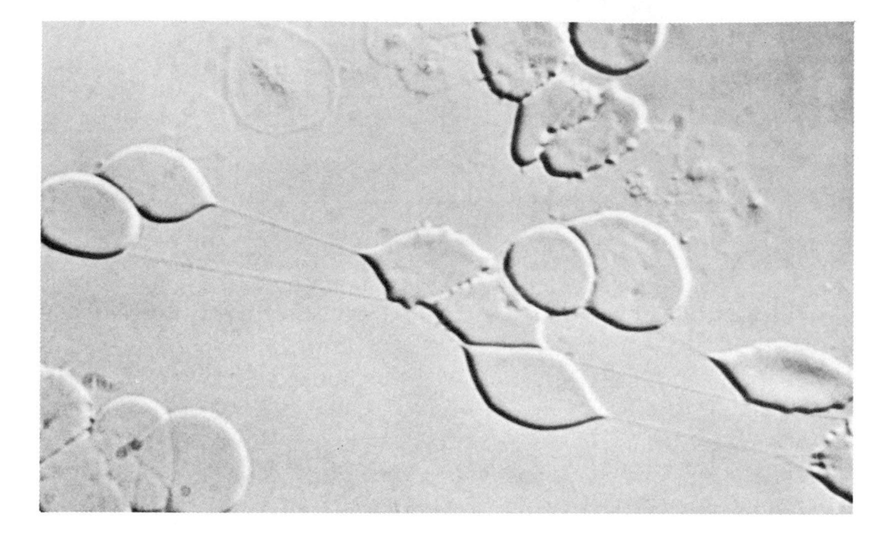

**Fig. 12.** Intercellular filaments and spindle-shaped deformation of red cells, seen when agglutinated red cells are forcibly separated (Nomarski interference microscope; × 2000).

predisposes to erythrophagocytosis. Two facets of this type of injury merit comment: the manner by which the antibody injury predisposes to phagocytosis and the phenomenon of partial phagocytosis. A variety of cellular functional abnormalities have been reported as a consequence of antibody binding to various red cell antigens (Rosse and Lauf, 1970), including the suggestion that the development of echinocytes was the result of the antibody effect. In fact, many of the reported observations must be considered in light of the fact that commercial antisera have been used, some containing preservative and others perhaps old enough to contain sufficient lysolecithin to produce the echinocytic shape change (Bessis and Brecher, 1971; Brecher and Bessis, 1972; Bessis and Weed, 1972; Longster and Tovey, 1972).

Although the specific lesion which initiates the antibody-injured red cell's predisposition to phagocytosis is not agreed upon, it seems likely that the primary injury may be followed by loss of some cellular constituent which acts as a chemotactic stimulant. This type of chemotaxis has been designated "necrotaxis" (Bessis, 1973a). Thus, the first step of phagocytosis is initiated. Adhesion of the erythrocyte to the phagocyte is undoubtedly enhanced by the decrease in surface charge which accompanies antibody coating (Pollack *et al.*, 1965). The third component of complement (C3) appears to be essential for erythrophagocytosis, although the mechanism of its action remains to be defined. When a $\gamma$G-coated red cell comes into contact with a phagocyte, usually a monocyte, adhesion and reciprocal invagination of the membranes of the two cells occurs (LoBuglio *et al.*, 1967) producing an effective decrease in the surface area of the adherent erythrocyte, which becomes spherical. It has been suggested that γG1 and γG3 alone are responsible for the spherical deformation occurring on contact with the phagocyte, while phagocytosis itself will occur only in complement-containing systems.

In the process of erythrophagocytosis, the entire erythrocyte may be ingested or, commonly, partial phagocytosis (Policard and Bessis, 1953; Bessis, 1955; Bessis and Boisfleury, 1970) occurs with bisection of the erythrocytes, with the resultant production of a smaller cell, a spheroschizocyte that may be released back into the circulation (Fig. 13). Obviously, the adverse decrease in cell surface area to volume ratio now compromises the further survival of these spherocytes.

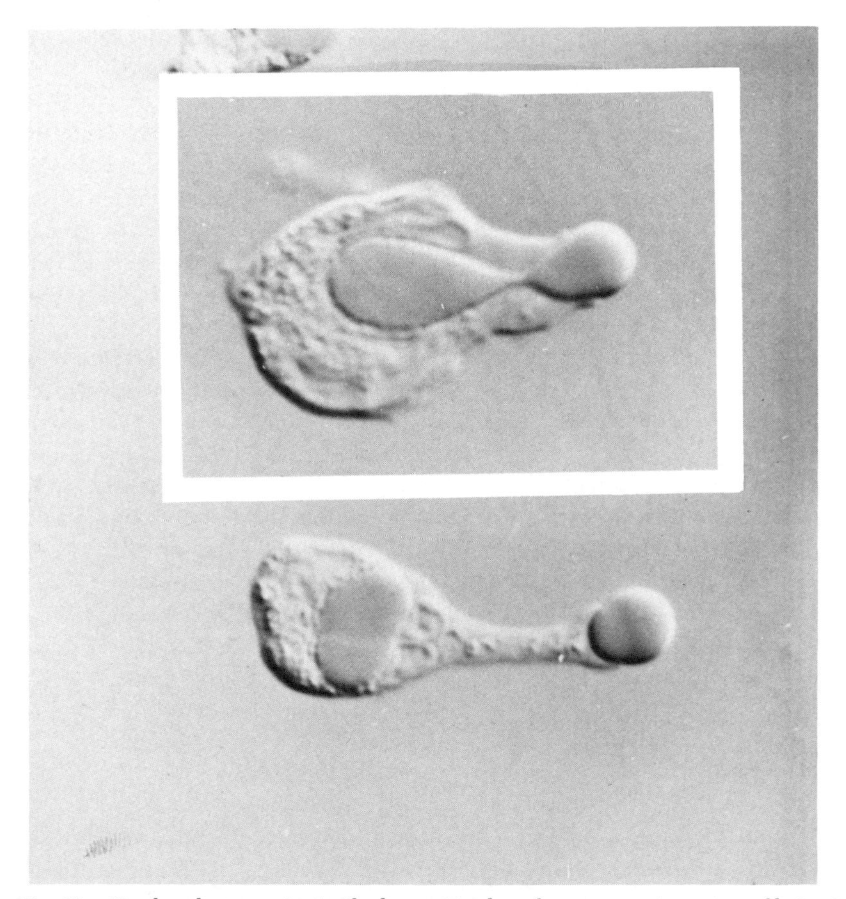

**Fig. 13.** Erythrophagocytosis. Antibody-sensitized erythrocytes are in process of being ingested by a phagocyte (Nomarski interference microscope).

*c. Heat Injury.* Temperatures between 48° and 65°C have a dramatic effect on red cell shape. The cells first lose their biconcave shape and then form multiple buds or microspherules (spheroschizocytes), each containing hemoglobin. These fragments are soft and labile at the elevated temperature, suggesting dissolution of the underlying membrane architecture. After minutes to hours, they will hemolyze. Although the chemical basis of the phenomenon remains to be established, it seems likely to shed light on other conformational changes which the membrane can undergo.

An interesting disorder, designated congenital pyropoikilocytosis has recently been described (Zarkowsky *et al.*, 1975). This abnormality is characterized by an apparent enhanced thermal sensitivity of the red cells. Budding, enhanced by temperatures of 50°C in normal cells, takes place at 44°C in these abnormal cells and even incubation at 37°C results in loss of myelin forms and microspherules over 17 hr. It is presumed that an abnormal membrane constituent, probably protein, underlies this peculiar abnormality which is associated with a hemolytic anemia.

*d. Mechanical Injury.* In addition to the antibody injury, the other most important type of extracellular damage is mechanical. Normal erythrocytes tolerate the mechanical stresses of passage through the normal microcirculation for 120 days remarkably well, deforming to negotiate narrow passages and returning to the discocytic shape. As has been outlined above, a variety of primary membrane and intracellular abnormalities may make the cell membrane rigid and unable to tolerate the physical stresses.

Disorders in which fragmentation of apparently normal erythrocytes is found include the various microangiopathic hemolytic states and iatrogenic situations (Brain *et al.*, 1962; Bull *et al.*, 1968; Bull and Kuhn, 1970) such as malfunctioning prosthetic heart valves in which a regurgitant jet of blood may subject the red cells to a sudden change in the rate of shear and fracture them. Under some circumstances, fragmentation of a red cell may produce two smooth and symmetrical smaller cells (both more spherical) (LaCelle, 1970) while at other times, the fragments or schizocytes (Bessis, 1972) are quite irregular in shape. What determines the postfracture shape is not clear.

### D. *Altered Immunological Reactivity*

#### 1. *Bombay Type*

The biological function of many red cell antigenic determinants is not at all clear. Antigenic constituents of the membrane such as the A and B antigens pose major problems in transfusion practice but their structural or biochemical contribution to normal cell function is not clear. This point is underlined by the erythrocytes from homozygous individuals, lacking the H gene. The H gene is essential for synthesis of the H antigen, to which is added the A and B antigenic determinants. Individuals of the Bombay type (lacking H determinants) have red cells without A, B, or H antigens. This poses major cross-matching problems for transfusion but the cells themselves are normal in their membrane (and intracellular) function (Levine *et al.*, 1973).

### 2. *Rh Null Cells*

In contrast to the Bombay type, there is a fascinating rare blood type characterized by nonreactivity to sera against all Rh-$H_2$ antigens and a hemolytic anemia characterized by stomatocytosis (Sturgeon, 1970). Most Rh null cases result from the action of a suppressor gene. The patients are homozygous for the gene; parents and offspring have Rh antigenic activity. The Rh antigen, unlike the glycoprotein ABO system (Green, 1968a,b; Whittemore *et al.*, 1969) appears to depend on integrity of membrane lipoprotein. Extraction of lipid with butanol will inactivate the antigen, although it can be restored by readdition of lipid. Rh antigenic activity, like membrane ATPase activity, is inhibited by sulfhydryl inhibitors and is pH dependent (Leddy *et al.*, 1970). Rh null cells appear to have some decrease in total membrane ATPase when corrected for the reticulocytosis found in the blood of these patients. All of these factors, i.e., the conditions for inhibition, the stomatocytosis with hemolysis and the possible lower ATPase activity suggest that the Rh antigen or its structural precursor is a functionally important membrane component.

### 3. *Paroxysmal Nocturnal Hemoglobinuria (PNH)*

This rare but fascinating disorder is characterized by an acquired defect of the red cell membrane that may result in increased binding of antibody or the membrane may be particularly susceptible to the hemolytic action of complement (Logue *et al.*, 1973; Rosse *et al.*, 1974) with significant hemolysis occurring at an acid pH. A great deal of work has been done on all aspects of these abnormal cells and the reader is referred to an excellent review by Rosse (1972) for an extensive bibliography.

The abnormal C-sensitive red cells make up a varying proportion of the population and these same cells appear to be deficient in membrane acetylcholinesterase activity, although the latter may not be functionally related to the hemolytic mechanism. A multitude of abnormalities has been suggested in this disorder, but no consistent alterations in membrane lipid composition (Bradlow, 1965) or membrane ultrastructure (Weinstein and McNutt, 1970) have been found, although treatment of normal cells with sulfhydryl-containing reagents renders their membranes sensitive to C lysis in a fashion similar to PNH cells (Sirchia and Dacie, 1967). This suggests that the fundamental defect is an acquired alteration in membrane protein (Jackson and Whittaker, 1972) which is of considerable interest since this disease is also associated with aplastic anemia and leukemia, which may involve erythroid, myeloid, and megakarocytic elements.

### 4. *Congenital Dyserythropoietic Anemias*

This is a group of three hereditary disorders characterized by ineffective erythropoiesis and morphological abnormalities of the erythroblasts (Heimpel *et al.*, 1971). To a variable extent, all three types manifest agglutination by anti-i and lysis by anti-i and anti-I.

Type I is characterized by loss of the nuclear envelope with chromatin bridges between nuclei of erythroblasts.

Type II, congenital dyserythropoietic anemia, has also been designated HEMPAS (hereditary erythrocyte multinuclearity with positive acidified serum test) (Crookston *et al.*, 1969). This is a disorder in which certain distinctive abnormalities of the erythrocyte membrane are found in association with hemolysis of the patient's cells in heterologous compatible serum at an acid pH (although not in the patient's own serum or plasma). The absence of lysis in the patient's own serum plus a negative sugar water test distinguishes this disorder serologically from PNH. Rosse *et al.* (1974) have presented evidence that HEMPAS cells bind either increased amounts of antibody, the fourth component of complement ($C_4$), or both. HEMPAS is also characterized by multinuclearity of the erythroblasts increases in certain erythrocyte enzyme activities (Valentine *et al.*, 1972; Rochant *et al.*, 1973; Verwilghen *et al.*, 1973) and significant reduplication of the plasma membrane arising from the endoplasmic reticulum (Heimpel *et al.*, 1970; Heimpel *et al.*, 1971; Wong, 1972; Breton-Gorius *et al.*, 1973) (Fig. 14). This membrane redundancy persists into the mature erythrocyte and can be identified with the phase microscope (Breton-Gorius *et al.*, 1973). There is also a dearth of polyribosomes in the erythroblasts. Recent studies of erythrocyte membrane protein have disagreed on whether any electrophoretic abnormality exists (Gockerman *et al.*, 1972; Boivin and Galand, 1974).

Type III, congenital dyserythropoietic anemia, is characterized by multinucleated and giant erythroblasts with tremendous variation in DNA content per erythroblast nucleus.

### 5. *Polyagglutinable Erythrocytes*

Certain red cells have alterations in membrane structure that render them polyagglutinable. This characteristic may pose significant problems in blood typing and cross matching (Mollison, 1967).

T polyagglutination is a phenomenon *in vitro* related to the presence of bacterial or viral neuraminidase activity that removes sialic acid, exposing a

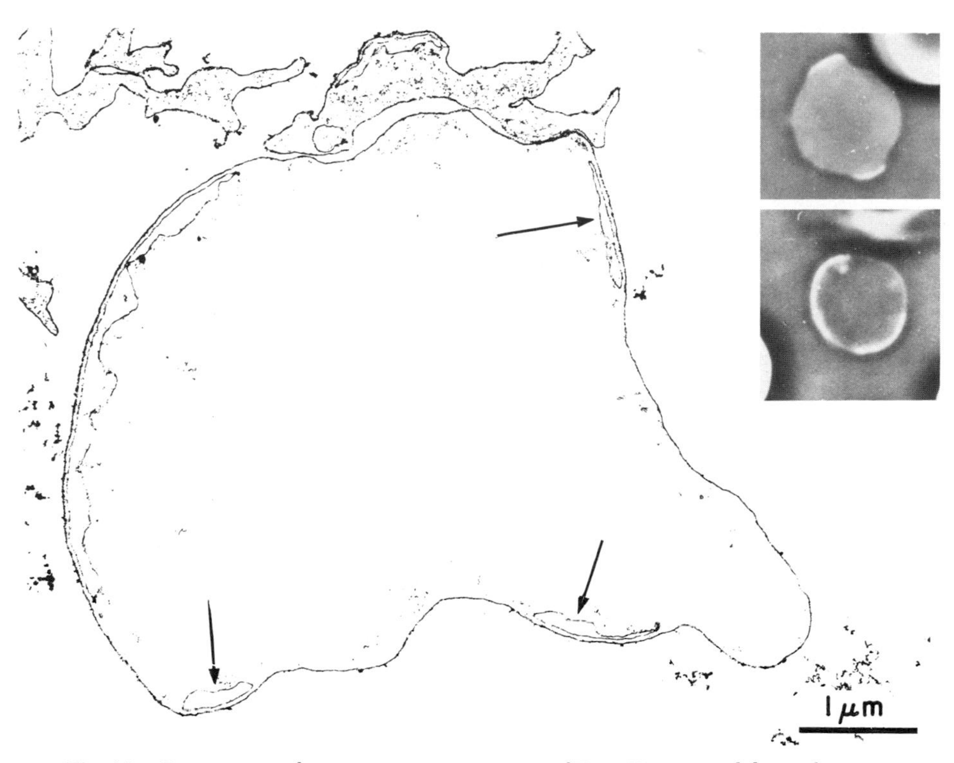


**Fig. 14.** Transmission electron microscopic section of Type II congenital dyserythropoietic anemia (HEMPAS). Notice the reduplication of the plasma membrane which can also be seen with phase contrast examination of freshly lysed erythrocyte ghosts (insert). (Reproduced from *Nouvelle Revue française d'Hématologie*, 1973, with permission from Mme. Breton-Gorius and the publisher.)

$\beta$-galactosyl residue. Since anti-T is present in most adult mammalian sera, these cells are polyagglutinable.

$T_n$ polyagglutination is an acquired condition *in vivo* in which *N*-acetyl-galactosamine residues are exposed in the face of decreased sialic acid content (Bird, 1972). The abnormality may be associated with leukopenia and thrombocytopenia and in some cases, shortened red cell life span (Dausset *et al.*, 1959), although several cases have had normal red cell survivals. $T_k$ (Bird and Wingham, 1972) and $T_{Cr}$ (Lalezari and Al-Mondhiry, 1973) variants have been described and in the $T_{Cr}$ case (Chien *et al.*, 1974), 90% of the cells were affected with increased rouleaux and polyagglutinability, but had normal survival.

Cad polyagglutination (Cazal *et al.*, 1968), by contrast, is an inherited dominant character. The erythrocytes in this case, have an antigen responsible for the polyagglutination because anti-Cad is present in most human adult sera. The sialic acid content of these cells is normal, however. By contrast to these polyagglutinable red cells, a genetically determined sialic acid deficiency (Furuhjelm *et al.*, 1969) with lack of the universal $En^a$ antigen is not associated with polyagglutinability.

### E. Alterations in Membrane Lipid Composition

Because of the lack of intracellular membranous structure, all of the lipid in mature erythrocytes can be found in the plasma membrane and recovered either from intact cells or ghosts (Weed *et al.*, 1963). Thus, in a variety of hemolytic disorders associated with fragmentation loss of membrane from the cell (Weed and Reed, 1966), a symmetrical decrease in all membrane lipids can be found and, in fact, such a finding is a good biochemical definition of fragmentation. The loss of membrane lipid from hereditary spherocytes (Langley and Felderhof, 1968; Weed and Bowdler, 1966; Reed and Swisher, 1966; Cooper, 1972) and from cells in autoimmune hemolysis and drug-induced hemolysis (Weed and Reed, 1966) are examples of the decreased membrane lipid values which appear secondary to loss of portions of the cell membrane.

Table IV summarizes abnormalities in the composition of red cell membrane lipids that have been reported to date, some of which are discussed in detail, below.

#### 1. *Abetalipoproteinemia (Plasma)*

In abetalipoproteinemia, the erythrocytes are essentially normal when released from the marrow but become increasingly abnormal in their lipid composition and morphology as they age *in vivo*. The evolution of the morphological change of the acanthocyte seen in this disorder parallels the exchange between plasma lipids and red cell lipids, although the mechanism of the shape change is not clear. The acanthocyte (Fig. 15) is an irregularly spiculated cell which, unlike the echinocyte, will not reverse its altered shape to a discocyte upon exposure to fresh plasma. In fact, (Fig. 15) the echinocytic shape change can be superimposed on an acanthocyte, e.g., by aged plasma, to give an echinoacanthocyte which will revert to an acanthocyte in fresh plasma. However, acanthocytes are not uniquely seen in abetalipoproteinemia, but they can also be found postsplenectomy in normal individuals (Brecher *et al.*, 1972) and in patients with liver disease. Although morphologically similar, these acanthocytes may differ from one another in the manner by which they become acanthocytes. Thus, the lipid changes in abetalipoproteinemia may only be one of several determinants leading to a

**TABLE IV**

*Abnormalities of Erythrocyte Lipids*

| *Disorder* | *Morphology* | *Red blood cell (RBC) abnormality* | *Mechanism* | *References* |
|---|---|---|---|---|
| Abetalipoproteinemia | Acanthocytosis | Progressive increase in RBC sphingomyelin/lecithin ratio | Faulty bowel absorption of triglyceride; exchange of plasma lipid with RBC | Ways *et al.* (1963) |
| High phosphatidylcholine hemolytic anemia | | Increased RBC lecithin, normal cholesterol and total phospholipid ↑ cation fluxes | Decreased transacylation of membrane lyso-PE | Jaffé and Gottfried (1968), Shohet *et al.* (1973) |
| Plasma LCAT deficiency | Target cells | $\alpha$-Lipoprotein deficiency; RBC have 50–80% increase in cholesterol; normal total phospholipid with elevated PC/PE+SM | Increased plasma-free cholesterol | Torsvik *et al.* (1968) |
| Liver disease | | | | |
| Obstructive | Target cells "spur" (crenated) | Elevated total cholesterol, phospholipid (lecithin) | Acquired LCAT deficiency may explain elevated cholesterol, excess lipid may explain target cells | Neerhout (1968b), Cooper (1970), and Cooper *et al.* (1972) |
| Zieve's syndrome (parenchymal) | Spherocytes | Same as obstructive disease | ?Elevated serum triglycerides, ?pancreatitis | Zieve (1958), Westerman *et al.* (1968) |
| *Clostridium welchii* infection | Spherocytes | Lysis of RBC lipids | Phospholipase C produced by bacteria | Shohet (1972) |
| Vitamin E deficiency | Pyknocytosis | Lipid peroxidation | Lack of protection by vitamin E against peroxidation | Oski and Barness (1967) |

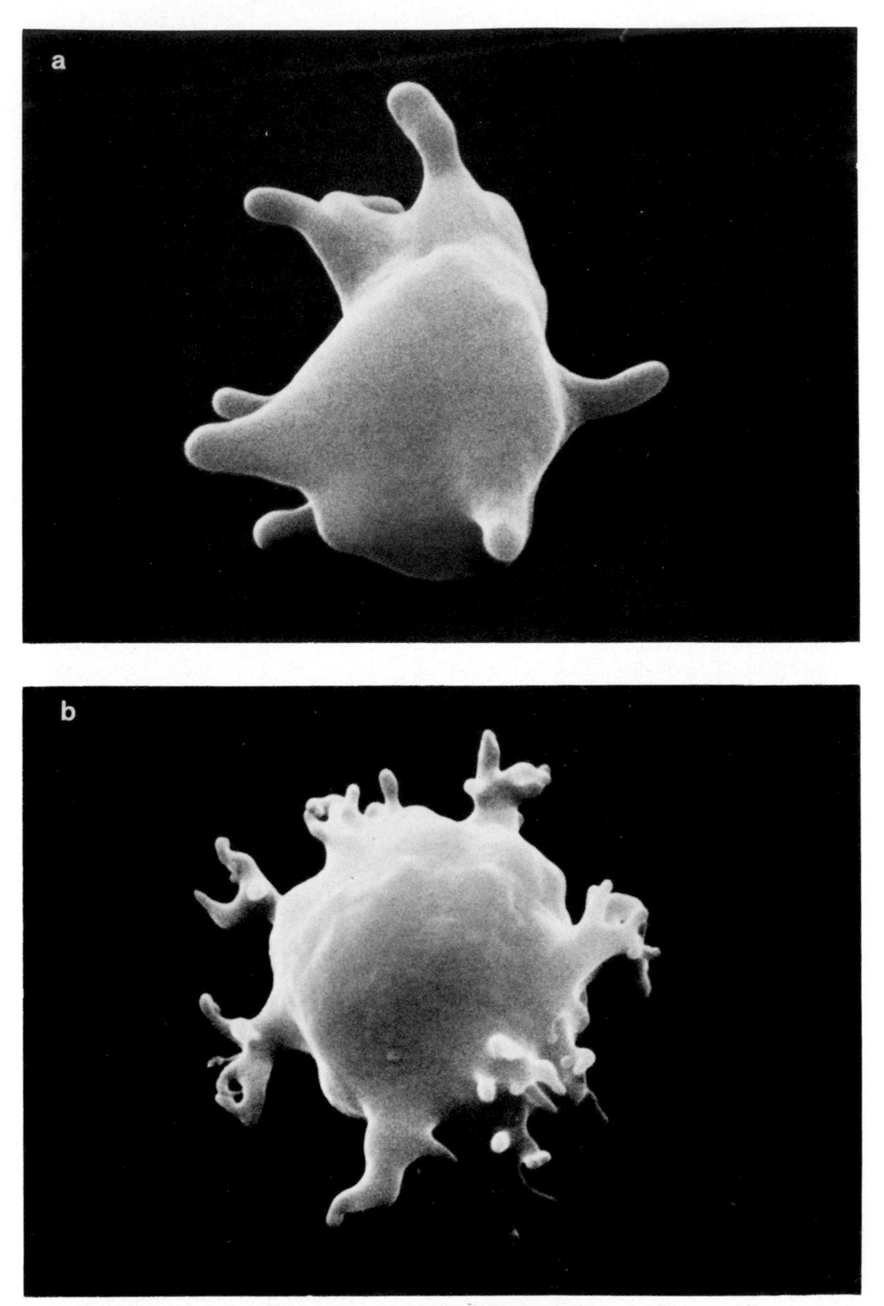


**Fig. 15.** (a) An acanthocyte (see text for definition). (b) An echinoacanthocyte.

final common mechanism producing irreversible, irregularly spiculated erythrocytes.

### 2. *High Phosphatidylcholine Hemolytic Anemia (HPCHA)*

A familial disorder that has been designated high phosphatidylcholine hemolytic anemia (HPCHA) has been characterized in some detail over the past five years (Jaffé and Gottfried, 1968; Shohet *et al.*, 1973). The defect appears to be faulty transfer of esterified fatty acid from phosphatidylcholine (PC) to phosphatidylethanolamine (PE) with resultant increase in PC and decrease in PE. The resultant alteration in lipid composition is associated with increased cation permeability and evidence of cell swelling which may well play a role in the pathogenesis of the hemolysis.

### 3. *Plasma Lecithin–Cholesterol Acyltransferase (LCAT) Deficiency*

This is a disorder which results in an increase in red cell cholesterol with target cell formation, bearing similarities to the situation in liver disease, described below.

### 4. *Liver Disease*

*a. Obstructive.* In obstructive liver disease, the erythrocytes accumulate cholesterol and lecithin. The depression of plasma LCAT activity due to the increased bilirubin may account for the elevated red cell cholesterol, but the explanation for the elevated red cell lecithin is not apparent. Presumably, the target shape reflects the excess of membrane lipid to volume, but the occasional crenated ("spur") cells cannot be explained solely on the basis of excess membrane. "Spur" cells or acanthocytes in liver disease have higher cholesterol to phospholipid and cholesterol to lecithin ratios. The morphology may represent an induced secondary phenomenon, related to the lipid abnormality.

*b. Zieve's Syndrome.* In Zieve's syndrome, a spherocytic hemolytic anemia in association with elevated plasma triglycerides is seen, yet the red cell lipids are similar to those in obstructive jaundice (Westerman *et al.*, 1968). There is no clear evidence, however, that the lipid composition per se is critical. The morphology and hemolysis may be related to some separate effect.

## Acknowledgments

This work was supported in part by Institutional National Research Service Award HL-07100 from the National Heart and Lung Institute, part by U. S. Public Health Service Grants

HL-06241 and AM 16095, the U.S. Atomic Energy Project at the University of Rochester and in part by INSERM, Unite 48 (Institut de Pathologie Cellulaire, Paris, France). It has been assigned publication #UR-3490-516.

## References

Allard, C., Mohandas, N., and Bessis, M. (1977). *Blood Cells* **3,** 2.
Ben-Bassat, I., Bensch, K., and Schrier, S. L. (1972). *J. Clin. Invest.* **51,** 1833.
Bertles, J. F. (1957). *J. Clin. Invest.* **36,** 816.
Bessis, M. (1955). *Ann N.Y. Acad. Sci.* **59,** 986.
Bessis, M. (1972). *Nouv. Rev. Fr. Hémat.* **12,** 721.
Bessis, M. (1973a). *In* "Living Blood Cells and Their Ultrastructure" p. 48. Springer-Verlag, New York.
Bessis, M. (1973b). *In* "Living Blood Cells and Their Ultrastructure", p. 166. Springer-Verlag, Berlin and New York.
Bessis, M. (1973c). *In* "Living Blood Cells and Their Ultrastructure" p. 247. Springer-Verlag, Berlin and New York.
Bessis, M., and Boisfleury, A. de (1970). *Nouv. Rev. Fr. Hematol.* **10,** 223.
Bessis, M., and Brecher, G. (1971). *Nouv. Rev. Fr. Hematol.* **11,** 305.
Bessis, M., and Mandon, P. (1972). *Nouv. Rev. Fr. Hematol.* **12,** 443.
Bessis, M., and Mohandas, N. (1975). *Blood Cells* **1,** 307.
Bessis, M., and Mohandas, N. (1977). *Blood Cells,* **3,** 229.
Bessis, M., and Weed, R. I. (1972). *Proc. Scanning Electron Microscope Symp. 5th* (Om Johari, ed.), p. 289.
Bessis, M., Dobler, J., and Mandon, P. (1970). *Nouv. Rev. Fr. Hematol.* **10,** 63.
Beutler, E. (1971). *Semin. Hematol.* **8,** 311.
Bird, G. W. G. (1972). *Acta Haematol.* **47,** 193.
Bird, G. W. G., and Wingham, J. (1972). *Br. J. Haematol.* **21,** 443.
Blum, J. R., Shohet, S. B., Nathan, D. G., and Gardner, F. H. (1969). *J. Lab. Clin. Med.* **73,** 980.
Boivin, P., and Galand, C. (1974). *Nouv. Rev. Fr. Hematol.* **14,** 355.
Bradlow, B. A., Lee, J., and Rubenstein, R. (1965). *Br. J. Haematol.* **11,** 315.
Brain, M. C., Dacie, J. V., and Hourihane, D. O'B. (1962). *Br. J. Haematol.* **8,** 358.
Branton, D. (1966). *Proc. Nat. Acad. Sci.* **55,** 1048.
Brecher, G., and Bessis, M. (1972). *Blood* **40,** 333.
Brecher, G., Haley, J. E., and Wallerstein, R. O. (1972). *Nouv. Rev. Fr. Hematol.* **12,** 751.
Breton-Gorius, J., Daniel, M. T., Caluvel, J. P., and Dreyfus, B. (1973). *Nouv. Rev. Fr. Hematol.* **13,** 23.
Bretscher, M. (1972). *Nature (London), New Biol.* **236,** 11.
Bull, B. (1973). *In* "Red Cell Shape, Physiology, Pathology, Ultrastructure" (M. Bessis, R. I. Weed, and P. F. Leblond, eds.), p. 125. Springer-Verlag, Berlin and New York.
Bull, B. S., Rubenberg, M. L., Dacie, J. V., and Brain, M. C. (1968). *Br. J. Haematol.* **14,** 643.
Bull, B. S., and Kuhn, I. N. (1970). *Blood* **35,** 104.
Butler, W. T., Alling, D. W., and Cotlove, E. (1965). *Proc. Soc. Exp. Biol. Med.* **118,** 297.
Carraway, K. L., Triplett, R. B., and Anderson, D. (1975). *Biochem. Biophys. Acta* **379,** 571.
Cazal, P., Morris, M., Caubel, J., and Brines, J. (1968). *Rev. Fr. de Transfus.* **11,** 209.
Chailley, B., Weed, R. I., Leblond, P. F., and Maigné, J. (1973). *Nouv. Rev. Fr. Hematol.* **13,** 71.

Charache, S., Conley, C. L., Waugh, D. F., Ugoretz, R. J., Spurrell, R. J., and Gayle, E. (1967). *J. Clin. Invest.* **46,** 1795.
Chau-Wong, M., and Seeman, P. (1971). *Biochim. Biophys. Acta* **241,** 473.
Chien, S., Cooper, G. W., Jan, K., Miller, L. H., Howe, C., Usami, S., and Lalezari, P. (1974). *Blood* **43,** 445.
Cooper, R. A. (1969). *J. Clin. Invest.* **48,** 1820.
Cooper, R. A. (1970). *Semin. Hematol.* **7,** 296.
Cooper, R. A. (1972). *J. Clin. Invest.* **51,** 16.
Cooper, R. A., Diloy Puray, M., Lando, P., and Greenberg, M. S. (1972). *J. Clin. Invest.* **51,** 3182.
Crookston, J. H., Crookston, M. C., Burnie, K. L., Francone, W. H., Dacie, J. V., Davis, J. A., and Lewis, S. M. (1969). *Br. J. Haematol.* **17,** 11.
Crosby, W. H. (1959). *Blood* **14,** 399.
Cutting, H. O., McHugh, W. J., Conrad, F. G., and Marlow, A. A. (1965). *Am. J. Med.* **39,** 21.
Davson, H., and Danielli, J. F. (1943). "The Permeability of Natural Membranes." Cambridge Univ. Press, London and New York.
Dausset, J., Moullec, J., and Bernard, J. (1959). *Blood* **14,** 1079.
DeBoisfleury, A., and Mohandas, N. (1977). *Blood Cells* **3,** 197.
Deuticke, B. (1968). *Biochim. Biophys. Acta* **163,** 494.
Döbler, J., and Bertles, J. F. (1968). *J. Exp. Med.* **127,** 711.
Dödge, J. T., Mitchell, C., and Hanahan, D. J. (1963). *Arch. Biochem. Biophys.* **180,** 119.
Dunham, P. B., and Gunn, R. B. (1972). *Arch. Int. Med.* **129,** 241.
Dunn, M. J. (1969). *J. Clin. Invest.* **43,** 674.
Eaton, J. W., Skelton, T. D., Swofford, H. S., Kolpin, C. E., and Jacob, H. S. (1973). *Nature (London)* **246,** 105.
Engelman, D. M. (1970). *J. Mol. Biol.* **47,** 115.
Evans, E. A. (1973a). *Biophys. J.* **13,** 926.
Evans, E. A. (1973b). *Biophys. J.* **13,** 941.
Evans, E. A., and Hochmuth, R. M. (1977). *J. Membr. Biol.* **30,** 351.
Feig, S. A., and Guidotti, G. (1974). *Biochem. Biophys. Res. Commun.* **58,** 487.
Feo, C. (1972). *Nouv. Rev. Fr. Hematol.* **12,** 757.
Feo, C., and Mohandas, N. (1977). *Nature (London)* **265,** 166.
Fidalgo, B. V., Katayama, Y., and Najjar, V. A. (1967). *Biochemistry* **6,** 3378.
Finean, J. B. (1962). *Circulation* **26,** 1151.
Firkin, B. F., and Wiley, J. S. (1966). *In* "Progress in Hematology" (E. B. Brown and C. V. Moore, eds.), Vol. V, p. 36. Grune & Stratton, New York.
Fung, Y. C., and Tong, P. (1969). *Biophys. J.* **8,** 175.
Furchgott, R. F., and Ponder, E. (1940). *J. Exp. Biol.* **17,** 117.
Furuhjelm, V., Myllylä, G., Nevanlinna, H. R., *et al.* (1969). *Vox. Sang.* **17,** 256.
Gilliland, B. C., Leddy, J. P., and Vaughan, J. H. (1970). *J. Clin. Invest.* **49,** 898.
Ginn, F. L., Hochstein, P., and Trump, B. F. (1969). *Science* **164,** 843.
Gockerman, J. P., Durocher, J., and Conrad, M. (1972). *Blood* **40,** 945.
Gomperts, E. D., Meta, J., and Zail, S. S. (1972). *Br. J. Haematol.* **23,** 363.
Gordesky, S. E., and Marinetti, G. V. (1973). *Biochem. Biophys. Res. Comm.* **50,** 1027.
Green, F. A. (1968a). *Nature (London)* **219,** 86.
Green, F. A. (1968b). *J. Biol. Chem.* **243,** 5519.
Greenquist, A. C., and Shohet, S. B. (1976). *Blood* **48,** 877.
Greenwalt, T. J., ed. (1973). "The Human Red Cell *in vitro.*"
Gunn, R. B., Silvers, D. N., and Rosse, W. F. (1972). *J. Clin. Invest.* **51,** 1043.
Haradin, A. W., Weed, R. I., and Reed, C. F. (1969). *Transfusion* **9,** 229.

Harris, J. R. (1969). *J. Mol. Biol.* **46**, 329.
Harris, E. J., and Prankerd, T. A. (1953). *J. Physiol.* **121**, 470.
Heimpel, H., Forteza-Vila, J., and Queisser, W. (1970). "Proc. XIII Int. Congr. Hematol." (J. F. Lehmanns, ed.), p. 391.
Heimpel, H., Forteza-Villa, J., Queisser, W., and Spiertz, E. (1971). *Blood* **37**, 299.
Heinz, R. (1890). *Virchows Arch. Pathol. Anat. Physiol.* **122**, 112.
Hochmuth, R. M., and Mohandas, N. (1972). *J. Biomech.* **5**, 501.
Hochmuth, R. M., Mohandas, N., and Blackshear, P. L., Jr. (1973). *Biophys. J.* **13**, 747.
Hoffman, J. F. (1966). *Am. J. Med.* **41**, 666.
Hoffman, J. F. (1972). *Nouv. Rev. Fr. Hematol.* **12**, 771.
Hsia, J. C., and Boggs, J. M. (1972). *Biochim. Biophys. Acta* **266**, 18.
Humphrey, J. H., and Dourmashkin, R. R. (1969). *In* "Advances in Immunology" (F. J. Dixon, Jr., and H. G. Kunkel, eds.), Vol. 11, p. 75. Academic Press, New York.
Jackson, P., and Whittaker, M. (1972). *Clin. Chim Acta* **41**, 299.
Jacob, H. S. (1970). *Semin. Hematol.* **7**, 341.
Jacob, H. S., and Jandl, J. H. (1964). *J. Clin. Invest.* **43**, 1704.
Jacob, H. S., Ruby, A., Overland, E., and Mazia, D. (1971). *J. Clin. Invest.* **50**, 1800.
Jacob, H. S., Amsden, T., and White, J. (1972). *Proc. Nat. Acad. Sci. USA* **69**, 471.
Jaffé, E. R., and Gottfried, E. L. (1968). *J. Clin. Invest.* **47**, 1375.
Jandl, J. H. (1965). *Blood* **26**, 367.
Jandl, J. H., Simmons, R. L., and Castle, W. B. (1961). *Blood* **18**, 133.
Jensen, M., Shohet, S. B., and Nathan, D. J. (1973). *Blood* **42**, 835.
Jensen, W. N. (1969). *Am. J. Med. Sci.* **257**, 355.
Jensen, W. N., and Lessin, L. S. (1970). *Semin. Hematol.* **7**, 409.
Jensen, W. N., Bromberg, P. A., and Bessis, M. (1967). *Science* **155**, 704.
Katsumato, Y., and Asai, J. (1972). *Arch. Biochem. Biophys.* **150**, 330.
Kirkpatrick, F. H. (1976). *Life Sci.* **19**, 1.
Kirkpatrick, F. H., Hillman, D. G., and LaCelle, P. L. (1975). *Experientia* **31**, 653.
Koyama, S., Aoki, S., and Deguchi, K. (1964). *Mie Med. J.* **14**, 143.
Kroes, J., and Ostwald, R. (1971). *Biochim. Biophys. Acta* **249**, 647.
Kuroda, K., Michiro, Y., and Morita, K. (1959). *Tokushima J. Exp. Med.* **6**, 10.
LaCelle, P. L. (1970). *Semin. Hematol.* **7**, 355.
LaCelle, P. L. (1971). *In* "Hematology for Internists" (R. I. Weed, ed.), p. 85. Little Brown, Boston, Massachusetts.
LaCelle, P. L., and Weed, R. I. (1969). *Blood* **34**, 858.
LaCelle, P. L., and Weed, R. I. (1971). *In* "Progress in Hematology" (E. C. Brown and C. V. Moore, eds.), Vol. VIII, p. 1. Grune Stratton, New York.
Lalezari, P., and Mondhiry, A. (1973). *Br. J. Haematol.* **25**, 399.
Langley, G. R., and Felderhof, C. H. (1968). *Blood* **32**, 569.
Leblond, P. F. (1972). *Nouv. Rev. Fr. Hematol.* **12**, 815.
Leblond, P. F. (1973). *Nouv. Rev. Fr. Hematol.* **13**, 771.
Leblond, P. F., LaCelle, P. L., and Weed, R. I. (1971a). *Nouv. Rev. Fr. Hematol.* **11**, 537.
Leblond, P. F., LaCelle, P. L., and Weed, R. I. (1971b). *Blood* **37**, 40.
Leblond, P. F., deBoisfleury, A., and Bessis, M. (1973). *Nouv. Rev. Fr. Hematol.* **13**, 873.
Leddy, J. P., Whittemore, N. B., and Weed, R. I. (1970). *Vox Sang.* **19**, 444.
Lessin, L. S. (1972). *Nouv. Rev. Fr. Hematol.* **12**, 871.
Lessin, L. S., Jensen, W. N., and Ponder, E. (1969). *J. Exp. Med.* **130**, 443.
Levine, P., Tripodi, D., Struck, J. Jr., Zmijenski, C. M., and Pollack, W. (1973). *Vox Sang.* **24**, 417.
Lichtman, M. A., and Marinetti, J. V. (1972). *Nouv. Rev. Fr. Hematol.* **12**, 775.

Lichtman, M. A., and Weed, R. I. (1973). *In* "Red Cell Shape, Physiology, Pathology, Ultrastructure" (M. Bessis, R. I. Weed, and P. F. Leblond, eds.), p. 79. Springer-Verlag, Berlin and New York.

Limber, G. K., Davis, R. F., and Bakerman, S. (1970). *Blood* **36,** 111.

LoBuglio, A. F., Cotran, R. S., and Jandl, J. H. (1967). *Science* **158,** 1582.

Logue, G. L., Rosse, W. F., and Adams, J. P. (1973). *J. Clin. Invest.* **52,** 29.

Longster, G. H., and Tovey, L. A. D. (1972). *Br. J. Haematol.* **23,** 635.

Lucy, J. A., and Glauert, A. M. (1964). *J. Mol. Biol.* **8,** 727.

Lux, S. E. (1979). *Semin. Hematol.* **16,** 21.

Lux, S. E., John, K. M., and Karnovsky, J. M. (1976). *J. Clin. Invest.* **58,** 955.

McMillan, P., and Luftig, R. B. (1973). *Proc. Nat. Acad. Sci. USA* **70,** 3060.

Maddy, A. H. (1964). *Biochim. Biophys. Acta* **88,** 448.

Maddy, A. H., and Malcolm, B. R. (1965). *Science* **150,** 1616.

Marchesi, V. T., and Palade, G. E. (1967). *J. Cell Biol.* **35,** 385.

Marchesi, V. T., and Steers, E., Jr. (1968). *Science* **159,** 203.

Marchesi, V. T., Steers, E., Tillack, T. W., and Marchesi, S. L. (1969). *In* "Red Cell Membrane, Structure and Function" (G. A. Jamieson and T. J. Greenwalt, eds.), p. 117. Lippincott, Philadelphia, Pennsylvania.

Mel, H. C., and Yee, J. (1975). *Blood Cells* **1,** 391.

Messer, M. J., and Harris, J. W. (1970). *J. Lab. Clin. Med.* **76,** 537.

Miller, D. R. (1972). *Ped. Clin. North Am.* **19,** 865.

Miller, D. R., Weed, R. I., Stamatoyannopoulos, G., and Yoshida, A. (1971a). *Blood* **38,** 715.

Miller, D. R., Rickles, F. R., Lichtman, M. A., LaCelle, P. L., Bates, J., and Weed, R. I. (1971b). *Blood* **38,** 184.

Mohandas, N., and Feo, C. (1975). *Blood Cells* **1,** 375.

Mohandas, N., Greenquist, A., and Shohet, S. B. (1976). *Blood* **48,** 991.

Mollison, P. L. (1967). "Blood Transfusion in Clinical Medicine", 4th edn. Davis, Philadelphia, Pennsylvania.

Murphy, J. R. (1968). *J. Clin. Invest.* **47,** 1483.

Nakao, M., Nakao, T., and Yamazoe, S. (1960). *Nature (London)* **187,** 945.

Nakao, M., Nakao, T., Yamazoe, S., and Yoshikawa, H. (1961). *J. Biochem.* **45,** 487.

Nakao, K., Wada, T., Kamiyani, T., Nakao, M., and Nagano, K. (1962). *Nature (London)* **194,** 877.

Nakao, K., Kurashina, S., and Nakao, M. (1967). *Life Sci.* **6,** 595.

Nathan, D. G., Oski, F. A., Sidel, V. W., Gardner, F. H., and Diamond, L. K. (1966). *Br. J. Haematol.* **12,** 385.

Nathan, D. G., and Shohet, S. B. (1970). *Semin. Hematol.* **7,** 381.

Nathan, D. G., Stossel, T. B., Gunn, R. B., Zarkowsky, H. S., and Laforet, M. T. (1969). *J. Clin. Invest.* **48,** 33.

Neerhout, R. C. (1968a). *Clin. Pediatr.* **7,** 451.

Neerhout, R. C. (1968b). *J. Pediatr.* **73,** 364.

Nicolson, G. L., Masouredis, S. P., and Singer, S. J. (1971). *Proc. Nat. Acad. Sci., USA* **68,** 1416.

Nightingale, D., Prankerd, T. A. J., Richards, J. D. M., and Thompson, D. (1972). *Q. J. Med.* **41,** 261.

Oski, F. A., and Barness, L. A. (1967). *J. Pediatr.* **70,** 211.

Oster, G., and Zalusky, R. (1974). *Biophys. J.* **14,** 124.

Overman, R. R., Hill, T. S., and Wong, Y. T. (1949). *J. Nat. Malaria Soc.* **8,** 14.

Palek, J. (1973). *Blood* **42,** 988.

Palek, J., and Liu, S. C. (1979). *Semin. Hematol.* **16,** 75.

Parker, J. C., and Welt, L. G. (1972). *Arch. Intern. Med.* **129,** 320.
Penniston, J. T., and Green, D. E. (1968). *Arch. Biochem. Biophys.* **125,** 339.
Peters, J. D., Rowland, M., Israels, L. G., and Ziparsky, A. (1966). *Can. J. Physiol. Pharmacol.* **44,** 817.
Policard, A., and Bessis, M. (1953). *C.R. Soc. Biol.* **147,** 982.
Pollack, W., Hoger, H. J., Reckei, R., Toren, D. A., and Singher, H. O. (1965). *Transfusion* **5,** 158.
Polley, M. J., Müller-Eberhard, H. J., and Feldman, J. O. (1971). *J. Exp. Med.* **133,** 53.
Ponder, E. (1948). "Hemolysis and Related Phenomena." Grune & Stratton, New York.
Ponder, E., and Ponder, R. (1962). *Nouv. Rev. Fr. Hematol.* **2,** 223.
Rebuck, J. W., and VanSlyk, E. J. (1968). *Am. J. Clin. Pathol.* **49,** 19.
Reed, C. F., and Swisher, S. N. (1966). *J. Clin. Invest.* **45,** 777.
Reed, C. F., Eden, E. G., Marinetti, G. V., and Swisher, S. N. (1960). *J. Lab. Clin. Med.* **56,** 281.
Rega, A. F., Weed, R. I., Reed, C. F., Berg, G. G., and Rothstein, A. (1967). *Biochim. Biophys. Acta* **147,** 297.
Robinson, J. R. (1980). *In* "Pathobiology of Cell Membranes" (B. F. Trump and A. U. Arstila, eds.), vol. II. Academic Press, New York.
Rochant, H., Minh, M. N., That, H. T., Henri, A., Basch, A., Sultan, C., and Dreyfus, B. (1973). *Nouv. Rev. Fr. Hematol.* **13,** 649.
Rosenthal, A. S., Kregenow, F. M., and Moses, H. L. (1970). *Biochim. Biophys. Acta* **196,** 254.
Rosse, W. F. (1972). *In* "Hematology" (W. J. Williams, E. Beutler, A. Erslev, and R. W. Rundles, eds.), p. 460. McGraw-Hill, New York.
Rosse, W. F., and Lauf, P. K. (1970). *Semin. Hematol.* **7,** 232.
Rosse, W. F., Logue, G. L., Adams, J., and Crookston, J. H. (1974). *J. Clin. Invest.* **53,** 31.
Rothman, J. E., and Lenard, J. (1977). *Science* **195,** 743.
Sachs, J. R. (1972). *J. Clin. Invest.* **51,** 3244.
Schatzmann, H. J. (1969). *In* "Biological Council Symposium on Drug Action" (A. W. Cuthbert, ed.), p. 85. Macmillan, London and New York.
Schatzmann, H. J., and Rossi, G. L. (1971). *Biochim. Biophys. Acta* **241,** 379.
Schwartz, S. O., and Motto, S. A. (1949). *Am. J. Med. Sci.* **218,** 563.
Sears, D. A., Weed, R. I., and Swisher, S. N. (1963). *J. Clin. Invest.* **43,** 975.
Seeman, P., and Iles, G. H. (1972). *Nouv. Rev. Fr. Hematol.* **12,** 889.
Selwyn, J. G., and Dacie, J. V. (1954). *Blood* **9,** 414.
Sheetz, M. P., and Singer, S. J. (1974). *Proc. Nat. Acad. Sci. USA* **71,** 4457.
Sheetz, M. P., Painter, R. G., and Singer, S. J. (1976). *Biochemistry* **15,** 4486.
Shohet, S. B. (1972). *New Engl. J. Med.* **286,** 638.
Shohet, S. B., and Lux, S. E. (1974). *In* "Hematology of Infancy and Childhood" (D. G. Nathan and F. A. Oski, eds.), p. 190. Saunders, Philadelphia, Pennsylvania.
Shohet, S. B., Nathan, D. G., Livermore, B. M., Feig, S. A., and Jaffe, E. R. (1973). *Blood* **42,** 1.
Singer, S. J., and Nicolson, G. L. (1972). *Science* **175,** 720.
Sirchia, G., and Dacie, J. V. (1967). *Nature (London)* **215,** 747.
Slater, L. M., Muir, W. A., and Weed, R. I. (1968). *Blood* **31,** 766.
Statius Van Eps, L. W., Schouten, J., Sloof, P. A. M., and Van Delden, G. J. A. (1971). *Clin. Chim. Acta* **33,** 475.
Sturgeon, P. (1970). *Blood* **36,** 310.
Takahara, S., and Miyamoto, H. (1948). *Otolaryngology (Tokyo)* **51,** 163.
Tanaka, K. R., Valentine, W. N., and Miwa, S. (1962). *Blood* **19,** 267.
Tillack, T. W., and Marchesi, V. T. (1970). *J. Cell Biol.* **45,** 649.

Torsvik, H., Gjone, E., and Norum, K. R. (1968). *Acta Med. Scand.* **183,** 387.
Tosteson, D. C., Shea, E., and Darling, R. C. (1952). *J. Clin. Invest.* **31,** 406.
Tosteson, D. C., Carlsen, E., and Dunham, E. T. (1955). *J. Gen. Physiol.* **39,** 31.
Tourtellotte, M. E., Branton, D., and Keith, A. (1970). *Proc. Nat. Acad. Sci. USA* **66,** 909.
Trump, B. F., and Arstila, A. U., eds. (1980). "Pathobiology of Cell Membranes," Vol. II. Academic Press, New York.
Udkow, M., LaCelle, P. L., and Weed, R. I. (1973). *Nouv. Rev. Fr. Hematol.* **13,** 817.
Usami, S., Chien, S., and Bertles, J. F. (1975). *J. Lab. Clin. Med.* **86,** 274.
Valentine, W. N., Crookston, J. H., Paglia, D. E., and Konrad, P. N. (1972). *Br. J. Haematol.* **23,** 107.
Vandenheuvel, F. A. (1963). *J. Am. Oil Chemists Soc.* **40,** 455.
Verwilghen, R. L., Lewis, S. M., Dacie, J. V., Crookston, J. H., and Crookston, M. C. (1973). *Q. J. Med.* **42,** 257.
Ways, P., Reed, C. F., and Hanahan, D. J. (1963). *J. Clin. Invest.* **42,** 1248.
Weed, R. I. (1968). "Proc. XII Congr. Int. Soc. Hematol. 1968" (E. R. Jaffe, ed.), p. 81. New York.
Weed, R. I. (1970a). *Semin. Hematol.* **7,** 249.
Weed, R. I. (1970b) *Am. J. Med.* **49,** 147.
Weed, R. I., and Bessis, M. (1973). *Blood* **41,** 471.
Weed, R. I., and Bowdler, A. J. (1966). *J. Clin. Invest.* **45,** 1137.
Weed, R. I., and Chailley, B. (1973). *In* "Red Cell Shape, Physiology, Pathology, Ultrastructure" (M. Bessis, R. I. Weed, and P. F. Leblond, eds.), p. 55. Springer-Verlag, Berlin and New York.
Weed, R. I., and LaCelle, P. L. (1969). *In* "Red Cell Membrane Structure and Function" (G. A. Jamieson and T. J. Greenwalt, eds.), p. 318. Lippincott, Philadelphia, Pennsylvania.
Weed, R. I., and LaCelle, P. L. (1973). *In* "The Human Red Cell in Vitro", The American National Red Cross 5th Annual Scientific Symposium (T. J. Greenwalt, ed.). Grune and Stratton, New York.
Weed, R. I., and Reed, C. F. (1966). *Am. J. Med.* **41,** 681.
Weed, R. I., and Weiss, L. (1966). *Trans. Assn. Amer. Physicians* **79,** 426.
Weed, R. I., Reed, C. F., and Berg, G. (1963). *J. Clin. Invest.* **42,** 581.
Weed, R. I., LaCelle, P. L., and Merrill, E. W. (1969). *J. Clin. Invest.* **48,** 795.
Weinstein, R. S. (1969). *In* "Red Cell Membrane Structure and Function" (G. A. Jamieson and T. J. Greenwalt, eds.), p. 36. Lippincott, Philadelphia, Pennsylvania.
Weinstein, R. S., and Bullivant, S. (1967). *Blood* **29,** 780.
Weinstein, R. S., and McNutt, N. S. (1970). *Semin. Hematol.* **7,** 259.
Weiss, L. (1974). *Blood* **43,** 665.
Weiss, L., and Tavassoli, M. (1970). *Semin. Hematol.* **7,** 372.
Wennberg, E., and Weiss, L. (1969). *Annu. Rev. Med.* **20,** 29.
Westerman, M. P., Balcerzak, S. P., and Heinle, E. W., Jr. (1968). *J. Lab. Clin. Med.* **72,** 663.
White, J. G., and Krivit, W. (1967). *J. Cell Biol.* **35,** 141A.
Whittam, R., and Wheeler, K. P. (1970). *Annu. Rev. Physiol.* **32,** 21.
Whittemore, N. B., Trabold, N. C., Reed, C. F., and Weed, R. I. (1969). *Vox Sang.* **17,** 289.
Wiley, J. S. (1970). *J. Clin. Invest.* **49,** 666.
Wiley, J. S. (1972). *Br. J. Haematol.* **22,** 529.
Wilkins, M. H. F., Blaurock, A. E., and Engleman, D. M. (1971). *Nature (London), New Biol.* **230,** 72.
Wins, P., and Schoffeniels, E. (1966). *Arch. Int. Physiol. Biochim.* **74,** 812.
Wong, K. Y. (1972). *Blood* **39,** 23.

Zail, S. S., and Joubert, S. M. (1968). *Br. J. Haematol.* **14**, 57.
Zarkowsky, H. S., Oski, F. A., Shaafi, R., Shohet, S. B., and Nathan, D. G. (1968). *New Engl. J. Med.* **278**, 573.
Zarkowsky, H. S., Mohandas, N., Speaker, C. B., and Shohet, S. B. (1975). *Br. J. Haematol.* **29**, 537.
Zeive, L. (1958). *Ann. Intern. Med.* **48**, 471.
Zwaal, R. F. A., Roelofsen, B., and Colley, G. M. (1973). *Biochim. Biophys. Acta* **300**, 159.

# EDITORS' SUMMARY TO CHAPTER II

This chapter is very important in that it sets the stage for many of the other chapters in this volume as well as reviewing some of the fundamentals of membrane conformational changes in the important erythrocyte model. The erythrocyte has many obvious advantages for studying membrane movements in simplified form. However, more importantly, changes in red cell membranes occur in a variety of diseases including not only those within the hematopoietic system but in other systemic diseases. Alterations in red cell metabolism may reflect molecular level membrane disease in such organs. An example of the possible use of red cell membrane metabolism to study systemic diseases is the suggestion by Rose *et al.* (1976) that spectrin phosphorylation may eventually serve to increase the diagnostic precision in the detection of silent carriers of Duchenne muscular dystrophy.

The present chapter also discusses the relationship between membranes and cytoplasmic components, such as contractile microfilaments and microtubules. These components apparently play a significant role in the maintenance of cell shape in all types of cells, and changes such as the phosphorylation of spectrin or other contractile proteins may induce some type of conformational change in the protein network throughout the membrane as it adjusts cell shape. Certainly, modification of microfilament proteins in many nucleated cells with the compound cytochalasin may result in dramatic alterations in cell shape such as the formation of large protoplasmic buds. These buds then pinch off and some contain the nucleus.

The echinocytic and stomatocytic transformations of red cells provide a dramatic model for the demonstration and study of the fundamental membrane movements which we have termed *esotropy and exotropy* (Arstila *et al.*, 1971; Trump and Arstila, 1971). Esotropy (from the Greek: "a turning in") is the definition we have given to the membrane movement in which the membrane turns in, so that the original outer (exoplasmic) surface now becomes the inner surface of a new membrane-bound vesicle. Esotropy may occur in either a forward or a reverse direction. Since the space within the cytocavitary network is topologically equivalent to the extracellular space and esotropy may occur by budding from either the cell membrane or from the cytocavitary membrane into the cytoplasm, then topologically, an esotropic vesicle will always contain extracellular material or the contents of the cytocavitary network. When the process occurs in the opposite direction, reverse esotropy, the vesicle fuses with a membrane and extracellular space or with the cytocavitary network space. Phagocytosis, pinocytosis, elaboration of transitional vesicles from endoplasmic reticulum at the forming face of the Golgi apparatus, elaboration of secretory vesicles from the maturing face of the Golgi apparatus, and the formation of microbodies are examples of forward esotropy. Some examples of reverse esotropy include fusion of secretory granules with the cell membrane, fusion of Golgi vesicles with phagosomes, and fusions between various secondary lysosomes.

We have defined exotropy (from the Greek: "a turning out") as the opposite membrane movement of esotropy in which the membrane turns out from the cytoplasm into the cytocavitary network or extracellular space so that the inner surface of the membranes faces toward one another. Therefore, an exotropic vesicle always contains cell sap. Exotropy into the endoplasmic reticulum appears to be an important mechanism of autophagocytosis and at the extreme a mitotic division could also be considered a gross example of exotropy. Exotropy may also occur in either a forward or a reverse direction. In reverse exotropy, the fusion of an exotropic structure brings two portions of cell sap into contact. Examples of reverse exotropy include the entry of viruses into cells, where the viral envelope fuses with the cell membrane. At the cellular level, the fusion of cells to produce syncytial giant cells and fertilization of the egg are also examples of reverse exotropy. Fusions between exotropic and esotropic vesicles have not yet been reported. These presumably relate to the "inside"–"outside" polarity of cell membranes.

Although there is much that remains to be learned about the basis of these membrane movements, available evidence strongly indicates that they depend upon high-energy compounds such as ATP, and that they are modified by cyclic AMP and $Ca^{2+}$, possibly through mechanisms involving cell filaments and microtubules. Also the role of calcium in these membrane movements is emphasized by the fact that calcium modifies either the stomatocyte or the echinocyte transformation especially during aging of red cells where echinocytic formations occur; ATP is depleted while calcium levels increase. This may be related to the membrane movements that occur in autophagocytosis where hormones such as glucagon interact with the cell membrane inducing or activating adenylate cyclase with the formation of cyclic AMP, which is presumed to act through mechanisms involving or related to calcium released from mitochondria. Calcium-mediated changes in the shape of endoplasmic reticulum seem to result in the sequestration of organelles.

The classification and nomenclature of erythrocyte shape changes by Bessis and co-workers has begun to greatly strengthen the study and classification of erythrocyte disorders. Some red cell disorders appear to involve primary membrane defects. Progress is being rapidly made in defining the molecular alterations involved. Interestingly, as in several other types of cell injury, altered ionic regulation, especially involving calcium, appears to be of key importance. In hereditary spherocytosis the changes may be greatly augmented in the environment of decreased pH and $O_2$ tension that exists in the spleen. It is of interest that cell membrane alterations may also be of some importance in sickle-cell disease. In other red cell disorders, cell surface changes appear to be of pivotal importance in determining the red cell's susceptibility to phagocytosis. Further studies in this area should clarify not only red cell disorders per se but also shed new light on cell membrane structure and function in general and on general cellular pathobiology in particular.

## References

Arstila, A. U., Jauregui, H. O., Chang, J., and Trump, B. F. (1971). *Lab. Invest.* **24**, 162.

Trump, B. F., and Arstila, A. U. (1971). *In* "Principles of Pathobiology" (M. F. LaVia and R. B. Hill, eds.), p. 9. Oxford Univ. Press, London and New York.

Rose, A. D., Roses, M. J., Miller, S. E., Hull, K. L., Jr., and Appel, S. H. (1976). *New Engl. J. Med.* **294**, 193.

# CHAPTER III

# FUNCTIONS AND ALTERATIONS OF CELL MEMBRANES DURING ACTIVE VIRUS INFECTION

P. M. Grimley and A. Demsey





PATHOBIOLOGY OF CELL MEMBRANES, VOL. II

ISBN 0-12-701502-7

## I. Introduction

Molecular and macromolecular alterations in host cytomembranes occur as direct or secondary consequences of virus assault in animal cells, and active interactions with celluar membranes characterize at least some portion in the life cycle of nearly all animal viruses. These may include steps of virus uptake, nucleic acid replication, protein synthesis, macromolecular assembly, virion maturation, and emergence or release. Indeed, the success of animal virus infections at a cellular level can be largely attributed to an efficient harnessing of the versatile molecular and biophysical properties of host membrane systems. Even the central process of membrane biogenesis by the host cell may become indentured to a general virus strategy of competitively dominating or subverting the preexisting regulatory and synthetic machinery (Mosser *et al.*, 1972b; Blough and Tiffany, 1975).

In addition to direct participation in virus developmental processes, the membrane surfaces in virus-infected cells may retain their specialized transport, protective, and metabolic functions. Membrane-limited host compartments also maintain spatial segregations favorable to regulation of positive-entropic biochemical processes, and confinement of DNA virus genomes within the nuclear envelope represents a relatively sophisticated deployment of this advantage (Gautschi *et al.*, 1976). Control of viral nucleic acid and protein synthesis within the intranuclear microenvironment can remarkably simulate host cell regulatory mechanisms (Chantler and Stevely, 1973; Seebeck and Weil, 1974; Su and DePamphilis, 1976). This may even involve selective molecular transport through the nuclear envelope (Kozak and Roizman, 1974).

In the cytoplasm, virus adaptation of host membranes facilitates the initiation of viral nucleic acid or protein synthesis (Caliguiri and Mosser, 1971; Mosser *et al.*, 1972a; Friedman *et al.*, 1972; Wagner *et al.*, 1972; Hay, 1974; Wirth *et al.*, 1977) and glycosylation of integral virus envelope proteins can depend entirely upon membrane-associated host enzymes involved in oligosaccharide synthesis (Sefton, 1976). These processes and the final maturation of enveloped viruses by budding through modified host membrane can be viewed as developmental mechanisms which minimize the requirement for direction of new macromolecular synthesis by viruses with a relatively limited genetic-coding capacity (see Portner and Pridgen, 1975). Indeed, an ample provision of host membrane material may be critical to the final steps of virion maturation (Choppin *et al.*, 1971; Blough and Tiffany, 1975). Infections by relatively well-adapted viruses need not grossly disrupt normal membrane biosynthesis (Quigley *et al.*, 1972; Luftig, *et al.*, 1974).

From a pathobiological viewpoint, host reactions to virus infection may be conditioned by cell membrane alterations. Insertion of viral proteins or modification of host membrane proteins may modulate immune recognition antigens at the cell surface (Edelman, 1976; Zinkernagel and Oldstone, 1976), modify intercellular adhesive properties (Wallach, 1972), promote cell fusion or aggregation (Feldman *et al.*, 1968; Scheid and Choppin, 1974; Larke *et al.*, 1977), or influence differentiation (Aoki, 1974). The possible significance of such virus-related phenomena in neoplastic, neurologic, and other chronic diseases offers a fertile area for future investigation.

With the exponential growth of experimental virologic observations and data, it becomes necessary within the practical confines of a subject review to choose between thorough analysis of specialized topics or a more general survey. The burgeoning interest of pathologists, immunologists, and other "nonvirologists" in cell membrane phenomena during virus infections suggested that a broad scope might be timely. Since molecular and biochemical features of virus structural membranes and assembly processes have been amply treated in a number of recent and comprehensive reviews (e.g., Klenk, 1973; Rifkin and Quigley, 1974; Blough and Tiffany, 1975), this presentation aims toward providing a somewhat broader perspective on those membrane functions and alterations which ultimately relate to the pathobiology of virus infections. We attempt to compensate for some unavoidable superficiality by directing attention to comprehensive current articles and earlier reviews at appropriate points in the text.

## II. Investigation of Viral Membrane Constituents

### A. *Membrane Isolation and Composition*

With the exception of naturally pure membranes obtained from erythrocyte ghosts or nerve sheaths, accurate biochemical analysis of native cell membranes awaited refinement of cell fractionation techniques. A number of reproducible methods now available for isolation of plasma membrane or cell "ghosts" (Warren and Glick, 1969; Steck, 1972; Atkinson, 1973; Neville and Kahn, 1974), internal cytomembranes (Bosmann *et al.*, 1968), and nuclear membranes (Berezney, 1974; Aaronson and Blobel, 1975) can be applied to the study of membranes in virus-infected cells (e.g., Spear *et al.*, 1970; Caliguiri and Tamm, 1970a; Friedman *et al.*, 1972; Heine and Roizman, 1973; Buck *et al.*, 1974; LeBlanc and Singer, 1974; Hay, 1974; Atkinson *et al.*, 1976). In general, these methods rely on mechanical cell disruption in a hypotonic buffer or nitrogen cavitation (Blough *et al.*, 1977) followed by differential velocity sedimentation to remove nuclei and large cell fragments and to prepare crude membrane supernatants or pellets. Specific membrane fractions are then separated by isopycnic centrifugation in discontinuous and continuous sucrose or ficoll gradients. The significance of biochemical analyses depends largely on the consistency of these preparations. Thin-section electron microscopy of pelleted cytomembrane fractions is an indispensible monitor (Bosmann *et al.*, 1968; Spear *et al.*, 1970; Caliguiri and Tamm, 1970a; Friedman *et al.*, 1972).

The predominant molecular species in animal cell biomembranes, regardless of source, are lipids and proteins in roughly balanced proportions. The lipids may be characterized as phospholipids, cholesterol, fatty acids, and glycolipids or glycosphingolipids (Law and Snyder, 1972; Klenk, 1973; Brady, 1975). Values of 30–40% lipid and 60–70% protein are typical for many clean mammalian cell membrane isolates (Stoeckenius and Engelman, 1969); however, considerable variation is recognized and differences in the organic composition of the naturally pure membranes of erythrocyte ghosts and in myelin dramatize this diversity: erythrocyte membranes contain lipid to protein ratio of about 1:1, whereas the ratio in myelin is approximately 4:1 (Singer, 1974). HeLa cell surface membranes contain about 40% lipid and 60% protein, and the proportion in most enveloped animal viruses is grossly similar (Klenk, 1973). The lipid pattern of highly purified virus particles in effect characterizes their envelope composition since the core (nucleocapsid) is a nucleic acid–protein structure (see reviews by Klenk, 1973; Rifkin and Quigley, 1974; Blough and Tiffany, 1975). Nevertheless, alertness to artifacts of preparation is necessary (Blough and Tiffany, 1975). Loosely bound membrane proteins can solubilize, while cellular microproteins may contaminate the surface of virions (Pinter and Compans, 1975).

Small amounts of carbohydrate usually represent no more than 10% of the dry weight in cell membranes or virus particles (Rifkin and Quigley, 1974) and this is distributed in glycoslated proteins as well as in glycolipids. Since carbohydrate chains are completed stepwise through the action of specific glycosyltransferases present in smooth and rough cytomembranes, glycosylated proteins may show microheterogeneity depending on the degree to which oligosaccharides are completed (Heath, 1971). With the exception of some large viruses, the nature of sugar residues in viral envelope glycoproteins is determined by the function of host cell glycosyltransferases rather than viral-encoded enzymes (Keegstra *et al.*, 1975; Sefton, 1976).

The molar ratio of cholesterol to phospholipid in certain virus envelopes approximates unity, whereas the ratio in whole cells is closer to 1:5 (Klenk, 1973). This apparently reflects a predominance of virus budding through the cell surface which contains a larger proportion of cholesterol than the internal cytomembranes; however, preferential selection of host lipids during the process of virion assembly is also possible (Blough and Tiffany, 1975). Viral phospholipids may contain a significantly higher proportion of sphingomyelin and a lower proportion of phosphatidylcholine than the plasmalemma (Quigley *et al.*, 1971, 1972). In general, the pattern of lipids in virus envelopes qualitatively resembles that of the host cell source (Klenk, 1973; Lenard and Compans, 1974; Rifkin and Quigley, 1974; Hirschberg and Robbins, 1974). A striking illustration is provided by togaviruses grown in baby hamster kidney cells and in mosquito cells: lipid composition of the progeny virions resembles that of the mammalian or insect host respectively, so that virions from the two sources contain only 36% of their lipids in common (Renkonen *et al.*, 1972). Microviscosity of togavirus membranes, measured by fluorescence depolarization, is also influenced by the host cell lipid (Moore *et al.*, 1976). In contrast, envelope proteins of viruses are largely specified by the virus genome and displace host proteins in the membrane plane (Hay, 1974; Birdwell and Strauss, 1974a; Dubois-Dalcq *et al.*, 1976b; Demsey *et al.*, 1977). This dichotomy in content of lipids and proteins is accounted for by a fluid-mosaic molecular construct of biological membranes (Section III,B).

### B. *Topographic and Topological Localization of Membrane Proteins*

Dramatic progress in the elucidation of virus membrane structure since a review by Allison (1971) owes largely to a multidisciplinary convergence of techniques both for topographical and topological localization of membrane proteins. Conventional electron microscopy remains a basic tool in this armamentarium, and high resolution of negatively stained samples, or even thin sections, permits recognition of asymmetrical macromolecular structures such as glycoprotein spikes in a virion envelope (e.g., Cartwright *et al.*, 1969; Klenk *et al.*, 1970; Garoff and Simons, 1974). At the same time, new ultrastructural techniques such as surface replication (Birdwell *et al.*, 1973; Demsey *et al.*, 1978; Dubois-Dalcq *et al.*, 1976a), freeze-fracture (Bächi *et al.*, 1969; Brown *et al.*, 1972; Sheffield, 1974; Haines and Baerwald, 1976), and secondary electron scanning (Wong and MacLeod, 1975; Dubois-Dalcq *et al.*, 1976b) offer a more integrated view of virus membrane topography. Freeze-drying of intact cells is a simple, but elegant method, which allows direct visualization of virus macromolecular elements and has provided resolution of at least two distinct virus components at a host surface (Demsey *et al.*, 1976, 1977). Figure 1 illustrates results of these procedures.

During the last decade, emphasis has also shifted toward ultrastructural localization of specific virus envelope constituents, sometimes combined with experimental manipulations to dissect individual steps in the complex chain of maturational events (e.g., Tiffany and Blough, 1971; Klein and Adams, 1972; Bächi *et al.*, 1973; Birdwell and Strauss, 1974a; Lampert *et al.*, 1975). Powerful tools for this work have been provided by biochemists and immunologists who pioneered the development of techniques in which antibodies or macromolecular probes are coupled to markers appropriate for electron microscopy and immunofluorescence or autoradiography. In general, the application of particulate electron-dense markers (e.g., ferritin, small viruses) coupled to specific antibodies or macromolecular probes has proven most suitable for high-resolution localization of virus-specific constituents. For a full discussion of technical details the reader is referred to Wagner (1973). Enzyme-labeled antibody can serve as a very sensitive marker for intracellular or surface cites of virus protein attachment. Hemocyanin or small virus particles are useful in conjunction with surface replication (Birdwell and Strauss, 1974a) or secondary scanning electron microscopy of cells prepared by critical point drying (Hämmerling *et al.*, 1975).

Table I summarizes representative samples of several major approaches which have been successfully applied for high-resolution immunolabeling of virus-infected systems. Perhaps the most versatile procedures are based upon construction of hybrid antibody molecules from Fab′ fragments (Hämmerling *et al.*, 1973). Dual monovalent specificities are directed both against the IgG of a species in which a specific antibody probe is raised and against a protein particle such as ferritin or a virus nucleocapsid. An example of this technique is illustrated in Fig. 2.

Although much emphasis has been placed upon antibody localization of specific viral proteins, the uses of colloidal iron to identify neuraminic acid residues (Klenk *et al.*, 1970) or of lectins with selective oligosaccharide affinities (Klein and Adams, 1972; Birdwell and Strauss, 1973) have been successfully exploited in the study of virus infection. Some novel technical approaches remain on the horizon. These include use of antibody fragments (Mannik and Downey, 1973; Kraehenbuhl *et al.*, 1974) and haptenic conjugates (Lamm *et al.*, 1972). Refinement of methods for labeling ultrathin sections of frozen (Tokuyasu and Singer, 1976) or embedded tissues (e.g., Kraehenbuhl and Jamieson, 1972) would be most advantageous in studying intracellular cites of virus protein biosynthesis without gross disruption of the membrane systems by subcellular fractionation.

---

**Fig. 1.** (A) Replica of a freeze-dried mouse $JLSV_9$-RLV cell surface revealing three budding Rauscher leukemia viruses (arrows). Note viral knoblike components concentrated on the buds and distributed randomly over the rest of the cell surface. Inset is a thin section of a virus particle budding from the membrane of a $JLSV_9$-RLV cell. Densities probably representing knobs are evident (arrows), as is the forming crescent-shaped viral nucleocapsid underneath the membrane. (B) Mid, and (C) late stages of Friend leukemia virus budding from STU-Eveline cells as seen in freeze-fracture replicas. Intramembranous particles are excluded from those regions of the host membrane enveloping the viral nucleocapsid. As virus budding progresses, release occurs by a pinching off of the membrane (arrows). ×95,000 (from Demsey *et al.*, 1977).

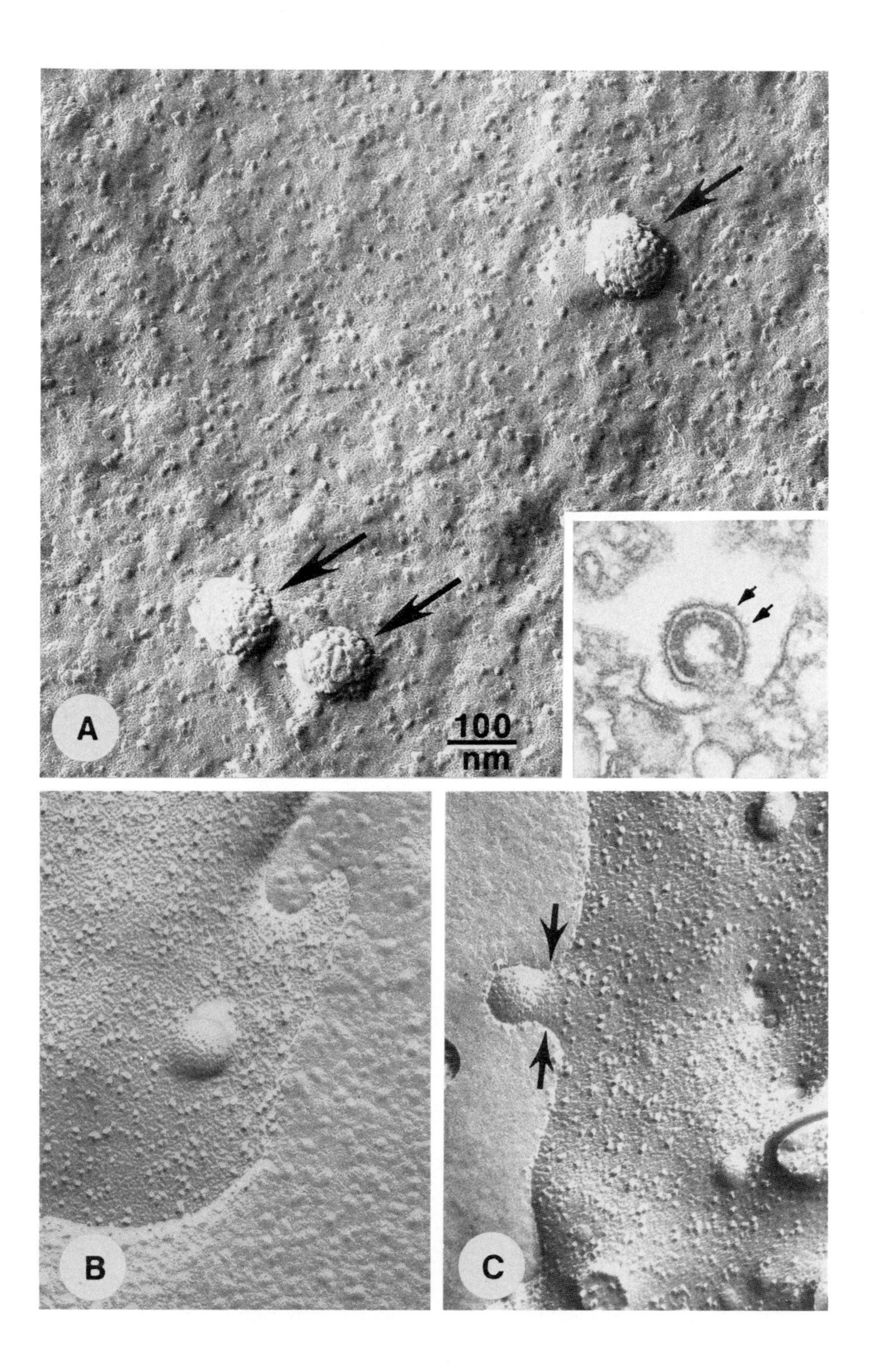
A
100
nm
B
C

**TABLE I**

*Examples of High Resolution Immunolabeling Procedures for Localization of Membrane-Associated Viral Antigens*[c]

| *Method* | *First reagent* | *Second reagent* | *Third reagent* | *Selected references* |
|---|---|---|---|---|
| Direct ferritin | $\gamma$/Ferritin | | | Howe *et al.* (1969), Pedersen and Sagik (1973), Bächi and Howe (1973) |
| Direct enzyme | $\gamma$/HPO | DAB + $H_2O_2$ | | Hoshino *et al.* (1972), Suzuki (1972) |
| Indirect ferritin | $\gamma$* | Anti-$\gamma$/ferritin | | Levinthal *et al.* (1969), Shigematsu *et al.* (1971), Coward *et al.* (1972) |
| Indirect hemocyanin | $\gamma$* | Anti-$\gamma$/hemocyanin | | Birdwell and Strauss (1974a) |
| Indirect enzyme | $\gamma$* | Anti-$\gamma$/HPO | DAB + $H_2O_2$ | Ciampor *et al.* (1974), Lampert *et al.* (1975), Hiraki *et al.* (1974) |
| Hybrid antibody | IgG | $F(ab')_2$ (*anti-*$\gamma$/anti-ferritin) | Ferritin | Wagner *et al.* (1971), Gelderblom *et al.* (1972) Aoki and Takahashi (1972 Heine and Schnaitman (1971) |

[a] $\gamma$, Total globulin fraction from specific antibody (first animal species); anti-$\gamma$, total globulin fraction from antibody raised in second animal species against globulin of first species; HPO, horseradish peroxidase; DAB, diaminobenzidine; SBMV, southern bean mosaic virus; and $\gamma$*, use of IgG fraction of Fab′ fragments may improve localization.

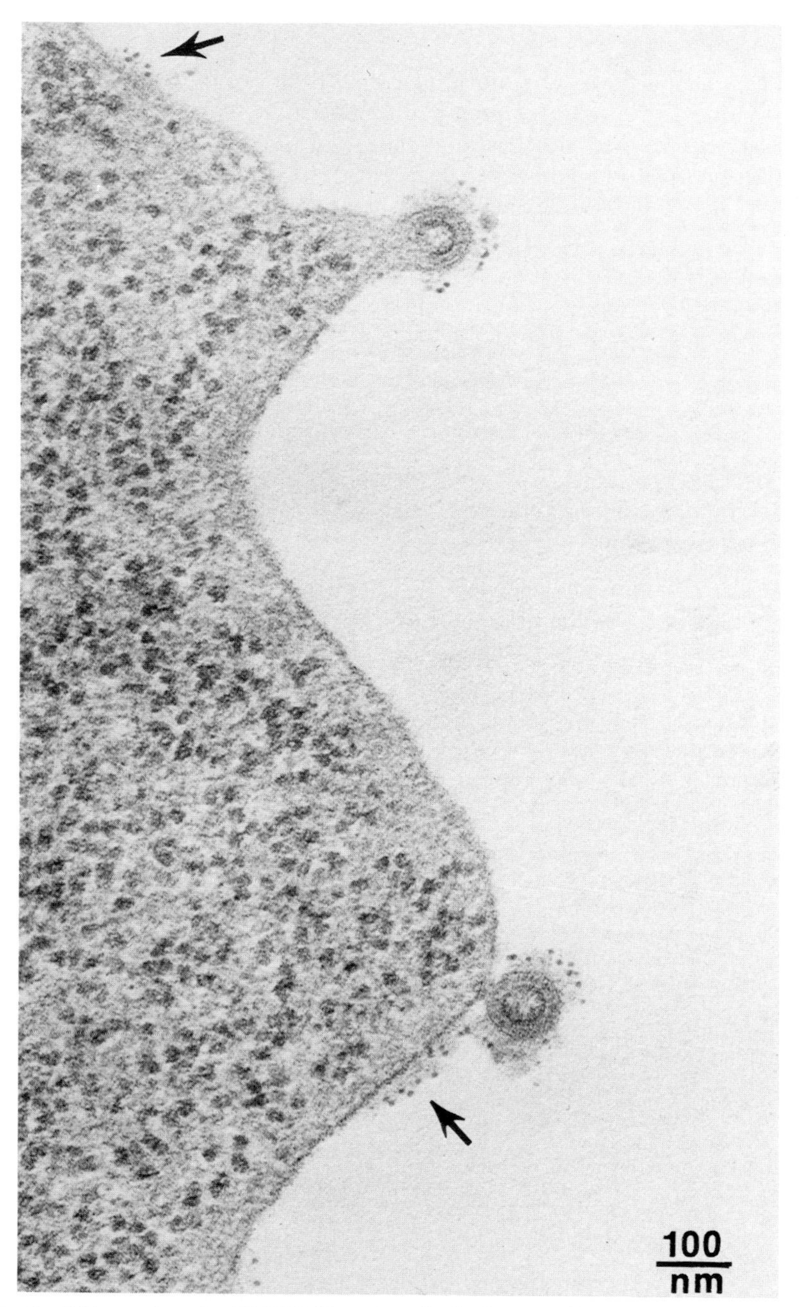


**Fig. 2.** Thin section of a FLC745 (Friend erythroleukemia) cell after labeling sequentially with rabbit anti-gp71 serum, hybrid sheep anti-rabbit IgG/anti-ferritin antibodies, and ferritin. Ferritin label is found on the surfaces of the Friend leukemia virus buds, as well as on other parts of the cell surface (arrows). ×95,000.

At a molecular level, the subcellular distribution and topological positioning of viral polypeptides within membranes may be elucidated by a number of relatively refined enzymatic techniques including gentle proteolytic cleavage on isolated membrane vesicles (Katz *et al.*, 1977; Wirth *et al.*, 1977; Witte *et al.*, 1977) or virus envelopes (Cartwright *et al.*, 1969; Schloemer and Wagner, 1974) and catalytic radiolabeling (see Juliano and Behar-Bannelier, 1975) followed by physical separation of radioiodinated molecules (Witte and Weissman, 1976; Knipe *et al.*, 1977c). More specific references to these approaches appear in appropriate sections below.

## III. Membrane Dynamics Relating to Virus Infection

### A. *Continuous Circulation and Biogenesis of Cell Membranes*

Cellular membranes exist in a state of dynamic equilibrium, both physical and metabolic, with continuous active synthesis and degradation (see Schimke and Dehlinger, 1972; Warren, 1972; Singer and Rothfield, 1973). Nascent membranes, synthesized in continuity with the endoplasmic reticulum (Higgins, 1974) eventually incorporate into the Golgi region through an "assembly-line" process (Morré *et al.*, 1971). Cell surface membrane is generated or replaced by assimilation of modified membrane vesicles which move peripherally from the Golgi region (Hicks, 1966; Grove *et al.*, 1968; Singer, 1974). The cell membranes utilized by many groups of enveloped viruses evidently arise in a similar fashion by continuous amplification of preexisting templates (Amako and Dales, 1967b; Mosser *et al.*, 1972b; Luftig *et al.*, 1974; Blough and Tiffany, 1975).

Exchanges of material between the cell surface and internal cytomembranes (see Whaley *et al.*, 1971) compensate for macromolecular migrations involved in the vectorial transport of fluids (phagocytosis) to the cell interior as well as the centrifugal export of secretions or waste products (exocytosis). Pinocytosis, for example, can result in internalization of cell surface at the rate of up to 20% per minute (Gosselin, 1967). The intimate linkage of virus uptake to membrane movements will be discussed below (Section IV,B). While this necessitates no gross interruption of the membrane circulatory dynamics, production and emergence of enveloped virus eventually taxes the process of membrane biogenesis. In the absence of compensatory membrane production, prolonged virus infection leads to a chronic depletion of surface membrane resources (Choppin *et al.*, 1971; Quigley *et al.*, 1972).

In bacterial models there is some degree of feedback control between protein and fatty acid synthesis (Fox, 1972), but coordination of protein and lipid synthesis during vertebrate membrane biogenesis appears more complex. Production of lipid constituents depends upon the availability of appropriate enzymes which are subject to metabolic or genetic regulatory controls (Majerus and Kilburn, 1969; Brady, 1975), but the synthesis of membrane proteins is not necessarily concerted, either temporally or spatially. In liver microsomes, for example, phospholipid synthesis occurs *in situ* in smooth membranes, while some complementary proteins may be synthesized on the ribosome-bound endoplasmic reticulum and later transferred to cites of insertion (Higgins and Barrnett, 1972). Phospholipid to protein ratios in smooth and rough microsomal fractions can vary independently (Higgins, 1974) and the half-life of barbiturate-stimulated microsomal enzymes may be shorter than the membrane half-life (Orrenius and Ericsson, 1966). Schimke and Dehlinger (1972) reproted differential rates of membrane protein turnover, with larger molecules renewing more rapidly.

Changes in the patterns of protein and lipid biosynthesis during animal virus infections is providing some useful clues to normal controls regulating membrane biogenesis. Early in the course of virus development, for example, there can be direct and selective inhibition or stimulation of membrane biogenesis (Plagemann *et al.*, 1970; Ben-Porat and Kaplan, 1971; Mosser *et al.*, 1972b; Willis and Granoff, 1974; Makino and Kenkin, 1975; Vance and Lam, 1975). The poxviruses offer unique models for further studies of biomembrane assembly mechanisms (Dales and Mosbach, 1968; Grimley *et al.*, 1970; Moss *et al.*, 1971a; Stern and Dales, 1974), since they synthesize membrane *de novo* from degraded host lipids. Investigations of other enveloped viruses are illuminating the subcellular pathways of membrane specialization (see Section V,C).

### B. *Fluid-Mosaic Structure*

It is now well established that viral and cellular proteins occupy contiguous and mobile domains in the cell surface membrane (Heine and Roizman, 1973; Birdwell and Strauss, 1973, 1974a; Hay, 1974; Blough and Tiffany, 1975; Edelman, 1976). Indeed, the capacity of host membranes to accommodate virus-specified proteins into a preexisting molecular framework is a biophysical property critical to the processes of virus maturation. A "recycling" of host cytomembranes by attachment or insertion of new viral gene products underlies the survival of virus groups which lack enzymatic resources to synthesize their own structural membranes.

Before 1960, hypotheses of biomembrane structure usually assumed a relatively uniform bilayer. The phospholipid molecules were considered to be relatively rigid while the protein was typically represented in an extended configuration applied to the exterior surfaces. This model appeared to explain the "railroad-track" image of cell and virus membranes observed in ultrathin sections after osmium tetroxide fixation (with a profile thickness in the range of 60–100 Å). Critical analysis of the essentially static bilayer concept revealed many discrepancies with experimental observations, and in the late 1960s evidence rapidly mounted for the presence of macromolecular subunits within various types of membranes (see Branton and Deamer, 1972). The conceptual developments have been lucidly recapitulated by Stoeckenius and Engleman (1969).

Freeze-fracture studies of plasma membranes by Branton and others were most revealing when they disclosed the presence of 80-Å diameter particles within the membrane plane. Internal localization was established by marking the membrane exterior with ferritin (Pinto da Silva and Branton, 1970) or with F-actin (Tillack and Marchesi, 1970). Independent evidence for the globular conformation of proteins was obtained with physical methods (see Branton and Deamer, 1972). These also indicated extensive "bareness" of the extended phospholipids. In favor of hydrophobic-protein–phospholipid interactions Singer (1974) found that neutral salt solutions do not dissociate large amounts of protein from membranes, and that cytochrome *b*5 isolated from liver microsomes and an erythrocyte surface glycoprotein each appeared to possess hydrophobic regions which could be cleaved from polar regions by limited proteolytic digestion. Additional experimental evidence consistent with penetration of protein macromolecules into the membrane lipid (Singer and Nicolson, 1972; Steck and Fox, 1972), phospholipid fluidity (Scandella *et al.*, 1972; Lee *et al.*, 1973), and protein mobility within the membrane plane (Frye and Edidin, 1970; DePetris and Raff, 1973) culminated in the proposal of a "fluid lipid-globular protein mosaic" structure (Singer and Nicolson, 1972).

This "fluid-mosaic" concept currently offers the most attractive synthesis of biophysical, biochemical, immunological, and ultrastructural observations in virus-infected and virus-transformed cells. Principal features of the model have been concisely summarized by Nicolson (1975) and Fig. 3 is based on prevalent conceptions: the membrane is composed of lipid molecules (principally phospholipid) arranged in an extended bimolecular configuration. Hydrocarbon, nonpolar tails of the lipid molecules are directed inward, away from the bulk aqueous phase. Thus, they form a semifluid matrix for integral membrane proteins which are stabilized within the hydrophobic plane. Hydrophilic moieties of these integral proteins, such as the glycopeptide portions of glycoprotein macromolecules, thrust outward into the aqueous environment. The integral proteins may comprise either single macromolecules or macromolecular complexes formed by weak interaction with peripheral proteins. The latter do not penetrate through the lipid bilayer and

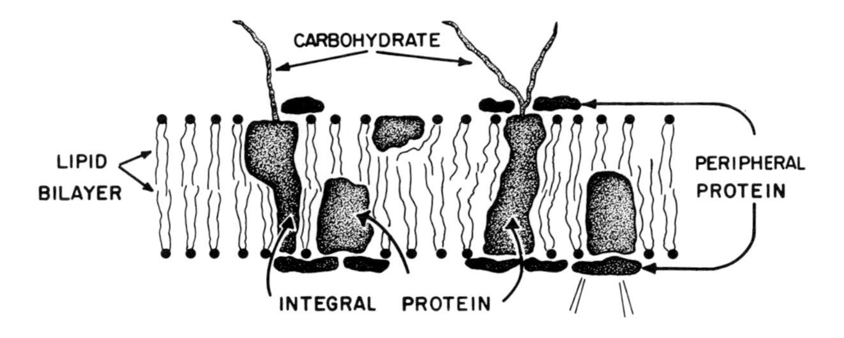


**Fig. 3.** Membrane structure based upon fluid-mosaic concept. Globular proteins are embedded in the plane of the lipid bilayer. "Integral" glycoproteins, penetrating into the hydrophobic region of the bilayer, could represent the G proteins in a virion envelope. The peripheral proteins, more loosely attached to the membrane surface, could be represented by the M proteins (see Section V).

are believed to account for up to 30% of all membrane-associated proteins. They often serve as membrane attachment cites to the cytoskeleton. Both the "integral" and "peripheral" membrane proteins have counterparts in virus envelopes (Blough and Tiffany, 1975; Knipe *et al.*, 1977b).

In recent studies, some of the dynamic techniques which led to devleopment of the fluid-mosaic concept have been applied to studies of virus-infected cells and virion envelopes. These techniques include visual or ultrastructural observation of virus antigen movements within the cell surface (Birdwell and Strauss, 1974a; Lampert *et al.*, 1975), incorporation of spin-labeled phospholipids for electron paramagnetic resonance spectroscopy (Sefton and Gaffney, 1974; Sharom *et al.*, 1976), or incorporation of a diphenylhexatriene probe for fluorescence depolarization analysis (Lenard *et al.*, 1974; Moore *et al.*, 1976; Levanon *et al.*, 1977). Detailed discussion of the biophysical techniques may be found in Branton and Deamer (1972) and Fox and Keith (1972).

### C. *Freeze-Fracture Observations*

Freeze-fracture methods for ultrastructural observation of membranes were devised in order to avoid potential artifacts associated with extensive dehydration and plastic embedding prior to thin sectioning. The approach has been successfully applied to examine morphogenesis of enveloped viruses in several groups: herpesvirus (Haines and Baerwald, 1976), myxoviruses (Bächi *et al.*, 1969; Bächi and Howe, 1973; Dubois-Dalcq *et al.*, 1976a), togaviruses (Brown *et al.*, 1972; Demsey *et al.*, 1974), visnavirus (Dubois-Dalcq *et al.*, 1976b), and oncornavirus (Sheffield, 1974; Demsey *et al.*, 1977). It can be profitably integrated with scanning electron microscopy (Dubois-Dalcq *et al.*, 1976b).

Biological material for freeze-fracture study is usually fixed briefly and infiltrated with a cryoprotectant solution to prevent ice crystal formation during the rapid freezing. The freeze-fracture process begins with a "snap freezing" at about −150°C, followed by cleavage (e.g., with a cold knife) and replication of the surface topography by deposition of fine molecular layers under high vacuum. Typically, platinum is deposited at an angle (about 45°), and "backed" perpendicularly by carbon. The carbon helps maintain integrity of the replica during digestion of the subjacent cellular material (e.g., with sodium hypochlorite or chromic acid). This is necessary to clean the replica for examination by transmitted electrons. Ice or glycerol–ice can be sublimed away from exposed biological components immediately after fracturing by allowing the cleaved specimen to remain in high vacuum for a brief time at an increased temperature (e.g., 2 min at −100°C). The controlled process is commonly known as freeze-etching.

With few exceptions, cell membranes are either (a) cross-fractured—the fracture plane cuts through the plasma membrane and enters the cell contents; or (b) split through the region of hydrophobic bonding of the lipid bilayer. The latter split is most informative and reveals either of two fracture faces: the PF (also A or +) face is characterized by numerous particles as seen from an exterior view; the EF (also B or −) face normally exhibits a sparser particulation as seen from an interior view (Branton and Deamer, 1972). These intramembranous particles (IMP) or membrane-associated particles (MAP) seen on freeze-fractured membrane faces are evidently large proteins (Tillack *et al.*, 1972; Marchesi *et al.*, 1976) and asymmetry of the inner- and outer-facing leaflets is anticipated in the fluid-mosaic model. Figure 4 illustrates planes in a freeze-fracture of cultured HeLa cell surface membranes. The PF face of an intramembranous fracture of one cell surface comprises most of the visible surface. A few patches of attached EF belong to a neighboring cell. The concept may be clarified by imagining the PF surface covers the remainder of the first cell which has lost only the outer half of its

---

**Fig. 4.** (Top) Replica of freeze-fractured HeLa cells. Mostly PF surface is revealed, although some small patches of EF surface from an adjacent cell are seen (arrows). ×40,000.

**Fig. 5.** (Bottom) Replica of freeze-fractured LLC-$MK_2$ (monkey kidney) cell infected with Dengue virus. Arrows indicate "inside-out" membrane polarity of vacuoles, some of which contain virus. ×40,000.

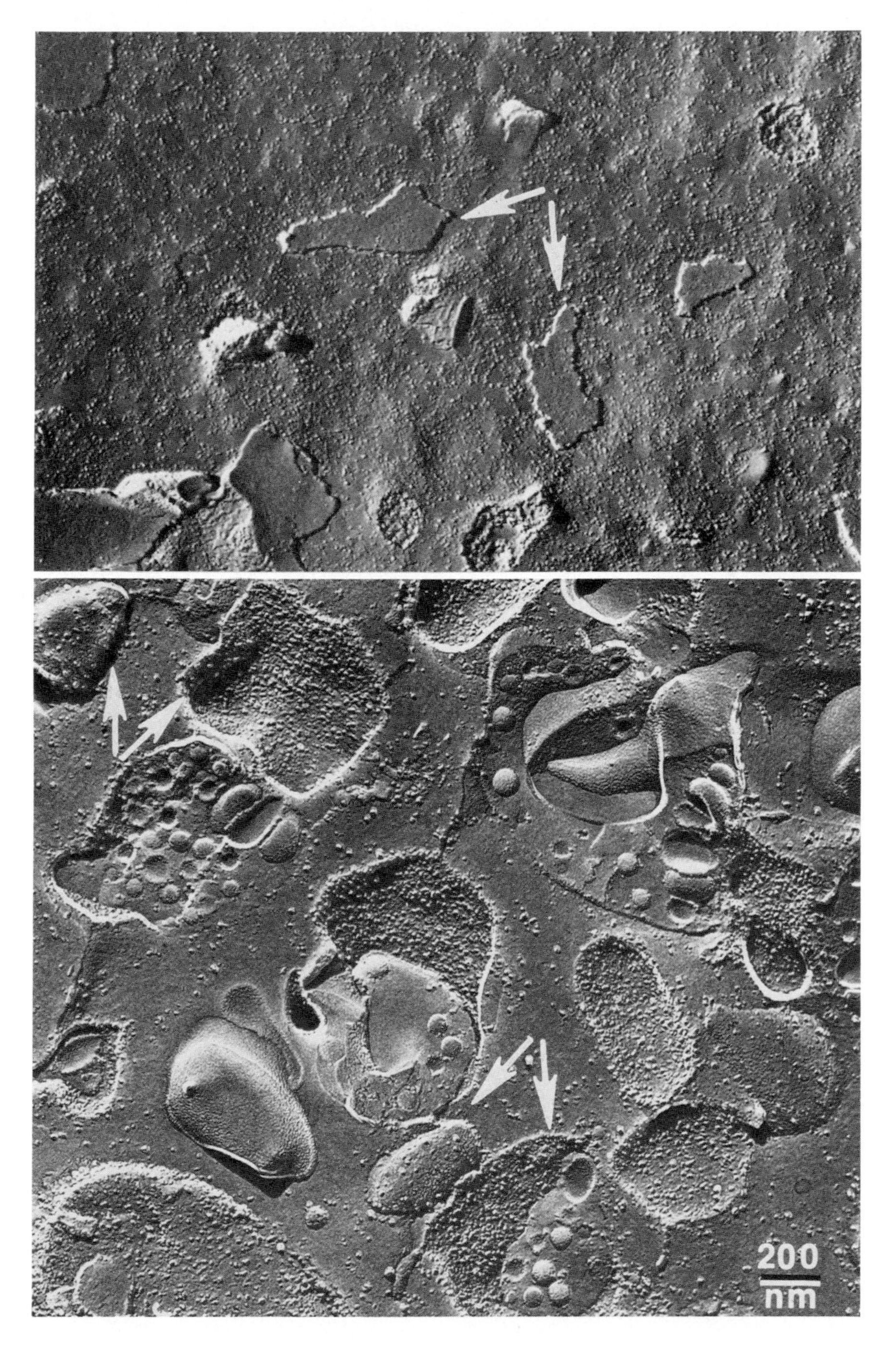
200
nm

membrane, whereas the EF surface has been left behind by tearing the remainder of the second cell away. Between the PF and EF faces would remain the outer half of the first cell's membranes, intercellular space, and the outer half of the second cell's membrane. In principle it is possible to replicate and retain the material that was fractured away, and such a complementary replica can be very useful for orientation purposes.

### D. *Internal Cytomembranes*

The fluid-mosaic structure evidently extends to internal cytomembranes which continuously fuse to the cell surface (Scandella *et al.*, 1972). Exocytic cytoplasmic vacuoles, which discharge certain viruses, demonstrate a similar internal structure to the cell surface after freeze-fracturing and etching (Demsey *et al.*, 1974); however, the intramembranous polarity is oriented with respect to the vacuolar lumen rather than to the cell surface (Fig. 5).

The intramembranous structure of the endoplasmic reticulum (ER) is of special interest, since certain RNA viruses, including togaviruses (Filshie and Rehacek, 1968; Blinzinger, 1972), coronaviruses (Caul and Egglestone, 1977), and intracisternal A-type oncornaviruses (Perk and Dahlberg, 1974) mature predominantly by budding through the ER or derivative cytomembrane systems. Furthermore, ER membranes indirectly support the envelopment of viruses which bud from the cell surface, since they are the source of new membrane biogenesis (see Section V,C). Macromolecular dynamics of the ER membranes, however, have only lately been examined. Lateral movements of membrane-bound ribosomes resemble those of surface proteins and suggest comparable internal fluidity with temperature-dependent phase transitions (Ojakian *et al.*, 1977). Freeze-fracture of the nuclear envelope which is homologous to the ER also provides evidence for structural similarities (Wunderlich *et al.*, 1974).

Certain cytomembranes exhibit a more regular and stable internal structure than the fluid-mosaic cell surface. For example, mitochondrial inner membranes, retinal photoreceptor membranes, and chloroplast membranes are highly enriched in enzymes which promote energy conversion by means of electron transfer or ATP synthesis. Efficiency of these processes is promoted by a tightly packed organization of protein subunits, perhaps involving polypeptide cross-linkage (Hendler, 1974). Even under such conditions, however, a potential for intramembranous movement of protein "particles" appears to persist (Apel *et al.*, 1976; Staehelin, 1976). Although there has been at least one report of virus envelopment within mitochondrial membranes (Lunger and Clark, 1973), there is yet insufficient information for any specific discussion of mitochondrial membrane structure, function, or turnover in relation to virus infections.

## IV. Virus Uptake and Penetration of Cytomembrane Barriers

### A. *Techniques for Investigating Virus Uptake*

Development of plaque assay and hemagglutination techniques encouraged early experimental efforts to quantitate virus uptake by host cells (e.g., Joklik and Darnell, 1961). This was accomplished by measuring the adsorption of infectious particles from an inoculum of known biological potency. Since the ratio of physical particles to infectious units ranges up to 1000:1 in a virus inoculum, only a fraction of the total virions can be measured by biological titration. Production of virus with radiolabeled nucleic acid and determination of cell-associated radioactivity as a percentage of radioactivity applied provides a more precise tool for quantitation of virion binding (e.g., Philipson, 1967; Sussenbach, 1967; Schloemer and Wagner, 1975). Investigations at a molecular level became more feasible as procedures for purification and fractionation of virus particles were refined (see Fraenkel-Conrat, 1969). This encouraged analysis of the virus envelope by selective extraction of lipids (see Klenk, 1973; Rifkin and Quigley, 1974) or electrophoretic separation of proteins (Cords *et al.*, 1975). Adaptation of methods for subcellular fractionation and separation of host membrane systems (e.g., Chan and Black, 1970; Roesing *et al.*, 1975) further expanded opportunities for experimental analysis.

Analysis of virus–cell interactions has become increasingly sophisticated with use of proteolytic or lipolytic enzymes (Cartwright *et al.*, 1969; Friedman and Pastan, 1969; Tillack *et al.*, 1972; Scheid and Choppin, 1974; Schloemer and Wagner, 1974), plant lectins (Ito and Barron, 1974), individual glycoproteins (Schloemer and Wagner, 1975), or purified virion subunits (Philipson *et al.*, 1968) to explore the surface properies of virus envelopes and host cells. Reconstitution of virus envelopes (Shimizu *et al.*, 1972) or construction of artificial membrane models offer further means for probing the molecular roles of specific phospholipid or protein components in virus attachment and penetration (Tiffany and Blough, 1971; Haywood, 1974; Mooney *et al.*, 1975; Sharom *et al.*, 1976). Virus–cell receptor complexes (Philipson *et al.*, 1968) can be isolated by buoyant density, and analysis of "cytotrophic" subunits of viruses or "viroceptive" proteins removed during the elution of virions from cell surfaces has been accomplished by polyacrylamide gel electrophoresis (see Crowell and Philipson, 1971; Philipson *et al.*, 1976).

Transmission electron microscopy has proven to be a uniquely valuable tool for examining virus uptake despite limitations in sampling and an inherent inability to discriminate the fate of infective and noninfective virus particles (Dales, 1973; Lonberg-Holm and Philipson, 1974; Smith and de Harven, 1974). High-resolution autoradiography has proven particularly rewarding since it can localize early events in virus replication (Amako and Dales, 1967b; Silverstein and Dales, 1968; Grimley *et al.*, 1968; Hummeler *et al.*, 1969, 1970; Willis and Granoff, 1974; Mackay and Consigli, 1976). Freeze-cleaving of membranes, described in Section III,C, provides a new and essential approach to the ultrastructural examination of virus receptors without treatment by organic solvents (Tillack *et al.*, 1972; Dubois-Dalcq *et al.*, 1976a). Improvement of resolution in secondary mode scanning electron microscopy (cf. deHarven, 1974; Panem and Kirsten, 1975; Wong and MacLeod, 1975; Dubois-Dalcq *et al.*, 1976b), should ultimately permit a more representative and topographical view of virus entry events, complementing conventional ultrastructural techniques. Other potentially valuable approaches are the use of high-voltage electron microscopy with stereoscopy (Grimley, 1971; Stokes, 1976) and a technique of freeze-drying intact cells (Demsey *et al.*, 1978). The latter method also avoids organic solvents and affords a relatively high resolution (10–15 Å). Most recently Levanon *et al.* (1977) used fluorescent polarization analysis with molecular probes to explore changes in intramembranous lipid fluidity of the cell surface during virus adsorption.

### B. *Functional Routes for Virus Invasion of Host Cells*

The obligate dependence of viruses upon host machinery for protein synthesis requires that instructions encoded by the virus genome gain direct access to intracytoplasmic or intranuclear compartments. Thus, virus *penetration* as defined by Dulbecco (1965) is consummated only when nucleic acid of the inoculum virus has escaped from its protective wrappings (uncoating) and reached intracellular sites where expression and replication of the parental genes can ensue. These virus entry mechanisms have proven to be unexpectedly complex and specialized for each major group of animal viruses. From an evolutionary perspective, perhaps they can be viewed as a process in which gross and molecular membrane dynamics of potential host cells have been adopted to assist the virus invasion. Examples of membrane-associated phenomena, to be discussed more fully in appropriate sections, include processes of membrane fusion and repair, interiorization and digestion of particulates (see Jacques, 1975), and nuclear cytoplasmic exchange (see Goldstein, 1974). In two comprehensive reviews, Dales (1965, 1973) cogently summarized the extensive evidence that virus penetration is a resultant of interacting virus and host influences. More recently, Lonberg-Holm and Philipson (1974) provided another lucid analysis of these events.

Since there normally exist no direct openings between the cell sap and exterior environment, viral components must traverse a membrane barrier. Physical disruption occurs when some bacteriophage penetrate their microbial hosts, and this may be accompanied by transient leakage of cell contents (Luria and Darnell, 1967). Transient dissolution of plasma membrane continuity has also been suggested by thin section observations of mammalian cells during penetration of a small nonenveloped adenovirus (Brown and Burlingham, 1973), and a murine leukemia virus (Miyamoto and Gilden, 1971). While local breaches of animal cell surfaces produced by physical, chemical, or immunological means can be rapidly restored as shown by the entry and resealing of microparticulates (Seeman, 1974), cytosol leakage is not notable during animal virus infections, and transmembranous entry is generally considered to involve more subtle molecular mechanisms. These include the attachment of the virus nucleocapsid or virus envelope to host membrane by means of mutually specific receptor molecules, with molecular translocation through the fluid-mosaic membrane structure (nucleocapsids) or membrane fusion (virus envelopes). In addition, there is a potential for active virus uptake by engulfment. This phenomenon may either resemble nonselective inhibition of small particulates (Epstein *et al.*, 1964; Morgan *et al.*, 1969) or occur subsequent to specific virus binding. The latter process was termed *viropexis* by Fazekas de St. Groth (1948).

In the sense of virus engulfment combined with specific membrane-binding mechanims, viropexis is probably very common. The microvesicular-limiting membrane around internalized virions remains topologically and functionally homologous to the plasma membrane from which it originates (Abodeely *et al.*, 1970; Choppin, 1976), so that penetration of intact virions engulfed within microvesicles can proceed by specialized mechanisms including nucleocapsid translocation or membrane fusion as observed at the cell surface (e.g., Smith and de Harven, 1974; Crowell, 1976).

Operationally, it is important to distinguish between mere interiorization of an infectious virion and actual genomic penetration. In either case, the infective element is no longer accessible to neutralizing antibody and particles cannot be eluted from the membrane exterior (Mandel, 1967). Nevertheless, interiorized virus particles may be sequestered for several hours within cytoplasmic vesicles (phagosomes) (Smith and de Harven, 1974; Ogier *et al.*, 1977), functionally quarantined from the cytoplasmic compartment by a limiting membrane. In the special case of nonenveloped nucleotropic viruses (adenoviruses and papoviruses), electron microscopic studies show an almost immediate attachment and engulfment of inoculum virions which are swiftly transported to the periphery of the nuclear envelope. Intact nucleoids of papovavirus arrive in the nuclei of host cells within 30 min

(Hummeler *et al.*, 1970; Mackay and Consigli, 1976), and centripetal migration of adenovirus is equally rapid (Morgan *et al.*, 1969; Dales and Chardonnet, 1973). This movement appears to be guided by cytoplasmic microtubules (Dales and Chardonnet, 1973). Virion uncoating may occur at orifices of the nuclear envelope (Hummeler *et al.*, 1970).

Dales has marshalled the strongest arguments in favor of virus infections being initiated by nonspecific engulfment (1973). Dales and Pons (1976) observed that enveloped viruses such as influenza can be infectious under conditions of neutralization or aggregation in which specific attachment and fusion to membrane surfaces was unlikely. A similar conclusion was drawn earlier in studies by Mandel (1967) which showed that neutralized poliovirus retained ability to penetrate HeLa cells, even though the released viral RNA was abnormally labile. Ultrastructural studies of a rhabdovirus suggested that enveloped virions were swallowed in pits at the cell surface which are believed to be coated by a sticky substance that traps particulates (Simpson *et al.*, 1969).

The molecular aspects of viral genome penetration from within phagosomes has not been fully elucidated, but except in the case of diplornaviruses which have a nuclease-resistant genome, the role of lysosomal hydrolases is questionable (Dales, 1973; Choppin, 1976). Even without formation of a phagolysosome, evolution of the phagosome by dehydration may increase permeability of the limiting membrane up to 100-fold (Jacques, 1975). Free adenovirus DNA appears to penetrate the plasma membranes of KB cells (Groneberg *et al.*, 1975) and recent investigations of DNA bacteriophage provide a basis for speculating that an amphiphilic virus coat protein could become the carrier to draw nucleic acid polyectrolytes through the hydrophobic fluid–lipid bilayer of host membrane (Marco *et al.*, 1974). This parallels the "permions" concept which postulates a class of large globular membrane proteins forming energetically favorable channels lined by polar residues within the lipid bilayer to facilitate permeation of charged macromolecules (Rothschild and Stanley, 1972).

In considering the biological and functional significance of various sets of experimental observations on virus entry mechanisms, limitations of tissue culture and other systems *in vitro* which have been widely employed for these investigations should be fully appreciated. Each experimental approach and method of examination is subject to its own set of arbitrary conditions and limited means of observation, introducing a biological "uncertainty principle." Cells in culture may gain or lose susceptibility to particular viruses (Holland, 1961) and changes induced by serum concentration (Allison, 1971) or temperature shifts conventionally employed during virus adsorption have no obvious parallel in the natural host. Furthermore, in the "unnatural universe of the laboratory," propagated virus strains may conceivably lose specific genetic characteristics which would be essential to their survival in nature (Reanney, 1974). Except for hematogenous dissemination, the physicodynamics of virus–cell interactions in the intact animal must differ remarkably from events in monolayer or suspension cultures. Direct cell–cell contacts and localized concentrations of virus particles in cell processes can be expected to amplify the efficiency of virus transfer in solid tissues by an exponential factor (Cohen, 1963; Grimley and Friedman, 1970a; Iwasaki and Koprowski, 1974). This can be observed even in organ cultures (Leestma *et al.*, 1969).

### C. *Molecular Mechanisms of Virus Attachment to Cell Membranes*

Direct contact and eventually irreversible attachment of inoculum particles to the host cell surface, its projections or invaginations, is an essential and probably universal first step in virus infections (Kohn and Fuchs, 1973; Lonberg-Holm and Philipson, 1974; Dales *et al.*, 1976). When virus particles and host cells are bathed in a generous volume of fluid and free to follow random Brownian movements, the kinetics of collision are predictably proportionate to relative concentrations. The rate of binding to cell surfaces, however, is exponentially lower than values expected if each virus–cell contact were to produce a firm attachment. Taken alone, these observations provide circumstantial evidence for the existence of specific virus-binding mechanisms, requiring steric orientation of macromolecules in addition to electrostatic forces (see Cohen, 1963). Attachment also can be significantly influenced by pH (Mooney *et al.*, 1975; Schloemer and Wagner, 1975), ionic strength (Pierce *et al.*, 1974), and other conditions modifying the charge environment (Vogt, 1967; Hochberg and Becker, 1968; Cartwright *et al.*, 1970; Miyamoto and Gilden, 1971).

Virus attachment is mediated by an interaction of natural cell membrane components ("viroceptive" molecules) with complementary elements of the virus envelope or capsid ("cytotropic" subunits). While some viroceptive molecules may represent common phospholipid or glycolipid constituents of animal cell membranes, others represent genetically specified host proteins. The genetic control of viroceptor specificity for picornaviruses has been elegantly revealed in experiments with cell hybrids (Medrano and Green, 1973). Since poliovirus type 1 can infect primate cells but not murine cells, Miller *et al.* (1974) exploited human–mouse heterokaryons with observed deletions of human chromosomes to assign the gene for the viroceptor of poliovirus to human chromosome 19.

Cytotropic subunits of the virion are obviously specified by virus genes. Mutations involving cell-attachment ability have been recognized for influenza virus (Palese *et al.*, 1974) and noninfectious Rous sarcoma virus (DeGiuli *et al.*, 1975).

The specificity of virus–cell interaction varies. Similar neuraminic acid-bearing viroceptive molecules may be recognized by more than one major virus group (Mori *et al.*, 1962). At the same time cross-competition (an attachment interference assay) and cell hybridization experiments have demonstrated the existence of distinct viroceptive molecules for individual virus serotypes within major virus subgroups (Philipson *et al.*, 1968; Chardonnet and Dales, 1970; Crowell, 1976). Such observations prompt the concept of virus receptor families (Lonberg-Holm and Philipson, 1974). Biologically, the species and tissue distribution of virus receptors can be an important determinant of host range and possibly even disease predilection (see Crowell, 1976; Weiss, 1976). Phenotypically mixed viruses, i.e., with mixed envelope antigens, may exhibit a host range similar to that of both parental strains (Weiss, 1976). Nevertheless, host susceptibility also involves genetically regulated intracellular factors unrelated to virus binding (Tucker and Docherty, 1975; Rey *et al.*, 1976).

The concept of a virus "receptor" complex ("viroceptor") is useful in the abstract, recognizing that the dynamics of virus attachment involves intramembranous mobilization of individual viroceptive molecules which aggregate and cross-link in loci subjacent to the virus particle (Howe *et al.*, 1970; Tillack *et al.*, 1972; Philipson *et al.*, 1976). Indeed, independent saturation of the cell surface by more than one class of virus particle indicates the formation of discrete viroceptor domains. Conceivably, more than a single species of viroceptive molecule may mobilize to provide a multivalent viroceptor configuration (Lonberg-Holm and Philipson, 1974; Choppin, 1976). Differences in affinity of molecules could account for experimental observations of relatively "loose" and "tight" degrees of virus particle binding (Pierce *et al.*, 1974). Evidence for viroceptor mobility was provided in experiments of Birdwell and Strauss (1974b); when Sindbis virus was adsorbed to glutaraldehyde-fixed cells, the particles counted by electron microscopy of surface replicas were evenly distributed; however, in unfixed cells the virus clustered in arrays of variable size, and lateral diffusion of viroceptive molecules evidently occurred even at 4°C.

Destruction of viroceptors by nonpenetrating treatment of intact cells with surface-active enzymes (see Kohn and Fuchs, 1973) such as neuraminidase (Tillack *et al.*, 1972), subtilisin (Philipson *et al.*, 1968), or trypsin (Levitt and Crowell, 1967), supports a model of asymmetric viroceptive molecules projecting outward from the cell surface (Tiffany and Blough, 1971; Marchesi *et al.*, 1976). Presumably, it is only peripheral moieties of these molecules which actually engage the cytotropic subunits of the virus particles.

Ultrastructurally, the intramembranous portions of viroceptive molecules for myxoviruses appear as 75-Å diameter particulates within frozen-cleaved erythrocyte surfaces (Tillack *et al.*, 1972). These are typical of the IMP mentioned in Section III,C. Estimates of viroceptor numbers, based upon saturation of the cell surface with virus particles at low temperature, range from $10^3$ per HeLa cell exposed to poliovirus type 1 (Lonberg-Holm and Philipson, 1974) to $10^4$ per HeLa cell exposed to adenovirus type 2 (Philipson *et al.*, 1968) and $10^5$ for chicken embryo cells exposed to Sindbis virus (Birdwell and Strauss, 1974b). From the latter figure, Birdwell and Strauss (1974b) calculated 20–160 Sindbis virus receptors for each $\mu m^2$ of cell surface. In view of probable aggregation dynamics, the actual numbers of individual viroceptive molecules should be severalfold greater than the numbers of attached particles, depending upon the virus circumference and valency (Philipson *et al.*, 1976).

Idiosyncracies in the attachment properties of enveloped viruses are as striking as their similarities and much remains to be learned. For example, some myxoviruses bind to artificial membranes (liposomes) only in the presence of sialylated fetuin (Tiffany and Blough, 1971), while Sendai virus and arboviruses bound to protein-free liposomes (Haywood, 1974, Mooney *et al.*, 1975). In contrast, removal of neuraminic acid from the surface of mouse cells actually increased the adsorption and infectivity of a rhabdovirus (Schloemer and Wagner, 1975). The host cell neuraminic acid evidently marks viroceptor molecules or prevents the formation of auxillary electrostatic bonds. Since rhabdoviruses have a remarkably wide host range, they presumably attach to a cell membrane component of relatively high frequency. Although the identity of the viroceptive molecules is not known, some functional properties were explored by Schloemer and Wagner (1975) in experiments in which goose erythrocyte or mouse L cell receptors for VSV were effectively blocked by fully sialylated fetuin or purified VSV oligoglycopeptides. If small VSV glycopeptides were generated by excessive trypsinization, the inhibition of hemagglutination was less effective. Either the viroceptor cites recognized sialoglycoprotein only in a restricted size range or cross-linking of sites—mimicking attachment of whole virus—is necessary for inhibition.

Attachment of nonenveloped viruses also involves a specific interaction with the cell surface which may be quite rapid (Mackay and Consigli, 1976; Philipson *et al.*, 1976). The viroceptors for adenoviruses and picornaviruses appear to be lipoproteins (McLaren *et al.*, 1968; Philipson *et al.*, 1968). When nonenveloped virus is adsorbed at low temperature, it is common for a large proportion of bound particles to elute spontaneously at 37°C (Crowell, 1976). This phenomenon has offered an advantageous tool for dissecting the interaction of cytotropic subunits and viroceptors, since the complex is often sufficiently firm to extract one or the other from its foothold in the membrane. Thus, cells lose their capacity for agglutination by fresh virus and the virus is rendered noninfectious (Mandel, 1967; Crowell and Philipson, 1971). Polyacrylamide gel electrophoresis of eluted picornavirus particles demonstrates loss of a rapidly migrating polypeptide VP 4 which is located on the surface of the virion (see Crowell, 1976) and presumably represents the cytotropic subunit. This is consistent with inactivation by removal of the VP 4 in low ionic strength solutions (Cords *et al.*, 1975). Further evidence of the firm attachment between nonenveloped viruses and host viroceptors comes from studies of adenoviruses and papovaviruses. Dissociated virions undergo a shift in buoyant density which may be explained by extraction of a cell-bound cytotropic subunit or addition of an excised segment of the lipoprotein viroceptor from the host plasma membrane (Philipson *et al.*, 1968).

Ultrastructural studies have suggested that an antennalike fiber in the capsid penton of adenoviruses represents the cytotropic subunit which anchors to host viroceptors (Chardonnet and Dales, 1970). Blockage of adenovirus attachment by purified fiber preparations supports this contention (Philipson *et al.*, 1968). The adenovirus fiber is a 183,000 MW polypeptide complex with two glycosylated chains (Dorsett and Ginsburg, 1975).

Cytotropic subunits in the envelopes of myxoviruses, rhabdoviruses and arboviruses also provide specific moieties for attachment to the cell surface. The envelope of myxoviruses and of paramyxoviruses is rich in neuraminidase (Kendal and Kiley, 1973; Scheid and Choppin, 1974). In paramyxoviruses, the hemagglutinating and neuraminidase activity are combined in a single molecule (HN) on the virus envelope (Scheid, 1976) and are represented in the virus spikes visible by negative staining and in thin sections. This macromolecule is evidently the cytoropic subunit which combines with sialic acid-bearing viroceptor components on the cell surface. In contrast, the hemagglutinating and neuraminidase activities of myxoviruses are associated with separate proteins of the virus envelope (see Kendal and Kiley, 1973). Palese *et al.* (1974) suggest that influenza neuraminidase may prevent self-aggregation of virus inoculum particles or progeny, while Scheid (1976) postulates a role in the inactivation of molecules which might compete for attachment cites.

The envelope of a rhabdovirus, vesicular stomatitis virus, is formed from lipids and glycolipids of host cell origin and contains two viral proteins (cf. Shimizu and Ishida, 1975; Atkinson *et al.*, 1976; Knipe *et al.*, 1977a). One of these is a sialoglycoprotein associated with the envelope spikes visualized by negative staining. Selective hydrolysis of the sialic acid moieties with neuraminidase almost totally eliminates infectivity, but activity can be restored by resialylation (Schloemer and Wagner, 1974). Sialoglycolipids of host derivation found in the virus envelope did not appear to be involved in virus attachment. Schloemer and Wagner (1975) confirmed that the positively charged polycation DEAE-dextran enhances the adsorption of VSV (Cartwright *et al.*, 1969) by twofold, but this did not increase infectivity and the sialoglycoprotein of the virus was still essential. Their results indicated that the virus neuraminic acid functioned as a true cytotropic subunit in the attachment process and did not merely influence the charge environment.

The biochemical nature of herpesvirus and poxvirus cytotropic proteins or viroceptors has not yet been extensively investigated. Inactivation of herpes simplex virus by the plant lectin concanavalin A (Ito and Barron, 1974) suggests the presence of a cytotropic glycoprotein.

### *D. Fusion of Enveloped Viruses with the Cell Surface*

There are many examples of complex biological systems arising by evolutionary superimposition (e.g., Silverstein, 1964) and it should not be surprising if the long course of virus evolution in animal cells (see Kurstak and Maramorosch, 1974) had accumulated a plurality of potential mechanisms for critically vulnerable steps in the transmission of virus. Virus entry into host cells by fusion of the virion envelope with the host surface membrane possibly represents an evolutionary superimposition upon the more generalized route of particulate entry by engulfment. As one example, enveloped herpesviruses display a relatively sophisticated capacity to fuse selectively with host cells (Miyamoto and Morgan, 1971), yet herpesvirus cores or even free nucleoproteins also may initiate infection, presumably by the mechanisms of endocytosis or phagocytosis (Spring and Roizman, 1968; Abodeely *et al.*, 1970). Nucleocapsids of a rhabdovirus evidently also retain cytotropic subunits distinct from those on the virus envelope (Cartwright *et al.*, 1969).

The formation of multinucleate syncytia during infections by enveloped viruses such as influenza and varicella long suggested that these viruses might produce a factor which promotes membrane fusion. In 1962, Hoyle reported evidence for the incorporation of myxovirus envelopes into the cell surface using radiolabeled virus particles. Direct observation of the fusion of enveloped viruses with the cell surface was later achieved by electron microscopy of thin sections (Morgan and Howe, 1968; Heine and Schnaitman, 1971; Miyamoto and Morgan, 1971; Granados, 1973). This mechanism might be particularly advantageous for RNA viruses such as the myxoviruses, rhabdoviruses, and oncornaviruses in which the virion genome is complementary to the messenger RNA involved in virus polypeptide synthesis, so that a virion associated transcriptase is essential for initiation of the replicative cycle (Fenner *et al.*, 1974). Direct "transfusion" of relatively large and intact nucleocapsid structures into the host cytosol would minimize the risk of exposure to potentially damaging lysosomal nucleases.

Several elegant studies with hybrid antibody have demonstrated that specific virus envelope proteins marked by a ferritin label are actually engrafted and diffused into the host surface within minutes after virus attachment (Heine and Schnaitman, 1971; Wagner *et al.*, 1971; Bächi *et al.*, 1973) (Fig. 6). Consistent with virological and biochemical studies of virus attachment, this process is temperature independent. There is now substantial evidence that a specific glycoprotein of myxoviruses, distinct from the hemagglutinin or neuraminidase, is responsible for fusion of virion envelopes to the cell surface (Scheid, 1976). While fusion is dependent on an intact mechanism for virus attachment (Scheid and Choppin, 1974; Seto *et al.*, 1974), treatment with specific antibodies directed against individual virus envelope compo-

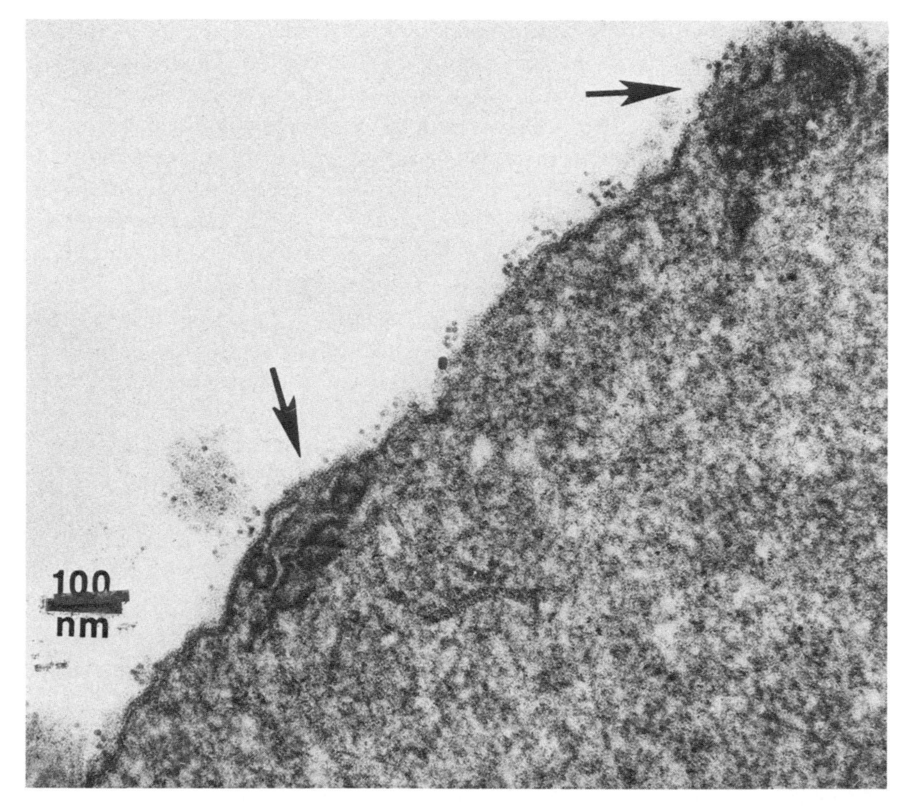


**Fig. 6.** Human erythrocyte, exposed to a paramyxovirus (Sendai) for 2 min, then reacted with ferritin-labeled antiviral antibody. Two aggregates of viral nucleocapsid material (arrows) indicate regions in which fusion of virions with cell surface occurred. Note that viral antigens have apparently diffused into the host membrane during the incubations. ×100,000 (from Bächi *et al.*, 1973; courtesy of Dr. Thomas Bächi).

nents confirms a functional dissociation of the attachment and fusion steps (Seto *et al.*, 1974).

### E. *Transmembranous Penetration and Uncoating of Nonenveloped Viruses*

The complexity of steps in virus uncoating varies. Release of infectious materials from some of the small RNA viruses occurs almost simultaneously to translocation across the cell membrane. This may involve a relatively simple configurational rearrangement of the capsid molecules (Luria and Darnell, 1967; Philipson *et al.*, 1976) which allows them to unravel and release the contained genome. While this process is temperature dependent it may not require energy generation, active host cell metabolism, or new protein synthesis (Luria and Darnell, 1967; Morgan *et al.*, 1969). Changes in the symmetry of adenovirus nucleocapsids in transit through the cytoplasm have been interpreted as a reflection of this "configurational" uncoating process (Morgan *et al.*, 1969; Brown and Burlingham, 1973). Experiments with picornavirus suggest that uncoating and attachment can have a common locus on the plasma membrane, although they are sequentially distinct steps. Uncoating is selectively inhibited by low pH, glutathione or a microsomal factor (Roesing *et al.*, 1975).

Uncoating of virus nucleic acid is measured by release of acid-soluble polynucleotides prelabeled with $^{32}P$ (Chan and Black, 1970) and by their increased sensitivity to nucleases. This can be gauged by the proportion of soluble counts remaining after treatment of $^{32}P$-labeled virus with TCA. The TCA soluble material is nuclease sensitive, indicating release from the protective nucleocapsid casing or protein "shell" of the virus (Chan and Black, 1970; Roesing *et al.*, 1975). The intracellular fate of released virus genetic material can also be pursued with radiolabeled inoculum (Dahl and Kates, 1970).

## V. Membrane-Dependent Steps in Virus Maturation

### A. *Membrane Complexes in Replication of Viral Nucleic Acids*

Production of virus progeny after penetration of parental virus into a host cell commences with transcription of the nucleic acid genome or its translation to polypeptides. In the case of intranuclear DNA viruses which have a large capacity to code for polypeptides (see Portner and Pridgen, 1975), the processes of DNA replication and RNA synthesis may parallel those of uninfected mammalian cells and expression of the viral genome may be similarly regulated by complex control mechanisms (see Sambrook, 1977). The cytoplasmic RNA viruses employ two major strategies for synthesis of viral polypeptides and replication of the genome (a) the genome RNA can itself be messenger RNA which is directly translated to viral polypeptides or act to replicate new genomes through an intermediate RNA strand with complementary base sequence; (b) the virion contains a transcriptase enzyme which enables its RNA genome to be transcribed to a messenger RNA of complementary base sequence. Complementary RNA strands then also serve as template for synthesis of new genome. The manner in which function of any one complete transcript is determined, i.e., message for viral protein synthesis or employment as a replicative intermediate, remains unresolved (see Portner and Pridgen, 1975; Spector and Baltimore, 1975).

Both picornaviruses and togaviruses employ a strategy in which virion genomes encode message and can be translated directly to polypeptides by host ribosomes and transfer RNA. For reasons not yet clear, replicative events in these infections also share an intimate and possibly unique association with internal cell membranes. The cytomembranes evidently provide a stable orientation for nascent viral enzymes, nucleic acids or proteins, thereby increasing the efficiency of interactions and facilitating assembly into nucleocapsids. At the ultrastructural level, this is manifested in a proliferation of smooth cytoplasmic membranes in picornavirus infections (Dales *et al.*, 1965; Amako and Dales, 1967b; Skinner *et al.*, 1968) (Fig. 7) and the morphogenesis of novel membrane structures in arbovirus infections (Blinzinger, 1972; Grimley *et al.*, 1972) (Figs. 8–11).

The development of poliovirus RNA begins in association with sedimentable cytoplasmic structures termed a *replication complex* (Girard *et al.*, 1967). Caliguiri and Tamm (1970a) dissected this process by biochemical analysis combined with subcellular fractionation. The membrane fractions were obtained with a modification of the technique developed by Bosmann *et al.* (1968) for isopycnic centrifugation in discontinuous sucrose density gradients. Pulse-labeling of nascent viral RNA with [$^3$H]uridine and of nascent viral proteins with [$^3$H]leucine demonstrated that replication of the poliovirus RNA genome and the transcription of genomes acting as messenger occurred in association with physically separable membrane elements (Caliguiri and Tamm, 1970a) resembling those described as membranous proliferations in thin sections (Fig. 7), whereas translation was associated with membrane-bound polyribosomes (granular ER). Smooth membrane iso-

---

**Fig. 7.** Proliferation of smooth cytoplasmic membranes and membrane-limited channels in cultured human cell infected by a picornavirus (polio type 1). Cluster of virus nucleocapsids in cytoplasmic matrix is prominent (arrow). ×60,000 (from Dales *et al.*, 1965; courtesy of Dr. Samuel Dales).

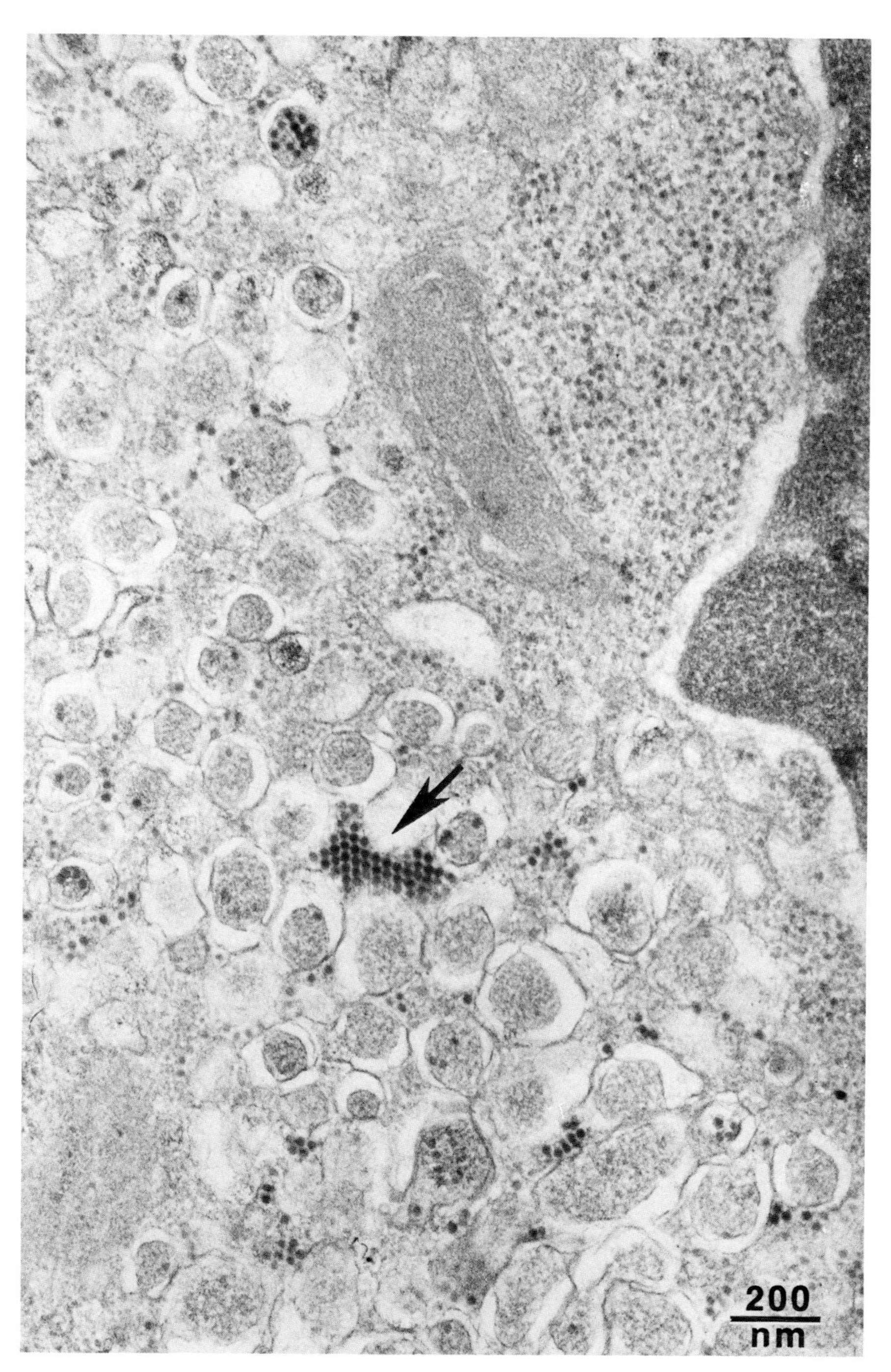
200
nm

lates in the density range of 1.12–1.18 gm/cm$^3$ contained RNA species identified by velocity sedimentation and acrylamide-agarose gel electrophoresis as single-stranded virion RNA, and as replicative forms with complementary strands (Caliguiri and Tamm, 1970b). Most of the viral RNA polymerase activity also is associated with the smooth microsome fraction (Caliguiri and Mosser, 1971), and it has more recently been shown that solubilized poliovirus RNA polymerase spontaneously associates with phospholipid membrane bilayers *in vitro* (Butterworth *et al.*, 1976). Further experiments suggest that initiation of virus particle assembly in the form of "procapsids" is closely coupled to the membrane-associated replication complex (Caliguiri and Mosser, 1971). Assembly *in vitro* of poliovirus also can be mediated by membranes isolated from infected HeLa cells (Perlin and Phillips, 1973).

Picornavirus infection dramatically stimulates cellular incorporation of [$^3$H]choline into membrane lipid (Amako and Dales, 1967b; Mosser *et al.*, 1972b). In cells infected by mengovirus, Plagemann *et al.* (1970) reported a doubling or tripling of choline incorporation into membrane phosphatidylcholine. Using high-resolution autoradiography after pulse-labeling with [$^3$H]choline, Amako and Dales (1967b) localized this lipid precursor over the smooth membrane proliferations. Biochemical analyses of Mosser *et al.* (1972a) disclosed a higher phospholipid to protein ratio in smooth membrane produced after poliovirus infection and a corresponding decrease of the enzyme NADH diaphorase which normally is associated with the endoplasmic reticulum. This indicated an alteration of protein constituents in nascent membranes produced during virus infection. Presumably, viral replicase is inserted into these membranes (Caliguiri and Mosser, 1971).

The mechanism for regulation of new membrane synthesis in picornavirus infection is not known, but studies of Mosser *et al.* (1972a) suggest a shift of lipid precursors from the host rough endoplasmic reticulum (RER) to new smooth membranes. This seems analogous to formation of new smooth membranes from the RER in stimulated hepatocytes (Higgins, 1974). The control of membrane biosynthesis during virus infection evidently involves a complex interaction between virus and host (see Blough and Tiffany, 1975). Since picornavirus-induced membrane proliferation proceeds in the presence of actinomycin D (Caliguiri and Tamm, 1970a), it apparently does not rely upon transcription of new host message. Neither is the membrane proliferation necessarily coordinated with the rates of viral RNA (Mosser *et*

---

**Fig. 8.** (Top) Cytoplasmic vacuole with membranous spherules (CPV-1) in a chicken embryo cell culture infected for 8 hr with the alpha togavirus, Semliki Forest virus. ×48,000.

**Fig. 9.** (Bottom) Nucleocapsids of an alpha togavirus (Semliki Forest virus) surrounding sarcoplasmic reticulum in mouse skeletal muscle cell infected for 12 hr. Note some mature virions within cisternae. ×60,000.

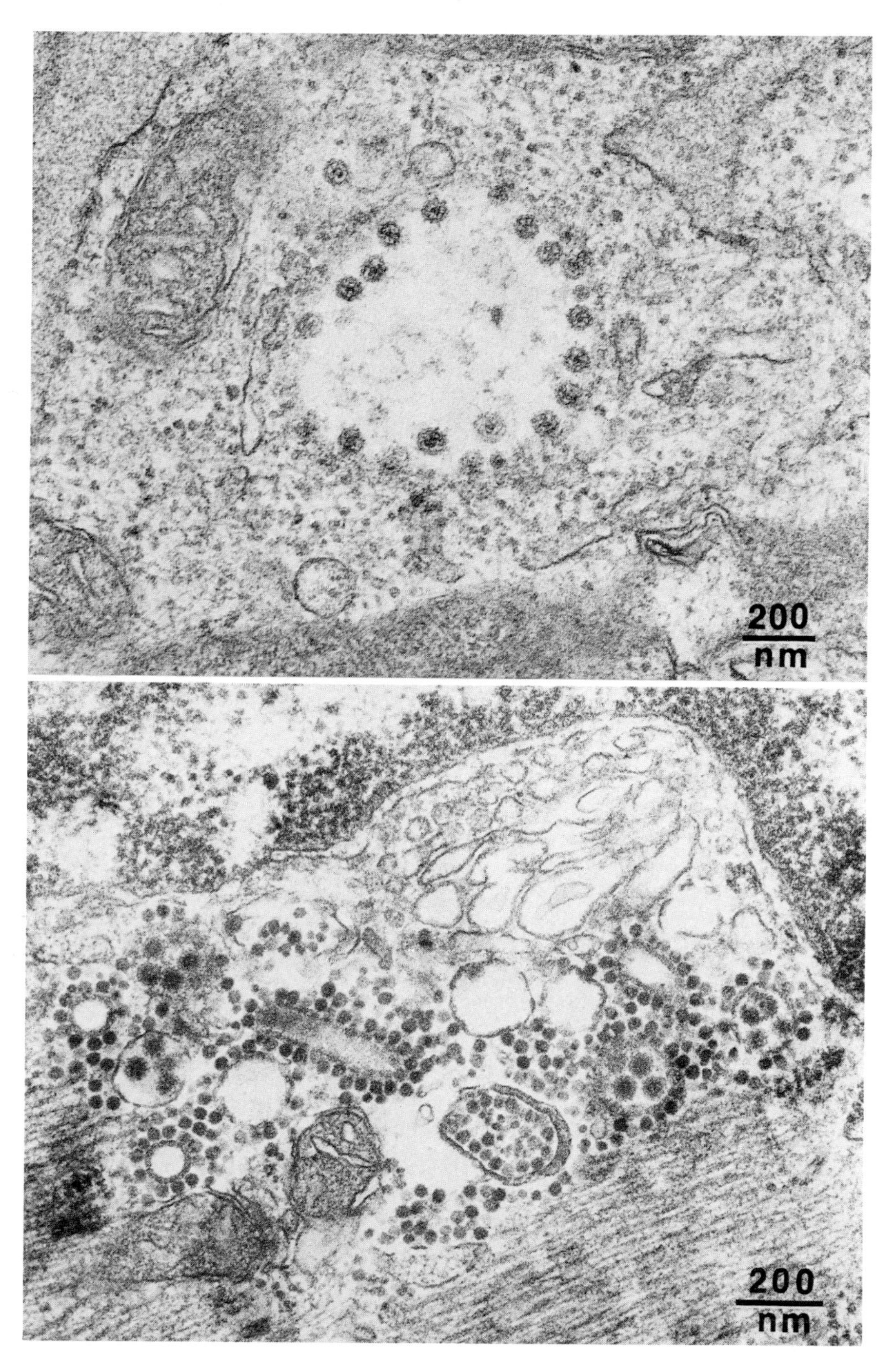
200
nm
200
nm

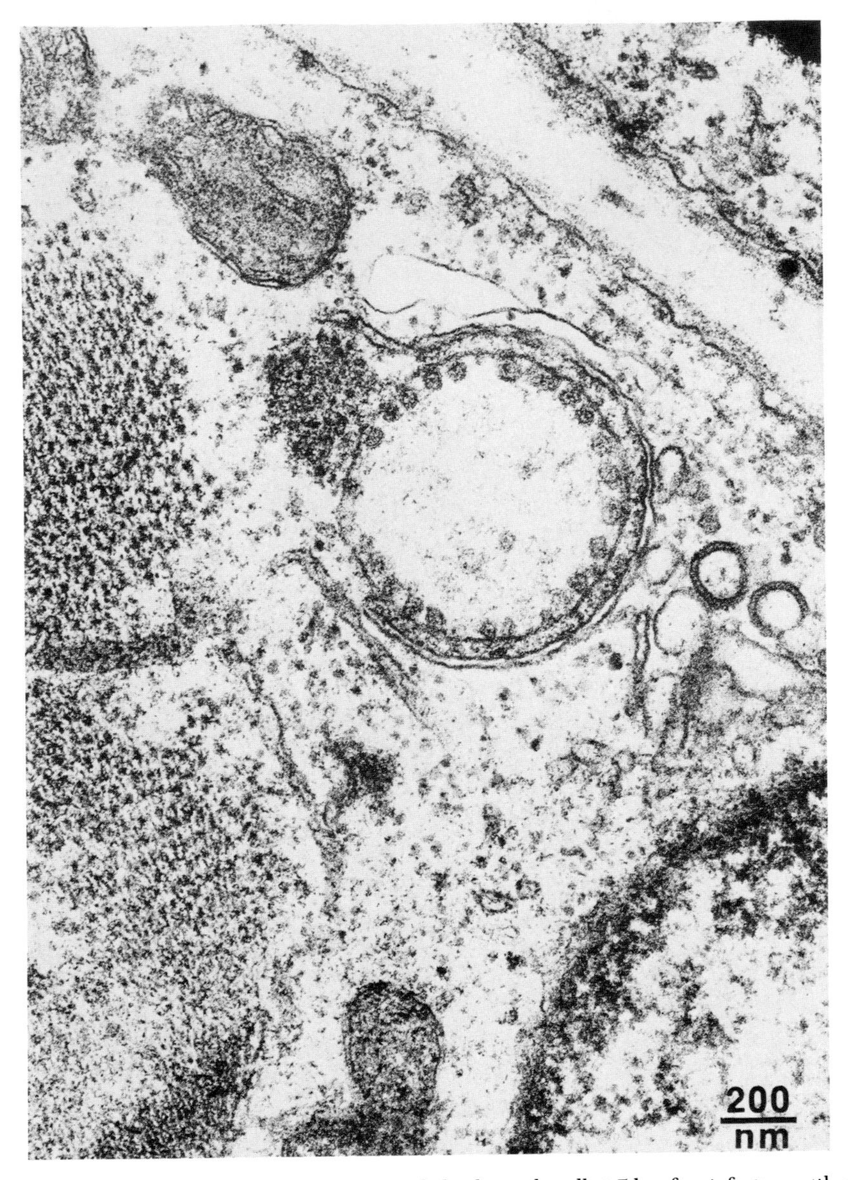


**Fig. 10.** Formation of CPV-1 in a mouse skeletal muscle cell at 7 hr after infection with an alpha togavirus (Semliki Forest virus). Note encircling profile of endoplasmic reticulum. Virus nucleocapsids are not evident at this time. ×48,000.

---

**Fig. 11.** Mosquito cell (*Aedes albopictus*) infected with a flavivirus (yellow fever). Tangential plane through the rough endoplasmic reticulum exposes numerous membranous spherules. Note mature virions within cisternae (arrow). ×33,000.

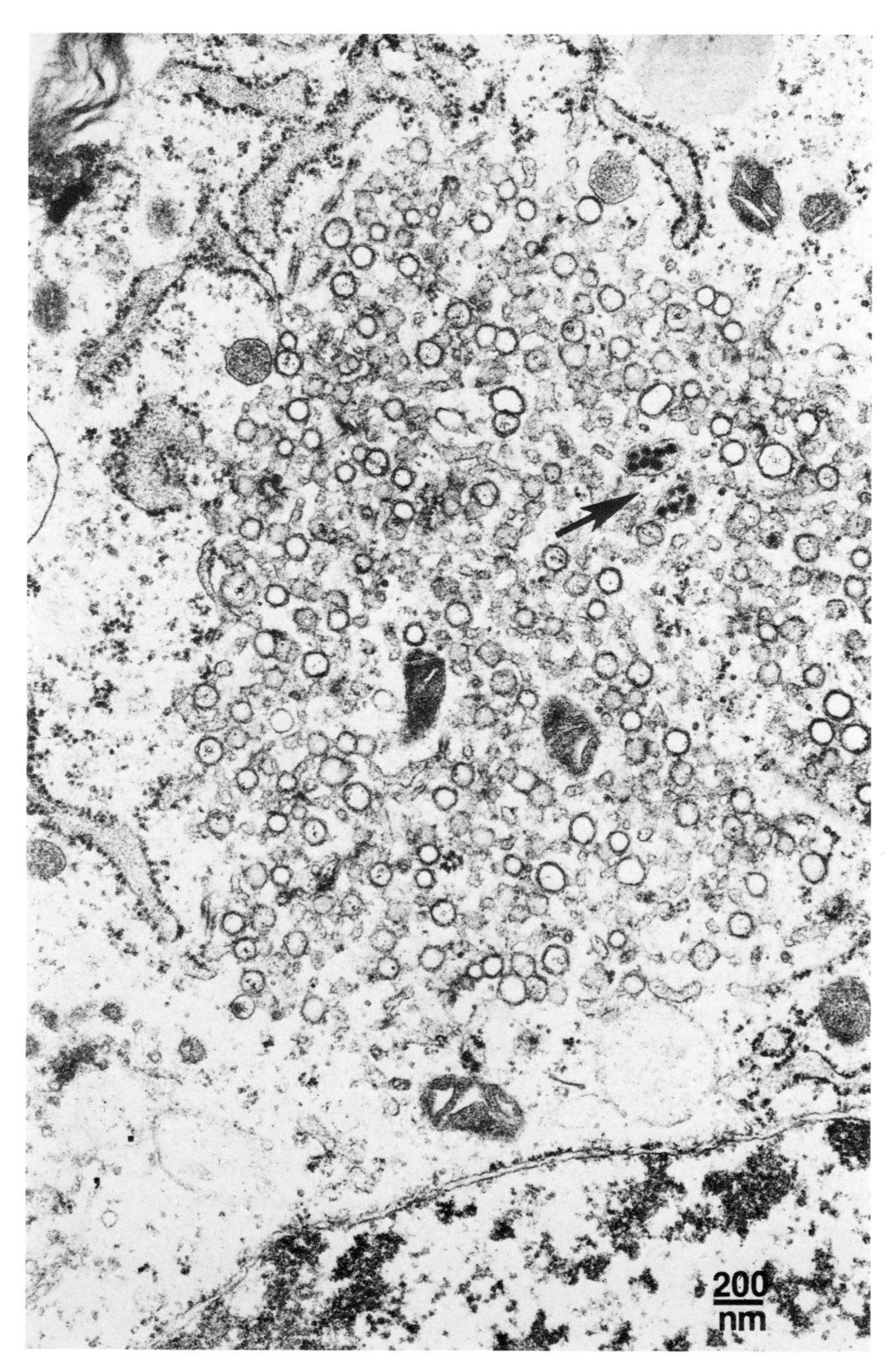
200
nm

*al.*, 1972b) or protein synthesis (Plagemann *et al.*, 1970). Conceivably, the interference of picornavirus infection with normal direction of cellular protein synthesis by host messenger RNA (see Spector and Baltimore, 1975) permits a selective proliferation of membranes, uninhibited by normal control mechanisms. A precedent for this concept is the evidence that smooth membrane proliferation in hepatocytes can be enhanced by low concentrations of actinomycin D, probably just sufficient to "derepress" normal controls (Orrenius and Ericsson, 1966).

Membrane structures also play a vital role in the development of togaviruses (arboviruses), although a net stimulation of membrane biosynthesis as noted in picornavirus infections may not be observed (Waite and Pfefferkorn, 1970b; Vance and Lam, 1975). The earliest ultrastructural evidence of infection with alpha togaviruses is often the appearance of 50-nm diameter membranous spherules with a fine central density (Fig. 8). These spherules are clearly distinguished from the smaller and denser virus nucleocapsids of enveloped virions (Fig. 9) and they become most numerous during the period of exponential virus growth (Grimley *et al.*, 1972). The spherules may be observed both at the cell surface and within large cytoplasmic vacuoles designated CPV-1 (Grimley *et al.*, 1972). The origin of CPV-1 membrane is uncertain. Thorotrast tracer from the growth medium may be included in CPV-1 suggesting origin by surface invagination. Cytochemical studies show the presence of acid phosphatase similar to that observed in the Golgi region (Grimley *et al.*, 1972), but another Golgi enzyme, TTPase was absent. The CPV-1 lack IDP associated with membranes of the RER, but they are frequently surrounded intimately by profiles of ER (Fig. 10). Freeze-fracture of CPV-1 membranes shows a lack of normal intramembranous particles within cleavage planes (Virtanen and Wartiovaara, 1974).

The possibility that membranous spherules represent a defective form of alpha togavirus was excluded by a series of experiments with alpha togaviruses passaged under different biological conditions and inoculated at low multiplicities of infection (Grimley *et al.*, 1972). The CPV-1 arise in tissue culture cells from several sources including human fibroblasts and HeLa cells and in mouse brain (Grimley and Friedman, 1970a) or striated muscle (Grimley and Friedman, 1970b).

Several lines of experimental evidence point toward a direct participation of CPV-1 in togaviral RNA replication. A large proportion of input virion RNA binds to host membranes within 1 hr (Friedman and Sreevalsan, 1970; Sreevalsan, 1970). Sedimentable membranous structures are associated both with nascent virus RNA (Friedman and Berezesky, 1967) and virus-induced viral RNA polymerase (Sreevalsan and Yin, 1969). When the cytoplasmic membranes of cells infected with an alpha togavirus were subjected to fractionation utilizing the approach of Caliguiri and Tamm (1970a), CPV-1 separated with smooth membranes and mitochondria at a density of 1.16 gm/cm$^3$. This fraction was enriched in viral RNA forms and viral RNA polymerase (Friedman *et al.*, 1972). High-resolution autoradiography after pulse labeling of infected cells with [$^3$H]uridine during the time of exponential virus growth showed development of silver grains over CPV-1 (Friedman *et al.*, 1972). Membranous spherules were decreased in numbers or absent in cells infected by temperature-sensitive alpha togavirus mutants grown at a temperature where RNA synthesis was restricted to no more than 24% of normal levels (Tan, 1970).

The role of membrane structures in group B togavirus (flavivirus) infections has been less fully documented, but Qureshi and Trent (1972) reported that a membranous structure with an average sedimentation coefficient of 250 S associated with viral RNA forms, RNA polymerase and viral-specific proteins in St. Louis encephalitis infection. Stohlman *et al.* (1975) reported localization of dengue virus RNA synthesis to the rough endoplasmic reticulum, and suggested a role of smooth membranes in virus capsid maturation. Boulton and Westaway (1976) also noted a predominance of Kunjin virus RNA and protein synthesis on internal cytomembranes. Membranous spherules measuring 100–120 nm in diameter are prominent in the ER of cells infected with flaviviruses (Fig. 11) and their morphogenesis typically precedes the appearance of mature virions (P. M. Grimley and N. A. Young, unpublished). These intracisternal spherules have been found in flavivirus-infected mammalian or arthropod cell cultures (Filshie and Rehacek, 1968; Matsumura *et al.*, 1971), and brain tissue (Boulton and Webb, 1971; Blinzinger, 1972). The similarity to spherules lining CPV-1 in alpha togavirus infections is intriguing and suggests homologous functions but there is yet no direct evidence.

In flavivirus infections, formation of new membrane structures may be associated with increased incorporation of [$^3$H]choline (Zebovitz *et al.*, 1974). Surprisingly, this does not occur in alpha virus infection. Waite and Pfefferkorn (1970b) showed that Sindbis virus infection of chick embryo cells causes a progressive and indiscriminate reduction of phospholipid synthesis. Temperature shift experiments with a temperature-sensitive Sindbis mutant showed that limited replication of RNA at the nonrestrictive temperature was essential for inhibition of choline incorporation, but that structural proteins of the virus were not involved. Their experiments with chick embryo and baby hamster kidney (BHK) cells indicated that the virus inhibition of phospholipid synthesis paralleled effects of the metabolic inhibitors actinomycin D and cycloheximide. Vance and Lam (1975) showed that Sindbis virus infection inhibits incorporation of choline into cellular phospholipids and inhibits an enzyme involved in biosynthesis of phosphatidycholine. It can only be concluded that the striking membrane changes which occur at the ultrastructural level in alpha togavirus infections must involve a highly selective redirection of residual membrane biosynthesis or a reorganization of preexisting membrane macromolecules. The latter was suggested by continued formation of CPV-1 for up to 3 hr in the presence of cycloheximide (Grimley *et al.*, 1972). This problem requires further investigation with individual cell membrane fractions.

### *B. Membrane-Associated Synthesis of Viral Proteins*

As obligate intracellular life forms, viruses rely totally upon host cell support for translation of the viral genetic code and synthesis of viral polypeptides. The latter includes structural polypeptides which are incorporated into progeny virions and nonstructural polypeptides such as viral enzymes employed during nucleic acid transcription or steps in progeny virion maturation (i.e., polypeptide phosphorylation or posttranslational macromolecular cleavages). Size of the viral genome determines its capacity to code for polypeptides. For example, the relatively large DNA genomes of herpesvirus ($82 \times 10^6$ MW) can theoretically code about 50 polypeptides of which over half have actually been identified (see Portner and Pridgen, 1975). Most RNA viruses have a more limited coding capacity (6–10 polypeptides). Thus, they provide relatively simple models to investigate the dynamics of membrane-associated protein synthesis, membrane integration, and virion maturation. Virus polypeptide biosynthesis can be specifically traced by introducing radiolabeled precursors at times after infection when host protein synthesis is restrained or totally inhibited by metabolic competition.

The site of messenger RNA translation may be on polysomes either within the cytosol or attached to membrane (RER). In uninfected mammalian cells, membrane-bound polyribosomes are typically engaged in synthesis either of proteins for secretion or of glycoproteins which become integrated into cell membrane. Lodish and Froshauer (1977) have reviewed evidence which suggests that attachment of ribosomes to the ER is mediated by the amino terminal sequence of nascent polypeptides. This "leader" may be very hydrophobic, so that the completed protein macromolecule immediately penetrates into the membrane bilayer near its origin of synthesis. Strong support for this train of events comes from studies of Sindbis virus (an alpha togavirus) by Wirth *et al.* (1977). In this very elementary model, the mature virion contains only three proteins: an internal core protein (C) and two envelope glycoproteins ($E_1$ and $E_2$). All three proteins are encoded by a single virus RNA (26 S), and synthesized by the same ribosome. The core protein is translated first in the cytosol. Enzymatic cleavage then exposes the amino terminus of protein $E_2$ which interacts with a proximate leaflet of ER and thereby binds the ribosome. After protein $E_2$ is cleaved it remains embedded in the membrane, becoming an integral protein of the future virus envelope. Synthesis of protein $E_1$, also membrane-associated, can then begin. Protease treatment of membrane vesicles isolated with the newly synthesized Sindbis proteins, suggests that protein $E_2$ has a cytoplasmic tail (carboxy terminus), thus totally traversing the membrane bilayer.

Studies of vesicular stomatitis virus have also been of special interest, since the rhabdovirus employs a more complex mechanism in the synthesis of envelope membrane proteins. Nascent chains of a protein penetrating the virion envelope (G) are closely associated with membrane-bound polyribosomes, whereas a nonpenetrating, matrix protein (M), associated with the inner aspect of the virion envelope, is evidently synthesized in the cytosol (David, 1977; Knipe *et al.*, 1977b). These G and M proteins may be considered analogous to the integral and peripheral membrane proteins defined in the fluid-mosaic concept (see Fig. 3). Again, linkage of ribosomes to membrane of the ER in rhabdovirus infection is mediated by the nascent G polypeptide (Lodish and Froshauer, 1977).

In addition to providing templates for insertion of virus gene products and ultimate conversion to virion envelopes, host cytomembrane systems exert a remarkable role in the biosynthesis of virus envelope glycoproteins. Small viruses lack sufficient genetic endowment to direct synthesis of the multiple enzymes which would be required for this task in the absence of host cell aid (Sefton, 1976). Thus, differences in the sugar composition of a togavirus (Sindbis) and a rhabdovirus (vesicular stomatitis) grown in different cell lines reflect disparities in host cell ability to complete the glycosylation of virus-coded polypeptides (Etchison and Holland, 1974; Keegstra *et al.*, 1975). Even more striking was the finding by Stollar *et al.* (1976) that a togavirus grown in cells from a potential mosquito vector lacks sialic acid, whereas the same virus grown in vertebrate cells contains sialylated protein. This was not associated with any detectable biological or antigenic differences.

In paramyxovirus infections, the host cell asserts a role in maturation of a fusion-related glycoprotein of the virion. This glycoprotein arises by cleavage from a larger precursor (Scheid, 1976). After replication in bovine kidney (MDBK) or mouse L cells virions emerge with abundant precursor but little active protein. Such virus adsorbs to host cells but is incapable of causing infection and fails in tests for hemolysis or cell fusion. Virions maturing in the chick embryo allantoic sac have little precursor and abundant cleavage product. They are fully capable of infection and manifest both cell fusion and hemolytic activities. Treatment of the MDBK-grown virus with trypsin *in vitro* can induce infectivity, evidently by cleaving the fusion factor precursor (Scheid and Choppin, 1974).

From the standpoint of virus replication, the role of transferase enzymes in glycosylation of membrane-associated proteins is the major focus of interest; however, viruses with a large genetic endowment, such as herpesvirus (Kaplan *et al.*, 1975) and poxvirus (Moss *et al.*, 1971b) may in fact produce nonstructural glycoproteins which conceivably modify cell behavior or interactions and thus play a pathobiological role in the infectious process at a tissue or organ level (Spear, 1975).

### C. *Origins of Virus Structural Membrane*

A number of biochemical and ultrastructural studies, cited earlier (Section III,A), have provided evidence for the biosynthesis of membrane elements in the cell interior with peripheral transport of prefabricated units to renew the cell surface or to form the envelope of secretory products. The kinetics of viroceptor regeneration on the cell surface after removal by enzymatic treatment indicate that receptor macromolecules are integrated with newly synthesized membrane in the cell interior, then move centrifugally (Zajac and Crowell, 1965; Levitt and Crowell, 1967; Marcus and Schwartz, 1968; Philipson *et al.*, 1968). Time-lapse cinematography and phase microscopy convincingly document the intermittent motion of endoplasmic reticulum in cultured cells with reversible connections, subdivisions, and regroupings of the tubular or pancake-shaped compartments (Buckley, 1965).

Biochemically, the endoplasmic reticulum evidently comprises a patchwork of preexisting and newly formed elements (see Higgins, 1974). Virus-induced membrane proliferations appear to be similarly constructed (Amako and Dales, 1967b; Mosser *et al.*, 1972a). An orderly succession of molecular modifications presumably accounts for specific differences observed in the protein distribution and enzymatic activities of interior and peripheral cytomembranes (cf. Heath, 1971; Hirano *et al.*, 1972; Meldolesi and Cova, 1972). For example, some monosaccharide residues are attached to secretory glycoproteins almost concurrently with termination of the polypeptide core; however, glycosylation is consummated in a stepwise process, and the sequential addition of monosaccharides also appears to propel secretory glycoproteins (such as immunoglobulin) through the Golgi region toward peripheral cites of exocytosis (Melchers, 1973). This progression is predetermined by the localization of specific glycosyltransferases which are concentrated in specialized membrane elements.

Evidence for a very similar progression in protein maturation has emerged from investigations of the "template viruses" which utilize internal cytomembrane elements of the host as a scaffold for addition of virus-specific products. The former include most groups of the RNA viruses and the DNA herpesviruses. In certain virus groups, covalent sulfation of membrane glycoproteins also involves multiple steps beginning in the rough endoplasmic reticulum, and continuing through the smooth membranes or even at the cell surface (Nakamura and Compans, 1977).

Thermodynamically, the direct transfer of newly synthesized membrane protein from membrane-bound polyribosomes to nascent segments of phospholipid (Higgins and Barnett, 1972) would appear to be more efficient than a release of proteins from polyribosomes and cytoplasmic diffusion to loci of membrane incorporation. A direct transfer process appears to characterize formation of the integral envelope proteins in several RNA virus infections (Hay, 1974; Klenk *et al.*, 1974; David, 1977; Katz *et al.*, 1977; Wirth *et al.*, 1977). Other proteins, considered equivalent to the interior peripheral proteins shown in Fig. 3, associate with the envelope just before virion emergence and evidently pass through a membrane-free cytosol phase (David, 1977; Knipe *et al.*, 1977b).

The rhabdovirus, vesicular stomatitis virus, contains five structural proteins, two of which are associated with the envelope. A glycosylated protein (G) penetrates the lipid bilayer (Katz *et al.*, 1977) and forms the outer envelope spikes. After a brief radioactive pulse, G protein is associated both with a smooth membrane fraction and with a fraction containing RER (Wagner *et al.*, 1972). Glycosylation evidently occurs only during association of the G protein precursor with membrane (David, 1977; Knipe *et al.*, 1977a). A second membrane protein (M) is associated with the inner aspect of the virus envelope, arrives after the G protein and apparently is synthesized on free polysomes (David, 1977; Knipe *et al.*, 1977b,c). This protein can bind strongly to isolated plasma membranes (Cohen *et al.*, 1971) and is conceivably responsible both for anchoring the G protein spikes and stabilizing the internal helix of ribonucleoprotein (Blough and Tiffany, 1975). In contrast to G and M proteins, the nucleocapsid ribonucleoprotein (N) is not membrane associated (Wagner *et al.*, 1970) and apparently is synthesized on free polyribosomes prior to complexing with viral RNA in the cytoplasm.

Protein constituents of myxovirus envelopes are more numerous. The envelope spikes represent four glycoproteins. The largest glycoprotein (HA) functions as a hemagglutinin and may cleave during maturation (Lazarowitz and Choppin, 1975). Myxovirus glycoproteins are similarly localized in the RER shortly after synthesis begins, and subsequently with smooth membrane (Compans and Caliguiri, 1973). Individual sugar residues on the viral glycoproteins are incorporated stepwise with glucosamine being added in the RER and fucose being added on smooth membrane (Compans, 1973). The precursors of the envelope hemagglutinins and the envelope neuraminidase polypeptides, both are synthesized in close association with the RER and at least partially glycosylated there (Hay, 1974; Klenk *et al.*, 1974). The macromolecular precursor of hemagglutinin is cleaved to yield two polypeptides sometime during the traverse of smooth membranes to the cell surface. The M protein is incorporated as one of the last steps in envelope assembly (Nagai *et al.*, 1976), thus very simialr to the sequence in rhabdovirus envelope formation discussed above.

The sequence of myxovirus antigen association with cell membranes has also been traced by immunoelectron microscopy. This shows labeling of the RER within 4 hr after infection (Hoshino *et al.*, 1972; Ciampor *et al.*, 1974) and of the smooth membrane or cell surface at later times (Ciampor *et al.*, 1974; Lampert *et al.*, 1975). In studies of surface budding viruses, viral antigens typically localize in discrete patches on the plasma membrane exterior, corresponding to a zone of internal membrane thickening where the nucleocapsid attaches (cf. Howe *et al.*, 1969; Aoki and Takahashi, 1972; Coward *et al.*, 1972; Bächi and Howe, 1973). This may also be seen directly as formation of a "fuzzy coat" on the cell surface exterior, corresponding to glycoprotein spikes visualized by negative contrast or in freeze-fracture preparations (Bächi *et al.*, 1969).

### *D. Virion Assembly, Emergence and Release*

In analyzing the terminal events of virus maturation, virions formed by budding can be divided conceptually along lines suggested by Blough and Tiffany (1975).

1. Nucleocapsids are fully assembled and tightly organized before budding into a tightly conforming envelope (e.g., herpesviruses, togaviruses).
2. Assembled nucleocapsids align beneath the cite of budding but are only loosely cloaked in an envelope of variable proportions (e.g., myxoviruses).
3. Final organization of assembled nucleocapsids occurs during emergence from the cell (e.g., rhabdoviruses).
4. Assembly of the nucleocapsid is concerted with emergence from the cell (e.g., oncornaviruses).

In the infections which form preassembled nucleoprotein, interaction of the cores with the inner membrane leaflet at a site of emergence appears to be the event which initiates outgrowth of a virus "bud" (Brown *et al.*, 1972; Dubois-Dalcq *et al.*, 1976a,b; Hashimoto *et al.*, 1975; Knipe *et al.*, 1977c). Intramembranous fluidity within this region is reduced (Sefton and Gaffney, 1974; Moore *et al.*, 1976) and the IMP normally detected by freeze-fracture typically disappear (Brown *et al.*, 1972; Sheffield, 1974; Dubois-Dalcq *et al.*, 1976a,b; Demsey *et al.*, 1977). At the same time, projecting spike proteins or globules characteristic of the mature virion envelope can be recognized ultrastructurally (Bächi *et al.*, 1969; Demsey *et al.*, 1977). In the case of togaviruses, myxoviruses, and rhabdoviruses, these projections represent the integral envelope-penetrating glycoproteins discussed in Section V, C. They typically arrive at the cell surface as long as 20 min in advance of virion emergence (cf. Lenard and Compans, 1974; Atkinson *et al.*, 1976), while posttranslational macromolecular alterations such as sulfation (Nakamura and

Compans, 1977), sialylation (Knipe *et al.*, 1977a), or cleavage (Lazarowitz and Choppin, 1975; Hay, 1974; Jones *et al.*, 1977) may still be in progress.

In myxovirus and rhabdovirus infections, attachment of a nonpenetrating "M protein" sets the stage for binding, positioning or shaping of the core (Blough and Tiffany, 1975; Shimizu and Ishida, 1975; Atkinson *et al.*, 1976). Recent studies of temperature-sensitive rhabdovirus mutants (Knipe *et al.*, 1977c) are consistent with this hypothesis. Mutations thought to involve the penetrating spike protein block membrane binding of both the nucleocapsids and M protein. On the other hand, mutations which cause failure in assembly of virus nucleocapsid do not block insertion of spike protein into membranes of the ER and migration to the cell surface. When the integral spike proteins are not stabilized by attachment of the peripheral M protein, structures of virion density fail to form. In myxovirus infection, the binding of M protein localizes to membrane cites in which penetrating hemagglutinin macromolecules are already present (Hay, 1974), while M protein is almost immediately incorporated into virions after attachment to the cell surface. Spike proteins are shed into virus from a larger pool of membrane-associated macromolecules and at a slower rate (Hay, 1974; Knipe *et al.*, 1977b).

A parallel disparity in the rates of shedding and pool sizes of newly synthesized core proteins and virion envelope precursors has been noted in oncornavirus infection (Witte and Weismann, 1976). An important mechanism in oncornavirus assembly is a terminal processing of large polyprotein intermediates (e.g., env-pr 85) to the macromolecular forms present in the virion envelope—gp 69/71, p15(E). This may involve sialylation and proteolytic cleavages (cf. Shapiro and August, 1976; Van Zaane *et al.*, 1976; Witte *et al.*, 1977).

Interaction of virus cores with envelope glycoproteins need not be specific. Virions of mixed phenotype (pseudotype particles) have been shown to emerge in a number of coinfections (Závada, 1972). In phenotypically mixed particles, spike glycoproteins of a paramyxovirus could be identified on virions with the morphology typical of a rhabdovirus (McSharry *et al.*, 1971). In this circumstance, however, the nonglycosylated M protein of the rhabdovirus segregated with the bullet-shaped nucleocapsids, further implicating an architectural interaction during rhabdovirus emergence.

Since arrival of M protein and nucleocapsids can occur just minutes prior to emergence and release of a mature virion (Atkinson *et al.*, 1976; Knipe *et al.*, 1977b), it remains difficult to explore the temporal sequence of organization. Evidence thus far can be interpreted to suggest: (a) a primary conformational or phase change in a membrane domain which promotes interaction with physically abutting core material (Hay, 1974; Lenard and Compans, 1974; Rifkin and Quigley, 1974; Blough and Tiffany, 1975; Brown and Smith, 1975; Dubois-Dalcq *et al.*, 1976a; Jones *et al.*, 1977; Witte *et al.*, 1977); (b) an active intermediation of "director" molecules such as the M protein to initiate the binding process (Shimizu and Ishida, 1975; Atkinson *et al.*, 1976; Knipe *et al.*, 1977c); and (c) a concerted "nucleation" of macromolecules triggered by physical contact and intrinsic properties of the nucleocapsid (Garoff and Simons, 1974; Hashimoto *et al.*, 1975). Interestingly, major shifts in the nucleci acid core composition of oncornavirus are not incompatible with budding of particles identified by density and negative staining (Levin *et al.*, 1974) and budding of Visna virus without "dense cores" has been observed (Coward *et al.*, 1972).

It is clear that any model of virion assembly must rest heavily upon the internal fluidity of membranes and lateral movements of viral or cellular proteins within and subjacent to the membrane (Birdwell and Strauss, 1974a; Hay, 1974; Blough and Tiffany, 1975). This assumption best explains observations that virus cores of different genotype and phenotype can emerge from physically proximate regions on the cell surface as observed by electron microscopy (Lunger and Clark, 1972; Grimley *et al.*, 1973). Furthermore many oncornaviruses incorporate host-specified glycoproteins within their relatively small and tightly fitted envelope. This is demonstrated both by immunoelectron microscopy and the capacity of purified virions to inhibit cytotoxic activity of various anti-H-2 sera. A curious aspect of this phenomenon, recently noted by Bubbers and Lilly (1977), is a selective incorporation of H-2 antigenic determinants into Friend leukemia virus, although the average proportion of such host surface molecules in each virion envelope probably is quite small (Aoki and Takahashi, 1972; Dorfman *et al.*, 1972). In general, experimental observations indicate that cell membranes of uninfected cells are far from saturated with integral proteins, and are thus able to accommodate virus-specified envelope glycoproteins by a simple process of lateral displacement perhaps involving steric segregation (Garoff and Simons, 1974). This is consistent with extensive "bare" regions in the phospholipid bilayer (see Section III,B), contiguity of host membrane and viral proteins observed in macrovesicles (Heine and Roizman, 1973), and quantitative persistance of host glycoproteins (Spear *et al.*, 1970; Birdwell and Strauss, 1973; Hay, 1974).

Depending on the virus group, virion emergence, and thereby envelopment, may proceed primarily at the inner nuclear membrane, through membranes of the ER, into cytoplasmic vacuoles, or at the cell surface (Fig. 12). While a functional interaction between the virus core and cell membrane at a cite of emergence depends upon complex local factors already discussed, general host cell conditions or the time after infection may also

**Fig. 12.** (A) Virions budding preferentially from filopodial projection on surface of a mouse L cell infected with an alpha togavirus (Semliki Forest virus). ×70,000. (B) Virions budding into channels of endoplasmic reticulum in neuron of a mouse infected with an alpha togavirus (Semliki Forest virus). ×60,000.

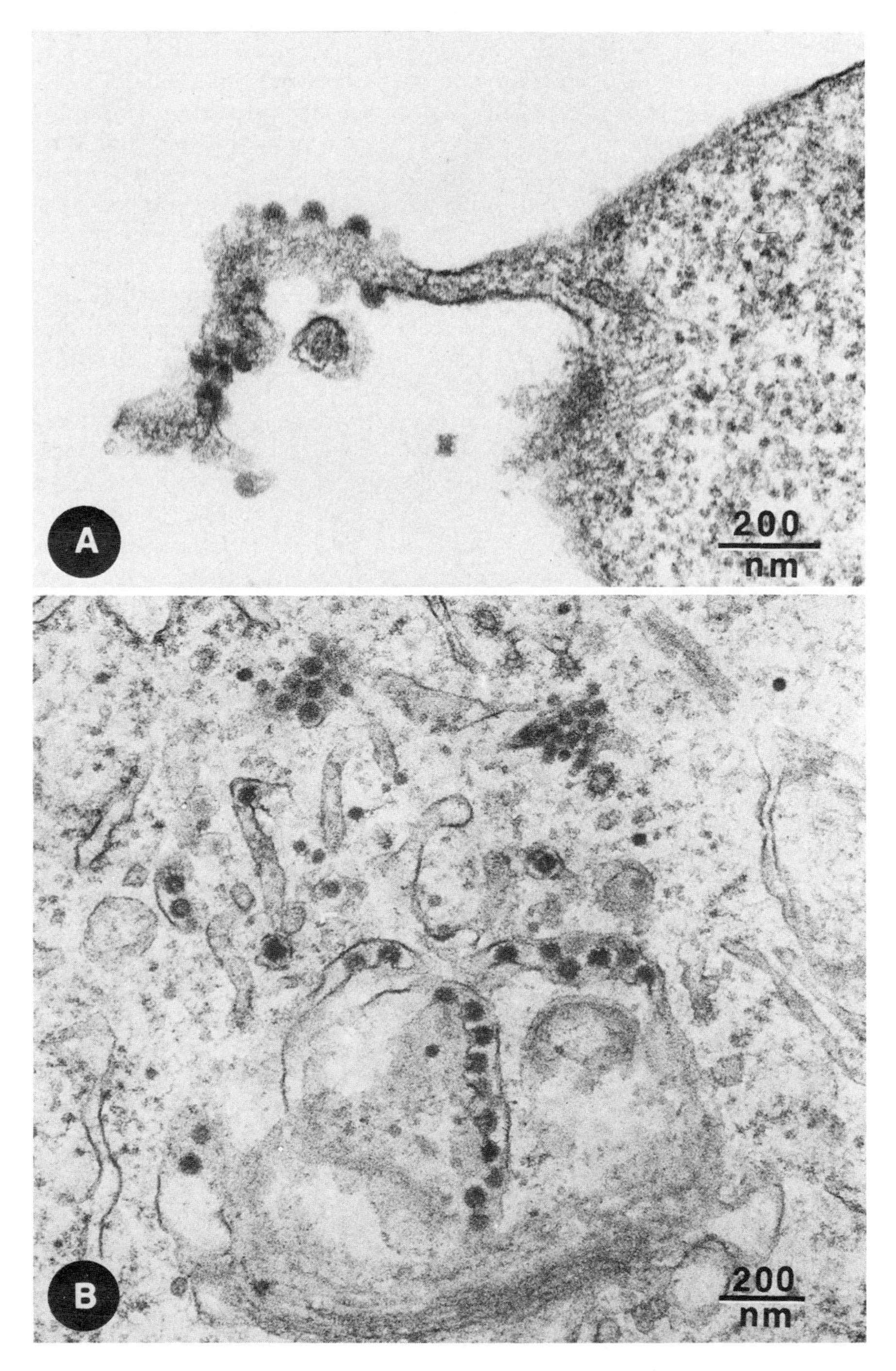
A
200
nm
B
200
nm

influence the pattern of virion emergence (Grimley and Friedman, 1970a,b; Gliedman *et al.*, 1975) or ability to emerge from carrier cells (Dubois-Dalcq *et al.*, 1976a; Ray *et al.*, 1976). Although physical collision of nucleocapsids with membranes probably is a relatively random process (Wong and MacLeod, 1975) microtubular movements or cytoplasmic flow dynamics may produce vectorial forces which direct virus emergence into surface projections (Fig. 12) or cell body extensions (Grimley and Friedman, 1970a; Birdwell *et al.*, 1973; Dubois-Dalcq *et al.*, 1976b).

Normally, the nucleocapsid alignment with specific envelope components at a membrane cite of emergence proceeds within minutes (Witte and Weissmann, 1974; Knipe *et al.*, 1977a) and the terminal steps in virion release may occupy no more than 20 sec (Waite and Pfefferkorn, 1970a). Some transitional events in this process can be dissociated in temperature-sensitive virus mutants or by experimental treatments. In cells infected with alpha togavirus, temperature-sensitive cleavage of one of the viral envelope proteins ($PE_2$) appears to be essential for the budding process (Jones *et al.*, 1977), and the preassembled virus cores do not bind to the plasma membrane in cells infected at nonpermissive temperature (Brown and Smith, 1975). Exposure of cells infected with togavirus to media of reduced ionic strength results in subplasmalemmal accumulation of nucleocapsids (Waite *et al.*, 1972). In studies of leukemia virus mutants, Wong and McCarter (1974) observed a stage in which virus partially emerged from the cell surface but where final constriction of the bud and release was not possible at the restrictive temperature. They ascribed this to a reversible defect in protein conformation.

A constriction mechanism presumably triggers the final steps of membrane fusion at the base of each particle (Brown *et al.*, 1972; Blough and Tiffany, 1975), and ultrastructural studies with freeze-etching indicate separate fusions of inner and outer membrane leaflets to complete the virion envelope (Brown *et al.*, 1972). Release from the cell surface apparently depends on the ionic or mucoprotein environment (Grimley and Friedman, 1970b; Waite *et al.*, 1972) as well as the proximity of viroceptive molecules. Ability of myxovirus progeny to separate from the cell surface may be mediated by the virion neuraminidase (Palese *et al.*, 1974; Scheid, 1976) and can be inhibited by antibody to viral surface antigens (Dowdle *et al.*, 1974).

## VI. The Nuclear Envelope in Virus Infection

### A. *Segregation and Replication of the Virus Genome*

Segregation and amplification of functions by means of internal membrane systems accompanied the evolution of eukaryocyte organization and genetic regulation (see Watson, 1976). The nuclear membrane separates the major genetic reservoir from the bulk of synthetic machinery which resides in the cytoplasmic compartment, and undirectional movement of informational transcripts (RNA) from the nucleus to the cytoplasm is basic to cellular control of protein synthesis (Goldstein, 1974; Watson, 1976). Intercompartmental flow of cations and polypeptides can also be vectorial (Goldstein, 1974). For example, histones and other nascent nucleoproteins preferentially enter the nucleus after synthesis in the cytoplasm and a rapid influx of molecules from the cytoplasm is associated with external stimulation of nuclear functions (Goldstein, 1974; Johnson *et al.*, 1974).

Intranuclear microenvironmental conditions which facilitate regulation of the cellular genome and processing of messenger RNA transcripts presumably sustain analogous functions during infection by viruses with a nuclear phase (cf. Honess and Roizman, 1974; Sambrook, 1977). Formation of nucleoprotein complexes occurs in several DNA virus infections (Chantler and Stevely, 1973; Seebeck and Weil, 1974; Meinke *et al.*, 1975; Gautschi *et al.*, 1976; Su and DePamphilis, 1976) and intranuclear assembly of viral nucleocapsids must depend upon appropriate physicochemical conditions for interaction of the viral proteins and nucleic acid (Gautschi *et al.*, 1976; Iida and Oda, 1975). Indeed, pseudovirions may contain host DNA (Qasba *et al.*, 1974).

A direct role of the nuclear membrane in initiating replication of viral nucleic acid, analogous to the role of cytoplasmic membranes discussed in Section V, A, has been debated for several years. In uninfected mammalian cells there appears to be no specific relationship of DNA replication to the nuclear membrane (Comings and Okada, 1973); however, the nuclear membrane does provide a major locus for orientation of DNA strands (see Comings, 1974; Franke and Scheer, 1974) and may selectively bind native DNA or synthetic polynucleotides (Kasper, 1974). In monkey cells infected by a papovavirus (SV40), LeBlanc and Singer (1974) found about 85% of DNA synthesis to be associated with a nuclear membrane fraction during a brief pulse-labeling with [$^3$H]thymidine. The mature viral DNA was evidently released after completion of replication. In adenovirus-infected human cells, Shiroki *et al.* (1974) found both parental virus DNA and nascent DNA in a nuclear membrane fraction. They identified DNA synthesized on nuclear membranes *in vitro* as a viral form by DNA–DNA hybridization. Almost coincidentally, Yamashita and Green (1974) reported the finding of new polypeptides closely associated with the isolated nuclear membranes of adenovirus-infected cells. These were interpreted as early viral gene products synthesized before replication of the adenovirus DNA. They appeared similar to DNA-binding polypeptides isolated from adenovirus-infected cells by Van der Vliet *et al.* (1975). Thus a nuclear membrane replication complex for initiation of adenovirus synthesis with a DNA polymerase capable of transcribing viral DNA sequences (Ito *et al.*, 1975) appears to exist.

### B. *Function of the Nuclear Pore Complex*

Despite some specific differences, both inner and outer nuclear membranes are fundamentally similar to the ER in terms of intrinsic macromolecular composition and enzymatic functions (Kasper, 1974; Berezney, 1974). Thin sections reveal that the nuclear envelope consists of paired membranes separated by a perinuclear cisternum which is intermittently continuous to cisternae of the granular ER (Blackburn, 1971). The outer nuclear membrane may have attached ribosomes and can participate in the synthesis of secretory proteins (Leduc *et al.*, 1968). The inner nuclear membrane appears asymmetrically thickened by a closely applied 300–600 Å deep layer of electron-dense material. This peripheral lamina consists primarily of polypeptides and is considered to provide a firm skeleton which supports and orients the nuclear pores (Aaronson and Blobel, 1975).

Nuclear pores or "channels" are the most significant microanatomic feature of the nuclear envelope (Fig. 13) since they probably represent the major pathway for nuclear–cytoplasmic macromolecular exchanges (Goldstein, 1974). Using the freeze-fracture technique and a computer program for calculation of topographic distributions, Maul *et al.* (1971) estimated a relatively constant pore to pore spacing, and the total numbers of nuclear pores were related to DNA content of the nucleus. Detailed fine structural studies of nuclear pores including negative staining of isolated nuclear envelopes indicate an orifice of relatively fixed diameter (ca. 60–80 nm) framed on inner and outer faces by annuli composed of symmetrical subunits which may be associated with fine fibrils (Franke and Scheer, 1974; Aaronson and Blobel, 1975). The center of the pore typically contains dense material or a granule (Blackburn, 1971). Electron images suggesting active extrusion of nuclear materials through pores have been obtained (Franke and Scheer, 1974), and some of the material occupying pore channels is evidently enriched in RNA, possibly representing messenger transcripts or ribosomal RNA in passage (Franke and Scheer, 1974; Goldstein, 1974). Migration is evidently restricted to the central part of the channel with a particle size range of 95–140 Å for

**Fig. 13.** Nuclear pores shown in complementary freeze-fracture replicas of the nucleus of an STU-Eveline cell. Most of the membrane surface revealed in the bottom micrograph is the EF surface of the outer nuclear membrane, although a step down to the PF surface of the inner membrane can be seen (arrow). The top micrograph reveals the apposing surfaces: i.e., mostly PF surface of the outer membrane, with some EF surface of the inner membrane (arrow). ×32,000 (from Demsey *et al.*, 1974).

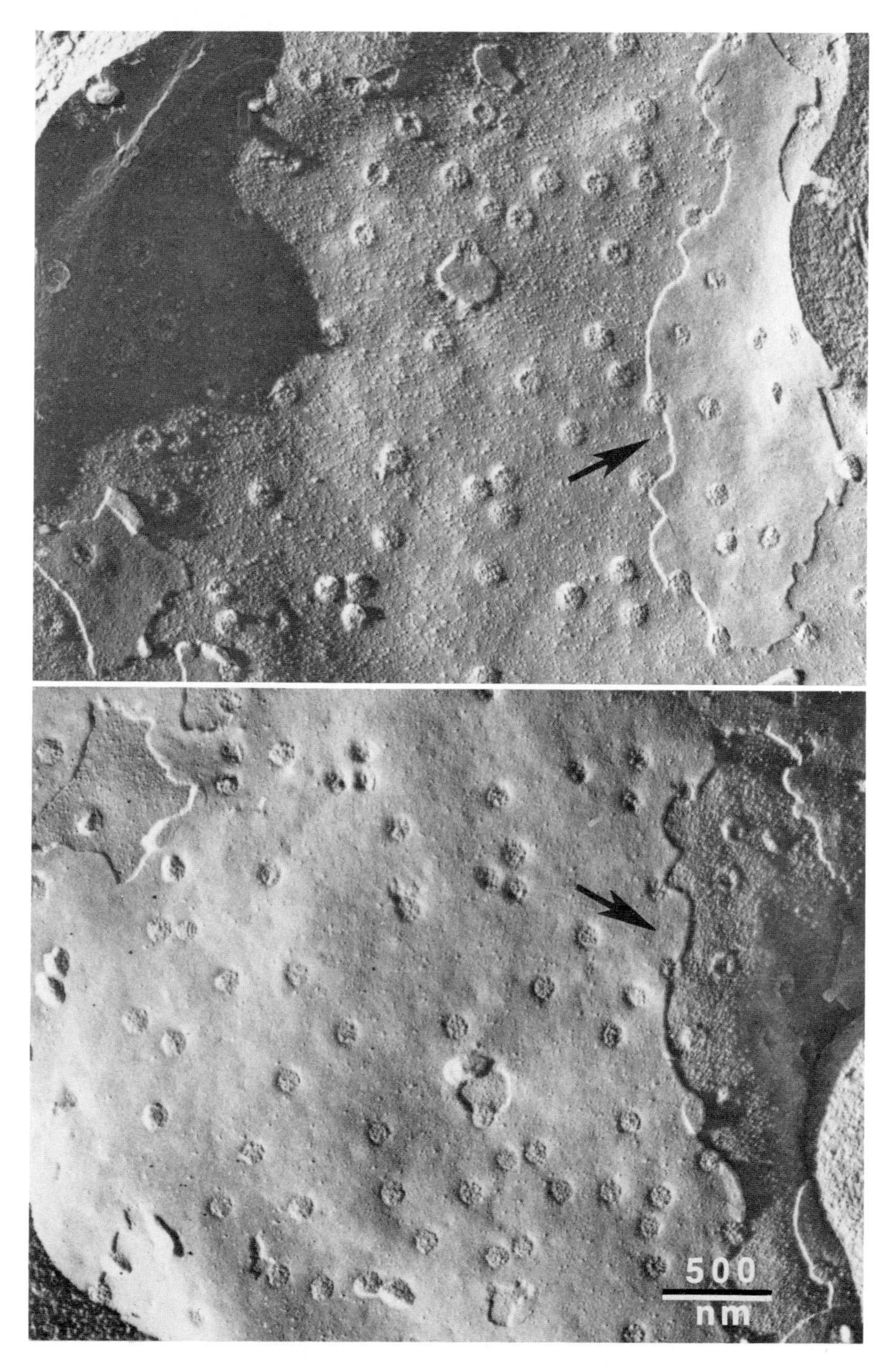
500
nm

ferritin and gold respectively (Feldherr, 1972). This is sufficient for egress of papovavirus nucleocapsids (Maul, 1976).

As in normal cellular processes, the replication of intranuclear DNA viruses also depends upon an active exchange of materials between the nucleus and the cytoplasmic compartments. This begins at the time of virus entry when cores of engulfed herpesvirus, adenovirus, or papovavirus rapidly approach the nuclear envelope (Morgan *et al.*, 1969; Barbanti-Brodano *et al.*, 1970; Hummeler *et al.*, 1970; Dales and Chardonnet, 1973; Mackay and Consigli, 1976). Final uncoating of the infectious nucleic acid genome may occur on the cytoplasmic outlet of the nuclear pore complex (Dales and Chardonnet, 1973) or within the nucleoplasm (Morgan *et al.*, 1969; Barbanti-Brodano *et al.*, 1970), possibly involving active transport (Dales and Chardonnet, 1973). The subsequent movement of viral RNA transcripts back into the cytoplasm is apparently regulated (Kozak and Roizman, 1974), and the proteins of DNA viruses which are synthesized on free or membrane-bound cytoplasmic polysomes must finally return to the nucleus for assembly of virus progeny (Ben-Porat *et al.*, 1969; Velicer and Ginsburg, 1970; Mark and Kaplann, 1971). This can be a selective process. In herpesvirus infection, capsid proteins required for intranuclear particle assembly processes evidently are released into the cytoplasmic sap, whereas glycoproteins destined for the cell surface can pass vectorially through or reside in membranes of the ER (Kaplan *et al.*, 1975). Similarly, in the case of the intranuclear nucleocapsids of the RNA myxoviruses, proteins destined for the virus envelope do not appear in the nuclear inclusions (Maeno and Kilbourne, 1970). Under certain experimental conditions viral proteins normally expected in the nucleus may fail to migrate (Duff *et al.*, 1970; Ishibashi, 1970). This could be due to a defect in active transport at the level of nuclear pore complex (see Blackburn, 1971).

### *C. Ultrastructural Pathology*

Gross configurational changes in the nuclear envelope occur in the course of normal cell growth, division, and maturation. Electron microscopy has shown that prophase dispersal of the nuclear envelope leads to formation of multiple individual membrane-bound cisternae and vesicles which lose asymmetry and become structurally indistinguishable from other constituents of the ER. Chromosomes are the organizing units responsible for reunification of the nuclear envelope during late anaphase and telophase when nuclear pores also reform (Franke and Scheer, 1974). Fluctuations in the numbers and distribution of nuclear pores also indicate that the nuclear envelope is a dynamic structure (Maul *et al.*, 1972).

Ultrastructural alterations of the nuclear membranes often are quite dramatic during herpesvirus infections and maturation of these viruses occurs to a large degree through the nuclear envelope. In fungal cells, herpes-type virus is temporarily enveloped by both inner and outer nuclear membranes in the passage from nucleoplasm to cytoplasm. These membranes are degraded in the cytoplasm and a final envelopment occurs by budding into cisternae of the Golgi apparatus or into cytoplasmic vacuoles (Kazama and Schornstein, 1973). Budding of herpesvirus nucleocapsids through the inner nuclear membrane can be readily observed in thin sections of vertebrate cells infected with herpes simplex (Darlington and Moss, 1969), herpes zoster, varicella (Achong and Meurisse, 1968), cavine herpes (Fong and Hsuing, 1977), Epstein-Barr virus (Glaser *et al.*, 1976), and cytomegaloviruses (Berezesky *et al.*, 1971). Herpes virions can traverse the nuclear membrane individually or in groups presenting a "peas in the pod" appearance (Fig. 14). The latter evidently segment into individual particles, since visualization of more than one nucleocapsid per virion envelope is very rare. The route of transfer from the perinuclear cisternum to the cell surface can only be surmised from thin-section observations. Before cytolysis or during infections by less virulent members of the herpes group, virions evidently can be transported to the cell surface within the system of ER cisternae and Golgi vesicles which eventually open to the exterior (Darlington and Moss, 1969). Nii *et al.* (1968) suggested that maturing nucleocapsids with a dense nucleic acid core bud preferentially through the nuclear envelope. On the other hand, Schaffer *et al.* (1974) observed that nucleocapsids of a temperature-sensitive herpes simplex mutant (ts022) with empty and partial cores could be preferentially enveloped at the restrictive temperature.

While relatively little is yet known about the molecular interactions at cites of nuclear virus budding, a freeze-fracture study (Haines and Baerwald, 1976) suggests events similar to those at the cell surface with loss of intramembranous particles (see Sections V, D). The envelope of herpes simplex virus contains new species of glycoproteins, but no appreciable loss of host membrane proteins is noted (Heine *et al.*, 1972). These observations suggest lateral displacement of host membrane proteins as discussed above (Section IV,E). Experiments of Ben-Porat and Kaplan (1971) indicated that herpes-type virions can assemble from newly synthesized segments of the inner nuclear membrane.

Breakdown of the nuclear envelope may occur early in herpesvirus replication, even prior to extensive disruption of normal cellular organization (Fig. 15). One possible explanation is a viral induction of host cell DNA synthesis (Melvin and Kucera, 1975) leading to an abortive prophase condition in which the nuclear envelope begins to fragment. Nucleocapsids then escape through gaps and mature at membrane surfaces in the cytoplasm rather than at the nuclear envelope (e.g., Fong and Hsuing, 1977). Studies of

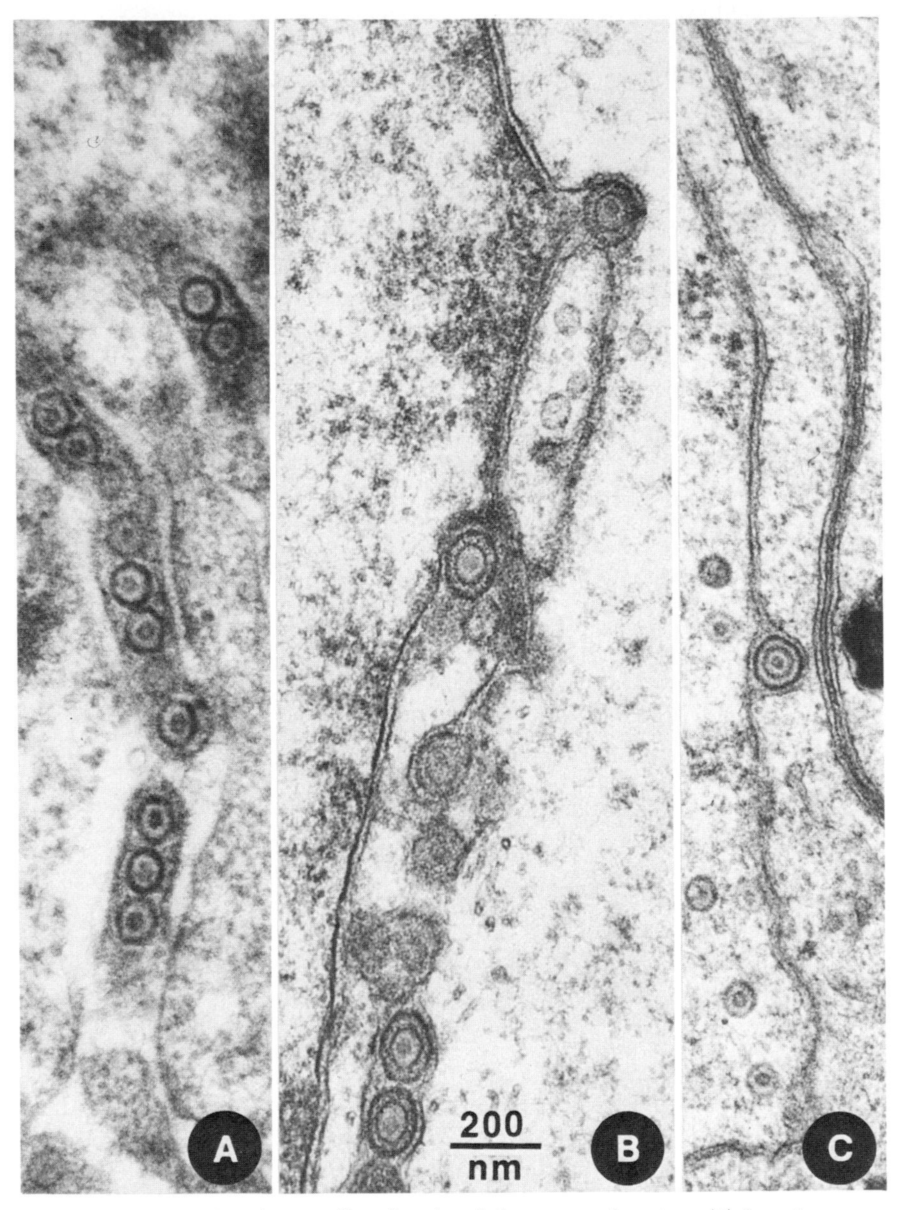


**Fig. 14.** Human lymphoma cells infected with herpes simplex virus. (A) Immature nucleocapsids extruded from nucleus in common outpouchings of the inner nuclear membrane. (B) Enveloped nucleocapsids which have budded through the inner nuclear membrane lie within endoplasmic reticulum cisternae connected to the perinuclear cisternum. (C) Comparison of immature intranuclear nucleocapsids and mature virion in the perinuclear cisternum. Reduplicated lamella of nuclear envelope appears to the right. ×60,000.

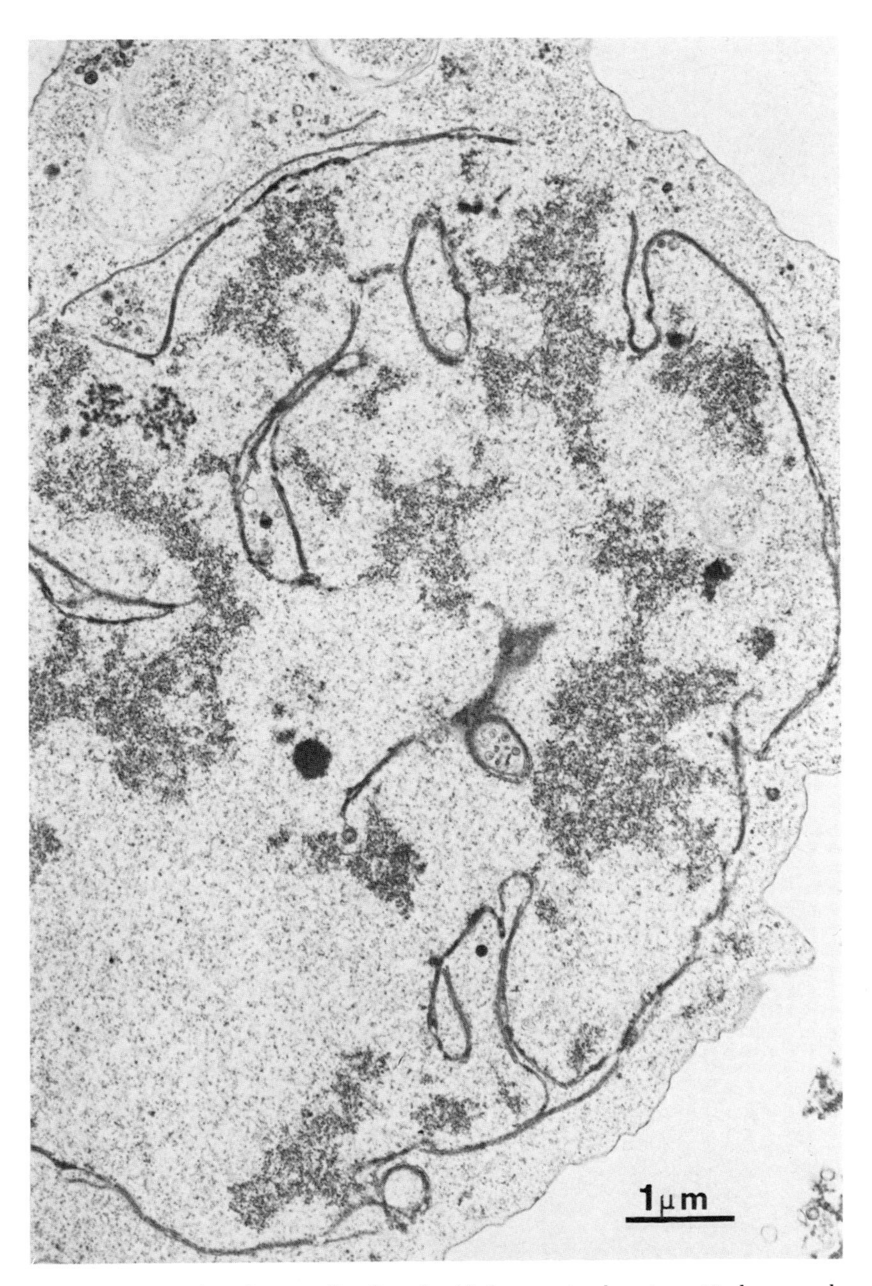


**Fig. 15.** Human lymphoma cell infected with herpes simplex virus. Nuclear envelope shows large discontinuities and abnormal reduplications of limiting membranes. Pattern of nuclear densities suggests chromosomal condensations. ×14,000.

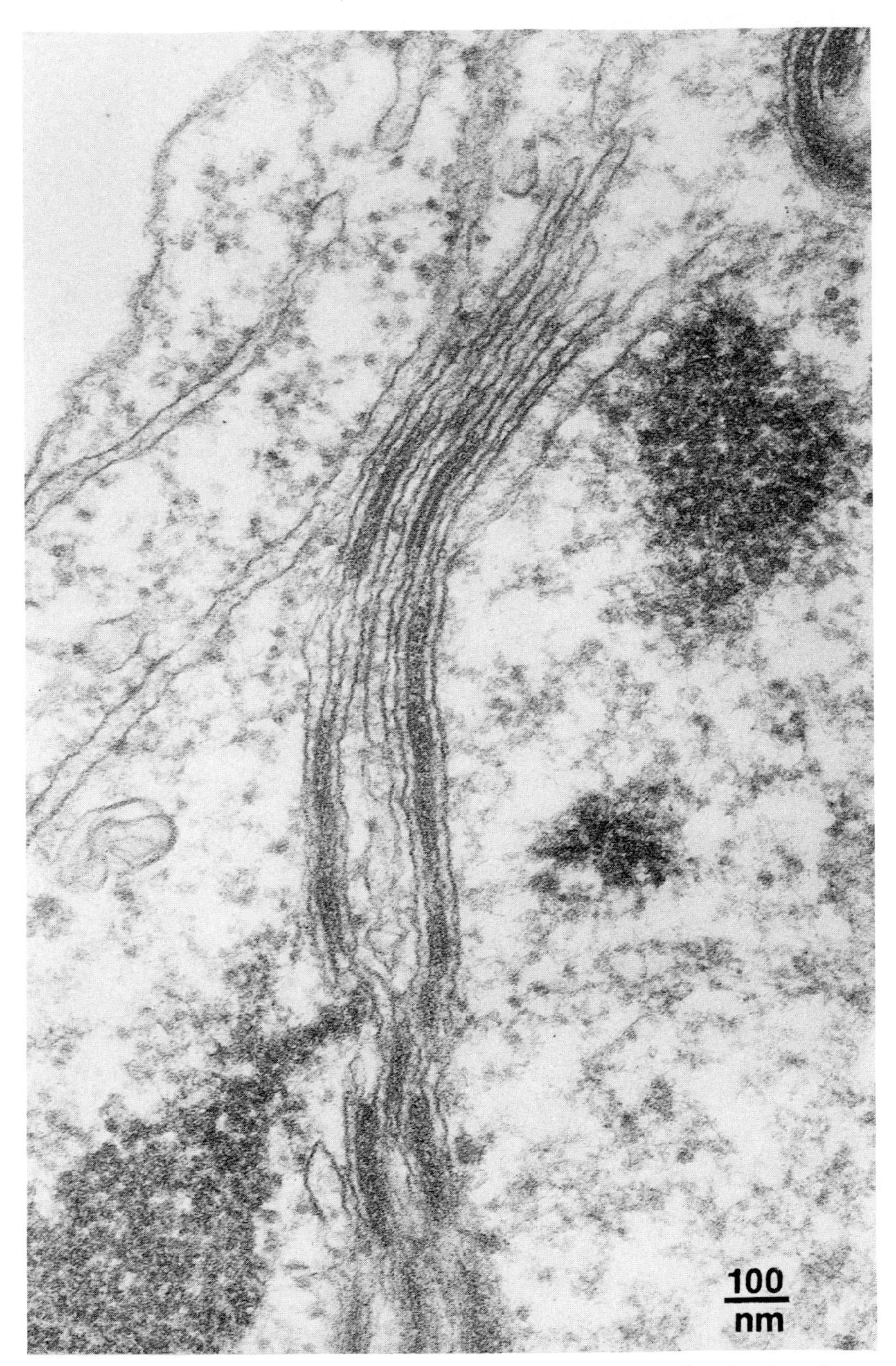


**Fig. 16.** Reduplications of nuclear envelope in human lymphoma cell 48 hr after infection with herpes simplex virus. ×78,000.

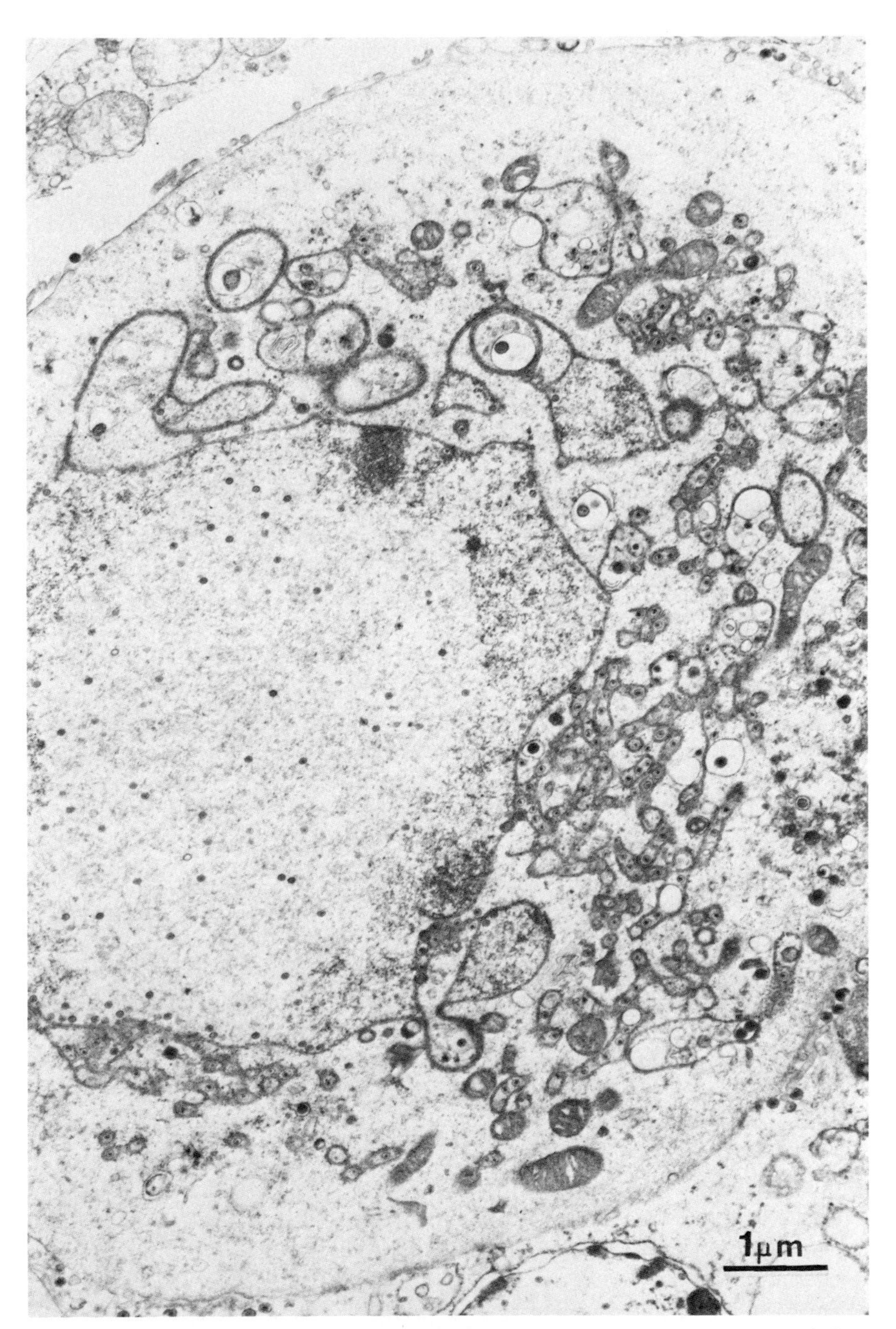


**Fig. 17.** Tortuous protrusions and reduplications of nuclear envelope in monkey kidney cells 24 hr after infection with herpes simplex virus. ×14,000.

a rat cytomegalovirus showed extensive accumulations of nucleocapsids in the cytoplasm of cells with apparently intact nuclear envelopes (Berezesky *et al.*, 1971). These nucleocapsids often lacked a dense core and were surrounded by an osmiophilic matrix. Some of them, however, appeared to represent complete nucleocapsids and budding into cytoplasmic vacuoles was observed. Similar observations have been made in other cytomegalovirus infections and could be explained by a maturation process in which a temporary nuclear envelope disintegrates (Kazama and Schornstein, 1973).

An almost pathognomonic feature of herpesvirus infection is a remarkable pairing or redundant stacking of nuclear membranes and the extrusion or drawing out of the nuclear envelope in tortuous projections, "blebs", and extended lamellae. Some of the envelope nuclear extensions may represent "short-circuit" bridge connections between regions of the outer nuclear membrane as described by Franke and Scheer (1974). In part, however, the ultrastructural appearance indicates that the inner nuclear leaflets and an associated 100–150 Å layer of dense granulofibrillar material or heterochromatin accompany the projections (Fig. 16). At low magnification, the overall pattern often suggests an elaborate lacework (Fig. 17). While generally related to an active process of virus envelopment, this membrane activity can also occur in abortive infection (Schaffer *et al.*, 1974; Glaser *et al.*, 1976). The finding of active phospholipid metabolism in pseudorabies virus-infected rabbit kidney cells (Ben-Porat and Kaplan, 1971) strongly suggests that the morphological changes reflect stimulated nuclear membrane synthesis analogous to the cytomembrane proliferations observed in picornavirus infections (Section V, A). Fusion or reduplications of paired envelope cisternae is noted also in adenovirus infection (Gregg and Morgan, 1959). These phenomena may be manifestations of pathobiological cytomembrane interactions due to synthesis of viral cytotropic or fusion proteins (see Sections IV, C and D) which insert indiscriminately. Redundancy of the nuclear envelope also occurs under conditions of active protein synthesis (Mollo *et al.*, 1969). Conversely, cell differentiation, as in spermatogenesis or leukopoiesis, often involves a contraction of the nuclear mass (Merriam, 1962) with formation of redundant nuclear envelope. Extreme examples of the latter are the polymorphonuclear leukocytes. Separation of envelope sheets in the form of annulate lamellae also can rapidly decrease nuclear surface area (Guylas, 1971). Formation of annulate lamellae noted in adenovirus-infected cells by Merkow *et al.* (1970) appeared to originate in close proximity to the nucleus within 1 hr after infection.

The frequent occurrence of nuclear bridges, projections, or pockets in lymphocytes (Pope *et al.*, 1968) is of some relevance to the discussion of herpesvirus infections, since members of this virus family, notably EBV, are lymphocytotropic and have been implicated in the pathogenesis of lymphomas (Miller *et al.*, 1977). No direct correlation should be drawn between these characteristics of lymphocyte nuclei, and the changes observed in experimental herpesvirus infections although they may appear superficially similar. Nuclear "pockets" projections or "blebs" are common in normal human leukocytes (Huhn, 1967; Smith and O'Hara, 1968) as well as in atypical and leukemic cells (Ahearn, 1967; Mollo and Stramignoni, 1967). Anomalous nuclear projections also have been described in a chromosomal triplication disorder (Lutzner and Hecht, 1966). These observations merely reinforce the concept that the nuclear membrane responds in an active manner to a variety of normal and pathological stimuli including virus infection.

## VII. Membrane-Related Cell Reactions

### A. *Activity of Lysosomes*

The disposition of hydrolytic enzymes and associated "inflammatory proteins" contained within primary lysosomes is governed by the limiting membranes of these microvesicular organelles. Fusions of the lysosomal membranes with phagosomes containing engulfed virions initiates a process of intracellular digestion similar to that observed with other particulate nucleic acids and proteins (see Friend *et al.*, 1969; Ericsson and Brunk, 1975). Intact virions may persist in phagolysosomes for several hours, but they are ultimately degraded and infectivity is abolished (Ogier *et al.*, 1977). As discussed in Section IV,B, infective virions probably escape the lytic pathway entirely, or liberate their nucleic acid within phagosomes before evolution to phagolysosomes. The proportion of inoculum virions which is shunted into the lytic pathway appears to vary with the virus group and even amongst serotypes (Dales, 1973; Ogier *et al.*, 1977). Only phagosomes with engulfed diplornaviruses appear to fuse preferentially with primary lysosomes (Dales, 1973). Since the diplornaviruses contain a nuclease-resistant (double-stranded) genome, lysosomal degradation can be advantageous in releasing their infectivity. Presumably, the single-stranded genomes of most other viruses are at least partially sensitive to lysosomal nucleases (Ogier *et al.*, 1977).

Evidence for an increase in lysosomal enzyme activity during virus infection comes from histochemical, cytochemical, and biochemical studies (Ruebner *et al.*, 1966; Allison, 1971; Greenham and Poste, 1971; Reeves and

Chang, 1971). During some virus infections, large numbers of multivesicular bodies appear (Fig. 18). These membranous structures are apparently related to the lysosomal system (see Anteunis, 1974) and may be particularly common in cytomegalovirus infections (Berezesky *et al.*, 1971).

In specialized defensive cells such as granulocytes and macrophages, lysosomal hydrolases and associated proteins accumulate and concentrate within secretory vesicles which transport them to the cell surface. Fusion of these lysosomal membranes with the plasma membrane is the normal route for egress. When release of these products is provoked under appropriate conditions, local destruction of foreign materials or microorganisms and stimulation of the inflammatory response can be beneficial to the host. When the release of lysosomal enzymes is disproportionate to the noxious stimulus or even inappropriate, harmful destruction of host tissues may ensue. Such over-reaction may propagate chronic inflammatory or connective tissue diseases (Hamerman, 1966). In certain virus infections, an increased production and extracellular secretion of lysosomal enzymes probably occurs (Ruebner *et al.*, 1966; Greenham and Poste, 1971; Reeves and Chang, 1971). In laryngotracheitis virus infection of fowl, this leads to a destructive chondrolysis in the turbinate cartilage beneath infected nasal mucosa (Schultz and Bang, 1977). Such extreme effects are probably rare; however, increased contact of lytic enzymes with the exterior of virus-infected cells may be an important factor in the process of cell to cell fusion (see Greenham and Poste, 1971).

A large number of experimental studies have been directed toward measuring changes in the permeability of lysosomal membranes during virus infection (see Allison, 1971; Tamm, 1975). This can lead to significant shifts in the intracellular distribution of acid hydrolases. In normal processes of cell division or differentiation, controlled and selective release of hydrolytic enzymes conceivably facilitates specific biochemical processes or moderates the environment for gene regulation (Allison, 1967). Pathological implications of incontinent hydrolytic enzyme release into the cell sap range from subtle effects on chromosomes (see Allison, 1967) to obvious weakening of the cell framework and cytolysis. In virus infections, particularly infections by nonbudding nucleotropic viruses, lysis of host membrane from within may facilitate the escape of progeny virions (see Tamm, 1975). A number of experiments have indicated that virus cytopathic effects occur more rapidly than expected on the basis of direct metabolic inhibition with drugs (see Tamm, 1975). The cytopathic effects in picornavirus infection appear to depend upon a virus-directed function and may even be retarded by treatment with metabolic inhibitors during the early stages of virus infection

**Fig. 18.** Multivesicular bodies in perinuclear cytoplasm of rat kidney cell infected with rat strain of cytomegalovirus. Note intranuclear herpes-type nucleocapsids (arrows). ×40,000.

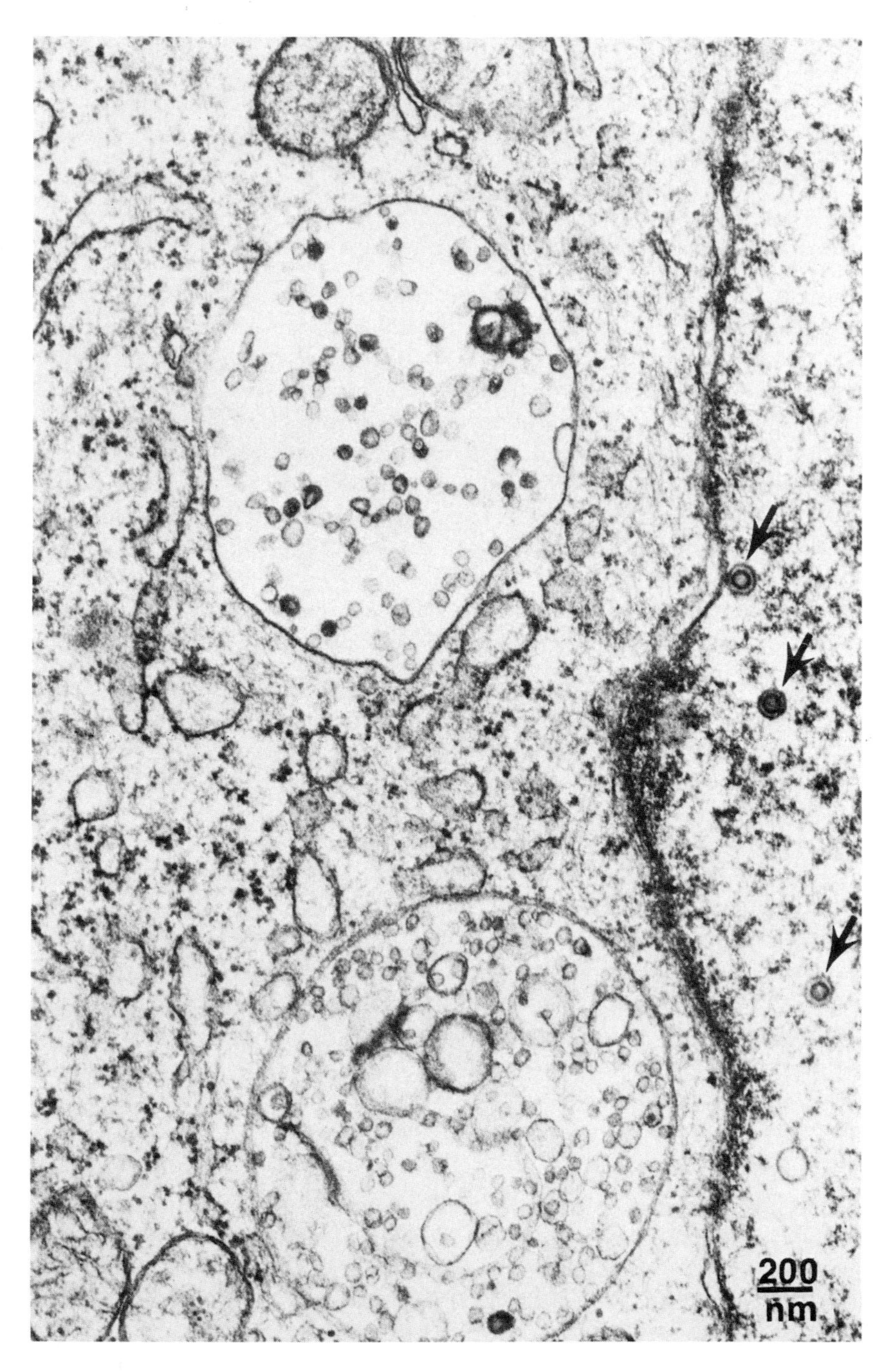
200
nm

(Amako and Dales, 1967a; Guskey and Wolff, 1974). Perhaps the most likely explanation for these phenomena is modification of lysosomal membranes by insertion of viral proteins (Allison, 1971); however, molecular mechanisms which account for labilization of lysosomal membranes by virus proteins remain to be elucidated.

### *B. Intracellular Membranous Proliferations*

Proliferation of smooth membranous elements associated with the ER or the nuclear envelope is a common subcellular response to virus infections. As previously detailed in Sections V,A and VI,C, some membrane alterations are integrally related to specific events in the virus replication cycle, including nucleic acid transcription, nucleocapsid assembly, or virion maturation. In this section, we draw attention to membranous proliferations which evidently represent a secondary subcellular reaction in virus-infected cells and characteristically appear after the onset of virion maturation. The functional significance, if any, of such cytopathologic phenomena remains obscure.

On an ultrastructural basis, nonspecific membrane proliferations which occur at a relatively late stage of virus replication may be subdivided into two general groups:

1. Arrays of smooth endoplasmic reticulum which closely resemble those observed in drug-treated hepatocytes (Hutterer *et al.*, 1969). The membrane profiles are relatively heterogeneous in dimensions (200–400 nm) and do not assume any organized pattern. This type of proliferation was observed by Heine and Dalton (1974) in cultured fibroblasts infected with varicella-zoster virus. We have observed similar smooth membrane elaborations in cultured human lymphoblasts or primate kidney cells infected with herpes simplex or cytomegalovirus.
2. Arrays of membranous tubules which share a relatively uniform dimension 200–300 nm), occupy dilated cisternae contiguous to or within the rough ER, and typically assume organized patterns. Membranous tubules which assume a reticular pattern (Fig. 19) typically occur in togavirus infections. They can be observed *in vitro* and *in vivo*. Membranous tubules with a more compact pattern are exemplified in Tana poxvirus infection (España *et al.*, 1971).

Of related interest is a family of membranous inclusions comprised of relatively electron-dense tubular elements with a similar range of dimensions (200–300 nm). These may be subdivided into compact and tubuloreticular forms (Schaff *et al.*, 1973). In contrast to any of the mem-

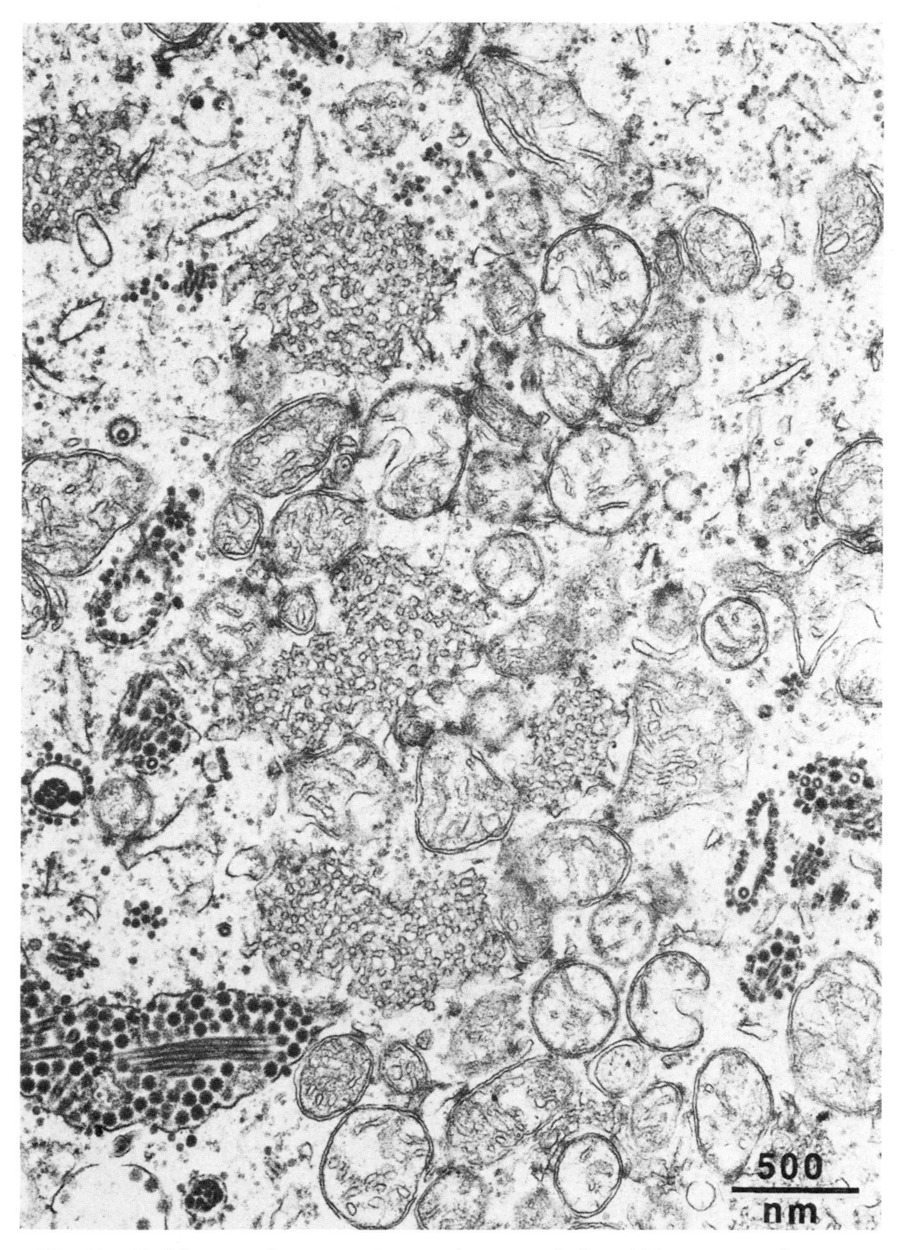


**Fig. 19.** Proliferation of anastomosing membranous tubules within a neuron of a suckling mouse infected with an alpha togavirus (Semliki Forest virus). Virus nucleoids surround profiles of endoplasmic reticulum. ×33,000.

branous proliferations described thus far, compact tubular inclusions and tubuloreticular inclusions are not usually found within the same cell sections as replicating virus. Rather, they appear to arise in cells adjacent to foci of infection, possibly in cells with occult virus. One hypothesis is that the tubular inclusions reflect a systemic or local immune response to virus. A comprehensive review of this subject has lately been published (Grimley and Schaff, 1976) and an experimental model of tubuloreticular inclusions has been developed in Burkitt lymphoma cells treated with halogenated pyrimidines (Hulanicka *et al.*, 1977).

### C. *Membrane Fusions*

Observation of multinucleated cells is a classical histopathological clue to the presence of a virus infection. With development of tissue culture techniques for growth of viruses *in vitro*, it quickly became obvious that this phenomenon was due primarily to aggregation and fusion of infected cells rather than to repeated endomitosis. In recent years the molecular basis of this phenomenon has been elucidated by ultrastructural, biochemical, and biological studies. In all cases, cell to cell fusion has been related to an insertion of virus gene products into host cell membranes.

There are two basic mechanisms for cell to cell membrane fusions—fusion "from within" and fusion due to external attachment of virions (Bratt and Gallaher, 1972). In myxovirus infections, the same molecules are involved in both processes and represent the cytotropic subunits of the virion which engage viroceptive molecules on the cell surface to create a firm attachment (see Section IV,C). Cell to cell fusion can be blocked by specific antibodies to envelope glycoproteins (Seto *et al.*, 1974) or by lectin binding (Ludwig *et al.*, 1974).

As discussed in Sections V, B and C, the cytotropic subunits responsible for cell fusions are viral proteins synthesized on the endoplasmic reticulum and translated to the cell surface by continuous peripheral movements of recycled or new membrane segments. Final maturation of these molecules requires cleavage of a larger glycoprotein precursor near or at the cell surface (Scheid, 1976). The delayed "activation" may prevent premature sticking of paired internal membranes before they can evert by fusion to the cell surface. Once at the cell surface, and before assembly into virions, cytotropic subunits can engage viroceptive molecules on adjacent uninfected cells. This may result in pathological fusions (Feldman *et al.*, 1968; Iwasaki and Koprowski, 1974) or cell aggregations (Larke *et al.*, 1977). In the central nervous system, cell to cell fusions induced "from within" may create a privileged pathway for spread of virus without exposure to humoral or cell-mediated immunological defenses (Iwasaki and Koprowski, 1974). In virus carrier cells which do not permit deployment of viral antigen at the plasma membrane, bud formation is defective and cell fusion is minimized (Dubois-Dalcq *et al.*, 1976a). The efficiency of fusion may also be subject to extracellular conditions such as pH (Gallaher and Bratt, 1974). Greenham and Poste (1971) suggest that activation of lysosomes is an important prelude to cell fusion and that acid hydrolases prepare the cell surface for attachment—perhaps by attenuating the glycocalyx.

In herpesvirus infections, membrane fusions are evidently mediated by nonvirion glycoproteins which lavishly coat interior cell membranes as well as the plasmalemma (Roizman and Kieff, 1975). Indeed, these virus gene products may be produced in such excess that they ultimately accumulate on the cytoplasmic aspect of cell membranes. A resultant "stickiness" probably explains the frequency of internal membrane fusions (Figs. 20 and 21). In contrast to myxovirus infections, hyperimmune serum directed against the virion antigens fails to inhibit polykaryocytosis (Ludwig *et al.*, 1974).

Poxvirus infection can also induce cell to cell fusion by production of a nonvirion protein. This evidently reaches the surface with mature virions by a unique mechanism. The virions become enwrapped by paired membrane elements in the Golgi membrane. The outer membrane element fuses to the cell surface while the inner membrane ruptures to effect release of the virion (Dales *et al.*, 1976). Poxvirus strains which produce hemagglutinin fail to induce fusion and vice versa. Dales *et al.* (1976) suggest that the fusion protein may actually contain the same polypeptide core as the hemagglutinin but lack an acceptor region for terminal glycosylation.

Cell to cell fusion mediated externally by viruses is probably of little consequence under natural biological conditions. Its main interest is as an experimental and diagnosic tool. In the field of cell biology, introduction of inactivated myxovirus (Sendai) for the purpose of producing heterokaryons had an almost revolutionary impact since it provided a reproducible working method applicable to a large variety of cell types (see Sidebottom, 1973). Fusion of XC cells by murine leukemia virus provides an important biological assay. A heat-labile virion protein appears to activate the syncytium formation (Johnson *et al.*, 1971).

The process of cell to cell fusion probably is essentially similar to that described during virion adsorption and entry into cells (Section IV,E). Despite a number of efforts to examine these processes at an ultrastructural level, however, the fine mechanism by which approximated continuous membranes open and reunite remains unclear (see Sidebottom, 1973; Okada *et al.*, 1975). New technical and experimental approaches will be required to resolve this question.

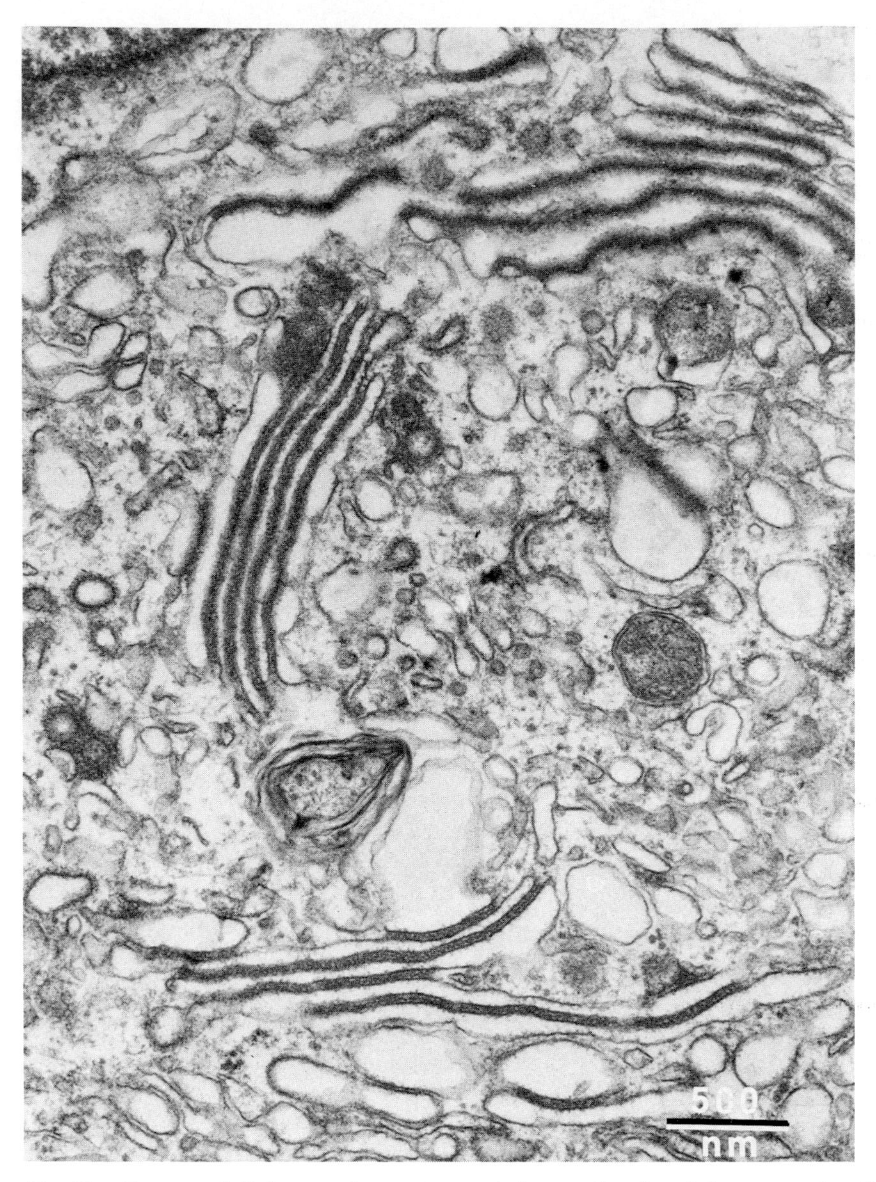


**Fig. 20.** Fusions of Golgi membranes in rat kidney cell infected by rat strain of cytomegalovirus. ×33,000.

---

**Fig. 21.** Fusion of endoplasmic reticulum (ER) with plasmalemma in human lymphoma cell infected by herpes simplex virus. Note ribosomes on cytoplasmic aspect of ER, and fuzzy coat (presumably virus glycoprotein) exterior to zone of fusion. ×82,000.

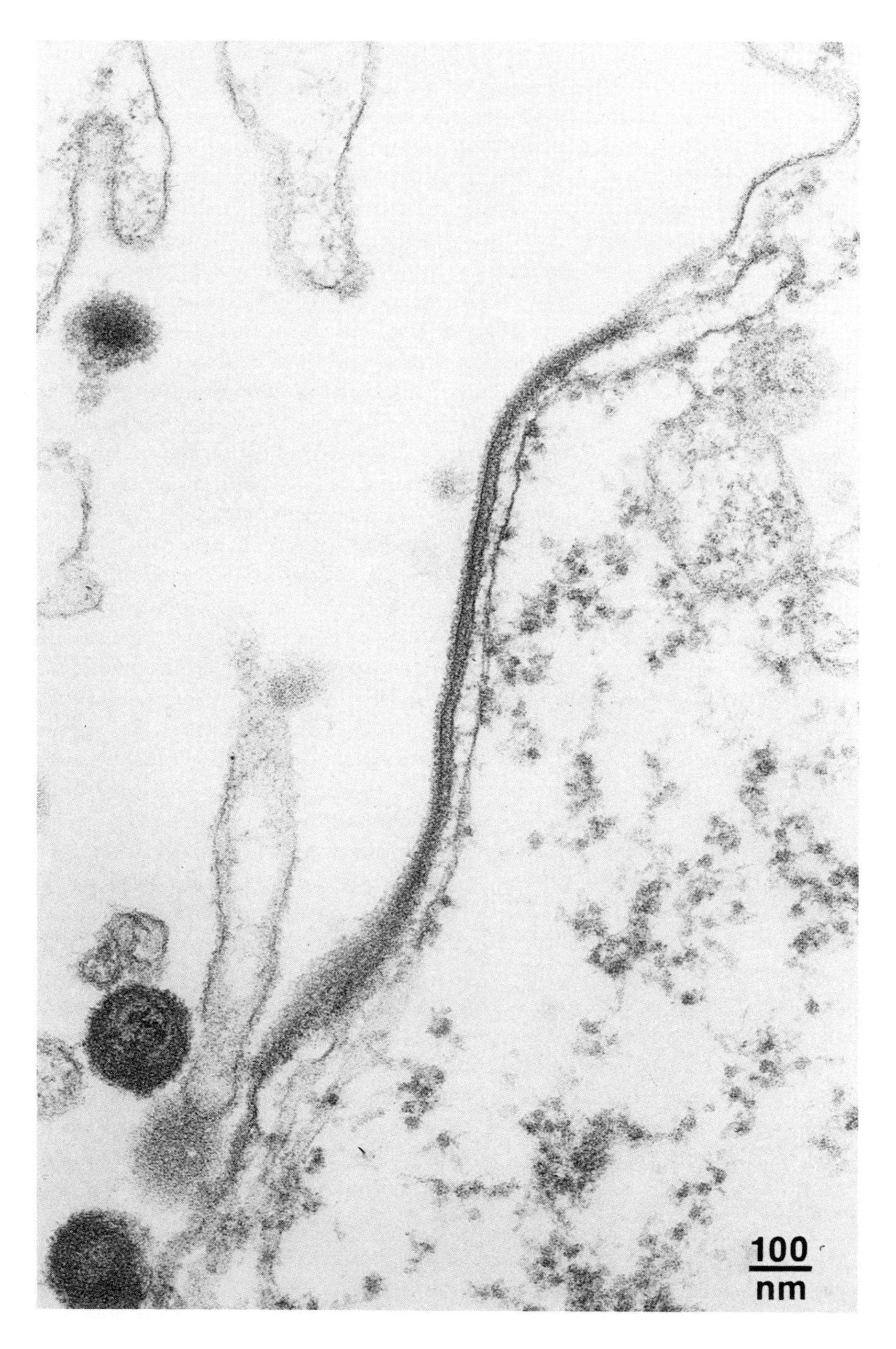
100
nm

### *D. Effects at the Cell Surface*

Insertion of virus-coded proteins into membranes which form the cell surface is an integral step in the maturation of most enveloped viruses (see Sections V, B and C). In contrast, changes in the pattern of host membrane lipid composition appear to be minimal during active virus infections (Blough *et al.*, 1977). A more profound reorganization of the cell surface, involving host membrane glycolipid patterns as well as glycoproteins, accompanies the process of phenotypic transformation by oncogenic viruses (see Glick *et al.*, 1974; Brady, 1975; Hynes, 1976), and may change the strength of normal cellular antigens (see Ting and Herberman, 1971). Acquisition of a neoplastic potential under these conditions involves a relatively stable change in cellular behavior, and the surface configuration also may reflect a more active metabolic state (see Sheinin, 1974). For example, in cells actively producing C-type virions surface amplification is typical (Fig. 22), but the extent of surface irregularity appears to depend upon a complex interaction of proviral and cellular genes (see Perecko *et al.*, 1973). The gene interactions may even result in an independent expression of virus envelope glycoprotein without concordant production of the virion core elements (Bilello *et al.*, 1974; Ledbetter *et al.*, 1977). From this perspective, the subject of cell surface reorganization after oncogenic virus transformation extends well beyond the scope of present discussion. Our attention is confined to just two subjects previously addressed by Allison (1971), both wih a potential influence on the pathobiology of active virus infections: (a) surface changes conditioning "social behavior" of cells; and (b) surface changes conditioning the immune response.

In the broadest sense, the "social behavior" of cells includes responsiveness to growth controls as well as cell movements and local interactions (Sheinin, 1974). Surface effects of active virus infection in tissues are limited to the latter phenomena. Relatively little is known of the immediate effects on membrane mobility, although scanning electron microscopy suggests increased activity at the cell surface in virus-producing cells (Fig. 22) and retraction or extension of cell processes may be observed in infected cell cultures. Considerably more is known about the effects of virus on local interactions. This has been treated in Section VII,C, as it relates to intercellular fusions of fixed cells and even aggregations of free-flowing platelets (Larke *et al.*, 1977). Effects of these processes on the pathobiology of infections, particularly chronic diseases of the central nervous system (Dubois-Dalcq *et al.*, 1976a), remain an important area for future investigation. Two

---

**Fig. 22.** Surface features of mouse cells in secondary electron scanning of samples prepared by the critical point method. (A) Primary culture of mouse embryo fibroblasts with relatively flat surfaces. ×1680. (B) Established mouse line JLSV$_9$ transformed by Rauscher virus. ×4000.

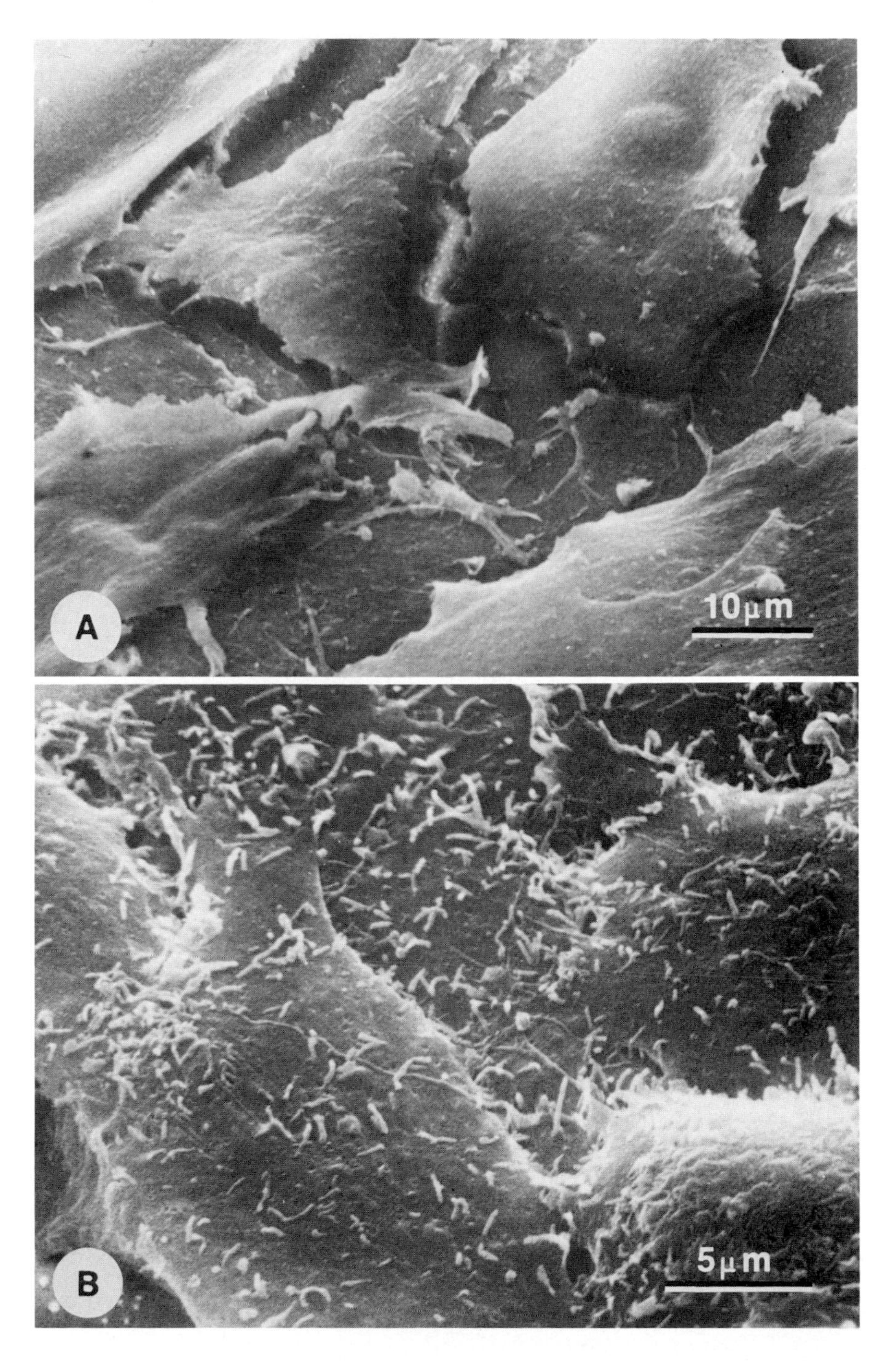
A
10μm
B
5μm

DNA viruses which produce an abundance of nonstructural glycoproteins appear to be capable of inducing changes in cell interactions under control of the virus genes. In poxvirus infection, there is an inverse relationship between stimulation of host cell fusion activity and induction of a surface hemagglutinin (Dales *et al.*, 1976). Strains of herpes simplex virus also differ in their ability to induce cell fusion or cell agglutination and these phenomena have been associated both with the quanity and quality of proteins associated with smooth membranes of the host cell (Roizman and Kieff, 1975; Spear, 1975).

Interest in the role of membrane macromolecular interactions on the immune response to virus infection has been growing rapidly. Both the cell type and virus are determinative factors (Brandt and Russell, 1975). There is now experimental evidence for an interaction of virus antigens and cellular histocompatibility determinants in cell-mediated immune cytolysis (Zinkernagel and Oldstone, 1976). Clinically, the cellular immune response to vaccinia virus immunization appears related to HLA type (De Vries *et al.*, 1977). Edelman (1976) postulates the formation of physical complexes between mobile histocompatibility antigens and other antigenic molecules (such as viral polypeptides) to form adaptor–antigen complexes which are recognized *in toto* by cytotoxic lymphocytes. Thus effects of viruses on the density of surface histocompatibility antigens may have important biological implications (see Weiss, 1977). It has been known for some time that maturing virions may incorporate histocompatibility antigens into the virus envelope, even on a selective basis (Bubbers and Lilly, 1977), while at the same time an absolute decrease in these antigens may occur on the infected cell surface (Ting and Herberman, 1971; Hecht and Summers, 1974). In oncornavirus infection, expression of host histocompatibility-antigens and antigens determined by the proviral genome may be closely interrelated (Cikes and Friberg, 1971).

Another rather intriguing phenomenon is the unexpected appearance of $F_c$ receptors on nonlymphoid cells infected by herpesviruses (Costa and Rabson, 1975; Westmoreland *et al.*, 1976). This has been demonstrated by binding of iodine-labeled purified IgG to the surface of cells infected by herpes simplex or cytomegaloviruses (Westmoreland *et al.*, 1976). The nature of the binding site or protein has not been resolved; however, a biological advantage for the propagation of herpesvirus infection has been postulated (Costa and Rabson, 1975). This may relate to findings of Stevens and Cook (1974) suggesting that antiviral IgG influences the maintenance of latent herpesvirus infection.

Thus we terminate this chapter on the rapidly advancing and converging frontiers of virology, membrane biology, and cellular immunology. Membrane-associated glycoproteins are proving to be strategic elements in each of these areas and further understanding of their structure, biosynthesis, and immunogenicity will provide important keys to unlocking the secrets of virus pathology and pathogenesis across the spectrum of active as well as oncogenic infections.

## References


Aaronson, R. P., and Blobel, G. (1975). *Proc. Nat. Acad. Sci. USA* **72**, 1007.
Abodeely, R. A., Lawson, L. A., and Randall, C. C. (1970). *J. Virol.* **5**, 513.
Achong, B. G., and Meurisse, E. V. (1968). *J. Gen. Virol.* **3**, 305.
Ahearn, M. J. (1967). *Nature (London)* **215**, 196.
Allison, A. (1967). *Sci. Am.* **217**, 62.
Allison, A. C. (1971). *Int. Rev. Exp. Pathol.* **10**, 181.
Amako, K., and Dales, S. (1967a). *Virology* **32**, 184.
Amako, K., and Dales, S. (1967b). *Virology* **32**, 201.
Anteunis, A. (1974). *Cell Tissue Res.* **149**, 497.
Aoki, T. (1974). *J. Nat. Cancer Inst.* **52**, 1029.
Aoki, T., and Takahashi, T. (1972). *J. Exp. Med.* **135**, 443.
Apel, K., Miller, K. R., Bogorad, L., and Miller, G. J. (1976). *J. Cell Biol.* **71**, 876.
Atkinson, P. H. (1973). *In* "Methods in Cell Biology" (D. M. Prescott, ed.), Vol. 7, pp. 157–188. Academic Press, New York.
Atkinson, P. H., Moyer, S. A., and Summers, D. F. (1976). *J. Mol. Biol.* **102**, 613.
Bächi, T., and Howe, C. (1973). *J. Virol.* **12**, 1173.
Bächi, T., Gerhard, W., Lindenmann, J., and Mühlethaler, K. (1969). *J. Virol.* **4**, 769.
Bächi, T., Aguet, M., and Howe, C. (1973). *J. Virol.* **11**, 1004.
Barbanti-Brodano, G., Swetly, P., and Koprowski, H. (1970). *J. Virol.* **6**, 78.
Ben-Porat, T., and Kaplan, A. S. (1971). *Virology* **45**, 252.
Ben-Porat, T., Shimono, H., and Kaplan, A. S. (1969). *Virology* **37**, 56.
Berezesky, I. K., Grimley, P. M., Tyrrell, S. A., and Rabson, A. S. (1971). *Exp. Mol. Pathol.* **14**, 337.
Berezney, R. (1974). *In* "Methods in Cell Biology" (D. M. Prescott, ed.), Vol. VIII, pp. 205–228. Academic Press, New York.
Bilello, J. A., Strand, M., and August, J. T. (1974). *Proc. Nat. Acad. Sci. USA* **71**, 3234.
Birdwell, C. R., and Strauss, J. H. (1973). *J. Virol.* **11**, 502.
Birdwell, C. R., and Strauss, J. H. (1974a). *J. Virol.* **14**, 366.
Birdwell, C. R., and Strauss, J. H. (1974b). *J. Virol.* **14**, 672.
Birdwell, C. R., Strauss, E. G., and Strauss, J. H. (1973). *Virology* **56**, 429.
Blackburn, W. R. (1971). *In* "Pathobiology Annual" (H. L. Ioachim, ed.), Vol. 1, pp. 1–31. Appleton, New York.
Blinzinger, K. (1972). *Ann. Inst. Pasteur, Paris* **123**, 497.
Blough, H. A., and Tiffany, J. M. (1975). *Curr. Top. Microbiol. Immunol.* **70**, 1.
Blough, H. A., Tiffany, J. M., and Aaslestad, H. G. (1977). *J. Virol.* **21**, 950.
Bosmann, H. B., Hagopian, A., and Eylar, E. H. (1968). *Arch. Biochem. Biophys.* **128**, 51.
Boulton, P. S., and Webb, H. E. (1971). *Brain* **94**, 411.
Boulton, R. W., and Westaway, E. G. (1976). *Virology* **69**, 416.
Brady, R. O. (1975). *Am. J. Clin. Pathol.* **63**, 685.
Brandt, W. E., and Russell, P. K. (1975). *Infect. Immun.* **11**, 330.

Branton, D., and Deamer, D. W. (1972). *In* "Protoplasmatologia" (M. Alfert, H. Bauer, W. Sandritter, and P. Sitte, eds.), pp. 1–70. Springer-Verlag, Berlin and New York.
Bratt, M. A., and Gallaher, W. R. (1972). *In* "Membrane Research" (F. Fox, ed.), pp. 383–406. Academic Press, New York.
Brown, D. T., and Burlingham, B. T. (1973). *J. Virol.* **12,** 386.
Brown, D. T., and Smith, J. F. (1975). *J. Virol.* **15,** 1262.
Brown, D. T., Waite, M. R. F., and Pfefferkorn, E. R. (1972). *J. Virol.* **10,** 524.
Bubbers, J. E., and Lilly, F. (1977). *Nature (London)* **266,** 458.
Buck, C. A., Fuhrer, J. P., Soslau, G., and Warren, L. (1974). *J. Biol. Chem.* **249,** 1541.
Buckley, I. K. (1965). *Protoplasma* **59,** 569.
Butterworth, B. E., Shimshick, E. J., and Yin, F. H. (1976). *J. Virol.* **19,** 457.
Caliguiri, L. A., and Mosser, A. G. (1971). *Virology* **46,** 375.
Caliguiri, L. A., and Tamm, I. (1970a). *Virology* **42,** 100.
Caliguiri, L. A., and Tamm, I. (1970b). *Virology* **42,** 112.
Cartwright, B., Smale, C. J., and Brown, F. (1969). *J. Gen. Virol.* **5,** 1.
Cartwright, B., Smale, C. J., and Brown, F. (1970). *J. Gen. Virol.* **7,** 19.
Caul, E. O., and Egglestone, S. I. (1977). *Arch. Virol.* **54,** 107.
Chan, V. F., and Black, F. L. (1970). *J. Virol.* **5,** 309.
Chantler, J. K., and Stevely, W. S. (1973). *J. Virol.* **11,** 815.
Chardonnet, Y., and Dales, S. (1970). *Virology* **40,** 478.
Choppin, P. W. (1976). *In* "Cell Membrane Receptors for Viruses, Antigens and Antibodies, Polypeptide Hormones, and Small Molecules" (R. F. Beers, Jr., and E. G. Bassett, eds.), pp. 271–284. Raven, New York.
Choppin, P. W., Klenk, H.-D., Compans, R. W., Caliguiri, L. A. (1971). *In* "The Gustav Stern Symposium—From Molecules to Man—Perspectives in Virology VII" (M. Pollard, ed.), pp. 127–158. Academic Press, New York.
Ciampor, F., Bystrická, M., and Rajeáni, J. (1974). *Arch. Gesamte Virusforsch.* **46,** 341.
Cikes, M., and Friberg, S., Jr. (1971). *Proc. Nat. Acad. Sci. USA* **68,** 566.
Cohen, A. (1963). *In* "Mechanisms of Virus Infection" (W. Smith, ed.), pp. 153–190. Academic Press, New York.
Cohen, G. H., Atkinson, P. H., and Summers, D. F. (1971). *Nature (London)* **231,** 121.
Comings, D. E. (1974). *In* "The Cell Nucleus" (H. Busch, ed.), Vol. 1, pp. 538–559. Academic Press, New York.
Comings, D. E., and Okada, T. A. (1973). *J. Mol. Biol.* **75,** 609.
Compans, R. W. (1973). *Virology* **55,** 541.
Compans, R. W., and Caliguiri, L. A. (1973). *J. Virol.* **11,** 441.
Cords, C. E., James, C. G., and McLaren, L. C. (1975). *J. Virol.* **15,** 244.
Costa, J. C., and Rabson, A. S. (1975). *Lancet* **1,** 77.
Coward, J. E., Harter, D. H., Hsu, K. C., and Morgan, C. (1972). *Virology* **50,** 925.
Crowell, R. L. (1976). *In* "Cell Membrane Receptors for Viruses, Antigens and Antibodies, Polypeptide Hormones, and Small Molecules" (R. F. Beers, Jr., and E. G. Basset, eds.), pp. 179–202. Raven, New York.
Crowell, R. L., and Philipson, L. (1971). *J. Virol.* **8,** 509.
Dahl, R., and Kates, J. R. (1970). *Virology* **42,** 453.
Dales, S. (1965). *Prog. Med. Virol.* **7,** 1.
Dales, S. (1973). *Bacteriol. Rev.* **37,** 103.
Dales, S., and Chardonnet, Y. (1973). *Virology* **56,** 465.
Dales, S., and Mosbach, E. H. (1968). *Virology* **35,** 564.
Dales, S., and Pons, M. W. (1976). *Virology* **69,** 278.
Dales, S., Eggers, H. J., Tamm, I., and Palade, G. E. (1965). *Virology* **26,** 379.

Dales, S., Stern, W., Weintraub, S. B., and Huima, T. (1976). *In* "Cell Membrane Receptors for Viruses, Antigens and Antibodies, Polypeptide Hormones, and Small Molecules" (R. F. Beers, Jr., and E. G. Bassett, eds.), pp. 253–270. Raven, New York.
Darlington, R. W., and Moss, L. H. (1969). *Prog. Med. Virol.* **11**, 16.
David, A. E. (1977). *Virology* **76**, 98.
DeGiuli, C., Kawai, S., Dales, S., and Hanafusa, H. (1975). *Virology* **66**, 253.
DeHarven, E. (1974). *Advan. Virus Res.* **19**, 221.
Demsey, A., Steere, R. L., Brandt, W. E., and Veltri, B. J. (1974). *J. Ultrastruct. Res.* **46**, 103.
Demsey, A., Calvelli, T. A., Kawka, D., Stackpole, C. W., and Sarkar, N. H. (1976). *Virology* **75**, 484.
Demsey, A., Kawka, D., and Stackpole, C. W. (1977). *J. Virol.* **21**, 358.
Demsey, A., Kawka, D., and Stackpole, C. W. (1978). *J. Ultrastruct. Res.* **62**, 13.
DePetris, S., and Raff, M. C. (1973). *Nature (London), New Biol.* **241**, 257.
DeVries, R. R. P., Kreeftenberg, H. G., Loggen, H. G., and Van Rood, J. J. (1977). *New Engl. J. Med.* **297**, 692.
Dorfman, N. A., Stepina, V. N., and Ievleva, E. S. (1972). *Int. J. Cancer* **9**, 693.
Dorsett, P. H., and Ginsberg, H. S. (1975). *J. Virol.* **15**, 208.
Dowdle, W. R., Downie, J. C., and Laver, W. G. (1974). *J. Virol.* **13**, 269.
Dubois-Dalcq, M., Reese, T. S., Murphy, M., and Fuccillo, D. (1976a). *J. Virol.* **19**, 579.
Dubois-Dalcq, M., Reese, T. S., and Narayan, O. (1976b). *Virology* **74**, 520.
Duff, R., Rapp, F., and Butel, J. S. (1970). *Virology* **42**, 273.
Dulbecco, R. (1965). *Am. J. Med.* **38**, 669.
Edelman, G. M. (1976). *Science* **192**, 218.
Epstein, M. A., Hummerler, K., and Berkaloff, A. (1964). *J. Exp. Med.* **119**, 291.
Ericsson, J. L. E., and Brunk, U. T. (1975). *In* "Pathobiology of Cell Membranes" (B. F. Trump and A. U. Arstila, eds.), Vol. 1, pp. 217–253. Academic Press, New York.
España, C., Brayton, M. A., and Ruebner, B. H. (1971). *Exp. Mol. Pathol.* **15**, 34.
Etchison, J. R., and Holland, J. J. (1974). *Virology* **60**, 217.
Fazekas de St. Groth, S. (1948). *Nature (London)* **162**, 294.
Feldherr, C. M. (1972). *Adv. Cell Mol. Biol.* **2**, 273.
Feldman, L. A., Sheppard, R. D., and Bornstein, M. B. (1968). *J. Virol.* **2**, 621.
Fenner, F., McAuslan, B. R., Mims, C. A., Sambrook, J., and White, D. O. (1974). "The Biology of Animal Viruses," 2nd ed. Academic Press, New York.
Filshie, B. K., and Rehacek, J. (1968). *Virology* **34**, 435.
Fong, C. K. Y., and Hsuing, G. D. (1977). *Fed. Proc. Fed. Am. Soc. Exp. Biol.* **36**, 2320.
Fox, C. F. (1972). *In* "Membrane Molecular Biology" (C. F. Fox and A. D. Keith, eds.), pp. 345–385. Sinaver Associates, Stamford, Connecticut.
Fox, C. F., and Keith, A. D. (1972). *In* "Membrane Molecular Biology" Sinauer Associates, Stamford, Connecticut.
Fraenkel-Conrat, H. (1969). "The Chemistry and Biology of Viruses." Academic Press, New York.
Franke, W. W., and Scheer, U. (1974). *In* "The Cell Nucleus" (H. Busch, ed.), Vol. 1, pp. 220–347. Academic Press, New York.
Friedman, R. M., and Berezesky, I. K. (1967). *J. Virol.* **1**, 374.
Friedman, R. M., and Pastan, I. (1969). *J. Mol. Biol.* **40**, 107.
Friedman, R. M., and Sreevalsan, T. (1970). *J. Virol.* **6**, 169.
Friedman, R. M., Levin, J. G., Grimley, P. M., and Berezesky, I. K. (1972). *J. Virol.* **10**, 504.
Friend, D. S., Rosenau, W., Winfield, J. S., and Moon, H. D. (1969). *Lab. Invest.* **20**, 275.
Frye, L. D., and Edidin, M. (1970). *J. Cell Sci.* **7**, 319.
Gallaher, W. R., and Bratt, M. A. (1974). *J. Virol.* **14**, 813.

Garoff, H., and Simons, K. (1974). *Proc. Nat. Acad. Sci. USA* **71**, 3988.
Gautschi, M., Siegl, G., and Kronauer, G. (1976). *J. Virol.* **20**, 29.
Gelderblom, H., Bauer, H., and Graf, T. (1972). *Virology* **47**, 416.
Girard, M., Baltimore, D., Darnell, J. E. (1967). *J. Mol. Biol.* **27**, 59.
Glaser, R., Farrugia, R., and Brown, N. (1976). *Virology* **69**, 132.
Glick, M. C., Rabinowitz, Z., and Sachs, L. (1974). *J. Virol.* **13**, 967.
Gliedman, J. B., Smith, J. F., and Brown, D. T. (1975). *J. Virol.* **16**, 913.
Goldstein, L. (1974). *In* "The Cell Nucleus" (H. Busch, ed.), Vol. 1, pp. 388–438. Academic Press, New York.
Gosselin, R. E. (1967). *Fed. Proc. Fed. Am. Soc. Exp. Biol.* **26**, 987.
Granados, R. R. (1973). *Virology* **52**, 305.
Greenham, L. W., and Poste, G. (1971). *Microbios* **3**, 97.
Gregg, M. B., and Morgan, C. (1959). *J. Biophys. Biochem. Cytol.* **6**, 539.
Grimley, P. M. (1971). *Proc. EMSA* **5**, 380.
Grimley, P. M., and Friedman, R. M. (1970a). *Exp. Mol. Pathol.* **12**, 1.
Grimley, P. M., and Friedman, R. M. (1970b). *J. Inf. Dis.* **122**, 45.
Grimley, P. M., and Schaff, Z. (1976). *In* "Pathobiology Annual" (H. L. Ioachim, ed.), pp. 221–257. Appleton, New York.
Grimley, P. M., Berezesky, I. K., and Friedman, R. M. (1968). *J. Virol.* **2**, 1326.
Grimley, P. M., Rosenblum, E. N., Mims, S. J., and Moss, B. (1970). *J. Virol.* **6**, 519.
Grimley, P. M., Levin, J. G., Berezesky, I. K., and Friedman, R. M. (1972). *J. Virol.* **10**, 492.
Grimley, P. M., Berezesky, I. K., and Levin, J. G. (1973). *J. Nat. Cancer Inst.* **50**, 275.
Groneberg, J., Brown, D. T., and Doerfler, W. (1975). *Virology* **64**, 115.
Grove, S. N., Bracker, C. E., and Morré, D. J. (1968). *Science* **161**, 171.
Guskey, L. E., and Wolff, D. A. (1974). *J. Virol.* **14**, 1229.
Guylas, B. J. (1971). *J. Ultrastruct. Res.* **35**, 112.
Haines, H., and Baerwald, R. J. (1976). *J. Virol.* **17**, 1038.
Hamerman, D. (1966). *Am. J. Med.* **40**, 1.
Hämmerling, U., Stackpole, C. W., and Koo, G. (1973). *In* "Methods in Cancer Research" (H. Busch, ed.), Vol. IX, pp. 255–282. Academic Press, New York.
Hämmerling, U., Polliack, A., Lampen, N., Sabety, M., and de Harven, E. (1975). *J. Exp. Med.* **141**, 518.
Hashimoto, K., Suzuki, K., and Simizu, B. (1975). *J. Virol.* **15**, 1454.
Hay, A. J. (1974). *Virology* **60**, 398.
Haywood, A. M. (1974). *J. Mol. Biol.* **87**, 625.
Heath, E. C. (1971). *Annu. Rev. Biochem.* **40**, 29.
Hecht, T. T., and Summers, D. F. (1974). *J. Virol.* **14**, 162.
Heine, J. I., and Dalton, A. J. (1974). *In* "Molecular Studies in Viral Neoplasia," pp. 63–96. Williams & Wilkins, Baltimore, Maryland.
Heine, J. W., and Roizman, B. (1973). *J. Virol.* **11**, 810.
Heine, J. W., and Schnaitman, C. A. (1971). *J. Virol.* **8**, 786.
Heine, J. W., Spear, P. G., and Roizman, B. (1972). *J. Virol.* **9**, 431.
Hendler, R. W. (1974). *In* "Biomembranes" (L. A. Manson, ed.), Vol. 5, pp. 251–273. Plenum, New York.
Hicks, R. M. (1966). *J. Cell Biol.* **30**, 623.
Higgins, J. A. (1974). *J. Cell Biol.* **62**, 635.
Higgins, J. A., and Barrnett, R. J. (1972). *J. Cell Biol.* **55**, 282.
Hiraki, S., Chan, J. C., Hales, R. L., and Dmochowski, L. (1974). *Cancer Res.* **34**, 2906.
Hirano, H., Parkhouse, B., Nicolson, G. L., Lennox, E. S., and Singer, S. J. (1972). *Proc. Nat. Acad. Sci. USA* **69**, 2945.

Hirschberg, C. B., and Robbins, P. W. (1974). *Virology* **61**, 602.
Hochberg, E., and Becker, Y. (1968). *J. Gen. Virol.* **2**, 231.
Holland, J. J. (1961). *Virology* **15**, 312.
Honess, R. W., and Roizman, B. (1974). *J. Virol.* **14**, 8.
Hoshino, M., Maeno, K., and Iinuma, M. (1972). *Separatum Experientia* **28**, 611.
Howe, C., Morgan, C., and Hsu, K. C. (1969). *Prog. Med. Virol.* **11**, 307.
Howe, C., Spiele, H., Minio, F., and Hsu, K. C. (1970). *J. Immunol.* **104**, 1406.
Hoyle, L. (1962). *Cold Spring Harbor Symp. Quant. Biol.* **27**, 113.
Huhn, D. (1967). *Nature (London)* **216**, 1240.
Hulanicka, B., Barry, D. W., and Grimley, P. M. (1977). *Cancer Res.* **37**, 2105.
Hummeler, K., Tomassini, N., and Zajac, B. (1969). *J. Virol.* **4**, 67.
Hummeler, K., Tomassini, N., and Sokol, F. (1970). *J. Virol.* **6**, 87.
Hutterer, F., Klion, F. M., Wengraf, A., Schaffner, F., and Popper, H. (1969). *Lab. Invest.* **20**, 455.
Hynes, R. O. (1976). *Biochim. Biophys. Acta* **458**, 73.
Iida, H., and Oda, K. (1975). *J. Virol.* **15**, 471.
Ishibashi, M. (1970). *Proc. Nat. Acad. Sci. USA* **65**, 304.
Ito, M., and Barron, A. L. (1974). *J. Virol.* **13**, 1312.
Ito, K., Arens, M., and Green, M. (1975). *J. Virol.* **15**, 1507.
Iwasaki, Y., and Koprowski, H. (1974). *Lab. Invest.* **31**, 187.
Jacques, P. J. (1975). *In* "Pathobiology of Cell Membranes" (B. F. Trump and A. U. Arstila, eds.), Vol. I, pp. 255–279. Academic Press, New York.
Johnson, G. S., Friedman, R. M., and Pastan, I. (1971). *J. Virol.* **7**, 753.
Johnson, E. M., Karn, J., Allfrey, V. G. (1974). *J. Biol. Chem.* **249**, 4990.
Joklik, W. K., and Darnell, J. E. (1961). *Virology* **13**, 439.
Jones, K. J., Scupham, R. K., Pfeil, J. A., Wan, K., Sagik, B. P., and Bose, H. R. (1977). *J. Virol.* **21**, 778.
Juliano, R. L., and Behar-Bannelier, M. (1975). *Biochim. Biophys. Acta* **375**, 249.
Kaplan, A. S., Erickson, J. S., and Ben-Porat, T. (1975). *Virology* **64**, 132.
Kasper, C. B. (1974). *In* "The Cell Nucleus" (H. Busch, ed.), Vol. 1, pp. 349–384. Academic Press, New York.
Katz, F. N., Rothman, J. E., Lingappa, V. R., Blobel, G., Lodish, H. F. (1977). *Proc. Nat. Acad. Sci. USA* **74**, 3278.
Kazama, F. Y., and Schornstein, K. L. (1973). *Virology* **52**, 478.
Keegstra, K., Sefton, B., and Burke, D. (1975). *J. Virol.* **16**, 613.
Kendal, A. P., and Kiley, M. P. (1973). *J. Virol.* **12**, 1482.
Klein, P. A., and Adams, W. R. (1972). *J. Virol.* **10**, 844.
Klenk, H.-D. (1973). *In* "Biological Membranes" (D. Chapman, and D. F. H. Wallach, eds.), Vol. 2, pp. 145–183. Academic Press, New York.
Klenk, H.-D., Compans, R. W., and Choppin, P. W. (1970). *Virology* **42**, 1158.
Klenk, H.-D., Wollert, W., Rott, R., and Scholtissek, C. (1974). *Virology* **57**, 28.
Knipe, D. M., Lodish, H. F., and Baltimore, D. (1977a). *J. Virol.* **21**, 1121.
Knipe, D. M., Baltimore, D., and Lodish, H. F. (1977b). *J. Virol.* **21**, 1128.
Knipe, D. M., Baltimore, D., and Lodish, H. F. (1977c). *J. Virol.* **21**, 1149.
Kohn, A., and Fuchs, P. (1973). *In* "Advances in Virus Research" (M. A. Lauffer, F. B. Bang, K. Maramorosch, K. M. Smith, eds.), Vol. 18, pp. 159–194. Academic Press, New York.
Kozak, M., and Roizman, B. (1974). *Proc. Nat. Acad. Sci. USA* **71**, 4322.
Kraehenbuhl, J. P., and Jamieson, J. D. (1972). *Proc. Nat. Acad. Sci. USA* **69**, 1771.
Kraehenbuhl, J. P., Galardy, R. E., and Jamieson, J. D. (1974). *J. Exp. Med.* **139**, 208.

Kurstak, E., and Maramorosch, K. (1974). "Viruses, Evolution and Cancer. Basic Considerations." Academic Press, New York.
Lamm, M. E., Koo, G. C., Stackpole, C. W., and Hämmerling, U. (1972). *Proc. Nat. Acad. Sci. USA* **69**, 3732.
Lampert, P. W., Joseph, B. S., and Oldstone, M. B. A. (1975). *J. Virol.* **15**, 1248.
Larke, R. P. B., Turpie, A. G. G., Scott, S., and Chernesky, M. A. (1977). *Lab. Invest.* **37**, 150.
Law, J. H., and Snyder, W. R. (1972). *In* "Membrane Molecular Biology" (C. F. Fox, and A. D. Keith, eds.), pp. 1–26. Sinauer Associates, Stamford, Connecticut.
Lazarowitz, S. G., and Choppin, P. W. (1975). *Virology* **68**, 440.
LeBlanc, D. J., and Singer, M. F. (1974). *Proc. Nat. Acad. Sci. USA* **71**, 2236.
Ledbetter, J., Nowinski, R. C., and Emery, S. (1977). *J. Virol.* **22**, 65.
Leduc, E. H., Avrameas, S., and Bouteille, M. (1968). *J. Exp. Med.* **127**, 109.
Lee, A. G., Birdsall, N. J. M., and Metcalf, J. C. (1973). *Biochemistry* **12**, 1650.
Leestma, J. E., Bornstein, M. B., Sheppard, R. D., and Feldman, L. A. (1969). *Lab. Invest.* **20**, 70.
Lenard, J., and Compans, R. W. (1974). *Biochim. Biophys. Acta* **344**, 51.
Lenard, J., Wong, C. Y., and Compans, R. W. (1974). *Biochim. Biophys. Acta* **332**, 341.
Levanon, A., Kohn, A., and Inbar, M. (1977). *J. Virol.* **22**, 353.
Levin, J. G., Grimley, P. M., Ramseur, J. M., and Berezesky, I. K. (1974). *J. Virol.* **14**, 152.
Levinthal, J. D., Dunnebacke, T. H., and Williams, R. C. (1969). *Virology* **39**, 211.
Levitt, N. H., and Crowell, R. L. (1967). *J. Virol.* **1**, 693.
Lodish, H. F., and Froshauer, S. (1977). *J. Cell Biol.* **74**, 358.
Lonberg-Holm, K., and Philipson, L. (1974). *In* "Monographs in Virology" (J. L. Melnick, ed.), Vol. 9, pp. 1–149 Karger, Basel.
Ludwig, H., Becht, H., and Rott, R. (1974). *J. Virol.* **14**, 307.
Luftig, R. B., McMillan, P. N., Bolognesi, D. P. (1974). *Cancer Res.* **34**, 3303.
Luria, S. E., and Darnell, J. E., Jr. (1967). "General Virology," 2nd ed. Wiley, New York.
Lunger, P. D., and Clark, H. F. (1972). *In Vitro* **7**, 377.
Lunger, P. D., and Clark, H. F. (1973). *J. Nat. Cancer Inst.* **50**, 111.
Lutzner, M. A., and Hecht, F. (1966). *Lab. Invest.* **15**, 597.
Mackay, R. L., and Consigli, R. A. (1976). *J. Virol.* **19**, 620.
McLaren, L. C., Scaletti, J. V., and James, C. G. (1968). *In* "Biological Properties of the Mammalian Surface Membrane" (L. A. Manson, ed.), pp. 123–136. The Wistar Institute Press, Philadelphia.
McSharry, J. J., Compans, R. W., and Choppin, P. W. (1971). *J. Virol.* **8**, 722.
Maeno, K., and Kilbourne, E. D. (1970). *J. Virol.* **5**, 153.
Majerus, P. W., and Kilburn, E. (1969). *J. Biol. Chem.* **244**, 6254.
Makino, S., and Jenkin, H. M. (1975). *J. Virol.* **15**, 515.
Mandel, B. (1967). *Virology* **31**, 248.
Mannik, M., and Downey, W. (1973). *J. Immunol. Methods* **3**, 233.
Marchesi, V. T., Furthmayr, H., and Tomita, M. (1976). *In* "Cell Membrane Receptors for Viruses, Antigens and Antibodies, Polypeptide Hormones, and Small Molecules" (R. F. Beers, Jr., and E. G. Bassett, eds.), pp. 217–222. Raven, New York.
Marco, R., Jazwinski, M., and Kornberg, A. (1974). *Virology* **62**, 209.
Marcus, P. I., and Schwartz, V. G. (1968). *In* "Biological Properties of the Mammalian Surface Membrane," pp. 143–147. The Wistar Institute Press, Philadelphia.
Mark, G. E., and Kaplan, A. S. (1971). *Virology* **45**, 53.
Matsumura, T., Stollar, V., and Schlesinger, R. W. (1971). *Virology* **46**, 344.
Maul, G. (1976). *J. Cell Biol.* **70**, 714.
Maul, G. G., Price, J. W., and Lieberman, M. W. (1971). *J. Cell Biol.* **51**, 405.

Maul, G. G., Maul, H. M., Scogna, J. E., Lieberman, M. W., Stein, G. S., Yee-Li Hsu, B., and Borun, T. W. (1972). *J. Cell Biol.* **55**, 433.
Medrano, L., and Green, H. (1973). *Virology* **54**, 515.
Meinke, W., Hall, M. R., and Goldstein, D. A. (1975). *J. Virol.* **15**, 439.
Melchers, F. (1973). *Biochemistry* **12**, 1471.
Meldolesi, J., and Cova, D. (1972). *J. Cell Biol.* **55**, 1.
Melvin, P., and Kucera, L. S. (1975). *J. Virol.* **15**, 534.
Merkow, L. P., Slifkin, M., Pardo, M., and Rapoza, N. P. (1970). *J. Ultrastruct. Res.* **30**, 344.
Merriam, R. W. (1962). *J. Cell Biol.* **12**, 79.
Miller, D. A., Miller, O. J., Dev, V. G., Hashmi, S., Tantravahi, R., Mediano, L., Green, H. (1974). *Cell* **1**, 167.
Miller, G., Shope, T., Coope, D., Waters, L., Pagano, J., Bornkamm, G. W., and Henle, W. (1977). *J. Exp. Med.* **145**, 948.
Miyamoto, K., and Gilden, R. V. (1971). *J. Virol.* **7**, 395.
Miyamoto, K., and Morgan, C. (1971). *J. Virol.* **8**, 910.
Mollo, F., and Stramignoni, A. (1967). *Br. J. Cancer* **21**, 519.
Mollo, F., Canese, M., and Stramignoni, A. (1969). *Nature (London)* **211**, 869.
Mooney, J. J., Dalrymple, J. M., Alving, C. R., and Russell, P. K. (1975). *J. Virol.* **15**, 225.
Moore, N. F., Barenholz, Y., and Wagner, R. R. (1976). *J. Virol.* **19**, 126.
Morgan, C., and Howe, C. (1968). *J. Virol.* **2**, 1122.
Morgan, C., Rosenkranz, H. S., and Mednis, B. (1969). *J. Virol.* **4**, 777.
Mori, R., Schieble, J. H., Ackermann, W. W. (1962). *Proc. Soc. Exp. Biol. Med.* **109**, 685.
Morré, D. J., Mollenhauer, H. H., and Bracker, C. E. (1971). *In* "Results and Problems in Cell Differentiation" (J. Reinert and H. Ursprung, eds.), pp. 82–126. Springer-Verlag, Berlin and New York.
Moss, B., Rosenblum, E. N., and Grimley, P. M. (1971a). *Virology* **45**, 123.
Moss, B., Rosenblum, E. N., and Garon, C. F. (1971b). *Virology* **46**, 221.
Mosser, A. G., Caliguiri, L. A., Scheid, A. S., and Tamm, I. (1972a). *Virology* **47**, 30.
Mosser, A. G., Caliguiri, L. A., and Tamm, I. (1972b). *Virology* **47**, 39.
Nagai, Y., Ogura, H., and Klenk, H.-D. (1976). *Virology* **69**, 524.
Nakamura, K., and Compans, R. W. (1977). *Virology* **79**, 381.
Neville, D. M., Jr., and Kahn, C. R. (1974). *In* "Subcellular Particles, Structures, and Organelles" (A. I. Laskin and J. A. Last, eds.), pp. 57–88. Dekker, New York.
Nicolson, G. L. (1975). *Am. J. Clin. Pathol.* **63**, 677.
Nii, S., Rosenkranz, H. S., Morgan, C., and Rose, H. M. (1968). *J. Virol.* **2**, 1163.
Ogier, G., Chardonnet, Y., and Doerfler, W. (1977). *Virology* **77**, 67.
Ojakian, G. K., Vreibich, G., and Sabatini, G. G. (1977). *J. Cell Biol.* **72**, 530.
Okada, Y., Koseki, I., Kim, J., Maeda, Y., Hashimoto, T., Kanno, Y., and Matsui, Y. (1975). *Exp. Cell Res.* **93**, 368.
Orrenius, S., and Ericsson, J. L. E. (1966). *J. Cell Biol.* **28**, 181.
Palese, P., Tobita, K., Ueda, M., and Compans, R. W. (1974). *Virology* **61**, 397.
Panem, S., and Kirsten, W. H. (1975). *Virology* **63**, 447.
Pedersen, C. E., and Sagik, B. P. (1973). *J. Gen. Virol.* **18**, 375.
Perecko, J. P., Berezesky, I. K., and Grimley, P. M. (1973). *In* "Proceedings of the Workshop on Scanning Electron Microscopy in Pathology" (O. Johari, ed.), pp. 521–528, IIT Research Institute, Chicago.
Perk, K., and Dahlberg, J. E. (1974). *J. Virol.* **14**, 1304.
Perlin, M., and Phillips, B. A. (1973). *Virology* **53**, 107.
Philipson, L. (1967). *J. Virol.* **1**, 868.
Philipson, L., Lonberg-Holm, K., and Pettersson, U. (1968). *J. Virol.* **2**, 1064.

Philipson, L., Everitt, E., and Lonberg-Holm, K. (1976). *In* "Cell Membrane Receptors for Viruses, Antigens and Antibodies, Polypeptide Hormones, and Small Molecules" (R. F. Beers, Jr., and E. G. Bassett, eds.), pp. 203–216. Raven, New York.
Pierce, J. S., Strauss, E. G., and Strauss, J. H. (1974). *J. Virol.* **13**, 1030.
Pinter, A., and Compans, R. W. (1975). *J. Virol.* **16**, 859.
Pinto da Silva, P., and Branton, D. (1970). *J. Cell Biol.* **45**, 598.
Plagemann, P. G. W., Cleveland, P. H., and Shea, M. A. (1970). *J. Virol.* **6**, 800.
Pope, J. H., Achong, B. G., and Epstein, M. A. (1968). *Int. J. Cancer* **3**, 171.
Portner, A., and Pridgen, C. (1975). *Prog. Med. Virol.* **21**, 27.
Qasba, P. K., Yelton, D. B., Pletsch, Q. A., and Aposhian, H. V. (1974). *In* "Molecular Studies in Viral Neoplasia," pp. 169–189. Williams & Wilkins, Baltimore, Maryland.
Quigley, J. P., Rifkin, D. B., and Reich, E. (1971). *Virology* **46**, 106.
Quigley, J. P., Rifkin, D. B., and Reich, E. (1972). *Virology* **50**, 550.
Qureshi, A. A., and Trent, D. W. (1972). *J. Virol.* **9**, 565.
Ray, U., Soeiro, R., and Fields, B. N. (1976). *J. Virol.* **18**, 370.
Reanney, D. C. (1974). *In* "International Review of Cytology" (G. H. Bourne, J. F. Danielli, and K. W. Jeon, eds.), Vol. 37, pp. 21–52. Academic Press, New York.
Reeves, M. W., and Chang, G. C. H. (1971). *Microbios* **4**, 167.
Renkonen, O., Kääriäinen, L., Gahmberg, C. G., and Simons, K. (1972). *Biochem. Soc. Symp.* **35**, 407.
Rifkin, D. B., and Quigley, J. P. (1974). *Annu. Rev. Microbiol.* **28**, 325.
Roesing, T. G., Toselli, P. A., and Crowell, R. L. (1975). *J. Virol.* **15**, 654.
Roizman, B., and Kieff, E. G. (1975). *In* "Cancer 2. A Comprehensive Treatise" (F. F. Becker, ed.), pp. 241–322. Plenum, New York.
Rothschild, K. J., and Stanley, H. E. (1972). *In* "Membranes and Viruses in Immunopathology" (S. B. Day and R. A. Good, eds.), pp. 49–80. Academic Press, New York.
Ruebner, B. H., Hirano, T., Slusser, R., Osborn, J., and Medearis, D. N., Jr. (1966). *Am. J. Pathol.* **48**, 971.
Sambrook, J. (1977). *Nature (London)* **268**, 101.
Scandella, C. J., Devaux, P., and McConnell, H. M. (1972). *Proc. Nat. Acad. Sci. USA* **69**, 2056.
Schaff, Z., Grimley, P. M., Michelitch, H. J., and Banfield, W. G. (1973). *J. Nat. Cancer Inst.* **51**, 293.
Schaffer, P. A., Brunschwig, J. P., McCombs, R. M., and Benyesh-Melnick, M. (1974). *Virology* **62**, 444.
Scheid, A. (1976). *In* "Cell Membrane Receptors for Viruses, Antigens and Antibodies, Polypeptide Hormones, and Small Molecules" (R. F. Beers, Jr., and E. G. Bassett, eds.), pp. 222–235. Raven, New York.
Scheid, A., and Choppin, P. W. (1974). *Virology* **57**, 475.
Schimke, R. T., and Dehlinger, P. J. (1972). *In* "Membrane Research" (C. F. Fox, ed.), pp. 115–134. Academic Press, New York.
Schloemer, R. H., and Wagner, R. R. (1974). *J. Virol.* **14**, 270.
Schloemer, R. H., and Wagner, R. R. (1975). *J. Virol.* **15**, 882.
Schultz, W. W., and Bang, F. B. (1977). *Am. J. Pathol.* **87**, 667.
Seebeck, T., and Weil, R. (1974). *J. Virol.* **13**, 567.
Seeman, P. (1974). *Fed. Proc. Fed. Am. Soc. Exp. Biol.* **33**, 2116.
Sefton, B. M. (1976). *J. Virol.* **17**, 85.
Sefton, B. M., and Gaffney, B. J. (1974). *J. Mol. Biol.* **90**, 343.
Seto, J. T., Becht, H., and Rott, R. (1974). *Virology* **61**, 354.
Shapiro, S. Z., and August, J. T. (1976). *Biochim. Biophys. Acta* **458**, 375.

Sharom, F. J., Barratt, D. G., Thede, A. E., and Grant, C. W. M. (1976). *Biochim. Biophys. Acta* **455**, 485.

Sheffield, J. B. (1974). *Virology* **57**, 287.

Sheinin, R. (1974). *In* "Viruses, Evolution and Cancer" (E. Kurstak and K. Maramorosch, eds.), pp. 371–400. Academic Press, New York.

Shigematsu, T., Dmochowski, L., and Williams, W. C. (1971). *Cancer Res.* **31**, 2085.

Shimizu, K., and Ishida, N. (1975). *Virology* **67**, 427.

Shimizu, Y. K., Hosaka, Y., and Shimizu, Y. K. (1972). *J. Virol.* **9**, 842.

Shiroki, K., Shimojo, H., and Yamaguchi, K. (1974). *Virology* **60**, 192.

Sidebottom, E. (1973). *In* "The Cell Nucleus" (H. Busch, ed.), Vol. 1, pp. 439–469. Academic Press, New York.

Silverstein, A. M. (1964). *Science* **144**, 1423.

Silverstien, S. C., and Dales, S. (1968). *J. Cell Biol.* **36**, 197.

Simpson, R. W., Hauser, R. E., and Dales, S. (1969). *Virology* **37**, 285.

Singer, S. J. (1974). *In* "Advances in Immunology" (F. J. Dixon and H. G. Kunkel, eds.), Vol. 19, pp. 1–66. Academic Press, New York.

Singer, S. J., and Nicolson, G. L. (1972). *Science* **175**, 720.

Singer, S. J., and Rothfield, L. I. (1973). *Neurosci. Res. Program Bull.* **11**, 1.

Skinner, M. S., Halperen, S., and Harkin, J. C. (1968). *Virology* **36**, 241.

Smith, G. F., and O'Hara, P. T. (1968). *J. Ultrastruct. Res.* **21**, 415.

Smith, J. D., and de Harven, E. (1974). *J. Virol.* **14**, 945.

Spear, P. G. (1975). *In* "Oncogenesis and Herpesviruses II." (G. de Thé, M. A. Epstein, H. zur Hausen, and W. Davis, eds.), pp. 49–61. International Agency for Research on Cancer, Lyon, France.

Spear, P. G., Keller, J. M., and Roizman, B. (1970). *J. Virol.* **5**, 123.

Spector, D. H., and Baltimore, D. (1975). *Sci. Am.* **232**, 24.

Spring, S. B., and Roizman, B. J. (1968). *J. Virol.* **2**, 979.

Sreevalsan, T. (1970). *J. Virol.* **6**, 438.

Sreevalsan, T., and Yin, F. H. (1969). *J. Virol.* **3**, 599.

Staehlein, L. A. (1976). *J. Cell Biol.* **71**, 136.

Steck, T. L. (1972). *In* "Membrane Molecular Biology" (C. F. Fox and A. D. Keith, eds.), pp. 76–116. Sinauer Associates, Stamford, Connecticut.

Steck, T. L., and Fox, C. F. (1972). *In* "Membrane Molecular Biology" (C. F. Fox and A. D. Keith, eds.), pp. 27–75. Sinauer Associates, Stamford, Connecticut.

Stern, W., and Dales, S. (1974). *Virology* **62**, 293.

Stevens, J. G., and Cook, M. L. (1974). *J. Immunol.* **113**, 1685.

Stoeckenius, W., and Engelman, D. M. (1969). *J. Cell Biol.* **42**, 613.

Stohlman, S. A., Wisseman, C. L. Jr., Eylar, O. R., and Silverman, D. J. (1975). *J. Virol.* **16**, 1017.

Stokes, G. V. (1976). *J. Virol.* **18**, 636.

Stollar, V., Stollar, B. D., Koo, R., Harrap, K. A., and Schlesinger, R. W. (1976). *Virology* **69**, 104.

Su, R. T., and DePamphilis, M. L. (1976). *Proc. Nat. Acad. Sci. USA* **73**, 3466.

Sussenbach, J. S. (1967). *Virology* **33**, 567.

Suzuki, I. (1972). *Gann* **63**, 629.

Tamm, I. (1975). *Am. J. Pathol.* **81**, 163.

Tan, K. B. (1970). *J. Virol.* **5**, 632.

Tiffany, J. M., and Blough, H. A. (1971). *Virology* **44**, 18.

Tillack, T. W., and Marchesi, V. T. (1970). *J. Cell Biol.* **45**, 649.

Tillack, T. W., Scott, R. E., and Marchesi, V. T. (1972). *J. Exp. Med.* **135**, 1209.

Ting, C-C., and Herberman, R. B. (1971). *Nature (London), New Biol.* **232**, 118.
Tokuyasu, K. T., and Singer, S. J. (1976). *J. Cell Biol.* **71**, 894.
Tucker, A. G., and Docherty, J. J. (1975). *Infect. Immun* **11**, 556.
Vance, D. E., and Lam, J. (1975). *J. Virol.* **16**, 1075.
Van der Vliet, P. C., Levine, A. J., Ensinger, M. J., and Ginsberg, H. S. (1975). *J. Virol.* **15**, 348.
Van Zaane, D., Dekker-Michielsen, M. J. A., and Bloemers, H. P. J. (1976). *Virology* **75**, 113.
Velicer, L. F., and Ginsburg, H. J. (1970). *J. Virol.* **5**, 338.
Virtanen, I., and Wartiovaara, J. (1974). *J. Virol.* **13**, 222.
Vogt, P. K. (1967). *Virology* **33**, 175.
Wagner, M. (1973). *Res. Immunochem. Immunobiol.* **3**, 185.
Wagner, R. R., Synder, R. M., and Yamazaki, S. (1970). *J. Virol.* **5**, 548.
Wagner, R. R., Heine, J. W., Goldstein, G., and Schnaitman, C. A. (1971). *J. Virol.* **7**, 274.
Wagner, R. R., Kiley, M. P., Snyder, R. M., and Schnaitman, C. A. (1972). *J. Virol.* **9**, 672.
Waite, M. R. F., and Pfefferkorn, E. R. (1970a). *J. Virol.* **5**, 60.
Waite, M. R. F., and Pfefferkorn, E. R. (1970b). *J. Virol.* **6**, 637.
Waite, M. R. F., Brown, D. T., and Pfefferkorn, E. R. (1972). *J. Virol.* **10**, 537.
Wallach, D. F. H. (1972). *In* "The Plasma Membrane: Dynamic Perspectives, Genetics and Pathology." Springer-Verlag, Berlin and New York.
Warren, L. (1972). *In* "Membranes and Viruses in Immunopathology" (S. B. Day and R. A. Good, eds.), pp. 89–104. Academic Press, New York.
Warren, L., and Glick, M. C. (1969). *In* "Fundamental Techniques in Virology" (K. Habel and N. P. Salzman, eds.), pp. 66–71. Academic Press, New York.
Watson, J. D. (1976). *In* "Molecular Biology of the Gene." Benjamin, New York.
Weiss, R. A. (1976). *In* "Cell Membrane Receptors for Viruses, Antigens and Antibodies, Polypeptide Hormones, and Small Molecules" (R. F. Beers, Jr., and E. G. Bassett, eds.), pp. 237–251. Raven, New York.
Weiss, R. (1977). *Nature (London)* **267**, 13.
Westmoreland, D., St. Jeor, S., and Rapp, F. (1976). *J. Immunol.* **116**, 1566.
Whaley, W. G., Dauwalder, M., and Kephart, J. E. (1971). *In* "Origin and Continuity of Cell Organelles" (J. Reinert and H. Ursprung, eds.), pp. 1–38. Springer-Verlag, Berlin and New York.
Willis, D., and Granoff, A. (1974). *Virology* **61**, 256.
Wirth, D. F., Katz, F., Small, B., and Lodish, H. F. (1977). *Cell* **10**, 253.
Witte, O. N., and Weissman, I. L. (1974). *Virology* **61**, 575.
Witte, O. N., and Weissman, I. L. (1976). *Virology* **69**, 464.
Witte, O. N., Tsukamoto-Adey, A., and Weissman, I. L. (1977). *Virology* **76**, 539.
Wong, P. K. Y., and McCarter, J. A. (1974). *Virology* **58**, 396.
Wong, P. K. Y., and MacLeod, R. (1975). *J. Virol.* **16**, 434.
Wunderlich, F., Wallach, D. F. H., Speth, V., and Fischer, H. (1974). *Biochim. Biophys. Acta* **373**, 34.
Yamashita, T., and Green, M. (1974). *J. Virol.* **14**, 412.
Zajac, R., and Crowell, R. L. (1965). *J. Bacteriol.* **89**, 1097.
Závada, J. (1972). *J. Gen. Virol.* **15**, 183.
Zebovitz, E., Leong, J. K. L., and Doughty, S. C. (1974). *Infect. Immun.* **10**, 204.
Zinkernagel, R. M., and Oldstone, M. B. A. (1976). *Proc. Nat. Acad. Sci. USA* **73**, 3666.

# EDITORS' SUMMARY TO CHAPTER III

This chapter on relationships of virus–cell interaction to cell membrane changes is very important as it sheds light not only on reactions of cells to these injurious agents, but also considerable information about topology, movement, permeability, and binding properties of normal membranes as well as about the regulation of membrane synthesis and turnover. The cell membranes in virus-infected cells not only continue their specialized functions in the preinfection stage, but also become involved by several means such as compartmentalization, direction of synthesis, etc. in virus developmental processes themselves. Indeed, viruses being obligate intracellular parasites are completely dependent on the modifications of membrane functions for their life cycle and the cell, in turn, expresses the cytopathic effects of virus infection by numerous changes in the cell membrane, including such things as changes of surface recognition antigens, modified intracellular adhesion, promotion of cell fusion, and modification of differentiation.

In addition to providing insights into the cytopathic effects of viruses, this chapter also illustrates the numerous significant ways in which studies of these phenomena help explain and understand the significance of normal cell membrane structure and function. In this regard, the reader is referred to Volume I in this series of treatises where several general concepts related to cell membranes and disease processes are discussed, for example, in Chapters I and VII.

In the case of membrane synthesis and degradation, the changed patterns seen during animal virus infection have provided useful clues and unique models. This is especially true with the poxviruses which are able to induce synthesis *de novo* of membranes from degraded host lipids. Such studies also have lent support to the fluid-mosaic structure for membranes by demonstrating that virus-specified proteins can be incorporated into a preexisting molecular framework. The fluid-mosaic concept which is reviewed in Chapter I of Volume I, offers a most attractive synthesis with which to understand the changes in virus–cell interactions.

The problem of virus entry into the cells and penetration of the genome through the membrane either through the surface of the plasmic membrane or of a phagosome, has been a most difficult one on which progress is currently being made. Since no direct channels that we know of exist, this penetration is a considerable problem. In the case of bacteriophages, there is evidence of physical disruption with transient leakage of cell contents in the case of entry into microbial hosts. Some evidence for transient dissolution of cell membranes in mammalian cells has also been seen, again with leakage of constituents. A very important problem in general pathology elsewhere, for example, is the question of whether cytoplasmic enzymes can escape from a still viable cell across a cell membrane. However, more subtle mechanisms are probably also involved, including attachment of virus to specific receptor molecules with molecular translocation through the fluid-mosaic membrane structure or actual membrane fusion between the membrane of the virus and that of the host, which would obviously also bring the contents into continuity (per so-called reverse exotropy see Chapter I, Volume I). In many cases these events of fusion, disruption, or molecular penetration are preceded by engulfment or pinocytosis of the virus, so-called viropexis, which also serves to protect the virus from extracellular circulating antibodies. However, it is important to realize that following engulfment, penetration still must occur.

With regard to attachment to the cell, an important first step, the current evidence and theories are similar to that being developed for other informational macromolecules, namely adhesion to specific receptors which are under genetic and other controls and which may markedly effect host cell susceptibility. The concept ot specific viro-receptors is currently being further developed and numerous species of these with various binding affinities probably exist. Intramembranous portions of viro-receptors have been visualized as 75 Å diameter particulates within freeze-cleaved surfaces where they appear to fit into the lipid mosaic. Various interesting chemical specificities which can influence the virus binding, even in cell-free systems, are being studied and may shed much more light on normal informational molecule interactions. Ultrastructural studies of some viruses of the adenovirus class suggest morphological identification of antenna-like areas which represent the binding portion of the virion.

The long controversial question of membrane fusion can also be approached through the study of viruses as these represent important tools for studying membrane fusion. Fusion of cells into syncytia is also an important concomitant of systemic virus infections in man. Viruses produce factors which stimulate membrane fusion, explaining some of the multicellular syncytia seen in certain virus infections. Amazingly, several studies indicate that certain virus envelope proteins can be actually inserted into the host cell membrane and diffuse within the membrane within a few minutes after attachment. This leads to instant changes in the characteristics of the host cell surface membrane. Such mechanisms may also relate to the mechanism involved in membrane fusion.

There are a number of membrane-dependent steps in virus–cell interactions which are important both to the virus cycle and also to the host cell response. Many of these are related to modification of host cell membrane synthesis and often marked proliferation of cytomembranes resembling smooth endoplasmic reticulum are seen. Such new membrane synthesis in the host cell may be fairly analogous to stimulation of membrane proliferation by chemicals. In togavirus infections the membrane spherules, possibly derivatives of phagocytic vacuoles may participate directly in togaviral RNA replication. These particles are called CPV-1.

As discussed by Drs. Grimley and Demsey, virus synthesis offers remarkable insights into distribution of proteins synthesized either on free or membrane-bound polysomes and going into the virus or into the cytomembranes which later become part of the virus envelope. In addition to providing templates, cytomembrane systems also play a role in the synthesis of virus envelope glycoproteins which are important in receptors, and for which sufficient genetic information is not necessarily available in the virus. Complex sequences in the virus synthesis which occur in the strict temporal succession lead to an amazing complexity of the virus as compared to their size utilizing both its and the cell's genomes.

Viruses also result in multiple and often dramatic effects on the nucleus, including changes in the pores, disruption, proliferation of nuclear envelope, all of which can be involved in the replicative cycle and some of which may be nonspecific.

The lysosomes also play an important role in virus interactions and intact virions may persist in lysosomes for several hours, ultimately to be degraded. In other instances labilization of lysosomes by viruses and increase in lysosomal activity is seen. Furthermore, during the budding process, viruses may be sequestered by a process akin to normal autophagocytosis. It has even been suggested that in infections by nonbudding nucleotropic viruses, lysis of host membranes from within may facilitate the escape of progeny virion. The formation of intracellular membranous proliferations is an interesting problem and appears to be of two types. One, a rise of endoplasmic reticulum which resemble those in drug-treated hepatocytes, and two, a raise of membranous tubules which occupy dilated cisternae contiguous to or within the rough endoplasmic reticulum. There are also varieties of membranous inclusions or tubular elements in the cytoplasm that have been noticed in a variety of viral diseases, and have been difficult to distinguish from virions themselves.

Reorganization of the cell surface by viruses is of particular interest with regard to the oncogenic viruses in which rather profound reorganizations involving host membrane glycolipid and glycoprotein composition accompanies the transformation by oncogenic viruses. These modifications are involved with regulation of the social behavior including cell movements, local interactions, formation of cell junctions, etc. Such surface modifications may also modify the immune response and relate to the pathophysiology of viral immune diseases.

In summary, studies of cell–virus interactions as they involve cell membranes provide models for the elucidation of many different types of changes in intracellular organelles and will therefore lead to a better understanding of both normal and abnormal cell biology.

# CHAPTER IV

# STRUCTURE AND FUNCTION OF PEROXISOMES AND THEIR ROLE IN DISEASE PROCESSES

Urs N. Riede, A. William Moore, and Walter A. Sandritter





PATHOBIOLOGY OF CELL MEMBRANES, VOL. II

ISBN 0-12-701502-7

## I. Introduction

Life is possible only when the elements of the organism are maintained in certain relationships to one another. This will only occur in the context of certain spatial arrangements. The proportional relationships among these elements are adapted by the organism precisely for its needs. This gives rise to a closed system in which each part is functionally attuned to each other part. No enlargement or diminution of this system is possible without a corresponding dysfunction. As a consequence, structure is a precondition for function.

Morphometry allows one to obtain exact measurements on the structures in the organism, from which we are tempted to draw conclusions as to its functional status. This often leads to false interpretations, however, because not every structural increase in an organelle corresponds to an increase in its function. Thus structure is not always proportional to function.

Peroxisomes are distributed widely in different cells, and take part in oxidative, carbohydrate, and fat metabolism. As a result, virtually every form of cellular damage leaves some visible or measureable footprint in the peroxisomes. The reaction patterns of these organelles can thus be viewed as indicators of cellular damage.

## II. Orthology of Peroxisomes

Peroxisomes are defined as "cytoplasmic particles (with specific sedimentation characteristics) which possess at least one hydrogen peroxide producing oxidase together with catalase" (de Duve, 1969; Novikoff *et al.*, 1972).

Peroxisomes seen in the epithelium of liver parenchyma and renal tubules are spherical, often vermiform organelles, surrounded by a single, triple-layer membrane. They contain inner bodies which are more electron dense than the surrounding finely granular matrix. The structure of these inner bodies, i.e., the core, is species dependent. Peroxisomes from different species may be classified as follows, after the suggestion of Hruban and Rechcigl (1969): Type I, coreless or anucleoid peroxisomes; Type II, peroxisomes with noncrystalloid cores; and Type III, peroxisomes containing crystalloid cores with regular structure (Figs. 1 and 2).

In addition to cores one also finds "marginal plates" in peroxisomes of different animals. These stretched out structures lie on the periphery of the peroxisomes. They are separated from the inner surface of the limiting membrane by a narrow space of lower density and are not considered part of the peroxisomal cores (Hruban and Rechcigl, 1969; Tiedemann, 1972). Besides an unknown matrix protein which makes up over half the peroxisomal mass (Reddy and Svoboda, 1973a), the liver peroxisomes of the rat contain the following characteristic enzymes: urate oxidase, catalase, D-amino acid oxidase, and L-$\alpha$-hydroxyacid oxidase. It is concluded that metabolism in mammalian peroxisomes consists of terminal flavine oxidation (de Duve, 1969). In addition to the above enzymes such as isocitrate dehydrogenase, glycerol phosphate dehydrogenase, and carnitine acetyltransferase (Leighton *et al.*, 1969; McGroarty and Tolbert, 1973).

Peroxisomes have the ability to reduce oxygen to water through the action of catalase and the oxidases. This process involves a two-step mechanism, in which hydrogen peroxide acts as an intermediate. Oxidase substrates, $RH_2$, such as uric acid, D-amino acids, L-amino acids, and L-$\alpha$-hydroxyacids serve as electron donors in the first step. The electron donors in the second step consist of either a second molecule of hydrogen peroxide, which is then oxidized back to oxygen, or an organic substrate, $R'H_2$, such as alcohols, aldehydes, or nitrite ions, which are subsequently oxidized by catalase (de

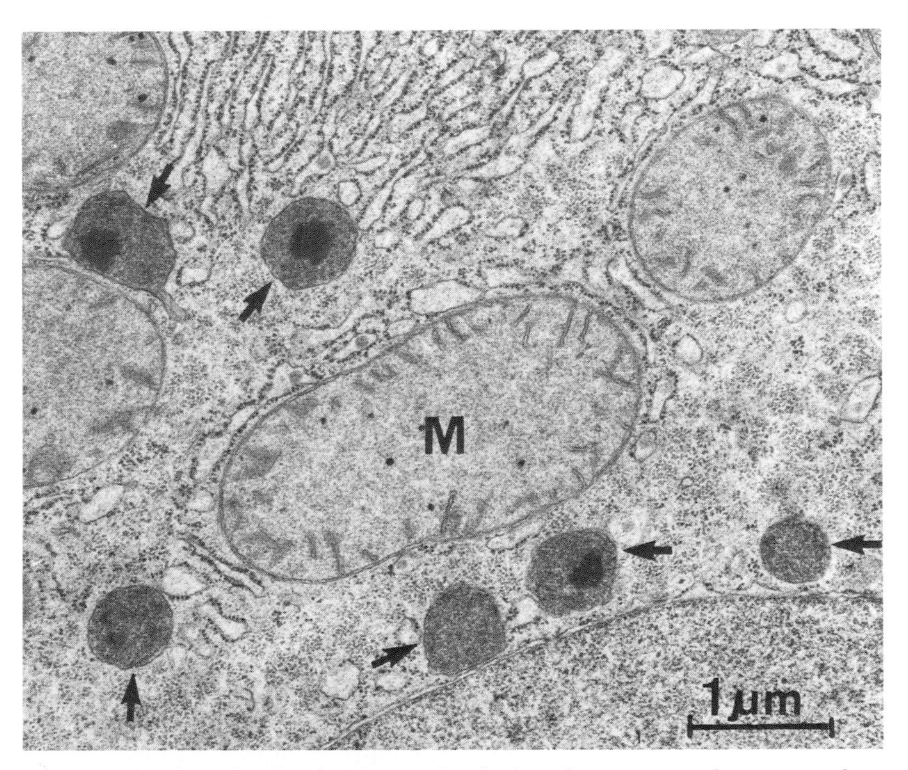


**Fig. 1.** Peroxisomes with and without nucleoids of a rat hepatocyte. Nucleus, N; mitochondrion, M; arrows, peroxisomes. ×1500.

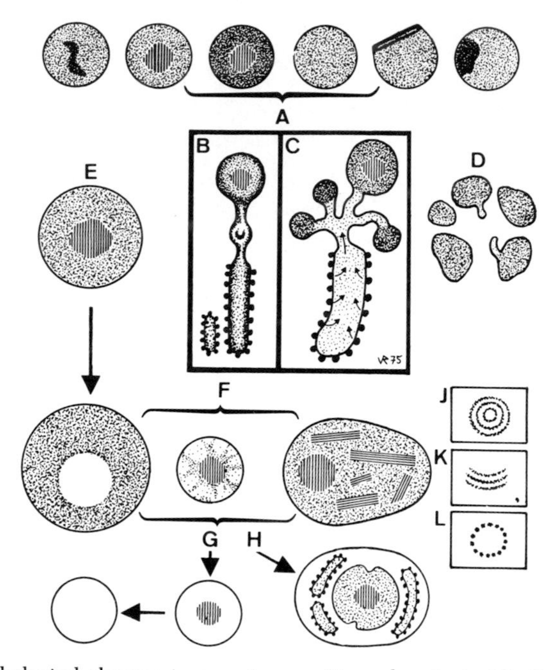


**Fig. 2.** Morphological changes in peroxisomes. Normal variants (A) of peroxisomes may show an absence or alteration of the nucleoid, and may show a homogeneous or localized matrix density. Microperoxisomes (D) may arise from the peroxisomal–endoplasmic reticulum complex (C) with the gastruloid cisternae connecting the two (B). These microperoxisomes may later enlarge to become megaperoxisomes (E). Degeneration of peroxisomes (F) involves loss of the nucleoid with the formation of punched-out peroxisomes, loss of matrix, or the appearance of matrical plates (K) or matrical tubules (J, L). Degradation of peroxisomes proceeds either without lysosomal participation by way of a process of autolysis (i.e., self dissolution) (G) ending as empty vesicles, or else with lysosomal participation by way of autophagy (H).

Duve and Baudhuin, 1966). The function of peroxisomes in cell economy is not yet fully understood. In 1966 de Duve and Baudhuin proposed a variety of metabolic roles for these organelles on the basis of numerous biochemical studies. They suggested that peroxisomes (a) serve as ancillary sites of carbohydrate oxidation; (b) participate in gluconeogenesis by providing an effective, exergonic pathway for the formation of $\alpha$-keto acids; (c) oxidize reduced NAD by way of an alternate pathway and thus play an important role in the aerobic support of cell sap metabolism; (d) protect the cell against the build up of intracellular $H_2O_2$; and (e) degrade D-amino acids from bacterial cell walls. In addition, peroxisomes participate in purine degradation in some species (Shnitka, 1966; Tsukada *et al.*, 1968). Finally peroxisomes are involved in lipid metabolism. They contain a fatty acetyl-CoA oxidizing system which employs oxygen and NAD as electron acceptors (Lazarow and de

Duve, 1976) as well as $\alpha$-glyceroid phosphate dehydrogenase and carnitine acetyltransferase. Recently Lazarow (1978) reported that rat liver peroxisomes are specialized for $\beta$-oxidation of long-chain fatty acids.

This metabolism role is further supported by the fact that all of the following cells contain peroxisomes: (a) cells which produce, store, or metabolize steroid hormones or androgenic hormones (Boeck, 1972; Reddy, 1973; Black and Bogart, 1973; Gulyas and Yuan, 1977); (b) cells which participate in cholesterol metabolism (Petrik, 1971; Schneeberger, 1972); and (c) cells which are involved in the absorption, transport, storage, or metabolism of lipids (Ahlabo and Barnard, 1971; Hruban *et al.*, 1972a,b; Novikoff and Novikoff, 1972). An important contributory argument is the peroxisomal proliferation seen following administration of hypolipidemic drugs (Reddy *et al.*, 1974a; Hruban *et al.*, 1974a) and the close spatial relationship between these organelles and lipid droplets and lipofuscin granules (A. B. Novikoff *et al.*, 1973a; Novikoff and Novikoff, 1972).

In liver parenchymal cells and renal tubular epithelium one finds both nucleoid containing peroxisomes and small, *nucleoidless microperoxisomes*. These microperoxisomes are widely scattered, and have been demonstrated in nearly all cell types (cf. Hruban *et al.*, 1972a; Novikoff and Novikoff, 1972; Svoboda and Reddy, 1974). They are especially numerous in the enterocytes of the small bowel (Novikoff and Novikoff, 1972), the epithelium of the sebaceous gland, yolk sac, renal cortex, thyroid, and retinal pigment (Hruban *et al.*, 1972a; Magalhaes and Magalhaes, 1971; Tice and Wollman, 1974; Beard, 1972), and finally in adipocytes of brown fat (Ahlabo and Barnard, 1971; Leuenberger and Novikoff, 1975). They contain catalase and D-amino acid oxidase (Novikoff and Novikoff, 1972), but have no urate oxidase activity and probably also lack L-$\alpha$-hydroxyacid oxidase activity (Novikoff and Novikoff, 1972). The enzymatic composition of microperoxisomes (especially, catalase content) tends to vary widely. This is especially true for microperoxisomes in the tubular epithelium of the kidney (Novikoff *et al.*, 1972).

Microperoxisomes lie in a transitional area between a central domain of the smooth endoplasmic reticulum (SER), consisting of dense layers of tubules with narrow lumina, and an exterior zone in which the SER tubules are distributed radially to the central domain (Novikoff *et al.*, 1972). Typically, microperoxisomes are heaped up in clusters, and lie in close relationship to SER, lipid droplets, and mitochondria (P. M. Novikoff *et al.*, 1973; Riede *et al.*, 1973a). Numerous, continuous transitions are seen between the membranes of the SER and those of the microperoxisomes, so that the microperoxisomes are regarded as "local swollen areas of the endoplasmic reticulum in which high concentrations of peroxisomal constituents are located" (Novikoff and Novikoff, 1972).

## III. Peroxisomal Topography

As discussed above, peroxisomes exhibit a close topographic and structural relationship to the endoplasmic reticulum (ER) and to lipofuscin granules (A. B. Novikoff *et al.*, 1973b). Peroxisomes may be regarded as special regions within the ER. Peroxisomes in the rat liver show a topographic relationship to the cellular membrane in the form of peroxisomal–desmosomal complexes (Tandler and Hoppel, 1970). In addition to pure peroxisomal–desmosomal complexes, one also sees hybrid complexes in which a peroxisome lies on one side of the desmosome and a mitochondrion on the other (Tandler and Hoppel, 1970). A relationship of peroxisomes to the cell membrane is also seen in cells of the pars recta of the renal proximal tubule. A variety of experimental conditions (especially inhibition of oxidative metabolism) gives rise to a proliferation of the SER and peroxisomes (Jacobsen and Joergenson, 1973; Riede *et al.*, 1973a). These aggregates of the SER, together with peroxisomes, usually have close contact with lateral and basal interdigitating cytoplasmic processes. These proliferated organelles lie in a strategically optimal position in the cytoplasm (Riede *et al.*, 1973a). A similar syntopy of peroxisomes and the sarcoplasmic reticulum can be seen in the myocardium and the striated muscle cells (Herzog and Fahimi, 1976).

Peroxisomes are found principally in the basal part of renal proximal tubule cells in all segments. Mitochondria in cells of the pars contorta are elongated and perpendicular to the basal membrane, whereas those in cells of the pars recta are round and distributed randomly in the cytoplasm (Beard and Novikoff, 1969; Jacobsen and Joergensen, 1973; Barret and Heidger, 1975). Peroxisomes are distributed principally in the supranuclear region of rat enterocytes in the small bowel (Novikoff and Novikoff, 1972), whereas the majority of mitochondria are seen in the lower half (Gonzales-Licea, 1970; 1971; 1972). This topographic distribution is remarkable in the sense that mitochondria drift toward the upper cellular pole in the resorption phase (Gonzales-Licea, 1972).

Mouse interstitial cells exhibit a rather clear-cut regional organization. Mitochondria are concentrated in areas where ER is scarce. Peroxisomes are found in both places in these cells and thus tend to cross regional lines (Black and Bogart, 1973).

## IV. Biogenesis of Peroxisomes

Originally, peroxisomes and microperoxisomes arise from the rough endoplasmic reticulum (RER) by budding (Novikoff and Shin, 1964; Essner, 1967; Novikoff and Novikoff, 1972). Thus microperoxisomes in the liver and kidney are precursors of peroxisomes (P. M. Novikoff *et al.*, 1973; Kramar *et al.*, 1974; Staeubli *et al.*, 1977; Reddy *et al.*, 1976). The majority of peroxisomes which have budded off from the membrane system of the ER are seen in transitional zones from rough to SER (Essner, 1967). A portion of the peroxisomes, especially the microperoxisomes, remains in connection with the membrane system of the ER, so that they communicate with the specialized ER region, where peroxisomal proteins are synthesized. Thus a continual or intermittent material exchange among individual peroxisomes is possible (Reddy and Svoboda, 1973a; de Duve, 1973). Although the turnover rate of catalase in liver peroxisomes has a half-life of 1.6–2.2 days (Poole, 1969; Poole *et al.*, 1969), the peroxisomes usually have no separate fate. They comprise a "peroxisomal compartment" in the cell, which acts as a functional complex with the ER (Novikoff *et al.*, 1972; de Duve, 1973; Kramar *et al.*, 1974; Reddy and Svoboda, 1973a).

The supply route for newly formed enzymes from ribosomes into peroxisomes is still hypothetical (cf. Staeubli *et al.*, 1977). The following three processes are discussed for catalase:

1. Following synthesis in membrane-bound ribosomes, catalase reaches the cisternae of the RER, and from there passes to the vesicles of the Golgi apparatus (Yokota and Nagata, 1974). Catalase is packaged and eventually transported to peroxisomes in the Golgi vesicles in a manner analogous to secretory proteins. Since catalase is not a secretory protein, however, this process is open to question (cf. Legg and Wood, 1970b).
2. Catalase reaches the peroxisomes directly after ribosomal synthesis and transport via the RER and Golgi apparatus (Wood and Legg, 1970). This route has been disproved by subsequent investigations (Novikoff and Novikoff, 1972; Lazarow and de Duve, 1973).
3. Incomplete catalase moieties are synthesized in membrane-bound ribosomes and pass from the cisternae of the RER directly into the peroxisomes, where completion of the catalase molecule eventually takes place. Since only a few traces of labeled catalase components are found in the cytosol soon after administration of labeled precursors, this enzyme transfer from RER into the peroxisomes must proceed with exceptional speed or by means of a continuous "RER–peroxisomal channel system" (de Duve, 1973). The role of the so-called gastruloid cisternae (Fig. 2), which lie between two segments of the RER and are in relation to the RER and peroxisomes, is unclear (Hruban *et al.*, 1974a; Lazarow, 1978).

Peroxisomes appear in the fetal liver of the mouse in the second embryonal week. By the 16th day one finds only a small population of anucleoid but peroxidase positive peroxisomes, whereas on the 20th day a large population of nucleoid-containing peroxisomes is present (Essner, 1970; Dvorak,

1971). From the third embryonal week to the first postnatal week the number of peroxisomes doubles (Tsukada *et al.*, 1968). A comparable numerical increase in peroxisomes is described in alveocytes Type II (Schneeberger, 1972) in adrenal cortical cells (Black and Bogart, 1973) and in the cells of proximal tubules of the kidney and small intestine (Pipan and Psenicnik, 1975; Goeckermann and Vigil, 1975). The proportional volume of peroxisomes per hepatocyte increases by tenfold in the perinatal period (Rohr *et al.*, 1971). After the first postnatal week the size of the nucleoids increases as well. Catalase activity progresses in a similar fashion. While the activity of urate oxidase increases from the first postnatal day, the D-amino acid oxidase activity increases only a week later. It is assumed that this is related to the overproduction of uric acid in neonatal animals.

Morphometric, autoradiographic, and histochemical studies on rat liver show, that mitochondria and peroxisomes proliferate after the mitotic cell cycle (Rohr and Riede, 1973; Goldenberg *et al.*, 1975).

During the course of evolution, purine metabolism has undergone progressive simplification, as seen in the loss of terminal links of the uricolytic enzymes. This is apparently the reason why adult birds retain anucleoid peroxisomes, whereas all mammals exhibit nucleoid-containing urate oxidase in the peroxisomes. The anthropoid primates, including man, constitute an

**TABLE I**

*Normal Values of Morphometric Parameters for 1 ml Liver Tissue in Three Different Rodents*[a]

| | *Rat (Wistar)* | *Mouse (NMRI)* | *Desert Rat (Meriones crassus)* | *Dimension* |
|---|---|---|---|---|
| Volume density of peroxisomes | 0.014 (0.0002) | 0.023 (0.0002) | 0.006 (0.0001) | $cm^3/cm^3$ |
| Numerical density of peroxisomes | $83.7 \times 10^9$ ($7 \times 10^9$) | $264 \times 10^9$ ($21 \times 10^9$) | $70.2 \times 10^9$ ($4 \times 10^9$) | $cm^{-3}$ |
| Volume density of RER | 0.11 (0.005) | 0.19 (0.004) | 0.18 (0.008) | $cm^3/cm^3$ |
| Volume density of SER | 0.056 (0.003) | 0.031 (0.002) | 0.061 (0.004) | $cm^3/cm^3$ |
| Volume density of mitochondria | 0.19 (0.005) | 0.17 (0.007) | 0.19 (0.011) | $cm^3/cm^3$ |
| Numerical density of mitochondria | $283 \times 10^9$ ($12 \times 10^9$) | $242 \times 10^9$ ($7 \times 10^9$) | $84 \times 10^9$ ($5 \times 10^9$) | $cm^{-3}$ |
| Surface density of mitochondrial cristae | 3.5 (0.28) | 1.95 (0.15) | 2.19 (0.13) | $m^2/cm^3$ |
| Volume density of lipid droplets | 0.003 (0.0005) | 0.003 (0.0002) | 0.022 (0.0003) | $cm^3/cm^3$ |

[a] All values are mean for five young adult animals; values in parentheses, standard error.

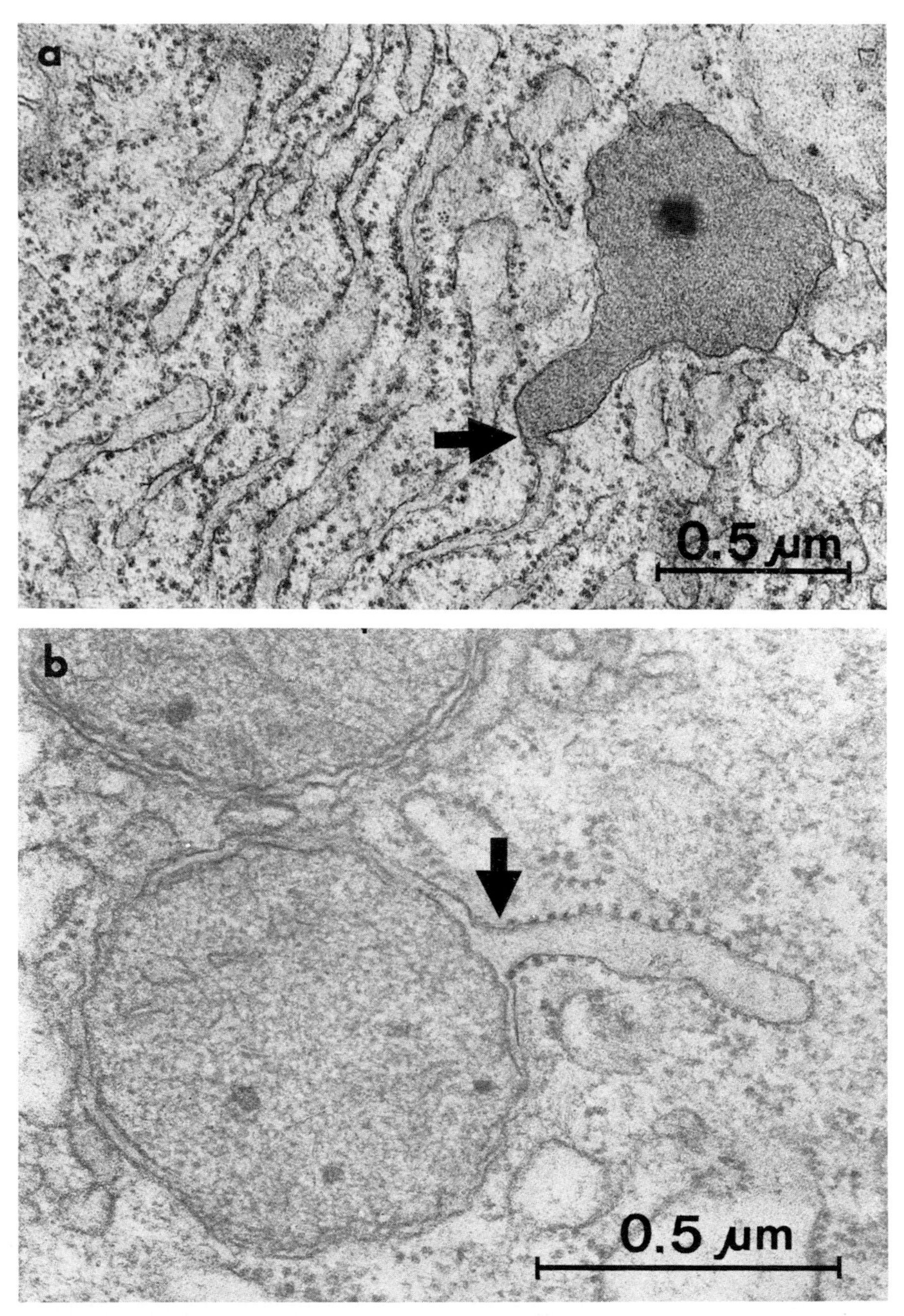


**Fig. 3.** Membrane continuum between a peroxisome and the rough endoplasmic reticulum (RER) (a) and a mitochondrion and the RER (b). Rat hepatocyte. (a) ×50,000; (b) ×80,000.

exception to this. They exhibit nucleoid peroxisomes in their livers with no urate oxidase, because uric acid is the end-product of purine degradation both in the anthropoid primates and in birds (Shnitka, 1966; Scott *et al.*, 1969; Tsukada *et al.*, 1968, 1971; Essner, 1970). Changes in nucleoid content are also seen in the course of ontogenesis. Although the fetal mouse liver contains anucleoid peroxisomes, one finds nucleoid-containing peroxisomes among the peroxisomes of human fetuses (Essner, 1969). Correlation of peroxisomal number with the ploidy level of the cell and its mitochondrial activity leads to the speculation that the peroxisomal control system may be related both to nuclear and to mitochondrial regulation (Szabo and Avers, 1969; Riede *et al.*, 1978. This working hypothesis is supported by the following observations:

1. In the presence of yeast cell nuclei with different ploidy levels, or in cells with the same ploidy level but different mitochondrial function, there were significant differences in peroxisome frequencies. This line of morphological evidence is supported by the observation that under certain conditions there are parallel changes in peroxisomal and mitochondrial activities, whereas under other conditions there are entirely different effects (Szabo and Avers, 1969).
2. Morphometric parameters for peroxisomes and mitochondria (Table I) have a constant, cell-specific and species-specific relationship (Loud, 1968; Weibel *et al.*, 1969; Riede *et al.*, 1972a; Rohr and Riede, 1973; Horvath *et al.*, 1971; Rohr *et al.*, 1973; Schmucker *et al.*, 1974).
3. Morphologically one sees a structural continuity (Fig. 3) on the one hand between peroxisomes and the nucleus (via the ER and perinuclear cisternae) and on the other hand between mitochondria and peroxisomes (via the ER) (Riede *et al.*, 1973c; P. M. Novikoff *et al.*, 1973).

## V. Breakdown of Peroxisomes

Although it is difficult to follow the fate of peroxisomes in mammalian cells (de Duve and Baudhuin, 1966; Pfeifer and Scheller, 1975; Reddy *et al.*, 1974b), studies on fat bodies of the moth larva (*Calpodes ethlius*) show that peroxisomes are eliminated from the cytoplasm either by dissolution (= autolysis) or autophagy (Locke and McMahon, 1971).

Peroxisomes can also be broken down either by autolysis, autophagy, or both in mammalian cells (Fig. 2).

1. Autolysis (Fig. 4): Peroxisomes dissolve after four days of continuous growth, in that their matrix is disrupted and they become empty, "burnt out" vesicles which are unidentifiable morphologically (Riede *et al.*, 1971e). Their membranes burst, and only the nucleoids persist (de Duve and Baudhuin 1966; Riede *et al.*, 1971e; Svoboda and Reddy, 1974; Staeubli

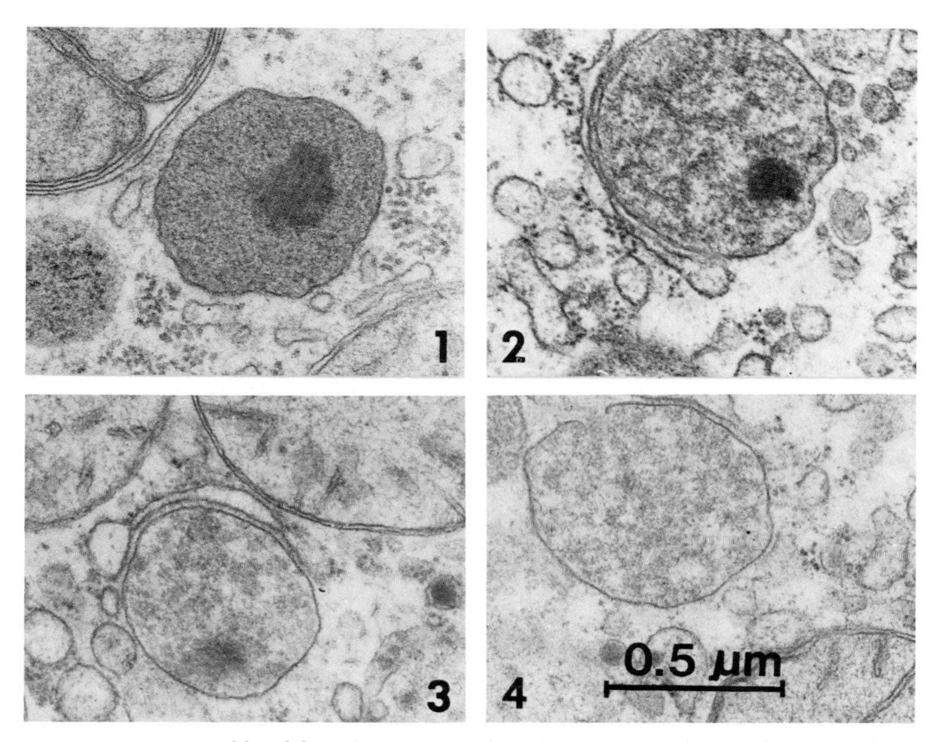


**Fig. 4.** Peroxisomal breakdown by way of self degradation (i.e., autolysis) ending as empty "burnt out" vesicles (1–4). ×47,500.

*et al.*, 1977). Eventually the individual components of the peroxisomes dissolve without leaving footprints in the hyaloplasm (Svoboda *et al.*, 1967).

2. Autophagy (Fig. 5): In contrast to the mitochondria, which usually lie in the autophagic vacuoles surrounded by only a narrow cytoplasmic zone (Arstila *et al.*, 1972), peroxisomes and microperoxisomes are always surrounded by a wide belt of cytoplasm (Riede *et al.*, 1971e; Novikoff *et al.*, 1972). Thus autophagy of peroxisomes presumably involves not simply an isolated organelle degradation, but rather the sequestration of a damaged cytoplasmic region. The activity of individual enzymes is still retained for a certain amount of time in autophagocytized peroxisomes, as in the mitochondria (Arstila *et al.*, 1972; Novikoff *et al.*, 1972). Lysosomal degradation of peroxisomes is seldom seen in the recovery phase following prior peroxisomal proliferation (Svoboda and Reddy, 1972; Reddy *et al.*, 1974b) or during severe liver atrophy (Riede *et al.*, 1973c, 1974b). It is thus assumed that autophagic degradation of peroxisomes proceeds very rapidly (Svoboda *et al.*, 1967). This is supported by determination of the average degradation time of peroxisomes (Pfeifer and Scheller, 1975).

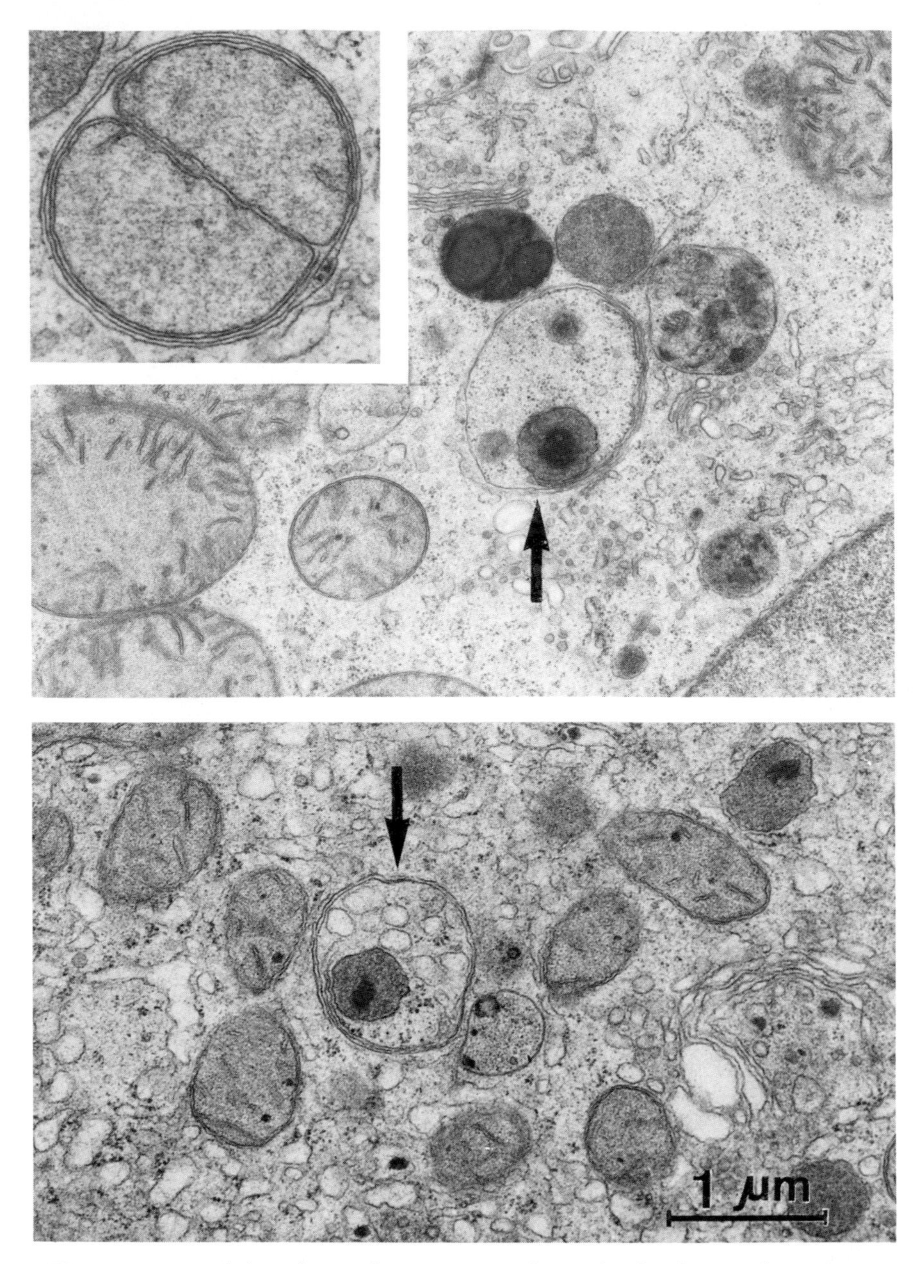


**Fig. 5.** Lysosomal degradation of peroxisomes and mitochondria by autophagy. In most cases of autophagic mitochondria one sees only a narrow band of cytoplasm (inset) between the autophagic vacuole membrane and the mitochondrial envelopes. By contrast, peroxisomes are always surrounded by a wide band of cytoplasm (arrow), which gives evidence of focal cytoplasmic degradation. ×20,000; inset ×40,000.

Peroxisomes in the periportal hepatocytes have a lifetime of about four days (cf. Poole *et al.*, 1969), and a degradation time of about 4 min. Compared to a degradation time of 16 min for mitochondria, this is substantially shorter. Autophagy of peroxisomes proceeds by a diurnal rhythm, in a manner similar to that of mitochondria, and exhibits a maximum at 2 P.M. and a minimum at 2 A.M. (Pfeifer and Scheller, 1975; Pfeifer, 1971, 1972, 1973).

## VI. Morphological Alterations in Peroxisomes

### A. *Peroxisomal Inner Bodies*

The majority of morphological alterations in peroxisomes under conditions of pathological metabolism involve the inner bodies (Fig. 2). Amorphous inner bodies are found in liver peroxisomes in patients with Reye's syndrome and Wilson's disease (Sternlieb and Quintana, 1977). Nucleoids appear in normally anucleoid human peroxisomes in patients with idiopathic, recurrent cholestasis (Biempica *et al.*, 1967). On the other hand, nucleoid-containing peroxisomes of the rat liver give rise to anucleoid peroxisomes after long-term administration of hypocholesterolemic agents or after irradiation (Svoboda and Azarnoff, 1966; Staeubli *et al.*, 1977; Christov *et al.*, 1974). Eccentric nucleoids are seen in the rat liver following $\beta$-3-furylalanine, and following an orotic acid rich diet (Novikoff *et al.*, 1966). In occasional single peroxisomes of the proximal tubule cell of the normal mouse kidney one finds marginal, amorphous matrix densities. Such changes in the peroxisomal matrix are more frequent after clofenapate stimulation (Reddy *et al.*, 1975b). Nucleoids in peroxisomes are considerably increased in size following azaserine administration, and in Reuber hepatomas (Hruban and Rechcigl, 1969). Marginal plates may appear in the peroxisomal matrix in man and monkey without pathological significance (Sternlieb and Quintana, 1977). Three weeks after discontinuation of long-term administration of clofibrate, spherical cavities appear in place of nucleoids, which are electron optically empty and sharply delimited from the normal peroxisomal matrix. Such "scooped out" peroxisomes may exceed mitochondria in size, and represent degenerative peroxisomal forms (Hartman and Tousimis, 1969). Cup-shaped hepatocytic peroxisomes were observed in male rat liver treated with (5-tetradecycloxy)-2-furancarboxylic acid (Svoboda, 1978).

### B. *Peroxisomal Matrix*

A flocculent rarefaction of the peroxisomal matrix appears in the rat liver 1 hr after ischemic necrosis (Trump *et al.*, 1965) or after portal vein ligature

(U. N. Riede and R. Rohrbach, unpublished). This degeneration stage also appears in liver peroxisomes after administration of actinomycin or allylisopropylacetamide to rats, or in man after viral hepatitis (Legg and Wood, 1972; De la Iglesia, 1969). After discontinuation of a 90-day period of clofibrate administration, the matrix is so rarefied in parts of the proliferated peroxisomes that only the nucleoids remain. Membrane ruptures are frequent. At the same time, the increased values for peroxisomal catalase activity return to normal (Svoboda and Reddy, 1972). Under certain metabolic loads, one sees tubulofibrillar structures in the peroxisomal matrix in the form of matrical tubules and matrical plates (Fig. 2). In these cases most of the peroxisomes are deformed and increased in size (Hruban *et al.*, 1974a).

1. The *matrical tubules* can be subdivided morphologically into two groups. (a) The double-walled tubules consist of two tubules constructed from helically arranged, grooved fibrils. This kind of matrical tubule contains no catalase, but probably contains isocitrate dehydrogenase (Hruban *et al.*, 1974a). They are seen in liver peroxisomes after dimethrine or meclizine treatment, in Morris hepatomas 7787 (Hruban *et al.*, 1974a), after feeding clofibrate to hypophysectomized rats, and in rats bearing aflatoxin hepatomas (Reddy and Svoboda, 1973b). (b) Granular tubules consist of spiral-like, layered granula. Catalase activity seems to be associated with the tubular wall (Reddy and Svoboda, 1972). They are present individually as normal variants in the rat kidney, in newt liver, in desert rat liver (*Meriones crassus*), and in rat Leydig cells, and are present in large numbers in neoplastic Leydig cells (Beard and Novikoff, 1969; Reddy and Svoboda, 1972, 1973b; Barrett and Heidger, 1975).
2. *Matrical plates* are usually arranged in stacks containing three plates (Hruban *et al.*, 1974a). They are seen in peroxisomes following administration of acetylsalicylate, clofibrate, nafenopine, and dimethrine (Hruban *et al.*, 1966, 1969, 1974a; Svoboda and Azarnoff, 1969; Reddy *et al.*, 1974b). The presence of these matrical plates does not correlate with serum cholesterol level.

The administration of citrate and propionate suppresses the deformation of peroxisomes and the development of matrical plates in animals fed with acetylsalicylate. This suppression does not occur with $\alpha$-ketoglutarate. If clofibrate-stimulated animals are given citrate, one sees a shift from predominantly round, dense peroxisomes to distorted peroxisomes filled with matrical plates. If catalase synthesis is inhibited by aminotriazole, peroxisomal proliferation induced by acetylsalicylate and dimethrine is not arrested, but matrical plates and tubules do not appear. It follows that these matrix enclosures may also contain catalase (Gotoh *et al.*, 1975).

## VII. Alterations in Peroxisomal Single Volume

The observation of sporadic giant peroxisomes does not give much of an idea as to their function, whereas enlarged single volume of these organelles is a parameter which can be correlated with functional change. In normal young adult rats, the peroxisomal compartment in the liver cell consists of about 70% microperoxisomes and 30% orthoperoxisomes. The same holds for human liver. Cellular damage or disease alters the composition of the peroxisomes population accordingly, and gives an idea as to the functional roles of these subpopulations. Fructose overload induces a glycogen storage liver with a marked reduction in organelles (Augustin *et al.*, 1978; Matthaei *et al.*, 1976). The peroxisome population here consists of only 50% microperoxisomes and 50% orthoperoxisomes. After three days of refeeding, however, the microperoxisomal population increases once again. This leads to the suspicion that regeneration of the peroxisomal compartment originates with the microperoxisomes, which are in turn the precursors of orthoperoxisomes. The reversal of this process is seen in the hypothyroid liver six months after total thyroidectomy, in which the peroxisomal compartment consists in large part of orthoperoxisomes and megaperoxisomes while age-specific nuclear polyploidization is essentially absent (Riede *et al.*, 1978). Thus the suppression of basic metabolism resulting from hypothyroidism is accompanied by superannuated peroxisomes in the form of megaperoxisomes (Fig. 6).

Similar alterations are observed in the peroxisomal compartment of the rat liver during and after nafenopine (Staeubli *et al.*, 1977). This hypolipidemic agent leads to a drastic peroxisomal hyperplasia, with a numerical increase in the orthoperoxisomes and megaperoxisomes. During the recovery phase it is largely the megaperoxisomes which are degraded, whence we may conclude that peroxisomal degradation in the cytoplasm is not an absolutely random process (cf. de Duve, 1973). The following processes have been considered as the cause of microperoxisomal accumulation:

1. Premature or precipitous separation of the peroxisomes from the ER (cf. Essner, 1967; Tsukada *et al.*, 1968).
2. Excessive budding of peroxisomes from the ER with subsequent shearing off of smaller "daughter peroxisomes" (cf. Legg and Wood, 1970a, 1972; Rigatuso *et al.*, 1970; de Duve, 1973).
3. Inhibition of synthesis in peroxisomal matrix protein or enzymes (cf. Legg and Wood, 1972; Svoboda and Reddy, 1972).

An increase in the mean peroxisomal single volume in rat liver, with the appearance of megaperoxisomes, must include discussion of:

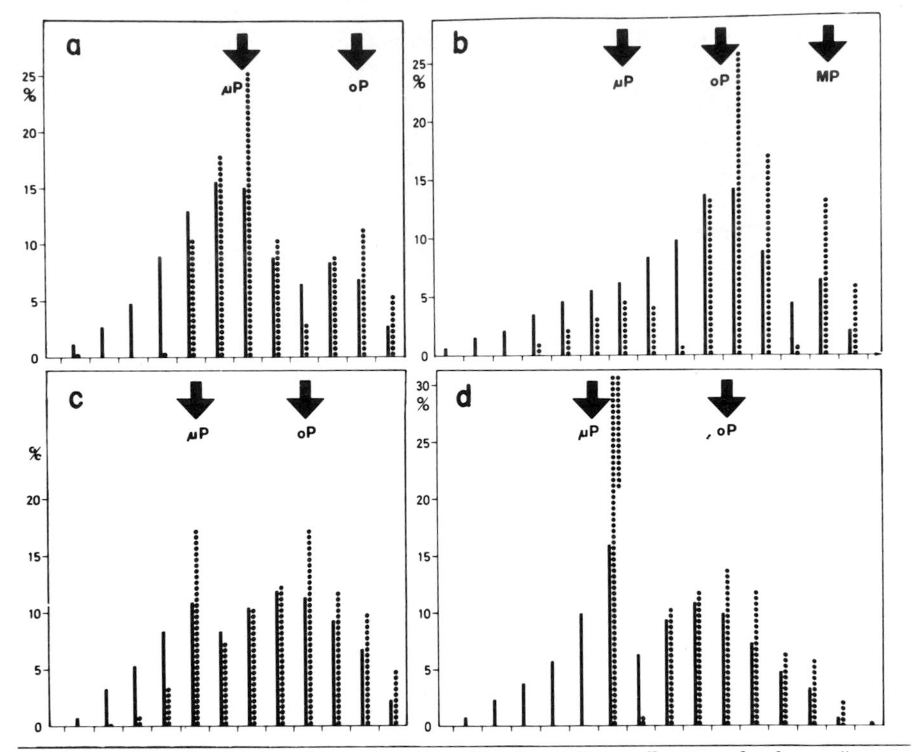


**Fig. 6.** Size distribution of peroxisomal profiles in rat liver cells. Dotted columns "true peroxisomal diameters" obtained by the Wicksell transformation. (a) Controls: 2 frequency peaks are seen, corresponding to micro- (µP) and orthoperoxisomes (oP). (b) Six months of hypothyroidism leads to an accumulation of megaperoxisomes (MP). (c) After 7 days of fructose loading, peroxisomal hypoplasia is seen. (d) Three days of refeeding leads to proliferation of microperoxisomes.

1. A delayed shearing process from the membrane system of the ER (cf. Essner, 1967; Tsukada *et al.*, 1968).
2. Swelling of peroxisomes through alterations in the physicochemical membrane properties of peroxisomes, analogous morphologically to mitochondrial swelling (cf. Trump *et al.*, 1965).
3. Increased synthesis rate of peroxisomal enzymes and/or matrix proteins, along with a normally occurring separation of peroxisomes from the membrane system of the ER (cf. Chiga *et al.*, 1971).

In this context, it is noteworthy that microperoxisomes (average diameter 0.45 µm) are typical in livers of patients with the following diseases (Sternlieb and Quintana, 1977):

1. copper deficiency (Menke's steel hair disease);
2. copper storage disease (symptomatic Wilson's disease); and
3. analbuminemia.

Megaperoxisomes (mean diameter 0.75 μm) may be found in livers of patients with:

1. copper storage disease (symptomatic Wilson's disease);
2. copper storage disease normalized after penicillamine treatment; and
3. chronic active hepatitis.

Two populations of peroxisomes can be distinguished morphologically and histochemically in the epithelium of the renal proximal tubule. Type I peroxisomes have a diameter of about 1 μm and are situated mostly at the base of the cell with some distribution in the paranuclear areas. Type II peroxisomes have a diameter of 2 μm, with a sparse matrix and marginal plates. This peroxisomal type is diffusely distributed throughout the cytoplasm, but exhibits a close spatial relationship with the ER and the mitochondria (Barret and Heidger, 1975).

The possible functional significance of such megaperoxisomes is suggested by a comparable morphometric investigation on the various parts of the renal tubule in the rat kidney (Beard and Novikoff, 1969; Orsoni *et al.*, 1969; Zimmerli, 1971). The peroxisomal volume per cell is 50% greater in the epithelium of the pars recta of the proximal tubule than in cells of the pars contorta. The peroxisomal single volume is 25% greater in the pars recta than in the pars contorta. Morphometric parameters of mitochondria are exactly reversed (Table II). Epithelial cells in the pars recta contain a 35% greater number of smaller mitochondria than cells in the pars contorta (Zimmerli, 1971). Cytochemical studies on these parts of the nephron show that tubule epithelial cells in the pars recta have nearly double the activity in α-hydroxyacid oxidase and peroxidase as do cells in the pars contorta. The activities of succinic dehydrogenase and NADH-cytochrome *c* reductase in mitochondria are reversed. These enzymes have a much greater activity in cells of the pars contorta than in cells of the pars recta. It is concluded that cells in the pars recta of the proximal tubule employ a different oxidative metabolism than cells in the pars contorta (Beard and Novikoff, 1969).

A morphometric investigation in the liver of the rat, mouse, and desert rat (*Meriones crassus*) suggests (Table I) that variation in the size of peroxisomes may have a special function in cellular metabolism. Desert rats (*Meriones crassus*) are required to spare water of oxidation, since they consume no drinking water. The fat content in desert rats is higher than that of ordinary rats. The enzymes of β-oxidation, succinic dehydrogenase, and NADH-cytochrome *c* reductase, exhibit higher activities in desert rat liver homogen-

TABLE II

*Morphometric and Cytochemical Comparison between Epithelial Cells in the Pars Recta and in the Pars Contorta of the Proximal Tubule*

| | *Pars contorta* | *Pars recta* |
|---|---|---|
| Morphometric parameters[a] | | |
| Peroxisomal single volume | 0.5 $\mu m^3$ | 0.75 $\mu m^3$ |
| Peroxisomal volume per cell | 27 $\mu m^3$ | 55 $\mu m^3$ |
| Mitochondrial single volume | 1.75 $\mu m^3$ | 1.0 $\mu m^3$ |
| Mitochondrial number per cell | 1200 | 1850 |
| Enzymatic activity[b] | | |
| α-Hydroxyacid oxidase | ++ | ++++ |
| DAB-oxidase, pH 9 | +++ | ++++ |
| Succinic dehydrogenase | +++ | ++ |
| NADH 2 cytochrome c reductase | +++ | ++ |
| DAB-oxidase, pH 6 | +++ | + |

[a] From Zimmerli (1971).
[b] From Beard and Novikoff (1969).

ates than in normal Wistar rats (R. Markstein, unpublished). Morphometric studies show that liver tissue in the mouse contains four times more small peroxisomes than that of the rat in a comparable morphometric configuration of the chondrioma. The desert rat liver, by contrast, contains an equal number of smaller peroxisomes, and the configuration of the chondrioma in these animals consists of triple-sized mitochondria in smaller numbers (Riede *et al.*, 1972a). Presumably this represents a genetically fixed adaptation of the peroxisomal compartment to the extreme metabolic load in the desert rat.

## VIII. Proliferation and Hyperplasia of Peroxisomes

### A. *Hypolipidemic Agents*

The most well-studied model for numerical increase in peroxisomes both at the biochemical and morphological level is the mouse and rat liver following several weeks treatment with the hypolipidemic agent, clofibrate (chlorophenoxiisobutyrate). Besides clofibrate analogues (nafenopine and clofenapate, acetylsalicylate, chlorcyclizine, l-benzylimidazole, N-(benzyloxy)-N, (3-phenylpropyl)acetamide, and biphenyl methylvalerianic acid can lead to a numerical increase in peroxisomes (cf. Reddy, 1974; Hruban *et al.*, 1974a; Kolde *et al.*, 1976; Staeubli *et al.*, 1977). (5-tetradecycloxy)-2-furancarboxylic acid represents a hypolipidemic agent which causes a hyperplasia of cup-shaped peroxisomes in liver cells, and an induction of several peroxisomal enzymes. The unusual enzyme responses were (a) elevation of catalase activity in liver and kidney in female rats; (b) increased activity of urate oxidase, L-α-hydroxyacid oxidase, and D-amino acid oxidase in the liver of both sexes; and (c) increased activity of L-α-hydroxyacid oxidase and D-amino acid oxidase in male kidney (Svoboda, 1978). Following administration of clofibrate, nafenopine, and methyl clofenapate, peroxisomal increase in the liver cell is accompanied by increased chondriogenesis. The RER is increased simultaneously (Svoboda and Azarnoff, 1966; Reddy *et al.*, 1974b; Moody and Reddy, 1978; Staeubli *et al.*, 1977). Reversal of an orotic acid-induced fatty liver is achieved by adding clofibrate (CPIB) to the diet (Novikoff and Edelstein, 1977). After 2 weeks of CPIB treatment, the vesiculated ER returns to its usual parallel configuration, and lipid droplets are not seen within its cisternae. At the same time hepatocellular peroxisomes show an increase in number and they possess eccentric nucleoids (Novikoff and Edelstein, 1977).

Increase in peroxisomal number is induced by administration of clofenapate in all three segments of the renal proximal tubule, but not in the cells of Henle's loop (Reddy *et al.*, 1975).

Peroxisomal hyperplasia with a numerical and volumetric increase was measured in enterocytes of the small intestine treated with nafenopine. The lack of any morphometric change in mitochondria in this study suggests that the hypolipidemic effect of nafenopine is a function of peroxisomes alone (Psenicnik and Pipan, 1977).

Increase in peroxisomal number brought about in liver cells by acetylsalicylate is inhibited by natural cell metabolites such as citrate, propionate, and α-ketoglutarate (Gotoh *et al.*, 1975). Increase in peroxisomal number induced by clofibrate is diminished only slightly by citrate, and that induced by dimethrin is not affected by citrate at all. The same holds for propionate and α-ketoglutarate. These substances do not influence peroxisomal increase induced by clofibrate and dimethrin. Citrate depresses the catalase activity even further in animals fed with acetylsalicylate or clofibrate, while propionate has no influence on the activity of this enzyme (Gotoh *et al.*, 1975). On the basis of these findings, it is thought that an increase in peroxisomal number must appear in the face of metabolic loads which lead to bottlenecks of certain metabolites in the Krebs cycle (Gotoh *et al.*, 1975).

In the case of a numerical increase of peroxisomes induced by clofibrate, it can be demonstrated that this cytoplasmic reaction is dependent on hypophyseal and/or sexual hormones (Svoboda *et al.*, 1969; Reddy *et al.*, 1974a). Hypophysectomy inhibits the increase in peroxisomal number induced by clofibrate. Nafenopine and methyl clofenapate, in contrast to clofibrate, un-

**TABLE III**

*Peroxisomal Proliferation and Hyperplasia*

| *Experimental design* | *Metabolic disorder* | *Tissues analyzed* | *Alterations of peroxisomal enzymes* | *Morphometric alterations of peroxisomes* | | *References* |
|---|---|---|---|---|---|---|
| | | | | $N_{VP}$[a] | $N_{VP}$[b] | |
| Acetylsalicylate | Hypolipidemia | Rat liver | Catalase ↑ | | | Huebner and Rohr (1969), |
| 12 days | | | | +150% | + 50% | Hruban and Rechcigl (1969) |
| 20 days | | | | +150% | +100% | |
| Clofibrate | Hypolipidemia | Rat liver | Catalase ↑ | | | Kolde *et al.* (1976) |
| 8 days | | | | +200% | +750% | |
| 21 days | | | | +400% | +100% | |
| Nafenopin | Hypolipidemia | Rat | Catalase ↑ | | | Staeubli *et al.* (1977), |
| 5 days | | | Carnitineacetyl-transferase ↑ | + 60% | + 70% | Moody and Reddy (1978) |
| Lactation period | Normalization of elevated serum lipid level | Rat liver | No data | | | Hope (1970) |
| 18 days | | | | No data | + 25% | |
| $GM_1$-gangliosidosis type I | β-Galactosidase deficiency | Human liver | No data | +200% | 0% | U. N. Riede (unpublished) |
| Phenobarbital | Microsomal enzyme induction | Rat liver | No data | | | Staeubli *et al.* (1969) |
| 5 days | | | | +100% | 0% | |

| | | | | | | |
|---|---|---|---|---|---|---|
| Folate<br>6 hr | Disturbance of protein metabolism | Rat liver | Urate oxidase ↓<br>Catalase ↓ | + 85% | + 40% | Riede *et al.* (1972d),<br>Vogel *et al.* (1964) |
| Adenine<br>1 day | Inhibition of purinesynthesis | Rat liver | No data<br>Urate oxidase ↓ | + 25% | + 10% | Riede *et al.* (1971e)<br>Hruban and Rechcigl (1969) |
| Cycloheximide<br>3 hr | Inhibition of protein synthesis | Rat liver | No data | + 75% | 0% | Riede *et al.* (1971b) |
| D-Penicillamine<br>7 weeks | Copper deficiency inhibition of cuproenzymes | Rat liver | No data | + 40% | +100% | Riede *et al.* 1971c) |
| Riboflavin deficiency<br>4 days | Inhibition of oxidative metabolism | Mouse liver | No data | + 50% | +150% | Rohr *et al.* (1974) |
| Manganese deficiency<br>12 months | Inhibition of oxidative and lipid metabolism | Rat liver | No data | No data | + 30% | Bell and Hurley (1973) |
| Thyroxine<br>10 days | Stimulation of oxidative metabolism | Rat liver | D-Amino acid oxidase ↑ | +300% | 0% | Reith *et al.* (1973),<br>Hruban and Rechcigl (1969) |
| D-Hypovitaminosis<br>4 weeks | Hypocalcemia, increased glycosis | Rat liver | No data | + 70% | +100% | Riede *et al.* (1973d) |
| Tocopherol deficiency<br>9 months | Lack of biologic antioxidants accumulation of peroxides | Rat liver | Induction of enzymes of $\beta$-oxidation | +100% | + 50% | Riede *et al.* (1971a; 1972c)<br>Markstein (1971) |
| X-ray irradiation<br>24 hr | Accumulation of peroxides | Rat liver | Catalase ↑ | + 30% | + 30% | Christov *et al.* (1974),<br>Hruban and Rechcigl (1969) |

[a] Significant changes in percentage of the controls. $N_{VP}$, numerical density of peroxisomes; $V_{VP}$, volume density of peroxisomes.

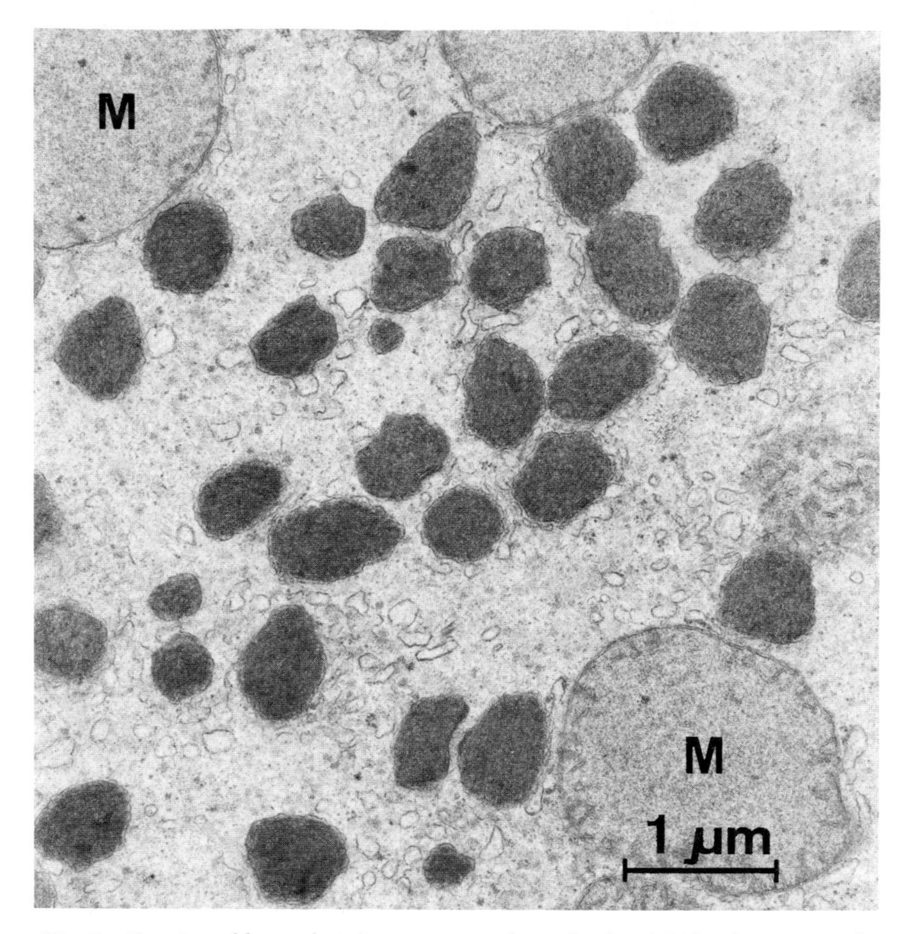


**Fig. 7.** Peroxisomal hyperplasia (increase in number and volume) 24 hr after x-ray irradiation. Rat liver; mitochondrion, M. ×19,000.

leash the same peroxisomal increase in both sexes of the mouse and rat (Reddy *et al.*, 1974a). An increase in peroxisomal number and an increase in catalase activity is brought about by clofibrate in intact males, but not in females (Svoboda *et al.*, 1969). Castrated males which are administered estrogen show no numerical increase in peroxisomes after clofibrate treatment (Svoboda *et al.*, 1969). Indeed, thyroidectomy inhibits the lipid-lowering effect of clofibrate, but has no effect upon peroxisomal increase and catalase activity. The same holds for adrenalectomy. This leads to the conclusion that an increase in peroxisomal number and increased catalase activity are both dependent upon androgen (Svoboda *et al.*, 1969). The maturity of the liver is also important. During the perinatal period, both male and female rats respond similarly to clofibrate stimulation with an increase in peroxisomal number (Svoboda *et al.*, 1969). This observation shows that the numerical increase in these organelles is not necessarily tied to an increase in enzymes. This is proven by the fact that peroxisomal increase can be induced by clofibrate, while at the same time catalase synthesis is blocked by the administration of allylisopropylacetamide (AIA), and the catalase content of the liver is drastically reduced. Only when protein synthesis is disturbed by administration of cycloheximide, does clofibrate fail to unleash further increase in peroxisomal number, while catalase content decreases (Reddy *et al.*, 1971). Thus one may conclude that the cell controls the synthesis of proteins required for peroxisomes independently of catalase synthesis. A similar phenomenon is seen in mitochondria (Reith *et al.*, 1973).

### B. *Morphometric Alterations in Different Cell Injuries*

Hepatocellular peroxisomes show a morphometrically measurable increase under the influence of a variety of metabolic disturbances. This can occur either as a result of inhibition (D, $B_2$, and E hypovitaminosis, Mn, and Cu deficiency; cf. Table III) or as a result of stimulation (thyroxine administration, lactation period; cf. Table III) of cell respiration. This finding is consistent with biochemical studies which show that peroxisomal respiration is always activated when oxygen tension in the cell increases because of insufficient mitochondrial respiration. This condition can result from a deficiency of oxidizable substrates (de Duve and Baudhuin, 1966; Baudhuin, 1969). Peroxisomal increase is also seen in the cytoplasmic adaptive reaction under conditions of increased natural substances and metabolites (folate, adenine, hydroxy radicals (cf. Table III; Fig. 7) in the organism.

From the morphometric viewpoint, one can distinguish two types of augmentation of the peroxisomal compartment:

1. proliferation with an increase in peroxisomal number only; and
2. hyperplasia with a numerical and volumetric increase of peroxisomes.

In cases with peroxisomal proliferation, we are dealing with a numerical increase of anucleoid microperoxisomes. Since this proliferation type occurs only a few hours after inhibition of protein synthesis (cycloheximide, folate; cf. Table III), it is unlikely that peroxisomal proliferation accompanies an increased synthesis of peroxisomal material. It is more reasonable that the enzymes and matrix proteins are exchanged among the individual peroxisomes by way of an existing channeling system (ER), and then be distributed among a large number of smaller peroxisomes. In this way the peroxisomal enzymes can be optimally utilized in the cytoplasm (cf. Goldenberg *et al.*, 1975).

When peroxisomal hyperplasia gives rise to an increase of the peroxisomal compartment within a cell, one is seeing a numerical increase of nucleoid-containing orthoperoxisomes. This process is presumably associated with an increased synthesis of peroxisomal material. Indeed in some cases the peroxisomes show hyperplasia but the activity of some peroxisomal enzymes decreases. Thus one must distinguish two further types of peroxisomal increase from a functional standpoint, similar to two types of cristal membrane increase (Riede *et al.*, 1974a,b).

1. Proportionate enzyme and matrical protein synthesis. In this case one can establish both peroxisomal hyperplasia as well as an increase in catalase or other proxisomal enzymes. Sometimes, however, the other enzymes in these organelles such as D-amino acid oxidase, L-$\alpha$-hydroxyacid oxidase, and urate oxidase, can show a considerable loss in activity (Reddy *et al.*, 1974b; Hruban *et al.*, 1974a,b). This type of peroxisomal proliferation is based upon a genuine cytoplasmic adaptive reaction.
2. Disproportionate enzyme and matrical protein synthesis. In this case, peroxisomal proliferation is seen but catalase does not increase (Goldenberg *et al.*, 1975; Kolde *et al.*, 1976; Staeubli *et al.*, 1977). This proliferation type is the result of a "false cytoplasmic adaptation" (cf. Wilson and Leduc, 1963), because the metabolic load of the cell is not compensated.

## IX. Hypoplasia of Peroxisomes

### A. *Catalase Inhibitors*

Allylisopropylacetamide (AIA) gives rise to an experimental porphyria, and inhibits the synthesis of liver catalase. Presumably this inhibition results either from a blockade of catalase apoprotein synthesis, or from a blockade of the association between heme and the protein moiety. The number of peroxisomes per cell remain remarkably unchanged following administration of this substance, and only the peroxisomal single volume becomes smaller. The matrix of these organelles is fluffy and less electron dense than the corresponding norm. There is a tendency for the peroxisomes to become aggregated in small areas throughout the cytoplasm. The SER is increased (Legg and Wood, 1972; Svoboda and Reddy, 1972). The loss of matrix in the peroxisomes induced by AIA presumably reflects a reduced synthesis of catalase and matrical proteins (Svoboda and Reddy, 1972).

Aminotriazole binds irreversibly with catalase, and inhibits its enzyme activity without blocking the synthesis of this enzyme (Wood and Legg, 1970). Only 2 hr after an aminotriazole injection one finds that catalase activity can no longer be demonstrated in the peroxisomes. The peroxisomal matrix becomes lighter without a change in peroxisomal number. After 4 hr, catalase activity returns in the peroxisomes, and after 6 hr this activity even overshoots the norm (Wood and Legg, 1970). These studies show that a specific suppression of catalase activity leads to hypoplasia of peroxisomes, manifested as a rarefaction of the peroxisomal matrix. Catalase, which is newly synthesized following removal of the blockade, distributes itself among the available peroxisomes.

### *B. Morphometric Alterations in Different Cell Injuries*

Numerical and volumetric reduction of peroxisomes signals hypoplasia in the peroxisomal compartment. This stage takes place under the influence of a variety of drugs before proliferation is unleashed in these organelles. We call this a preproliferative deficit in peroxisomes (clofibrate, adenine; cf. Table IV). Those interactions with carbohydrate metabolism, which lead to an accumulation of liver glycogen (cortisone, fructose; cf. Table IV) are also accompanied by a decrease in size of the peroxisomal compartment. Peroxisomal hypoplasia is observed together with a reduction of the synthetically active compartment during a generalized disturbance of cellular metabolism (starvation, antimetabolites, hypothyroidism; cf. Table IV; Fig. 8). It is remarkable that specific catalase inhibitors (AIA, aminotriazole) give rise only to a rarefaction of the peroxisomal matrix, not to a numerical reduction in the peroxisomal compartment.

It is not completely clear as yet the extent to which the size and morphometric configuration of the peroxisomal compartment is controlled by the cell nucleus. Nonetheless there are experiments which show that the size of the peroxisomal compartment is altered with the ploidy level (hypothyroidism; cf. Table IV).

Finally, the simultaneous appearance of fatty liver with a reduction in the peroxisomal compartment (orotate load, $B_6$ hypovitaminosis; cf. Table IV) leads one to speculate about a role of peroxisomes in lipid metabolism. The question as to whether the hypoplasia seen in these organelles is the cause or simply a consequence of fatty liver is as yet unanswered.

## X. Peroxisomes and Diseases

At the present time only two congenital metabolic disturbances are known which may be considered as peroxisomal diseases: acatalasemia and the cerebro-hepato-renal syndrome of Zellweger.

**TABLE IV**

*Peroxisomal Hypoplasia*

| *Experimental design* | *Metabolic disorder* | *Tissues analyzed* | *Alterations of peroxisomal enzymes* | *Morphometric alterations of peroxisomes* | | *References* |
|---|---|---|---|---|---|---|
| | | | | $N_{VP}$[a] | $V_{VP}$[a] | |
| Clofibrate 5 hr | Hypolipidemia | Rat liver | Catalase ↑ | − 20% | 0% | Riede *et al.* (1972)<br>Kaneko *et al.* (1969) |
| Orotate 7 days | Inhibition of purine nucleotide synthesis: fatty liver | Rat liver | No data | − 50% | − 50% | Riede *et al.* (1971) |
| Gestation 3 weeks | Increased serum lipid level (liver dysfunction) | Rat liver | No data | No data | − 75% | Hope (1970) |
| Starvation ketosis (lactation period) | Ketosis fatty liver | Cow liver | No data | − 30% | − 20% | Reid (1973) |
| Calcitonin 6 weeks | Hypocalcemia inhibition of lipolysis | Rat liver | No data | − 35% | − 30% | Riede *et al.* (1976b) |
| Cortisone 6 days | Stimulation of gluconeogenesis increased hepatic lipids | Rat liver | Catalase ↓<br>D-Amino acid oxidase ↑ | − 25% | −35% | Wiener *et al.* (1968)<br>Hruban and Rechcigl (1969) |

| | | | | | | |
|---|---|---|---|---|---|---|
| Frustose load 1 week | Stimulation of glycogen synthesis ATP deficiency (fatty liver) | Rat liver | No data | − 60% | − 45% | Riede *et al.* (1975), Augustin *et al.* (1978) |
| Glycogenosis type I | Glucose-6-phosphate deficiency | Human liver | No data | − 65% | − 20% | Riede *et al.* (1979) |
| $B_6$-hypovitaminosis 6 weeks | Disturbance of amino acid metabolism: fatty liver | Rat liver | No data | − 50% | − 25% | U. N. Riede (unpublished) |
| Bleomycin 3 days | Inhibition of protein synthesis | Rat liver | No data | − 70% | − 70% | Riede *et al.* (1973b) |
| Adenine 3 hr | Inhibition of purine synthesis | Rat liver | Urate oxidase ↓ | − 30% | − 30% | Riede *et al.* (1971e) |
| Food restriction 6 weeks | Inhibition of metabolism | Rat liver | No data | − 60% | − 60% | Riede *et al.* (1973c) |
| Hypothyroidism 9 months | Inhibition of metabolism | Rat liver | Catalase ↑ D-Amino acid oxidase ↓ | − 50% | − 15% | Riede *et al.* (1978), Hruban and Rechcigl (1969) |
| Ischemia 1 hr | Inhibition of cell respiration | Rat liver | No data | − 75% | − 5% | Riede *et al.* (1976a) |
| Morris hepatoma 66 | Disturbed cellular proliferation | Rat hepatoma | Catalase ↓ | − 80% | − 50% | Riede and Lorenz (1976), Tsukada *et al.* (1978) |

Significant changes in percentage of the controls. $N_{VP}$, numerical density of peroxisomes; $V_{VP}$, volume density of peroxisomes.

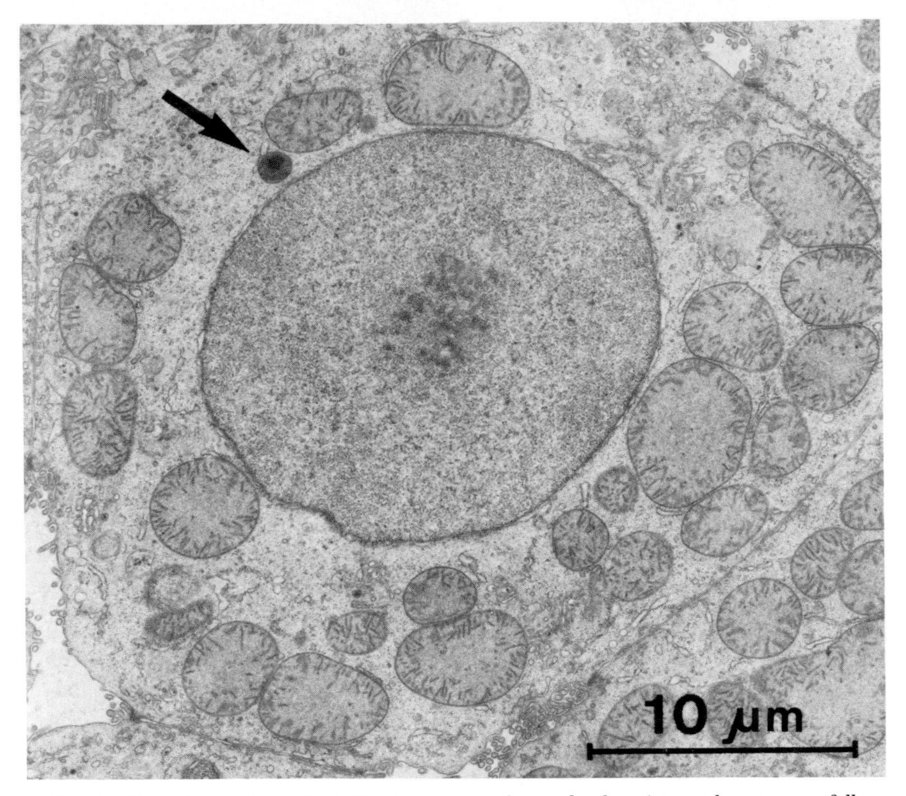


**Fig. 8.** Peroxisomal hypoplasia (decrease in number and volume) in rat hepatocytes following partial starvation for 6 weeks. ×3500.

### A. *Acatalasemia*

Acatalasemia is seen both in mutant mice and in humans (cf. Goldfischer *et al.*, 1971, 1973; Svoboda and Reddy, 1974). Tissue and blood catalase activity is lower than normal in these cases. This results from a greater heat lability of catalase. It is presumed that catalase breakdown products (= catalase peroxidase) arise *in vivo* in acatalasemic mice, and that they lead to hypolipidemia by way of inhibition of cholesterol synthesis (cf. Goldfischer *et al.*, 1973). Acatalasemia expresses itself in humans as an extremely low concentration of catalase in the liver and in other organs. The sole clinical symptoms in the few observed cases have been oral, gangrenous ulcerations (cf. Goldfischer *et al.*, 1973; Svoboda and Reddy, 1974).

### B. *Cerebro-Hepato-Renal Syndrome*

In the cerebro-hepato-renal syndrome of Zellweger, the liver peroxisomes are not demonstrable cytochemically, although the total content of liver catalase is not different from that in the normal human liver. One striking finding is the disturbance of oxidation by succinate and reduced nicotinamide dinucleotide. Presumably there is a blockade in the electron transport chain between flavoprotein and coenzyme Q (Goldfischer *et al.*, 1973). The liver epithelial cytoplasm is substantially altered, and resembles that of a fructose load-induced glycogen storage liver (Riede *et al.*, 1975a). The hepatocytes are full of glycogen and fat, and contain only a little endoplasmic reticulum. The few mitochondria are tubularly deformed (Goldfischer *et al.*, 1973).

### C. *Testicular Feminization*

Peroxisomal hypoplasia is also seen in the Leydig cells of the testis from mouse mutants with testicular feminization. The Leydig cells of normal mice contain an abundant SER and numerous peroxisomes. By contrast, peroxisomes are infrequent, the SER is hypoplastic and the mitochondria are numerous in Leydig cells from mice with X-linked testicular feminization (Reddy and Ohno, 1976).

### D. *Inflammation*

Peroxisomal enzymes also play a role in bacterial and abacterial inflammation (cf. Canonico *et al.*, 1977). In patients with pneumococcal sepsis, one observes a decreased synthesis of peroxisomal enzymes together with peroxisomal hypoplasia in the liver. The same holds for abacterial inflammation (Canonico *et al.*, 1977). It is particularly noteworthy that antihyperlipidemic agents have an inverse effect on fat metabolism as compared to inflammatory agents. Antihyperlipidemic agents increase fatty acid oxidation and inhibit fatty acid esterification and cholesterol synthesis. Inflammatory agents do the opposite.

After 8 weeks of infestation with *Schistosoma mansoni* hepatocytic peroxisomes show a 30–40% increase in number (Ramadan, 1971). In this type of infestation the peroxisomes could be involved (a) in the elimination of L-amino acids not metabolized by the injured liver, and (b) in the degradation or conversion of foreign D-amino acids from the parasite (Ramadan, 1971).

Since peroxisomes are so widespread, we may consider other idiopathic metabolic diseases under the scope of a peroxisomal defect. Included are hypercholesterolemia and hyperlipidemia (cf. Hruban *et al.*, 1972a,b). Metabolic insufficiency in peroxisomes could either mean that some peroxisomal enzymes exhibit altered physicochemical properties, or that the genesis of peroxisomes is disturbed. This could lead to an intracellular build-up of toxic metabolic products, such that the function of other metabolic compartments is altered as well.

## XI. Peroxisomal Reaction Patterns during Cell Injury

As discussed above, peroxisomes have an ancillary function in intermediate metabolism such that each metabolic alteration, regardless of whether it damages the cell, leads to a change in the peroxisomes. These peroxisome changes are in part morphologically recognizable, but may in some cases first be recognized only by morphometry, and eventually lead to a reorganization of the peroxisomal compartment. These peroxisomal changes, however, do not occur in isolation within the cell, but rather are accompanied by changes in the ER and in the mitochondria. These simultaneous organelle changes may be represented as *reaction patterns* of cellular injury, based upon a quantitative organelle pathology (Moore *et al.*, 1977), in which the patterns are expressed as symbolic logic and tested for internal consistency by electronic data processing.

### A. *Mitochondrial–Peroxisomal Patterns*

Studies on yeast cells suggest that mitochondrial regulatory substances control the enzyme activity in peroxisomes (Szabo and Avers, 1969). Whether or not this is also true for mammalian cells is still unclear. In any case, morphometric changes take place under conditions of cell damage in both peroxisomes and mitochondria, which appear in a typical fashion in the several phases of the cytoplasmic adaptive reaction (Fig. 9).

In the *alarm phase* of an adaptive reaction, the hepatocellular chondrioma exhibits an increased number of smaller mitochondria. Since the cristal membranes are not increased in this phase, we cannot attribute this change in the chondrioma to increased chondriogenesis, but must consider rather an increased mitochondrial fission. This process is a nonspecific response to any cellular insult (Chevremont and Chevremont-Comhaire, 1953; Rohr and Riede, 1973). Peroxisomes show proliferation or hyperplasia in most cellular insults during the alarm phase, during which presumably the ER sheers off increasingly many microperoxisomes, which are anucleoid at first, but subsequently mature into nucleoid-containing peroxisomes (P. M. Novikoff *et*

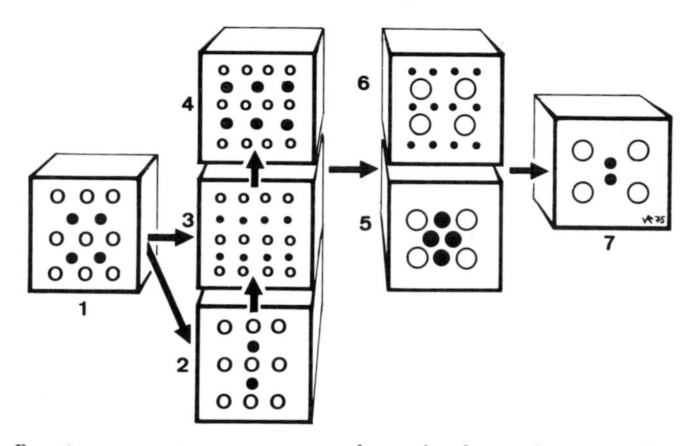


**Fig. 9.** Reaction pattern in peroxisomes and mitochondria in the course of a cytoplasmic adaptive reaction. Mitochondria and peroxisomes normally exhibit a constant, cell specific ratio (1). Microperoxisomes and mitochondria show a numerical increase during the alarm phase of cell adaptation (3). This process may be preceded by a preproliferative deficit of peroxisomes (2), which is followed by hyperplasia of ortho- or megaperoxisomes (4). In the resistance phase we find a transformation of the peroxisomal compartment and chondrioma such that megamitochondria production is accompanied by either a numerical increase of microperoxisomes (6, pattern of hyperfunctional resistance phase) or formation of megaperoxisomes (5, pattern of hypofunctional resistance phase). Both the peroxisomal and mitochondrial compartment are diminished during the exhaustion phase (7). White circles, mitochondria; black circles, peroxisomes.

*al.*, 1973). A preproliferative deficit may precede this peroxisomal proliferation.

During the *resistance phase*, the hepatocellular chondrioma switches over to a smaller number of enlarged mitochondria. This mitochondrial enlargement is based either upon the growth of individual mitochondria, upon the fusion of mitochondria, or upon the disturbed division of mitochondria (Rohr and Riede, 1973). These morphometric alterations in the chondrioma are in some cases accompanied by peroxisomal proliferation or hyperplasia, in other cases by peroxisomal dystrophy with the appearance of megaperoxisomes. Usually, the peroxisomal enzyme content is lower than normal.

In the *exhaustion phase*, the hepatocellular chondrioma is characterized by a drastically reduced number of abnormal megamitochondria. These mitochondria are either abnormally rich in matrix (Riede *et al.*, 1973c) or swollen (Rohr *et al.*, 1973). In either case, the cristal surface in such mitochondria is diminished. The peroxisomes in this phase always exhibit a drastic hypoplasia, ageneration or atrophy, leading to a numerical and/or volumetric reduction. This reaction pattern can also be seen in acute lethal cell injuries (Riede *et al.*, 1973b), and corresponds to Trump's stage 4 (Trump and Arstila, 1975).

### *B. Cristal-Peroxisomal Patterns*

The mitochondrial cristae are the bearers of the respiratory chain and contain oxidase which catalyze the final step of the biological oxidation process. In contrast to the mitochondria, peroxisomes contain oxidases, which do not participate in the salvage of energy-rich substrates. The mitochondria contain DNA and RNA, but like peroxisomes are directed by ribosomal protein synthesis (Borst *et al.*, 1967). In spite of the independence of these two organelles suggested by current understanding, certain morphometric parameters of the hepatocellular chondrioma are altered along with the size of the peroxisomal compartment. We can sketch out the following six reaction patterns (Fig. 10).

The numerical density of peroxisomes increases during the alarm phase because of increased neogenesis. There is no increase in the mean peroxisomal single volume. However, the cell shows an impoverishment of

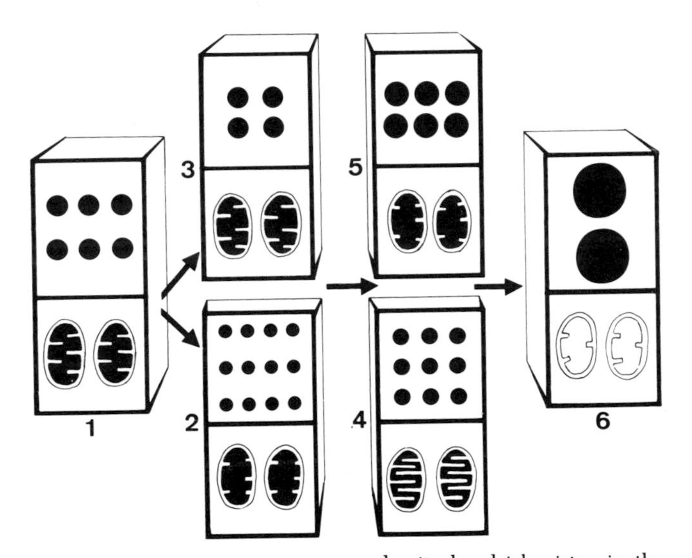


**Fig. 10.** Reaction patterns in peroxisomes and mitochondrial cristae in the course of a cytoplasmic adaptive reaction. In contrast to normal conditions (1) a numerical increase of peroxisomes and reduction of cristal membrane surface is seen in the alarm phase (2). A preproliferative peroxisomal deficit without morphometric changes of mitochondrial cristae may proceed this process (3). In the resistance phase the peroxisomal augmentation is either accompanied by an increased (4, hyperfunctional resistance phase) or a decreased surface density of mitochondrial cristae (5, hypofunctional resistance phase). The peroxisomal compartment and the cristal surface density is diminished in the exhaustion phase (6, vita minima).

cristae either as a result of pathologic increase in matrix or as a result of mitochondrial swelling with cristolysis (lit. see Rohr and Riede, 1973).

An alternative reaction pattern in the peroxisomes in the alarm phase is a decrease in peroxisomes in the cell with a corresponding volumetric decrease in the peroxisomal compartment. The membrane content of cristae in mitochondria typically experiences no losses in this situation. Thus the mitochondrial content of the cell predominates. In certain cases, however, insufficient, degenerated cristal membranes can accumulate in cells and mimic in increase in respiratorially active membranes (Riede *et al.*, 1975a,b). This cytoplasmic reaction represents a changeover of oxidative metabolism to an energetically less economic—evolutionarily more primitive?—constellation of cellular compartments which participate in oxidate metabolism (cf. de Duve, 1969). This reaction pattern in mitochondria and peroxisomes is typical for rapidly growing hepatomas (Hruban *et al.*, 1973) and for the adaptive response to extreme metabolic loads (Riede *et al.*, 1978). This pattern is also seen in acute, lethal cellular injuries (Riede *et al.*, 1976a) and corresponds to Trump's stage 2 (Trump and Arstila, 1975).

Two further reaction patterns arise in the resistance phase from these two reaction patterns of the alarm phase. One is characterized by a persistence of the cristal loss, the other by an increase in cristal membranes. In the case of the cristal membrane increase, the peroxisomal compartment shows the comparable morphometric transformation with a corresponding numerical increase. This pattern reflects an optimal adaptation reaction to sublethal cellular injury with increased function (i.e., hyperfunctional resistance phase). A concomitant induction of mitochondrial and peroxisomal enzymes is usually observed in this setting (Staeubli *et al.*, 1969; Hope, 1970; Reith, 1973). This pattern is seen only in the course of chronic, sublethal cellular injury, and corresponds to a cytoplasmic adaptive reaction against disturbed cell respiration.

In the situation in which the cristal membrane loss persists, the peroxisomal compartment experiences morphometric transformation by way of a corresponding volumetric increase. This pattern represents an insufficient cytoplasmic adaptation reaction with hypofunction (i.e., hypofunctional resistance phase). Megaperoxisomes predominate in the peroxisomal compartment, and are usually superannuated, functionally ineffective organelles (Riede *et al.*, 1978; Staeubli *et al.*, 1977). This reaction pattern is one of "vita reducta" ("reduced life") (cf. Riede *et al.*, 1971c).

If the cause of cellular injury continues further, the cell reaches the exhaustion phase. The mitochondria lose the greater part of their cristal membranes through swelling-induced cristal loss, and thus lose their respiratory potential. The proximal compartment suffers similar changes, and becomes substantially reduced either volumetrically or numerically. Eventually, only a few megaperoxisomes or occasional microperoxisomes remain behind in the cytoplasm.

This reaction pattern is also seen in acute, lethal cellular injury (Rohr *et al.*, 1974; Riede *et al.*, 1973b), and corresponds to Trump's stage 4 (Scarpelli and Trump, 1971). Simultaneous reduction of peroxisomal and mitochondrial volume and cristal membrane surface is likewise seen in chronic, sublethal cellular injury, and demonstrates that the cell has reached a "vita minima" ("minimum life"). At this point the possibilities for adaptation by the cytoplasm are exhausted, and the cell has reached the "point of no return."

### C. *Ergastoplasmic-Peroxisomal Pattern*

As already discussed, there is a close morphological and topographic relationship between peroxisomes and the ER. This is seen especially well in the rat liver and the proximal tubule of mouse kidney during the perinatal period, in which peroxisomal biogenesis is first established in the late phase of the fetal period with the development of the RER (Rohr *et al.*, 1971; Dvorak, 1971; Goeckermann and Vigil, 1975). This functional relationship also expresses itself in the morphometrically demonstrable reaction patterns of both organelles in the course of a metabolic load. (Fig. 11).

In the alarm phase of acute sublethal cellular injury, one sees a preproliferative peroxisomal deficit without volumetric change of the RER. The peroxisomal deficit in this phase is due more to a peroxisomal autolysis and less to increased peroxisomal autophagy. Alternatively, proliferation or hyperplasia of orthoperoxisomes or dysplasia of microperoxisomes can appear during the alarm phase of the cytoplasmic adaptive reaction. Typically this is accompanied by loss of membrane and volume in the ergastoplasm. Possibly, RER membranes are increasingly required for use as peroxisomal envelopes in the phase of cellular injury. The following observations support this idea:

1. RER membranes have the same structural properties as peroxisomal membranes (Reddy *et al.*, 1974a).
2. Peroxisomes may be interpreted as localized dilations of the ER, which retain multiple, membranous continuities (Novikoff and Novikoff, 1972; de Duve, 1973). Since an increase in catalase activity is not demonstrable in this situation, we are presumably dealing with a disproportionate synthesis of matrix proteins and peroxisomal enzymes in this type of peroxisomal increase.

Simultaneous proliferation or hyperplasia of the RER and peroxisomes may occur in the resistance phase. This peroxisomal disease may be pre-

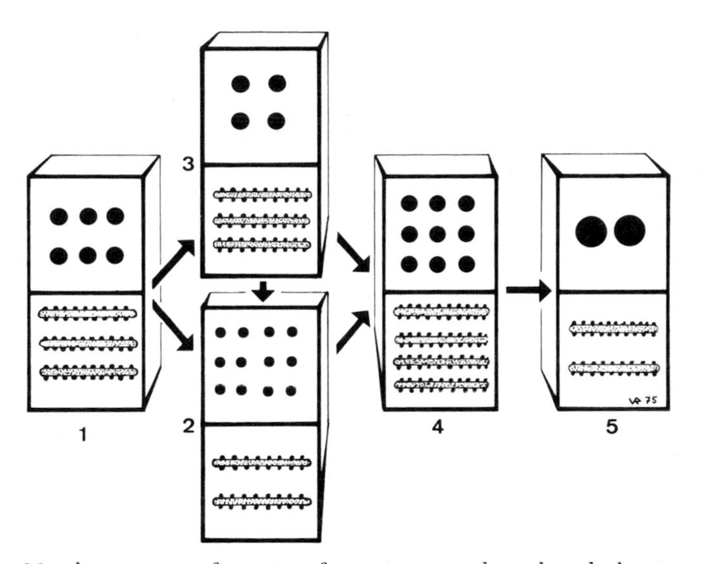


**Fig. 11.** Morphometric transformation of peroxisomes and rough endoplasmic reticulum (RER). In contrast to normal conditions (1), a numerical increase of peroxisomes and volumetric reduction of the RER is seen in the alarm phase (2). A preproliferative peroxisomal deficit without diminution in the RER may precede this process (3). The RER and orthoperoxisomes are augmented in the resistance phase (4). The RER and the peroxisomal compartment diminish during the exhaustion phase (5).

ceded by a preproliferative deficit. Both proportionate and disproportionate synthesis of peroxisomal enzymes and peroxisomal matrix proteins may occur, because in some cases catalase activity increases.

Finally in the exhaustion phase, peroxisomes are rarefied either as a result of hypoplasia, dystrophy, or ageneration. The RER usually exhibits a reduction both in volume and/or membrane surface. The increase in fat bodies, seldom absent in this phase of cellular injury, may be due to disintegrated lipoprotein synthesis and/or disintegrated function in the peroxisomal compartment. This reaction pattern corresponds to Trump's stage 3 (Scarpelli and Trump, 1971).

## XII. Peroxisomes and Tumors

### A. *Peroxisomal Number and Tumor Growth Rate*

Peroxisomes behave differently in neoplastic cells than in normal cells. Nucleoid peroxisomes are absent in rapidly growing Novikoff hepatomas.

These tumors may contain a limited number of anucleoid microperoxisomes, however (A. B. Novikoff *et al.*, 1973). The number and size of peroxisomes in Morris hepatomas (Fig. 12) reflect the growth rate of these tumors, (Mochizuki *et al.*, 1971; Riede and Lorenz, 1976). Thus fast-growing hepatomas contain only a few anucleoid microperoxisomes, hepatomas with intermediate growth rates have more abundant and larger peroxisomes which contain crystalloid, and slowly growing hepatomas have numerous, large, nucleoid peroxisomes.

Furthermore, the catalase activity is inversely related to the growth rate of Morris hepatomas. The urate oxidase activity of moderately well-differentiated hepatomas is greater than that of well-differentiated tumors, and is even higher in normal liver parenchymal cells. Tumor peroxisomes react to a proliferation stimulus differently than peroxisomes in the host

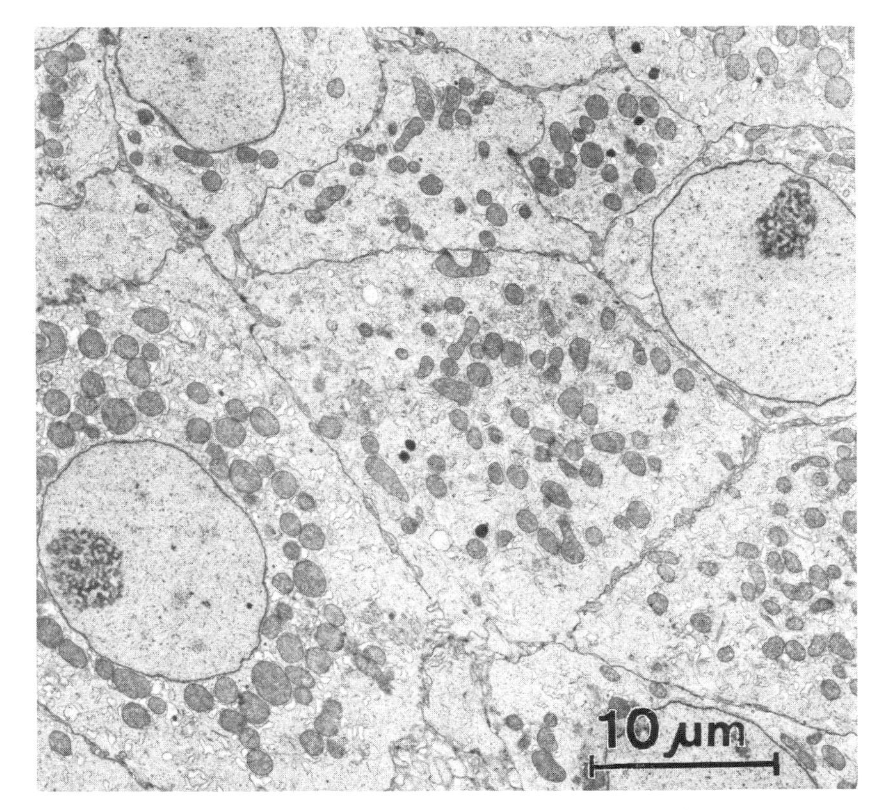


**Fig. 12.** Ultrastructure of tumor cells (Morris hepatoma 66). Only few peroxisomes are present. ×2500.

liver. Thus we see a loss of peroxisomes and catalase in the host liver in an animal with a hepatoma 7316 A after seven days of a protein-free diet, whereas the tumor peroxisomes remain biochemically and morphometrically unaltered (Tsukada *et al.*, 1978).

Transplantable BNL and $BH_3$ mouse hepatomas with intermediate growth rates have no peroxisomes and a poorly developed RER. BRL mouse hepatomas have peroxisomes, but they are small and few in number (Malick, 1972). Tiny peroxisomes are also seen in rapidly growing Reuber H-35 hepatomas (Hruban *et al.*, 1965).

Peroxisomal single volume increases, but peroxisomal proliferation is not seen in cells of Morris hepatoma 9618 A under the influence of clofibrate (Tsukada *et al.*, 1975b). The same holds for tumors induced by 3-methyl-4-diaminoazobenzene. Peroxisomes in these rat tumor cells are morphologically indistinguishable from hepatocellular peroxisomes. Undifferentiated carcinomas and adenocarcinomas which arise in the course of this carcinogenesis contain either few peroxisomes or none at all. These peroxisomes usually have no nucleoids. The catalase content in these rat liver tumors is substantially lower than in normal liver tissue (Itabashi *et al.*, 1975). Catalase activity in liver and kidney is depressed in tumor-bearing animals but may be drastically increased in conjunction with a peroxisomal proliferation by way of clofibrate stimulation (Reddy and Svoboda, 1971; Svoboda and Reddy, 1974).

At the 10th week of carcinogen (3′-methyl-4-dimethylaminobenzene) feeding hyperplastic liver lesions (hyperplastic foci and nodules) appeared and advanced to further stages. Most of the foci and some of the nodules showed very low catalase activity and a small number of peroxisomes. When rats were administered CPIB, most of the hyperplastic lesions showed an increase in catalase activity and peroxisomal number. However, there are hyperplastic lesions in which the cells proliferate without cytoplasmic maturation and without responsiveness to CPIB. These persistently altered cells serve as intimate precursors for subsequent liver carcinomas (Itabashi *et al.*, 1977).

Peroxisomes and mitochondria in transplantable Morris renal tumors, which presumably arise from cells of the proximal tubule, are both less numerous than in normal epithelium of the proximal tubule. Catalase activity and D-amino acid oxidase activity are both substantially lower in these adenocarcinomas than they are in the renal cortical tissue of normal rats (Hruban *et al.*, 1973).

Peroxisomes in cells of a human hepatic hamartoma are indistinguishable from those of normal human hepatocytes; they and the mitochondria are reduced in liver adenoma cells (Philipps *et al.*, 1973). The presence of peroxisomes in human primary liver cell carcinomas is an index of the degree of differentiation of the tumor cells (O'Conor *et al.*, 1972). Human hepatocarcinoma cells often have no peroxisomes at all (Woyke *et al.*, 1974).

### *B. Morphology of Tumor Cell Peroxisomes*

Peroxisomes are morphologically unchanged in pyrrolizidine-induced megalohepatocytosis, which is regarded as an early phase of hepatoma (Allen *et al.*, 1970).

A portion of the cisternae of the RER clings to peroxisomes in tumor cells of Morris hepatoma 5123 C. That portion of the ER in contact with the peroxisomes is always smooth on the concave face (Hruban *et al.*, 1972b). Peroxisomal proliferation appears among hepatocytes in nontumor areas in clofibrate-stimulated rats with hepatomas induced by aflatoxin B1. The tumor cells show no numerical increase in peroxisomes, but exhibit matrical tubules (Reddy and Svoboda, 1973b) such as appear spontaneously in Morris hepatomas 7787 (cf. Tsukada *et al.*, 1975a). The same matrical tubules are also seen in neoplastic rat hepatocytes during carcinogenesis induced by 2-acetylaminofluorene or by 3-methyl-*p*-dimethylaminobenzene (Tsukada *et al.*, 1975a). It appears that the production of such matrical tubules is only possible during preneoplastic hyperplasia caused by carcinogens (Tsukada *et al.*, 1975b). Presumably this strengthens the action of clofibrate. This drug seems to induce matrical plates at an earlier stage than matrical tubules during carcinogenesis (Tsukada *et al.*, 1975b, 1978).

In human trabecular hepatomas, the mitochondria show a decrease in number and a striking pleomorphism. Many of these mitochondria exhibit amorphous inclusions. The anucleoid peroxisomes are quite variable in size and shape, and some contain amorphous inclusions and partial myelin figures (Ruebner *et al.*, 1967).

Numerous, anucleoid peroxisomes are found in Leydig cell tumors of the rat. In contrast to normal Leydig cells, peroxisomes in these neoplastic cells contain numerous matrical tubules. The tissue of these highly differentiated and hormonally active tumors contains neither urate oxidase nor α-hydroxyacid oxidase, but does contain catalase (Reddy and Svoboda, 1973b).

Peroxisomes vary with respect to form and size in adenomas of the perianal glands of the dog. They contain fewer marginal plates and more amorphous nucleoids in comparison to normal perianal gland cells. The peroxisomes in these neoplastic cells contain L-α-hydroxyacid oxidase, but no urate oxidase or D-amino-acid oxidase (Kuhn, 1968).

A varying number of microperoxisomes occur in all cells of a virilizing adrenocortical adenoma (Gorgas *et al.*, 1976). These organelles are abundant in larger adenomatous cells. In this tumor, however, histochemical DAB reaction failed to reveal peroxidatic activities of catalase within the peroxisomes (Gorgas *et al.*, 1976).

Peroxisomal behavior in neoplastic cells is presumably due to a disturbance in the cellular mechanisms which control neogenesis of these organelles and the production of peroxisomal enzymes (cf. Itabashi *et al.*, 1975). Thus the peroxisomal compartment is transformed such that only anucleoid microperoxisomes are present. As for mitochondria, this disintegrated synthesis of matrical components leads to the production of paracrystalline formations in peroxisomes (cf. Riede and Nobmann, 1974; Reddy and Svoboda, 1972). The peroxisomal enzyme pattern is altered in parallel with this alteration (Hruban *et al.*, 1973; Reddy and Svoboda, 1972; Mochizuki *et al.*, 1971). The cytoplasm becomes poor in mitochondria and peroxisomes disappear with progressive anaplasia (Mochizuki *et al.*, 1971).

## XIII. Appendix

### A. *Symbolic Logic as a New Method in Organelle Pathology*

*Symbolic logic* is a mathematical and philosophical method in which descriptive statements, as well as quantitative data, are placed in a formal relationship to one another (Borkowski, 1977). Each *unit statement* is called an *element*, and corresponds to a single, declarative statement in a natural language (such as English). *Compound statements* are constructed from one or more unit statements, whose meaning has been qualified by one or more *operators*. There is one *unary operator*, namely *negation* (−), which qualifies the status of a single element, or *operand*; and five *binary operators* which express a qualitative relationship between two operands. The five binary operators are: *and* (&), *inclusive-or* (|), *implication* (>), *back-implication* (<), and *equivalence* (=). (These symbols are chosen for their availability on a standard keypunch.) If A is a statement, then −A is a statement with the opposite truth value; that is, −A is false when A is true and vice versa. If A, B are statements, then statement A&B is true if and only if both A *and* B are true; statement A|B is true if and only if *either* of A *or* B are true (or both); statement A>B is true if and only if whenever A is true, B is true; statement A<B is true if and only if B>A is true; and statement A=B is true if and only if *both* A>B and B>A are true.

Our procedure for evaluating expressions in symbolic logic depends upon the concept of *nullity*, introduced by Quine (1948). We convert each symbolic logic expression into its equivalent in nullities, then exhaustively perform an operation called *null-addition* upon all allowable pairs of nullities in the system (Moore *et al.*, 1977). A *nullity* is a set of elements which, taken in combination, are *false*. The nullity {A} expresses the condition, "it is false that A;" the nullity {−A}, on the other hand, expresses the condition, "it is false that not −A," in other words, "it is true that A." The nullity {A,B} expresses condition, "it is false that *both* A and B are true," or "either A is false or B is false (or both," or "not−A implies B," or "not−B implies A." Likewise, the nullity {A,B,C} expresses the condition, "it is false that *all* of A, B, and C are true," etc. Every superset of a true nullity is a true nullity. If {A,B} is a true nullity, then so is {A,B,C}. The empty nullity, {} or ∅, signals a contradiction in the system.

Each expression in symbolic logic is readily converted into its equivalent in nullities. As a first step, the initial symbolic logic expression is negated. Then every appearance of A>B is converted to −A|B; every appearance of A<B is converted to A|−B; and every appearance of A=B is converted to (−A|B)&(A|−B). Every appearance of A|B is separated into two expressions, A and B (Separation rule). Every appearance of −(A&B) is converted into −A|−B (DeMorgan rule). Every appearance of (A|B)&C is converted into (A&C)|(B&C) (Distributive rule). The Separation, DeMorgan, and Distributive rules are applied repeatedly to exhaustion. Each such string of elements, which are connected exclusively by &, then become the members of a single nullity. For example, the expression "A&−B&C," which cannot be further reduced, becomes the nullity {A,−B,C}.

The unit calculation is *null-addition*, denoted ⊕. Two nullities, say {A,B,C} and {−B,C,D}, are subject to this operation if and only if there is exactly one element, called the sign reversal element (here, element B), which is positive in one nullity and negative in the other. The *null-sum* is defined as the set containing the members of both sets *except* the sign reversal elements, i.e., {A,B,C} ⊕ {−B, C,D} = {A,C,D}. Null-addition of the nullity pair {A,B,C} and {B,C,D} is not allowed because there is no sign reversal element, and null-addition of the nullity pair {A,B,C} and {−B,−C,D} is not allowed because there is more than one sign reversal element. Null-addition is performed to exhaustion on all allowable pairs of nullities in the system. This calculation has been shown by rigorous mathematical proof to find all and only the valid nullities in the system (Moore *et al.*, 1977). Substantial computing economies are afforded (a) by first starting with null-additions in which one nullity contains exactly one member (such as {A} ⊕ {−A,B,C} = {B,C}), and (b) by casting out superset nullities, since they are superfluous. (For example, {A,B,C} is superfluous if {A,B} is already in the system.)

### *B. Quantitative Organelle Pathology*

The current model of quantitative organelle pathology employs a subdivision of the hepatocyte (H) into five compartments: rough endoplasmic reticulum (R), smooth endoplasmic reticulum (S), mitochondria (M), mitochondrial cristae (C), and peroxisomes (P). These organelles are further classified as either particulate organelles (H,M,P) or as tubulocisternal organelles (R,S,C). Particulate organelles, tubulocisternal organelles and the hepatocyte as a whole, are collectively designated as compartments. Compartments H, M, P have three morphometric measurements. Two of these are directly relevant to the pathological state: volume density (V) and numerical density (Z). The volume to number ratio of the particulate organelles Q corresponds to the mean single volume of the particulate organelles. Compartments R, S, C have three measurements: volume density (V), surface density (F) of the organelle membranes, and the volume surface ratio (Q). Measurement Z for compartments H, M, P is used interchangeably with measurement F for compartments R, S, C. Each morphometric measurement is quantified as significantly increased (H, "high"), normal (N), or significantly decreased (L, "low") using 95% confidence limits about Student's *t*-distribution. Each measurement was written out as a three-letter shorthand, with the first letter denoting the compartment (R, M, P, etc.), the second letter the measurement type (V, Z, F, etc.), and the third letter the amount (H, N, L). For example, PVH denotes "peroxisomal volume increased (high)." Morphometric parameters (volume density, numerical density, and surface density) for the hepatocyte have the unit volume of liver tissue (1 $cm^3$) as a reference system and correspond to the morphometric symbols $V_{VH}$, $N_{VH}$, $S_{VH}$ (Weibel *et al.*, 1969). The morphometric parameters for all cytoplasmic organelles, e.g., peroxisomes, have the volume density of cytoplasm ($V_{VC}$) as a reference system, i.e., volume fraction of cytoplasm per unit volume liver tissue (1 $cm^3$) and correspond to the morphometric symbols ($V_{VP}/V_{VC}$, $N_{VP}/V_{VC}$) (Rohr and Riede, 1973).

A system of nine pathological diagnoses and five morphological diagnoses was constructed for each organelle. The pathological diagnoses were: normal (N), proliferation (P), ageneration (E), hypertrophy (H), atrophy (A), hyperplasia (R), hypoplasia (O), dysplasia (D), and dystrophy (Y). The morphological states for particulate compartments were: unchanged single volume (U), microorganelle (C), small organelle (S), megaorganelle (M), and giant organelle (G). The morphological states for the tubulocisternal compartments were analogous. Each possible pathological and morphological diagnosis was written out as a two-letter shorthand, with the first letter denoting the compartment (R, M, P, etc.) and the second letter denoting the pathological (N, P, E, etc.) or morphological (U, C, S, M, G) diagnosis. For example, PR denotes "peroxisomes hyperplastic" and PC denotes "microperoxisomes."

Symbolic logic sentences expressing the relationship between the measured quantities (PVH, PVN, PVL, etc.) and pathological (PN, PP, etc.) and morphological states (PV, PV, etc.) were prepared according to the following descriptions (Fig. 13).

1. Proliferation (P) is defined either as a numerical increase without volume change in particulate organelles or as membrane increase (= "numerical increase of membrane units") without volume change in tubulocisternal organelles.
2. Ageneration (E) is the opposite of proliferation. It is characterized as a numerical reduction in particulate organelles or as a reduction in membrane surface in tubulocisternal organelles without simultaneous volumetric change of the organelles.
3. Hypertrophy (T) is defined as volumetric increase without numerical change in particulate organelle number or in membrane surface of tubulocisternal organelles.
4. Atrophy (A) is the opposite of hypertrophy. It is defined as volumetric decrease in organelles without change in number of particulate organelles or in membrane surface of tubulocisternal organelles.
5. Hyperplasia (R) is defined as a numerical and volumetric increase in particulate organelles or as an increase in volume and membrane surface in tubulocisternal organelles.
6. Hypoplasia (O) is the opposite of hyperplasia. It is defined as a numerical and volumetric decrease in particulate organelles or as a volume and membrane surface decrease in tubulocisternal organelles.
7. Dysplasia (D) corresponds to a numerical reduction with silmultaneous volumetric increase in particulate organelles, or to a membrane surface reduction with simultaneous volumetric increase in tubulocisternal organelles.
8. Dystrophy (Y) is the opposite of dysplasia. It is defined as numerical increase and volume decrease in particulate organelles, or as membrane surface increase and volume decrease in tubulocisternal organelles.

The morphological state of the particulate organelles may be characterized additionally in terms of the volume number ratio (Q). This parameter corresponds to the volume density per numerical density of the particulate organelles and, say, for peroxisomes, is denoted by the morphometric symbol $V_{VP}/N_{VP}$. This ratio (Q) allows estimation of the average single volume of the particular organelles (Weibel *et al.*, 1969).

Five morphological states are defined for the particulate organelles. The *unchanged* state (U) shows organelles with a normal single volume. The *small* organelle state (S) shows organelles with moderately decreased single

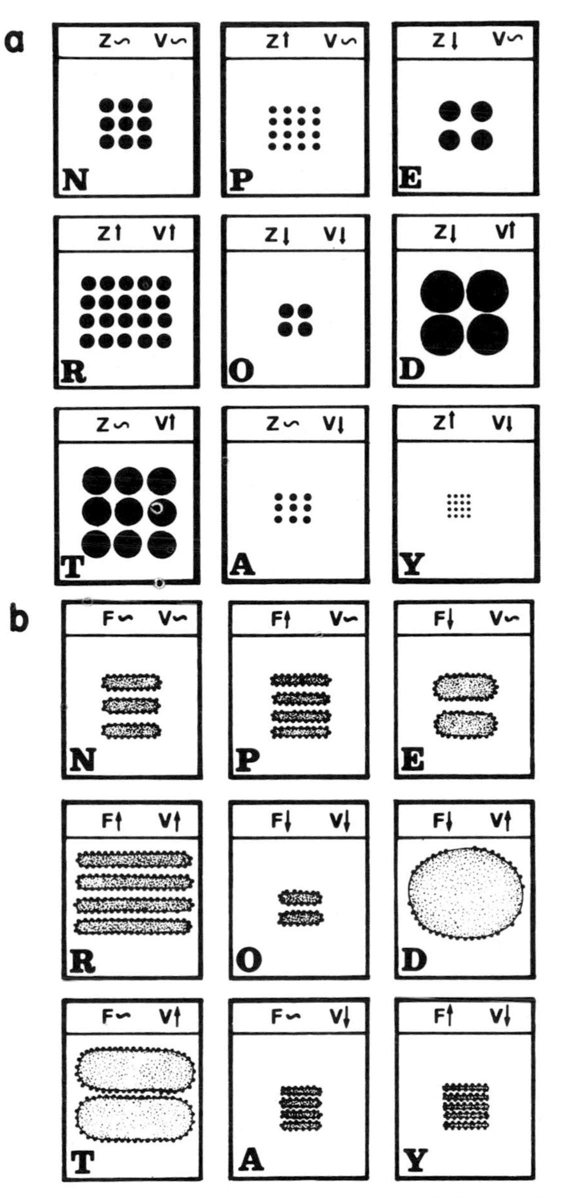


**Fig. 13.** The pathological states in quantitative organelle pathology. (a) Particulate organelles, e.g., peroxisomes and mitochondria; (b) tubulocisternal organelles, e.g., endoplasmic reticulum and mitochondrial cristae. Numerical density, Z; volume density, V; surface density, F. Normal (∼), increased (↑), decreased (↓) morphometric values. Normal, N; proliferation, P; ageneration, A; hypertrophy, T; atrophy, A; hyperplasia, R; hypoplasia, O; dystrophy, D; and dysplasia, Y.

volume, in which either the organelle volume is not low or the organelle number is not high. The *microorganelle* state (C) shows organelles with such a decreased single volume that *both* the organelle volume is low *and* the organelle number is high. The *megaorganelle* state (M) shows organelles with a moderately increased single volume, in which either the organelle volume is not high or the organelle number is not low. The *giant* organelle state (G) shows organelles with such an increased single volume that *both* the organelle volume is high *and* the organelle number is low.

Volume density of the mitochondria ($V_{VM}$) and surface density of the mitochondrial cristae (= inner membrane plus cristae = $S_{VMC}$) were employed for the determination of pathological and morphological states of the cristae mitochondriales.

## Acknowledgments

The authors wish to thank Prof. J. Staudinger and Prof. H. Tsukada for their discussion and critical review of this chapter. The personal research work performed by Dr. Riede has been supported by grants from the Deutsche Forschungsgemeinschaft (Az Ri 271/5).

## References

Ahlabo, I., and Barnard, T. (1971). *J. Histochem. Cytochem.* **19**, 670.
Allen, J. R., Carstens, L. A., Norback, D. H., and Loh, P. M. (1970). *Cancer Res.* **30**, 1857.
Arstila, A. U., Shelburne, J. D., and Trump, B. F. (1972). *Lab. Invest.* **27**, 317.
Augustin, P., Riede, U. N., and Sasse, D. (1978). *Path. Res. Pract.* **162**, 226.
Barrett, J. M., and Heidger, P. M. (1975). *Cell Tissue Res.* **157**, 283.
Baudhuin, P. (1969). *Ann. N.Y. Acad. Sci.* **168**, 214.
Beard, M. E. (1972). *J. Histochem. Cytochem.* **20**, 173.
Beard, M. E., and Novikoff, A. B. (1969). *J. Cell Biol.* **42**, 501.
Bell, L. T., and Hurley, L. S. (1973). *Lab. Invest.* **29**, 723.
Biempica, L., Kosower, N. S., and Novikoff, A. B. (1967). *Lab. Invest.* **17**, 171.
Black, V. H., and Bogart, B. J. (1973). *J. Cell Biol.* **57**, 345.
Boeck, P. (1972). *Z. Zellforsch. Mikrosk. Anat.* **133**, 131.
Borkowski, L. (1977). *Formale Logik*, p. 1. C. H. Beck, Munich.
Borst, P., Kroone, A. M., and Ruttenberger, G. J. C. M. (1967). Genetic elements: properties and function (D. Shuger, ed.), p. 81–116. Academic Press, New York.
Canonico, P. G., Rill, W., and Ayala, E. (1977). *Lab. Invest.* **37**, 479.
Chevremont, M., and Chevremont-Comhaire, S. (1953). *Arch. Biol.* **64**, 399.
Chiga, M., Reddy, J., and Svoboda, D. (1971). *Lab. Invest.* **25**, 49.
Christov, K., Riede, U. N., Helin, H., and Rohr, H. P. (1974). *Pathol. Eur.* **9**, 11.
De Duve, C. (1969). *Ann. N.Y. Acad. Sci.* **168**, 369.
De Duve, C. (1973). *J. Histochem. Cytochem.* **21**, 941.
De Duve, C., and Baudhuin, P. (1966). *Physiol. Rev.* **46**, 323.
De la Iglesia, F. A. (1969). *Acta Hepato-Splenol.* **16**, 141.

Dvorak, M. (1971). *Ergeb. Anat. Entwicklungsgech.* **45**, 4.
Essner, E. (1967). *Lab. Invest.* **17**, 71.
Essner, E. (1968). *J. Histochem. Cytochem.* **17**, 454.
Essner, E. (1969). *J. Histochem. Cytochem.* **17**, 454.
Essner, E. (1970). *J. Histochem. Cytochem.* **18**, 80.
Goeckerman, J. A., and Vigil, E. L. (1975). *J. Histochem. Cytochem.* **23**, 957.
Goldenberg, H., Hüttinger, M., Böck, P., and Kramar, R. (1975). *Histochemistry* **44**, 47.
Goldfischer, S., Roheim, P. S., and Edelstein, D. (1971). *Science* **173**, 65.
Goldfischer, S., Johnson, A. B., Essner, E., Moore, C., and Ritch, R. H. (1973). *J. Histochem. Cytochem.* **21**, 972.
Gonzales-Licea, A. (1970). *Lab. Invest.* **23**, 163.
Gonzales-Licea, A. (1971). *Lab. Invest.* **24**, 273.
Gonzales-Licea, A. (1972). *Lab. Invest.* **26**, 403.
Gorgas, K., Böck, P., and Wuketich, St. (1976). *Beitr. Pathol.* **159**, 371.
Gotoh, M., Griffin, C., and Hruban, Z. (1975). *Virchows Arch. B* **17**, 279.
Gulyas, B. J., and Yuan, L. D. (1977). *Cell. Tissue Res.* **179**, 357.
Hartman, H. A., and Tousismis, A. J. (1969). *Experientia* **25**, 1248.
Herzog, V., and Fahimi, H. D. (1976). *J. Mol. Cell Cardiol.* **8**, 271.
Hope, J. (1970). *J. Ultrastruct. Res.* **33**, 292.
Horvath, E., Kovacs, K., and Blaschek, J. A. (1971). *Acta Anat.* **79**, 44.
Hruban, Z., and Rechcigl, M. (1969). Microbodies and related particles. Morphology, Biochemistry and Physiology. *Int. Rev. Cytol.* (Suppl.) **1**.
Hruban, Z., Swift, H., and Rechcigl, M. (1965). *J. Nat. Cancer Inst.* **35**, 459.
Hruban, Z., Swift, H., and Slesers, A. (1966). *Lab. Invest.* **15**, 1884.
Hruban, Z., Mochizuki, Y., Morris, H. P., and Slesers, A. (1972a). *Lab. Invest.* **26**, 86.
Hruban, Z., Mochizuki, Y., Slesers, A., and Morris, H. P. (1972b). *Cancer Res.* **32**, 853.
Hruban, Z., Mochizuki, Y., Morris, H. P., and Slesers, A. (1973). *J. Nat. Cancer Inst.* **50**, 1487.
Hruban, Z., Gotoh, M., Slesers, A., and Chou, S. F. (1974a). *Lab. Invest.* **30**, 64.
Hruban, Z., Mochizuki, Y., Gotoh, M., Slesers, A., and Chou, S. F. (1974b). *Lab. Invest.* **30**, 474.
Huebner, H. H., and Rohr, H. P. (1969). *Beitr. Pathol.* **139**, 362.
Itabashi, M., Mochizuki, Y., and Tsukada, H. (1975). *Gann* **66**, 463.
Itabashi, M., Mochizuki, Y., and Tsukada, H. (1977). *Cancer Res.* **37**, 1035.
Jacobsen, N. O., and Jorgensen, F. (1973). *Z. Zellforsch. Mikrosk. Anat.* **136**, 479.
Kaneko, A., Sakamoto, S., Morita, M., and Onoe, T. (1969). *Tohoku J. Exp. Med.* **99**, 81.
Kolde, G., Roessner, A., and Themann, H. (1976). *Virchows Arch. B.* **22**, 73.
Kramar, R., Goldenberg, H., Böck, P., and Klobucar, N. (1974). *Histochemistry* **40**, 137.
Kuhn, C. (1968). *Z. Zellforsch. Mikrosk. Anat.* **90**, 554.
Lazarow, P. B., and de Duve, C. (1973). *J. Cell Biol.* **59**, 507.
Lazarow, P. B., and de Duve, C. (1976). *Proc. Nat. Acad. Sci. USA* **73**, 2043.
Lazarow, P. B. (1978). *J. Biol. Chem.* **253**, 1522.
Legg, P. G., and Wood, R. L. (1970a). *Histochemie* **22**, 262.
Legg, P. G., and Wood, R. L. (1970b). *J. Cell Biol.* **45**, 118.
Legg, P. G., and Wood, R. L. (1972). *Z. Zellforsch. Mikrosk. Anat.* **128**, 19.
Leighton, F., Poole, B., Lazarow, P. B., and de Duve, C. (1969). *J. Cell Biol.* **41**, 521.
Leuenberger, P. M., and Novikoff, A. B. (1975). *J. Cell Biol.* **65**, 324.
Locke, M., and McMahon, J. T. (1971). *J. Cell Biol.* **48**, 61.
Loud, A. V. (1968). *J. Cell Biol.* **37**, 27.
McGroarty, E., and Tolbert, N. E. (1973). *J. Histochem. Cytochem.* **21**, 949.
Magalhaes, M. M., and Magalhaes, M. C. (1971). *J. Ultrastruct. Res.* **37**, 563.

Malick, L. E. (1972). *J. Nat. Cancer Inst.* **49**, 1039.
Markstein, R. (1971). Ph.D. Thesis, Univ. of Basel, Switzerland.
Matthaei, C., Sasse, D., and Riede, U. N. (1976). *Beitr. Pathol.* **157**, 56.
Mochizuki, Y., Hruban, Z., Morris, H. P., Slesers, A., and Vigil, E. L. (1971). *Cancer Res.* **31**, 763.
Moody, D. E., and Reddy, J. K. (1978). *Am. J. Pathol.* **90**, 435.
Moore, G. W., Riede, U. N., and Sandritter, W. (1977). *J. Theor. Biol.* **65**, 633.
Novikoff, A. B., and Shin, W. Y. (1964). *J. Microsc.* **3**, 187.
Novikoff, A. B., Roheim, P. S., and Quintana, N. (1966). *Lab. Invest.* **15**, 27.
Novikoff, A. B., Novikoff, P. M., Davis, D., and Quintana, N. (1972). *J. Histochem. Cytochem.* **20**, 1006.
Novikoff, A. B., Novikoff, P. M., Nelson, Q., and Davis, C. (1973a). *J. Histochem. Cytochem.* **21**, 1010.
Novikoff, A. B., Novikoff, P. M., Quiantana, N., and Davis, C. (1973b). *J. Histochem. Cytochem.* **21**, 1010.
Novikoff, P. M., and Edelstein, D. (1977). *Lab. Invest.* **36**, 215.
Novikoff, P. M., and Novikoff, A. B. (1972). *J. Cell Biol.* **53**, 532.
Novikoff, P. M., Novikoff, A. B., Quintana, N., and Davis, C. (1973). *J. Histochem. Cytochem.* **21**, 540.
O'Conor, G. T., Tralka, T. S., Henson, E., and Vogel, C. L. (1972). *J. Nat. Cancer Inst.* **48**, 587.
Orsoni, J., Rohr, H. P., and Gloor, F. (1969). *Pathol. Eur.* **4**, 345.
Petrik, P. (1971). *J. Histochem. Cytochem.* **19**, 339.
Pfeifer, U. (1971). *Ergeb. Anat. Entwicklungsgesch.* **44**, 1.
Pfeifer, U. (1972). *Virchows Arch. B* **10**, 1.
Pfeifer, U. (1973). *Virchows Arch. B* **12**, 195.
Pfeifer, U., and Scheller, H. (1975). *J. Cell Biol.* **64**, 608.
Philipps, M. J., Langer, B., Stone, R., Fisher, M. M., and Ritchie, S. (1973). *J. Cell Biol.* **41**, 536.
Pipan, N., and Psenicnik, M. (1975). *Histochem.* **44**, 13.
Poole, B. (1969). *Ann. N.Y. Acad. Sci.* **168**, 229.
Poole, B., Leighton, F., and de Duve, C. (1969). *J. Cell Biol.* **41**, 536.
Psenicnik, M., and Pipan, N. (1977). *Virchows Arch. B* **25**, 161.
Quine, W. V. (1948) Theory of Deduction, parts I–IV pp. 54–81. Harvard Cooperative Study, Cambridge, Massachusetts.
Ramadan, M. A. (1971). *Virchows Arch. B* **9**, 1.
Reddy, J. (1973). *J. Histochem. Cytochem.* **21**, 967.
Reddy, J. (1974). *Am. J. Pathol.* **75**, 103.
Reddy, J., and Ohno, S. (1975). *XXVI Annu. Meeting Histochem. Soc., Atlantic City, New Jersey*, p. 14.
Reddy, J., and Svoboda, D. (1971). *Lab. Invest.* **24**, 74.
Reddy, J., and Svoboda, D. (1972). *J. Histochem. Cytochem.* **20**, 140.
Reddy, J., and Svoboda, D. (1973a). *Am. J. Pathol.* **70**, 421.
Reddy, J., and Svoboda, D. (1973b). *Virchows Arch. B* **14**, 83.
Reddy, J., Chiga, M., Bunyaratvej, S., and Svoboda, D. (1971). *J. Cell Biol.* **44**, 226.
Reddy, J., Tewari, J. P., Svoboda, D. J., and Malhotra, S. K. (1974a). *Lab. Invest.* **31**, 268.
Reddy, J., Azarnoff, D. L., Svoboda, D. J., and Prasad, J. D. (1974b). *J. Cell Biol.* **61**, 344.
Reddy, J., Rao, M. S., Moody, D. E., and Qureshi, S. A. (1976). *J. Histochem. Cytochem.* **24**, 1239.
Reid, I. M. (1973). *Exp. Mol. Pathol.* **18**, 316.

Reith, A., Brdiczka, D., Nolte, J., and Staudte, H. W. (1973). *Exp. Cell Res.* **77**, 1.
Riede, U. N., and Lorenz, H. (1976). *Beitr. Pathol.* **159**, 61.
Riede, U. N., and Nobmann, E. (1974). *Beitr. Pathol.* **153**, 319.
Riede, U. N., Markstein, R., Bianchi, L., and Rohr, H. P. (1971a). *Proc. R. Microsc. Soc.* **6**, 25.
Riede, U. N., Seebass, C., and Rohr, H. P. (1971b). *Virchows Arch. B* **9**, 16.
Riede, U. N., Roth, M., Molnar, J. J., Bianchi, L., and Rohr, H. P. (1971c). *Experientia* **27**, 794.
Riede, U. N., Straessle, H., Bianchi, L., and Rohr, H. P. (1971d). *Exp. Mol. Pathol.* **15**, 231.
Riede, U. N., Widmer, A. E., Bianchi, L., Molnar, J. J., and Rohr, H. P. (1971e). *Pathol. Eur.* **5**, 1.
Riede, U. N., Kuepfer, A., Rasser, Y., Rupp, S., and Rohr, H. P. (1972a). *Z. Zellforsch. Mikrosk. Anat.* **123**, 240.
Riede, U. N., Ettlin, Ch., von Allmen, R., and Rohr, H. P. (1972b). *Naunyn-Schmiedeberg's Arch. Exp. Pathol. Pharmakol.* **272**, 336.
Riede, U. N., Stitny, C., Althaus, S., and Rohr, H. P. (1972c). *Beitr. Pathol.* **145**, 24.
Riede, U. N., Konigsberger, H., Kaiser, B., Torhorst, J., and Rohr, H. P. (1972d). *Beitr. Pathol.* **147**, 175.
Riede, U. N., Torhorst, J., Rohr, H. P. (1973a). *Pathol. Eur.* **8**, 211.
Riede, U. N., Kaiser, W., Matt, C. von., and Rohr, H. P. (1973b). *Z. Krebsforsch.* **80**, 323.
Riede, U. N., Hodel, J., Matt, C. von, Rasser, Y., and Rohr, H. P. (1973c). *Beitr. Path.* **150**, 246.
Riede, U. N., Leibundgut, U., and Rohr, H. P. (1973d). *Beitr. Pathol.* **150**, 378.
Riede, U. N., Rasser, Y. M., and Rohr, H. P. (1974a). *Beitr. Pathol.* **152**, 383.
Riede, U. N., Kreutzer, W., Kiefer, G., and Sandritter, W. (1974b). *Beitr. Pathol.* **153**, 379.
Riede, U. N., Uhl, H., Seufer, G., Robausch, Th., Kiefer, G., and Sandritter, W. (1975a). *Beitr. Pathol.* **154**, 63.
Riede, U. N., Schmitz, E., Robausch, Th., Kiefer, G., Grünholz, D., and Sandritter, W. (1975b). *Beitr. Pathol.* **154**, 140.
Riede, U. N., Lobinger, A., Grünholz, D., Steimer, R., and Sandritter, W. (1976a). *Beitr. Pathol.* **157**, 391.
Riede, U. N., Vomstein, M., and Rohrbach, R. (1976b). *Beitr. Pathol.* **157**, 147.
Riede, U. N., Riede, P. R., Horn, R., Batthiany, R., Kiefer, G., and Sandritter, W. (1978). *Pathol. Res. Pract.* **162**, 398.
Riede, U. N., Spycher, M. A., and Gitzelmann, R. (1979). *Pathol. Res. Pract.* (in press).
Rigatuso, J. L., Legg, P. G., and Wood, R. L. (1970). *J. Histochem. Cytochem.* **18**, 893.
Rohr, H. P., and Riede, U. N. (1973). *Curr. Top. Pathol.* **58**, 1.
Rohr, H. P., Wirz, A., Henning, L. C., Riede, U. N., and Bianchi, L. (1971). *Lab. Invest.* **24**, 128.
Rohr, H. P., Brunner, H. R., Rasser, Y. M., Matt, C. von, and Riede, U. N. (1973). *Beitr. Pathol.* **149**, 347.
Rohr, U. A., Riede, U. N., and Rohr, H. P. (1974). *Beitr. Pathol.* **152**, 46.
Ruebner, B., Gonzales-Licea, A., and Slusser, R. J. (1967). *Gastroenterology* **53**, 18.
Scarpelli, D. L., and Trump, B. F. (1971). Cell Injury Research Pathology Education, Upjohn Co. Kalamazoo, Michigan.
Schmucker, D. L., Jones, A. L., and Mills, E. S. (1974). *J. Gerontology* **29**, 506.
Schneeberger, E. E. (1972). *Lab. Invest.* **27**, 581.
Scott, P. J., Visentin, L. P., and Allan, J. M. (1969). *Ann. N.Y. Acad. Sci.* **168**, 244.
Shnitka, T. K. (1966). *J. Ultrastruct. Res.* **16**, 598.
Staeubli, W., Hess, R., and Weibel, E. R. (1969). *J. Cell Biol.* **42**, 92.
Staeubli, W., Schweizer, W., Suter, J., and Weibel, C. R. (1977). *J. Cell Biol.* **74**, 665.

Sternlieb, I., and Quintana, N. (1977). *Lab. Invest.* **36**, 140.
Svoboda, D. J. (1978). *J. Cell Biol.* **78**, 810.
Svoboda, D. J., and Azarnoff, D. L. (1966). *J. Cell Biol.* **25**, 442.
Svoboda, D., and Reddy, J. (1972). *Am. J. Pathol.* **67**, 541.
Svoboda, D. J., and Reddy, J. K. (1974). *In* "Pathobiology Annual" (Ioachim, H. L., ed.), Vol. 4, p. 1. Appleton, New York.
Svoboda, D., Grady, H., and Azarnoff, D. (1967). *J. Cell Biol.* **35**, 127.
Svoboda, D., Azarnoff, D., and Reddy, J. (1969). *J. Cell Biol.* **40**, 734.
Szabo, S., and Avers, C. J. (1969). *Ann. N.Y. Acad. Sci.* **168**, 302.
Tandler, B., and Hoppel, C. L. (1970). *Z. Zellforsch. Mikrosk. Anat.* **110**, 166.
Tice, L. W., and Wollman, S. H. (1974). *Endocrinology* **94**, 1555.
Tiedemann, K. (1972). *Z. Zellforsch. Mikrosk. Anat.* **133**, 141.
Trump, B. F., and Arstila, A. U., eds. (1975). Pathobiology of Cell Membranes, p. 1. Academic Press, New York.
Trump, B. F., Goldblatt, P. J., and Stowell, R. E. (1965). *Lab. Invest.* **14**, 1946.
Tsukada, H., Mochizuki, Y., and Konishi, T. (1968). *J. Cell Biol.* **37**, 231.
Tsukada, H., Koyama, S., Gotoh, M., and Tadano, H. (1971). *J. Ultrastruct. Res.* **36**, 159.
Tsukada, H., Mochizuki, Y., and Gotoh, M. (1975a). *J. Nat. Cancer Inst.* **54**, 519.
Tsukada, H., Mochizuki, Y., Gotoh, M., and Morris, H. P. (1975b). *J. Nat. Cancer Inst.* **55**, 153.

Tsukada, H., Mochizuki, Y., and Gotoh, M. (1977a). *J. Nat. Cance Inst.* **54**, 519.
Tsukada, H., Mochizuki, Y., Itabashi, M., and Gotoh, M. (1977b). *J. Nat. Cancer Inst.* **55**, 153.
Tsukada, H., Mochizuki, Y., and Gotoh (1978). *In* "Morris Hepatomas" (H. P. Morris and W. R. Criss, eds.), p. 331. Plenum, New York.
Vogel, W. H., Snyder, R., and Schulman, M. P. (1964). *Biochim. Biophys. Acta* **85**, 164.
Weibel, E. R., Stäubli, W., Gnägi, H. R., and Hess, F. A. (1969). *J. Cell Biol.* **42**, 68.
Wiener, J., Loud, A. V., Kimberg, D. V., and Spiro, D. (1968). *J. Cell Biol.* **37**, 47.
Wilson, J. W., and Leduc, E. (1963). *J. Cell Biol.* **16**, 218.
Wood, R. L., and Legg, P. G. (1970). *J. Cell Biol.* **45**, 576.
Woyke, S., Domagla, W., and Olzewski, W. (1974). *Acta Cytol.* **18**, 130.
Yokota, S., and Nagata, T. (1974). *Histochemistry* **39**, 243.
Zimmerli, U. (1971). M.D. Thesis, Univ. of Basel, Switzerland.

# EDITORS' SUMMARY TO CHAPTER IV

Although it is only a little over 10 years since the presence of new organelles, microbodies, or peroxisomes were described in the liver and kidney already a vast amount of knowledge has been collected on the pathological alterations in these organelles. These organelles, interestingly, have also become useful as important markers in cells such as liver or kidney where they have a particular ultrastructure. This has become of great value in diagnostic electron microscopy in defining such cells with minimal clues. The microbodies represent an important clue to the identity of a proximal tubular cell or a hepatocyte. Furthermore, even well-differentiated tumors of these organs may retain the microbodies and, therefore, in a metastasis, can be extremely important in recognizing the organ of origin. This information on the pathology of peroxisomes is to a certain extent in sharp contrast to our knowledge on their normal functions since even now their exact role in cell metabolism is not well-defined. Only by examining these organelles in pathological processes will we discover clues about their role in normal cells. Thus knowledge in cellular pathology precedes knowledge in cell biology.

From the point of view of toxicologists and diagnostic pathologists these small organelles have interest and importance. As shown in this chapter by Dr. Riede and co-authors, the peroxisomes are extremely sensitive indicators on subtle changes in the metabolism in the cells which possess peroxisomes. Fortunately these include such important organs as liver and kidney. The sensitivity of these organelles depends upon their short half-life of only two days as well as their characteristic structure. Thus it is easier to observe changes in their structure without time-consuming morphometrical analysis changes in their number and volume fraction. Since these organelles can also be specifically stained by the electron microscopic histochemical diaminobenzidine method, it is also possible to get a fairly accurate estimate of the number of these organelles, for instance, in the human liver biopsy sample. Perhaps the sensitivity of these organelles to alterations is also due to the fact that their metabolism is closely coupled both with the metabolism of the mitochondria and endoplasmic reticulum. As pointed out by Dr. Riede and co-authors, the presence or absence of peroxisomes as well as changes in their structure also give important clues in cancer diagnosis since, for instance, even minor differentiation will affect these parameters.

From the point of practical medicine it is of considerable interest that all compounds which induce peroxisome proliferation, seemingly also possess hypolipidemic properties (Reddy *et al.*, 1976a). Since the drugs used in controlling hyperlipidemic states in man are of great importance and since many of them also induce peroxisome proliferation, much more research is needed in order to evaluate the biological and pathological significance of peroxisome proliferation. At least some of the compounds, such as nefenopin, have been shown by Reddy *et al.* (1976b) to induce hepatocellular carcinomas in acatalasemic mice.

## References

Reddy, J. K., Moody, D. E., Azarnoff, D. L., and Rao, M. S. (1976a). *Life Sci.* **18**, 941.
Reddy, J. K., Rao, M. S., and Moody, D. E. (1976b). *Cancer Res.* **36**, 1211.

# CHAPTER V

# ORGANELLE TURNOVER

**M. Locke and Janet V. Collins**



## I. Introduction—The Topological Problem Posed by Autophagy

The most striking feature of the arrangement of the membranes of cells is that there are no free edges. All the membranes are at interfaces separating cytoplasm from external compartments. Topologically a cell is composed of nucleocytoplasmic and mitochondrial compartments together with transiently isolated external spaces. The bounding membranes may be classified according to the following scheme (Fig. 1). The topological arrangement poses a problem for cells undergoing partial autolysis prior to reconstruction. Components about to be destroyed are free in the cytoplasm. During destruction many of them must be spatially segregated in an environment appropriate for lysis. How can doomed organelles pass into a lytic environment without releasing the lytic enzymes into the cell? A solution to this



ISBN 0-12-701502-7

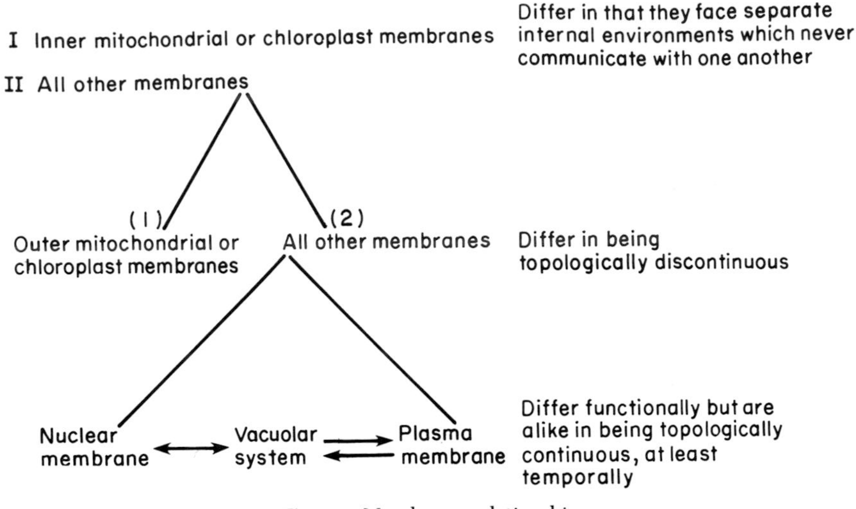


**Fig. 1.** Membrane relationships.

topological problem, (de Duve and Wattiaux, 1966) came from studies on metamorphosing insect fat body (Locke and Collins, 1965). Organelles are made topologically external to the cytoplasm by envelopment prior to the addition of lytic enzymes. The cells of *Calpodes ethlius* larvae (the caterpillar of a hesperiid or skipper butterfly) are particularly useful for studies on the life history of organelles. The precise and synchronous timing of events in development makes it possible to find and study evanescent stages only rarely observed in the more commonly studied vertebrate cells which are in a dynamic state of loss and replacement. This chapter describes the sequential remodeling of insect cells and suggests a general mechanism for the isolation of organelles which are lost through autophagy.

## II. Insects as Experimental Material—The Precision and Synchrony of Changes in *Calpodes*

The dramatic changes which take place during the life cycle of holometabolous insects—the progression from the egg through several larval stages to the pupa and adult—are partly brought about by the death and replacement of cell types and partly by partial cell autolysis and the replacement of organelles. Those cells which modify their structure and function for each stage provide ideal material for studying general problems of cell modeling and remodeling. With this in mind, several tissues were studied during molting and metamorphosis in *Calpodes ethlius* Stoll (Lepidoptera, Hesperiidae).

*Calpodes* is a skipper butterfly easily and quickly reared in the laboratory.

The life cycle takes about six weeks from egg to egg. One of its particular advantages is the precision with which it develops. At 22°C, 192 ± 13 hr elapse between the 4th to 5th ecdysis and the emergence of the pupa from the sloughed 5th-stage skin (Locke, 1970a).

This 5th stage is divided into three phases of development. The first phase spans the period from ecdysis to 66 hr into the 5th stage (M + 66 hr) by which time the prothoracic glands no longer need thoracotropic hormone from the brain to induce molting. The second phase ends at M + 156 hr when the tissues become independent of the prothoracic glands in completing the molt to the pupal stage. During the first phase (M + 0 to M + 66 hr) the cells in most tissues acquire a complement of organelles appropriate for the larval syntheses taking place during the second phase. At M + 156 hr there is a switch from larval syntheses to overt changes related to pupation, including the sequential autophagy of specific larval organelles.

Thus, we may follow several cell types through a sequence involving the loss of 4th-stage organelles, formation, growth, and function of 5th-stage organelles, followed by their destruction and the formation of pupal and adult structures. In particular, by concentrating on periods of a few hours out of the whole sequence, we may study the mechanism by which organelles are removed from a cell during autophagy. Figure 2 is a guide to the temporal sequence, particularly for events in the fat body. The fat body contains a single cell type which is functionally diverse: at various times it stores and mobilizes fat, glycogen, and protein which it may have synthesized or sequestered. It also plays a part in excretion and urate metabolism, for the

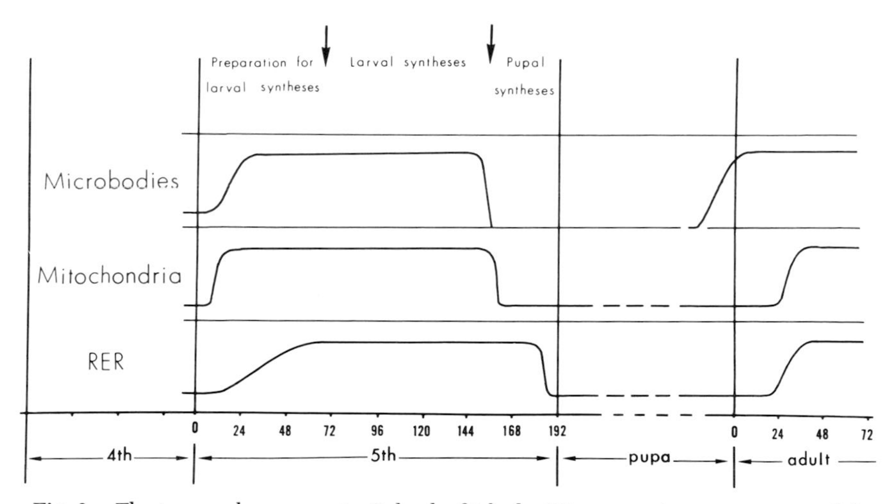


**Fig. 2.** The temporal sequence in *Calpodes* fat body. Diagrammatic representation of the timing of changes in the microbodies, mitochondria, and rough endoplasmic reticulum (RER) during the 4th and 5th larval stages, the pupa, and the adult. Changes in the adult from Larsen (1970, 1971).

cells have catalase and urate oxidase in microbodies. The fat body resembles vertebrate liver structurally and functionally, but differs in the temporal separation of hormonally induced phases of growth, function, and partial autolysis. This difference from the liver makes the fat body uniquely suitable for studying problems of cell development.

## III. Autophagy of Larval Organelles in the Fat Body

### A. *The Temporal Sequence*

The last 36 hr of development in the 5th stage (M + 156 to M + 192 hr) is intrinsically controlled from within the cells, that is, pupation continues even in the absence of the prothoracic glands. By this time, the fat body has begun to change from larval occupations such as the synthesis of blood proteins to overt preparations for pupation. The first sign of the switch is the envelopment of the microbodies (peroxisomes) by paired membranes, the *isolation membranes,* which takes place in about 6 hr (M + 150 to M + 156 hr). A vesicle appears to flatten at the surface of the microbody (Fig. 3) and isolation proceeds (Fig. 4) as more vesicles are added to the substance of the envelope until investment is complete (Figs. 5, 6, and 7). Such a granule, in which organelles are completely recognizable, and which is surrounded by two membranes is called an *isolation body.* Several isolation bodies so formed come together with a fusion of their outer isolation membranes. Acid phosphatase is not present initially between the isolation membranes but can be detected in small vesicles presumed to come from the Golgi complex. After their fusion with the isolation bodies, the inner isolation membranes break down and an autophagic vacuole is formed (Fig. 8).

At the beginning of the autophagic sequence prior to pupation, isolation membranes only envelop microbodies with few or no mitochondria being destroyed. However, after a few hours (M + 154 to M + 160 hr) many, but not all, mitochondria suffer a similar fate. A vesicle flattens at the surface of a mitochondrion and enlarges until the mitochondrion is completely isolated by two unit membranes and the space between them (Figs. 9–13). The frequent presence of microvesicles near the enlarging envelope suggests that they contribute to its growth (Fig. 9). The isolation bodies fuse together in aggregates of almost pure mitochondria with only rare traces of rough endoplasmic reticulum (RER) (Fig. 13). The contents break down within about

---

**Figs. 3 to 7.** The isolation of microbodies. (Figures 4 and 6 from Locke and McMahon, 1971.)

**Fig. 3.** A vesicle adheres to the side of the microbody. ×87,000.

**Fig. 4.** At this time only microbodies are enveloped even though the isolation membrane may be as close to mitochondria. ×64,000.

**Fig. 5.** The inner isolation membrane is very closely apposed to the microbody membrane. ×56,000. (From Locke and Sykes, 1975.)

**Fig. 6.** A microbody may be isolated with very little other material. ×64,000.

**Fig. 7.** The contents of the envelope react intensely after treatment with hot aqueous $OsO_4$, ×56,000. (From Locke and Sykes, 1975.)

**Fig. 8.** Autophagy of the microbodies. The events leading to microbody destruction from M + 150 to M +156 hr. Microbodies are isolated although mitochondria are present.

**Figs. 9 to 13.** The isolation of mitochondria.

**Fig. 9.** A vesicle adheres to a mitochondrion and grows to an envelope by the addition of microvesicles. ×112,000. (From Locke and Sykes, 1975.)

**Fig. 10.** The envelope is closely adherent to the outer membrane of the mitochondrion. ×56,000.

**Fig. 11.** The contents of the envelope react intensely after hot aqueous $OsO_4$. ×56,000. (From Locke and Sykes, 1975.)

**Fig. 12.** An isolated mitochondrion or mitochondrial isolation body. ×107,000. (From Locke and Sykes, 1975.)

**Fig. 13.** An autophagic vacuole almost entirely composed of isolated mitochondria. ×41,000. Isolation membranes, im; mitochondrial isolation body, ib. (From Locke and Collins, 1965.)

**Fig. 14.** Autophagy of the mitochondria. The events leading to mitochondrial destruction from M + 154 to M + 160 hr. Although there is an abundant rough endoplasmic reticulum (RER), only the mitochondria are invested. Not all the mitochondria are destroyed. Some survive to repopulate the cell in the adult.

**Figs. 15 to 19.** The isolation of the rough endoplasmic reticulum (RER) and the origin of the isolation membranes from vesicles. Golgi complex, g; isolation membranes, im; isolation body, ib. (Figures 16, 17, and 18 from Locke and Collins, 1965.)

**Fig. 15.** Isolation membranes segregating regions of RER to form isolation bodies of about the same size as mitochondria. The inner isolation membranes follow the contour of the RER. ×31,000.

**Fig. 16.** ×21,000.

**Fig. 17.** ×31,000. The isolation membranes appear to arise from vesicles in the transition vesicle zone of the RER-GC.

**Fig. 18.** The RER isolation bodies fuse together to give autophagic vacuoles containing almost nothing but RER. ×41,000.

**Fig. 19.** The contents of the isolating envelope react intensely after hot aqueous $OsO_4$. ×67,000. (From Locke and Sykes, 1975.)

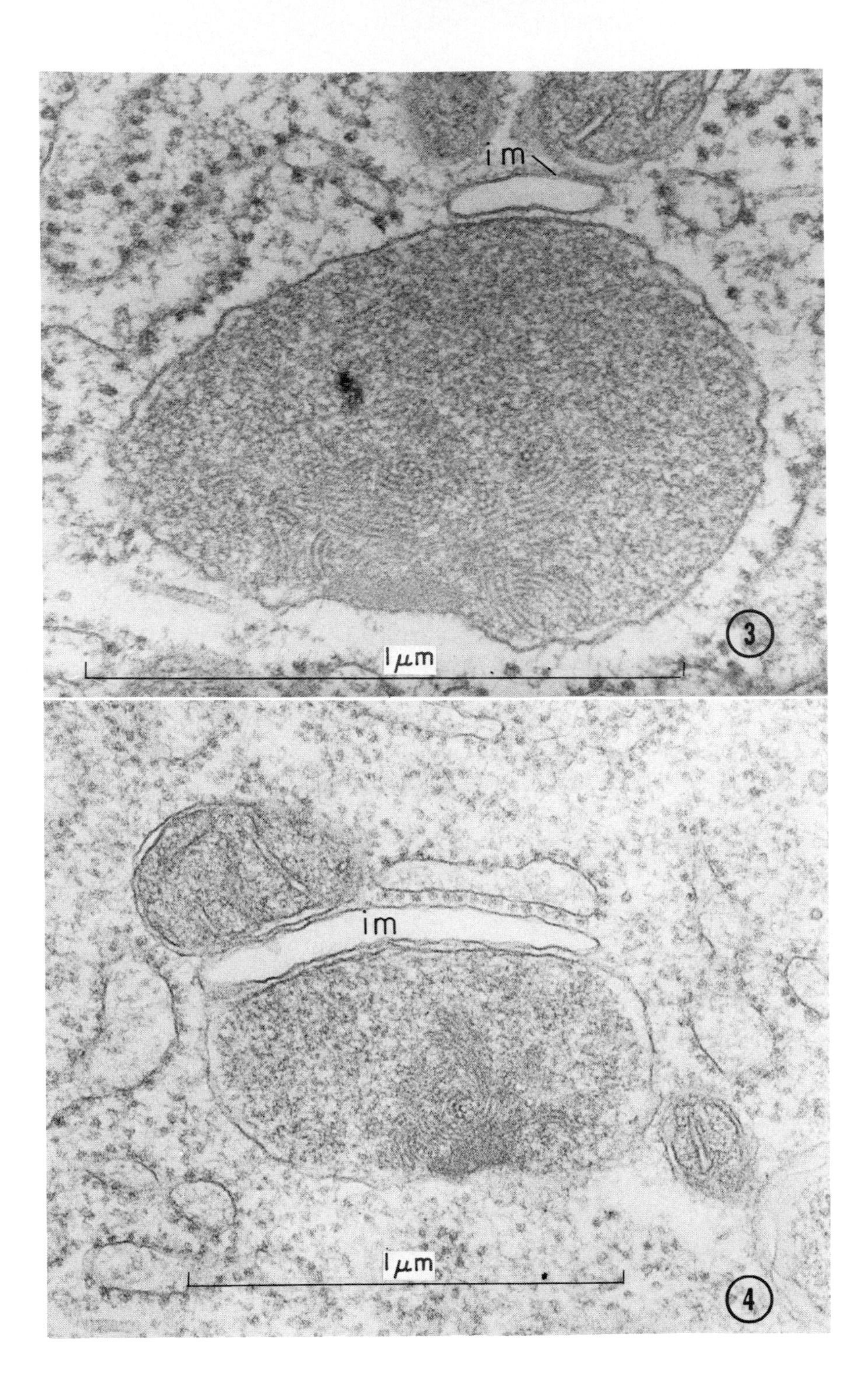
im
1 μm
3
im
1 μm
4

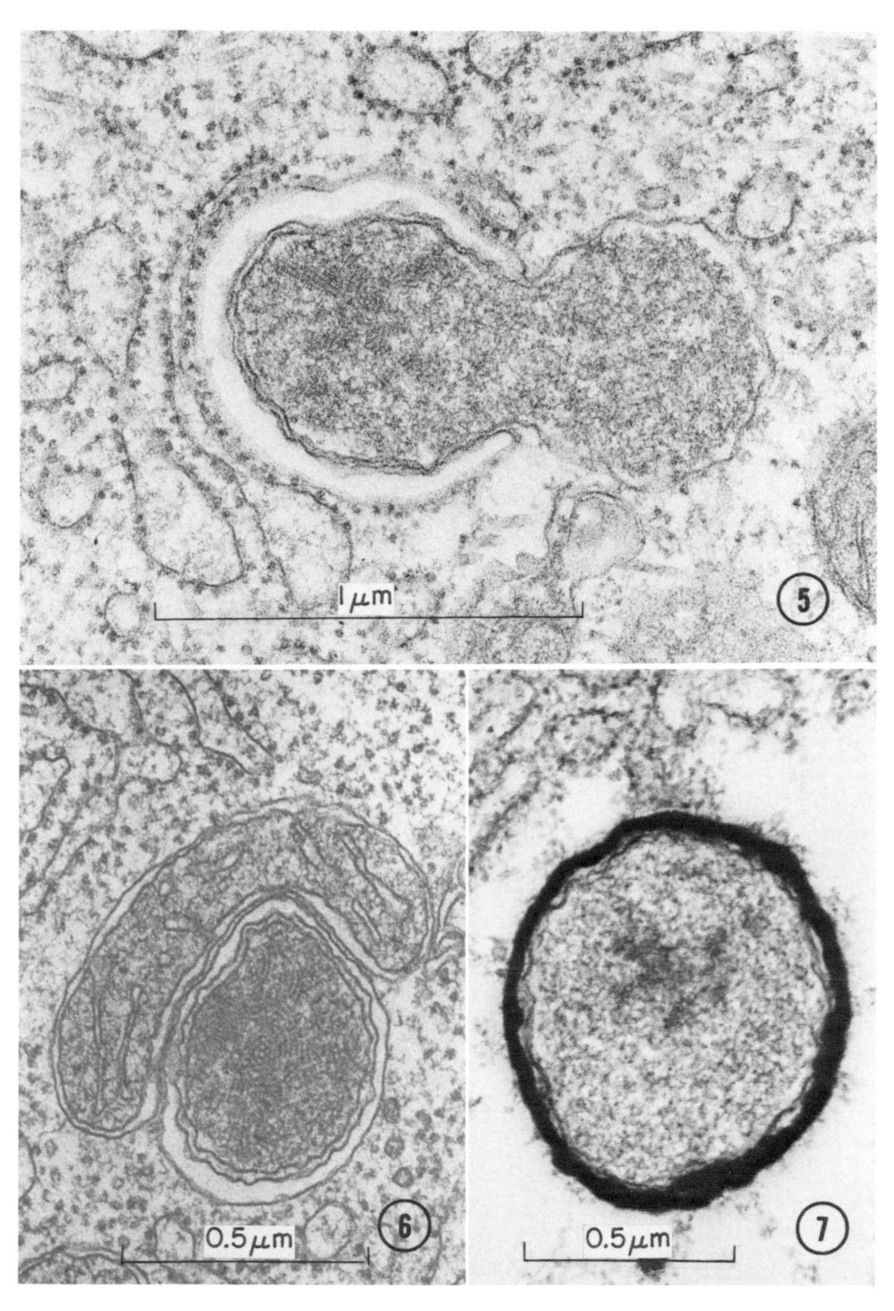


Figs. 3–7. Explanations on p. 227.

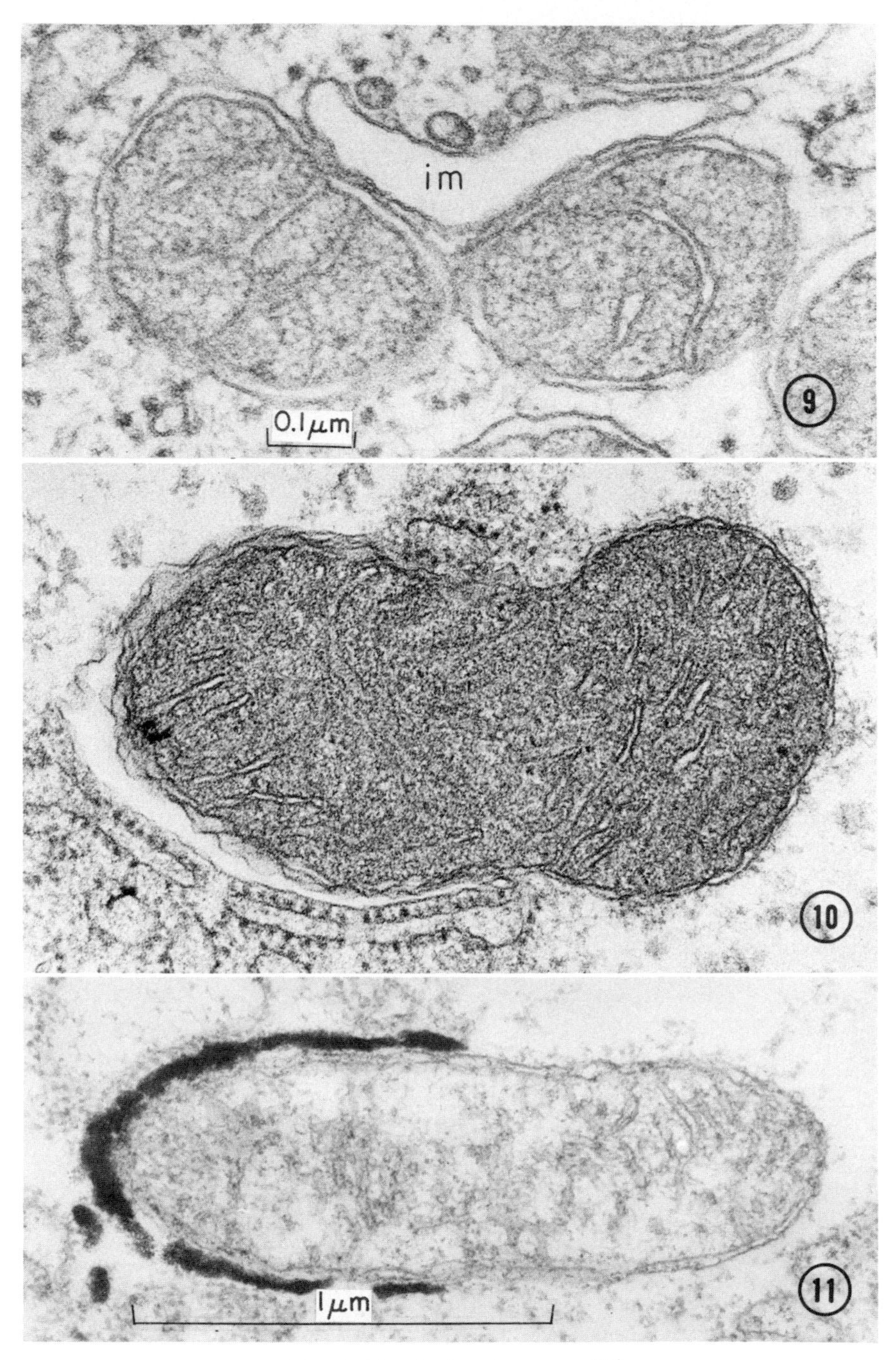


**Figs. 8–12.** Explanations on p. 227.

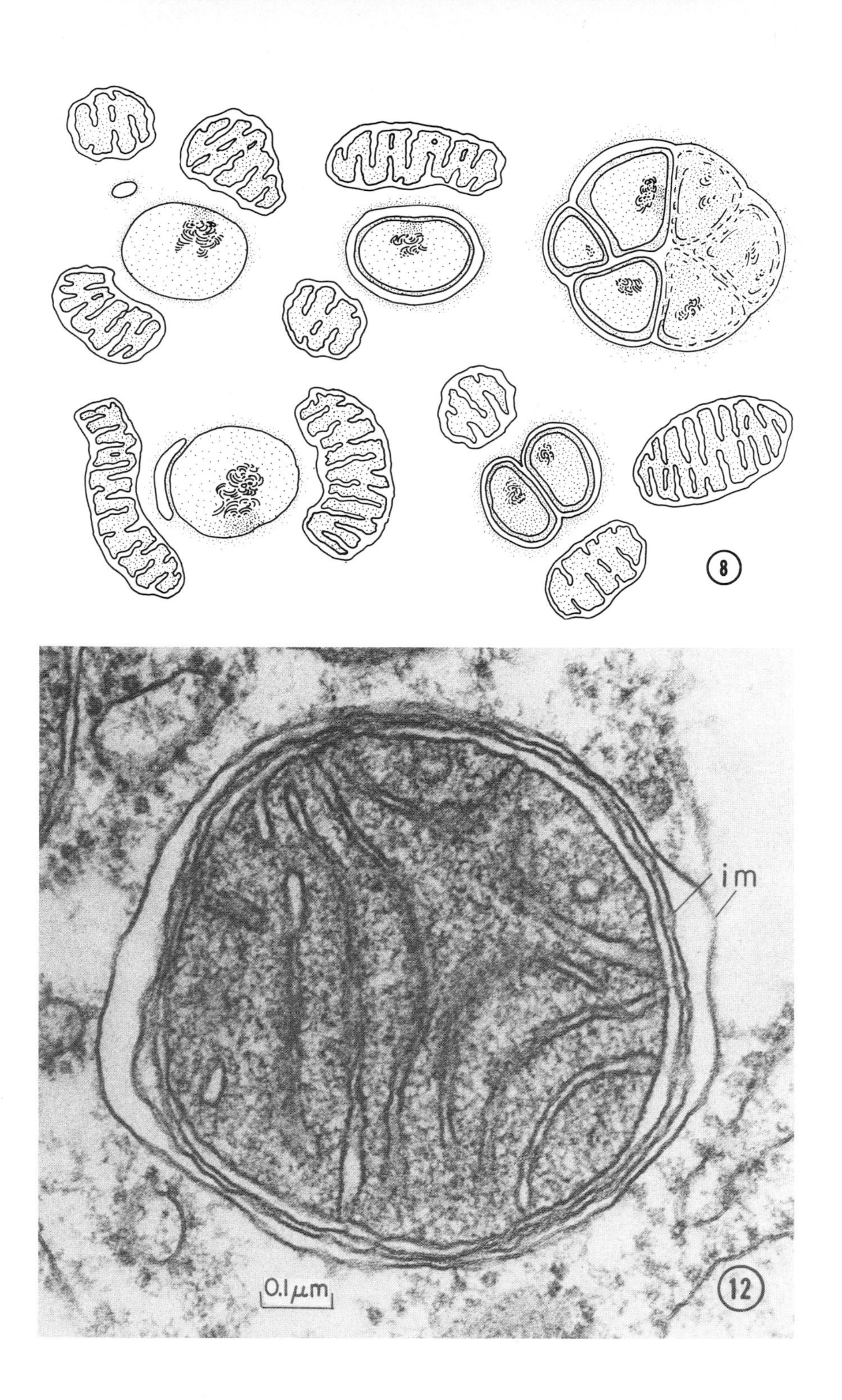

8
im
0.1 μm
12

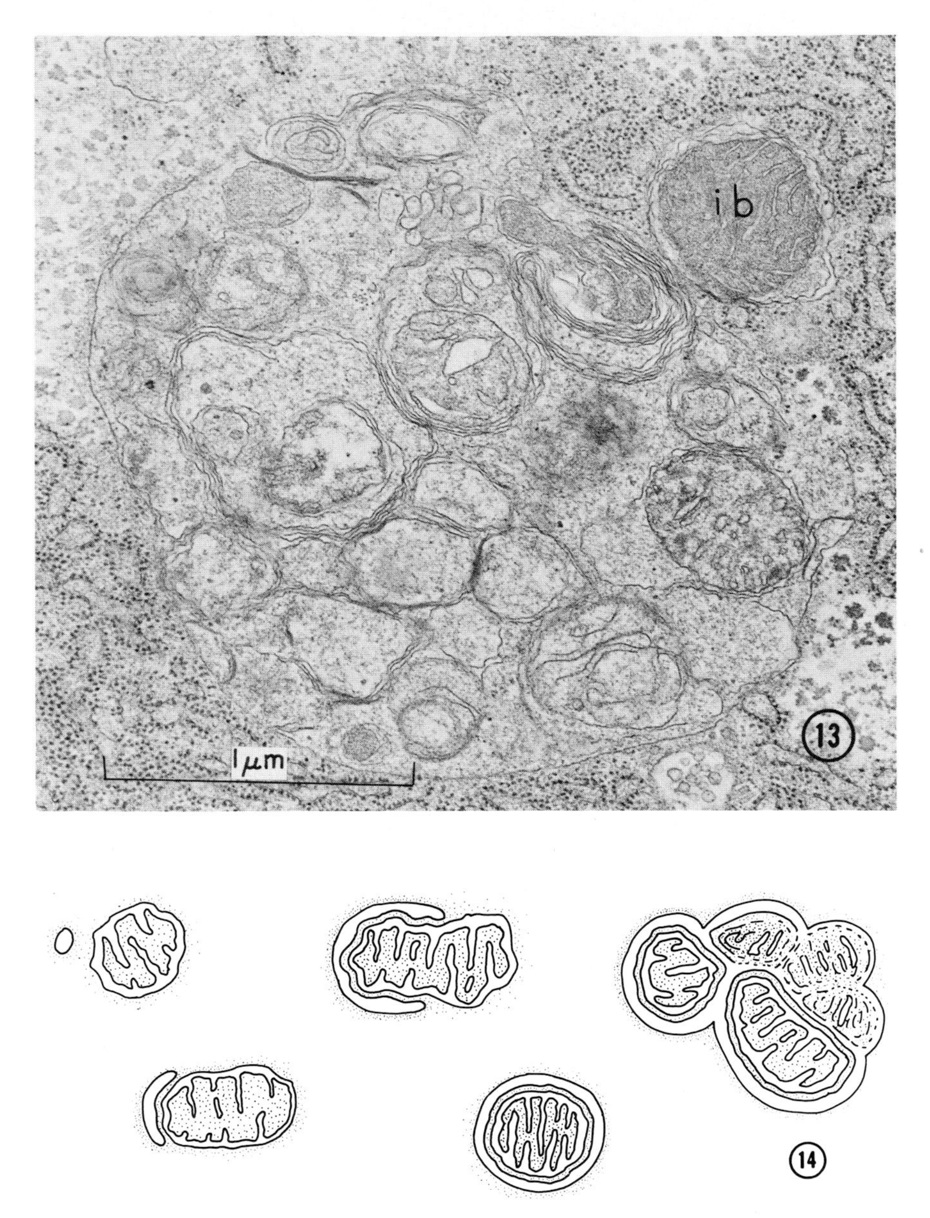


Figs. 13–17. Explanations on p. 227.

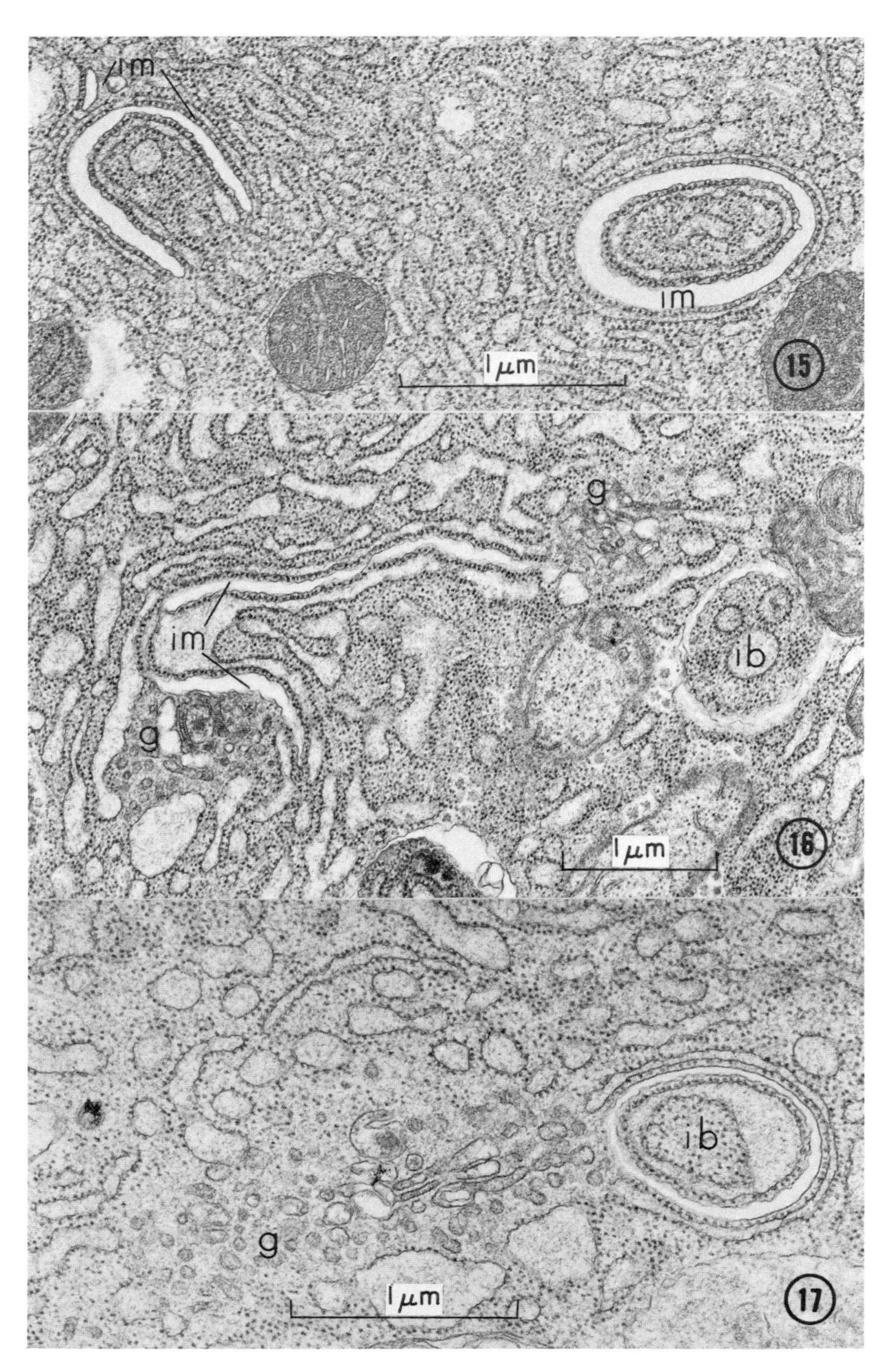
im
im
1 μm
15
g
im
ib
g
1 μm
16
ib
g
1 μm
17

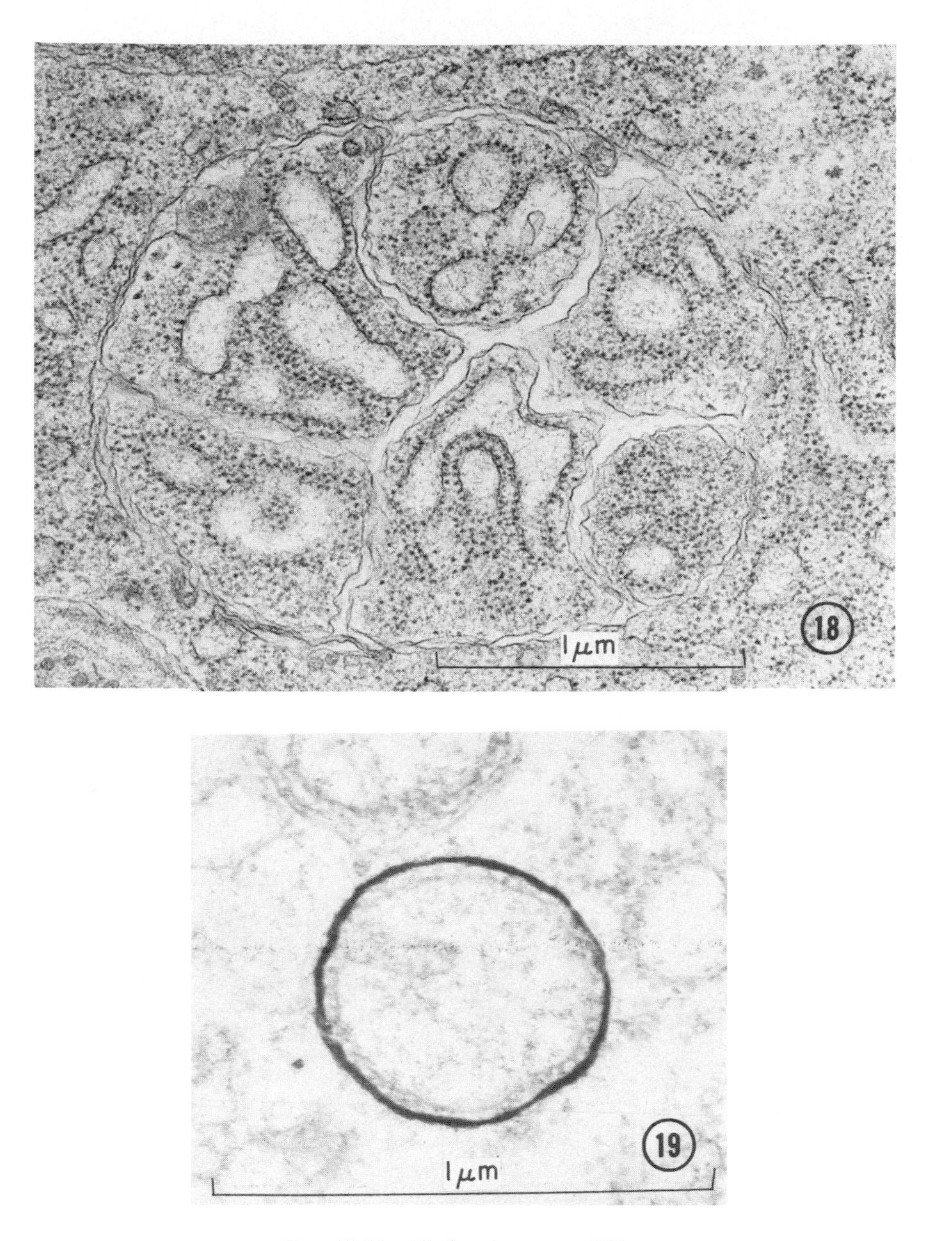


Figs. 18–19. Explanations on p. 227.

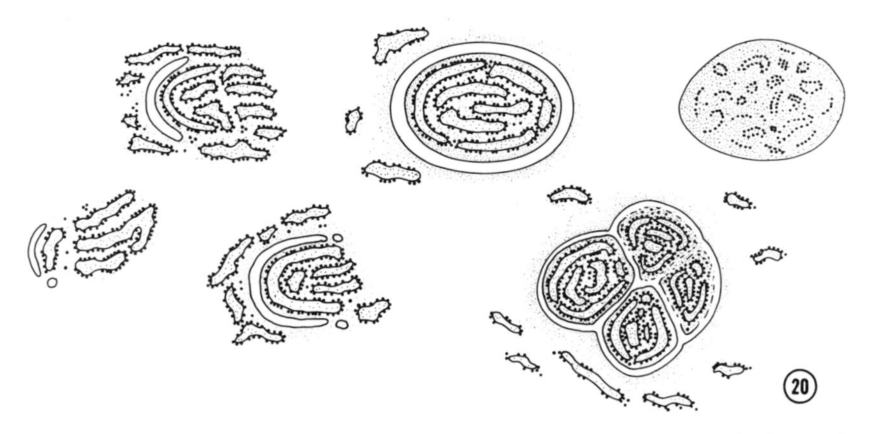


**Fig. 20.** Autophagy of the rough endoplasmic reticulum (RER). The events leading to the sequestration of the RER from M + 180 to M + 186 hr. The contents of the autophagic vacuoles formed from the RER at this time are stored.

8 hr so that few acid phosphatase positive granules are left by M + 162 hr (Fig. 14).

The fat body next undertakes a phase of heterophagy (M + 162 to M + 180) (Locke and Collins, 1968). Massive quantities of blood proteins are taken up and stored in granules occupying as much as 25% of the volume of the cell. Before and after this time some protein is pinocytosed and broken down in multivesicular bodies but not stored.

The formation of protein storage granules is followed by a phase of isolation and autophagy of most of the RER (M + 180 to M + 186 hr). The RER is isolated in masses of about the same size as the isolation bodies formed around mitochondria (Figs. 15, 16, 17, and 19). The inner isolation membrane closely follows the contours of the ribosomes on the RER while the outer isolation membrane has a smoother profile. These isolation bodies fuse in groups containing little but RER (Fig. 18). Some of the RER may break down rather quickly and become indistinguishable from the granules of sequestered protein with which RER autophagosomes often fuse, but more commonly the RER condenses leaving ribosomes embedded in a solid matrix. These granules are used up during adult development in the pupa (Fig. 20).

### B. *The Specificity of Isolation*

The changing composition of autophagic vacuoles during development implies that there is some specificity in the isolating mechanism. Mitochondria make up less than 2% of the cell volume at the beginning of autophagy and they outnumber the microbodies by 10 to 1. Nevertheless the earliest stages of autophagy involve the selective envelopment of microbodies with few or no mitochondria being destroyed. In the presence of abundant mitochondria, isolation membranes are only associated with microbodies. The isolation membranes may even touch nearby mitochondria but they still envelop only the microbodies. A few hours later many mitochondria are isolated and incorporated in autophagic vacuoles even though they are a minor component of the mass of the cells. The presence of only small amounts of RER in the autophagic vacuoles at this time is not because isolation membranes can only invest discrete membrane-bounded organelles. After heterophagy of blood proteins the RER is the main target for isolation. The selection of these three components for isolation might be less specific than appears from their relative proportions in the cell. The bulk of the cell is made up of lipid droplets, masses of glycogen and the nucleus which may all be precluded from investment by their size and texture. The remainder of the cell is mainly RER. At the time when microbodies and mitochondria are lost the RER might also be resistant to being broken up into small masses because of its continuity and texture. The discrete components like mitochondria and microbodies are usually associated with one another in spaces often related to Golgi complexes. They appear as if they were freely moving within fluid channels between the lipid and glycogen and through or at the edges of the RER. Within this system there might be ample opportunity for vesicles to approach organelles in the first stages of isolation. The problem is similar to the control of transition vesicle movement between the RER and Golgi complexes, or between secretory vesicles and the plasma membrane. The difference is that the vesicles flatten close to the membrane of the organelle to be invested, rather than fusing with it. There may be a parallel with some membranes of the smooth endoplasmic reticulum which confront regions of plasma membrane or mitochondria (Locke, 1970b). The interraction may be a property of the vesicle membrane or the organelle or both.

In the 5th-stage larva the main function of the RER in the fat body is the synthesis and secretion of the blood proteins which are later sequestered. Protein synthesis stops at the time of sequestration and the Golgi complexes change their form. The extended curtains of RER tend to break up into smaller islands, which perhaps permit isolation. The rarity of isolated mitochondria at this time might be related to the relative proportions of RER now available for autophagy compared with surviving mitochondria.

Thus, although there is an organelle-specific sequence of isolation and autophagy this could be accomplished at least partly by the sequential availability of organelles or fragments suitable for isolation, rather than only through a specific recognition process between isolation membrane and isolate. However, the preferential investment of microbodies by isolation membranes which may be as close to mitochondria implies some selective properties on the part of isolation membranes (Fig. 4).

### C. *The Form and Contents of Isolating Envelopes*

The most characteristic feature of the isolating envelope in a conventionally fixed and stained preparation is its empty appearance. It does not contain acid phosphatase. The lumen is free of any inclusions or wisps of density, making it appear even lighter than the resin of the section. This suggested that the lumen might not be empty but might be occluded with something which does not usually stain.

Observations on material reacted with osmium tetroxide (Friend, 1969; Friend and Brassil, 1970) confirmed this idea. Tissues were fixed in unbuffered 2–4% aqueous osmium tetroxide and reacted in 2% osmium tetroxide for 48 hr at 40°C. After this treatment the lumen of the isolating envelope is evenly and intensely stained (Figs. 7, 11, and 19). The reaction is precisely localized. In the fat body at this time the only other cell components which react are microvesicles believed to be transition vesicles between the RER and Golgi complex (Figs. 21 and 22), a few vesicles of RER adjacent to lipid droplets, and occasionally some saccules of the Golgi complex. We may conclude that the isolation envelopes have contents. If the contents have some rigidity it may help to explain the rather constant separation between inner and outer membranes and the flattened rather than spherical form. An alternative explanation for the flattened form would be that the cytoplasmic face of the membranes adheres to microbody or mitochondrial membrane or to ribosomes on RER. If most newly added isolation membrane has such an affinity, then the appearance of creeping around an organelle to cause investment could be explained. In favor of this hypothesis is the precision with which the profile of the inner isolation membrane duplicates that of the organelle which it is investing. Figures 5, 10, 15, and 17 show that a remarkably constant separation of about 120 Å is maintained between the densely stained edges of an isolating membrane and the object invested, be it membrane or ribosome. The absolute values mean little, since the measurement is only a record of the separation between points having a particular electron density which varies with the conditions of preparation. Separation does not imply a gap in the sense that nothing of interest is there. However, the constancy of the separation is probably significant. It is difficult to explain without supposing that rather uniform forces operate between the two surfaces.

### D. *The Origin of the Isolation Membranes*

The morphological criterion for calling paired membranes isolation membranes, is that they appear to separate an organelle or section of cytoplasm from the rest of the cell. The certainty of the identification may be confused by the plane of section and depends upon the degree of completion of the envelopment. In some planes a structure only partly enclosed could appear

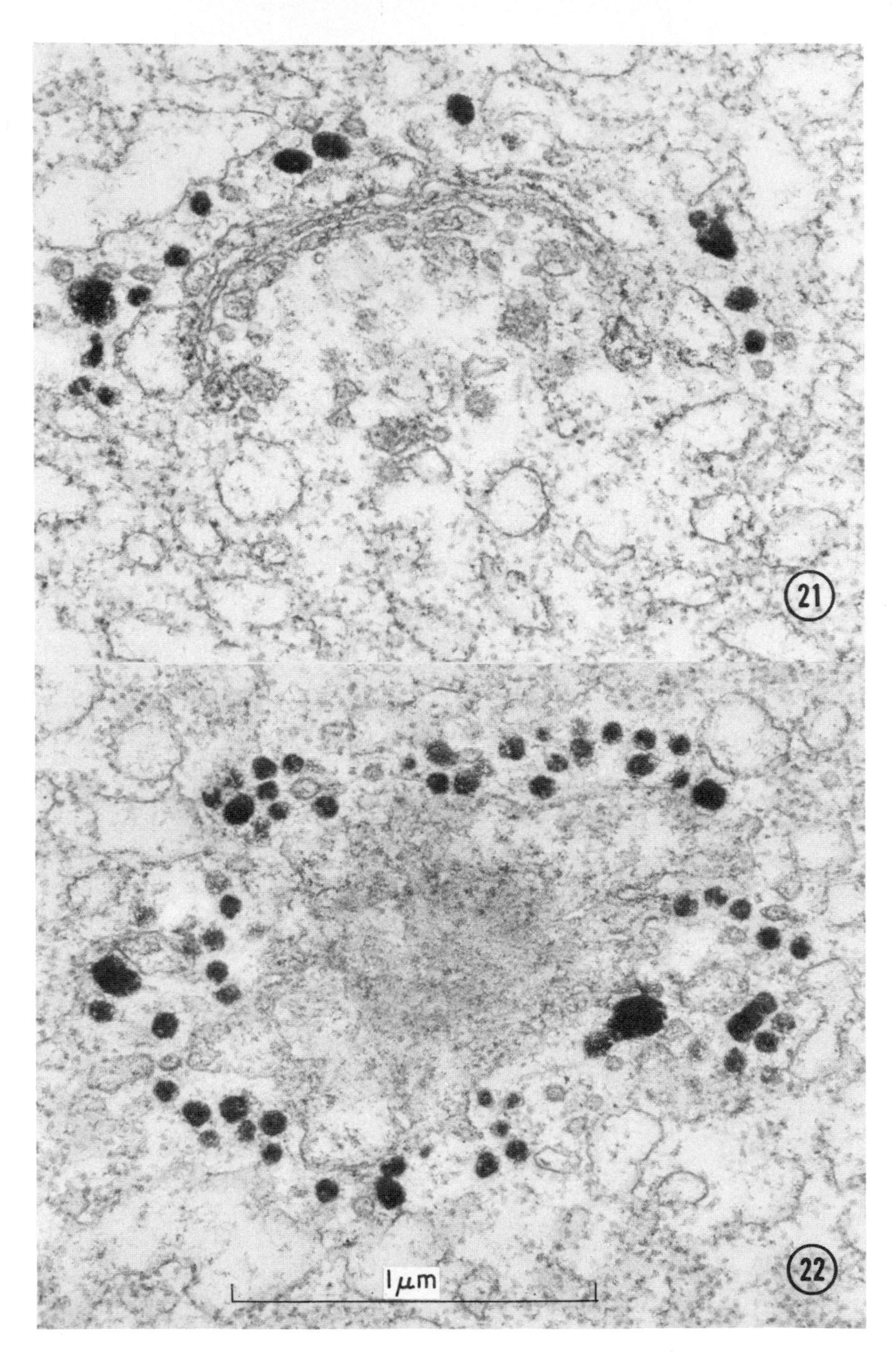


**Figs. 21 and 22.** The contents of the transition vesicles of the Golgi complexes react intensely after treatment with hot aqueous $OsO_4$ as do the contents of the isolation membranes. ×50,000. (From Locke and Sykes, 1975.)

completely isolated. It is assumed that incompletely enveloped organelles are stages in isolation. Some organelles may be partly enveloped transiently in relation to entirely different functions. For example, growing 5th-instar microbodies have confronting RER cisternae (Locke and McMahon, 1971), and mitochondria in the prothoracic gland have confronting SER cisternae (Locke, 1970b). The problem of identifying isolation-membrane antecedents before they have begun to enclose an organelle is even more difficult. Early stages of isolation membrane formation could be structurally indistinguishable from microvesicles and smooth endoplasmic reticulum. The only other criterion for isolation membranes is the osmiophilia of the lumen after the hot osmium tetroxide treatment. The two criteria for the recognition of isolation membranes agree as to their origin.

The morphology of the fat body suggests that isolation membranes arise from microvesicles (Fig. 9). There is often a continuous field of vesicles from an isolation membrane to the convex face of the Golgi complex where they appear to be similar to those in the position of the transition vesicles (Figs. 16 and 17). These vesicles, the cisterna of the isolation envelope and the transition vesicles all react with osmium tetroxide and are almost the only cell components to react in this way. We may tentatively summarize the relation between the Golgi complex and isolation membranes as in Fig. 23.

The localization of osmiophilic material in the lumen of the isolating envelope confirms that the membranes enclose some cell product and demonstrates their relationship to the Golgi complex and endoplasmic reticulum (ER) transitional elements. The ER has also been suggested as a membrane source in the formation of autophagic vacuoles in other tissues, notably rat liver (Arstila and Trump, 1968, 1969). These authors noted the similarity between early isolation membranes and the ER, and at a later stage, the presence of acid hydrolases in the space between membranes limiting autophagosomes.

The notion that in the fat body the isolation membranes arise from vesicles originating near the outer concave face of the GC has been confirmed (Locke and Sykes, 1975). Autophagy is a two-step process. Organelles are first isolated by membranes having osmiophilic contents but which are without acid hydrolases. Acid phosphatase is carried via microvesicles (primary lysosomes) in a second later step to the isolation bodies which then become autophagic vacuoles. The primary lysosomes arise from the inner, concave face of the GC.

## IV. Autophagy in Other Tissues

### A. *The Epidermis*

The epidermis in the 5th stage, like the fat body, undergoes a phase of growth (e.g., ER formation, M + 0 to M + 66 hr), a phase of larval syntheses (lamellate cuticle deposition, M + 66 to M + 156 hr) and a switch to pupal syntheses (M + 156 to M + 192 hr). The switch to pupal syntheses involves a loss and reorganization of the RER which is isolated and broken down in autophagic vacuoles in the same way as in the fat body. Some epidermal cells also lose most of their RER immediately after ecdysis to the 5th stage. The proleg spines and the cells which support them are most active in synthesis

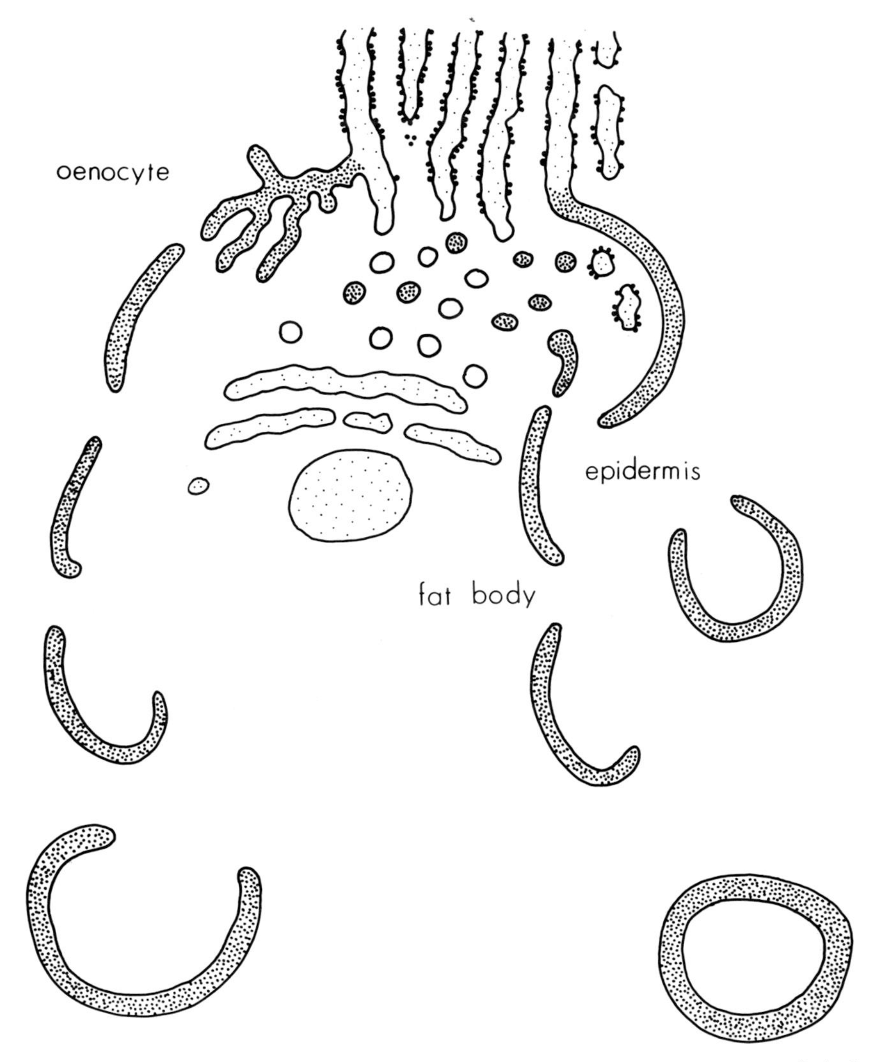


**Fig. 23.** The relation between isolation membranes and the RER-GC in oenocytes, fat body and epidermis. The localization of the osmiophilia is shown by dense stippling.

and secretion prior to ecdysis and undergo a dramatic loss of cytoplasm, mainly RER, as soon as the 5th-stage larval cuticle has been deposited. The isolation membranes seem to arise from the ER, perhaps in the transitional region. These observations suggest that in the epidermis also, isolation membranes arise from that specialized part of the ER which normally bleeds into the Golgi complex (Fig. 23).

### *B. The Oenocytes*

The oenocytes are characterized by the development of tubular smooth endoplasmic reticulum with free ribosomes but only traces of RER (Locke, 1969a). They undergo massive autophagy. At some times in the molt/intermolt cycle, one-third to one-half of the cell may be composed of giant autophagic vacuoles formed from hundreds of isolation bodies. Envelopment is typically not specific and all components have been seen isolated, including complete Golgi complexes. In this nonspecific autophagy the oenocytes resemble experimental atrophy in vertebrate tissues (Cole *et al.*, 1971). The isolation membranes arise from smooth tubular endoplasmic reticulum and as in the fat body, they cut off fragments of the cell of a rather uniform size not much larger than a mitochondrion. The cisternae of both the smooth endoplasmic reticulum and the isolation membranes react intensely after Friend's hot osmium procedure. Smooth endoplasmic reticulum may arise directly from the nuclear membrane but typically it connects with RER a little away from the transition region. This interrelation and the connection with isolation membranes is proposed in Fig. 23.

## V. Plasma Membrane Turnover and the Apical Multivesicular Bodies

The turnover of plasma membrane poses the same topological problem to a cell as a cytoplasmic organelle since the membrane must present itself to a lytic environment without the release of enzymes into the cytoplasm. This feat is probably accomplished by the formation of multivesicular bodies devoted primarily to membrane turnover. Numerous studies have demonstrated the uptake of macromolecules in pinocytosis vesicles and their fusion with multivesicular bodies (Friend and Farquhar, 1967; Arstila *et al.*, 1971; Locke and Collins, 1968.) The vesicle contents become part of the matrix and the membrane adds to the MVB membrane. MVBs have also been shown to be the site of digestion of the phagocytosed macromolecules. Osmium impregnation suggests that some vesicles within MVBs may have originated as Golgi complex vesicles (Friend 1969), but other vesicles may have budded inward from the surface. If the total volume of a MVB is derived from the fusion of micro-vesicles there would be a gross excess of MVB surface membrane. This seems to be eliminated by the budding inward of microvesicles and breakdown in the granule lumen (Locke and Collins, 1965). It has been suggested that the MVB's in HeLa cells may contain vesicles composed of plasma membrane, which have budded inward from the surface of the MVB, since only the matrix of the granule contained phagocytosed macromolecules (Arstila *et al.*, 1971; Fedorko *et al.*, 1968a,b). A single MVB may thus be a heterophagic vacuole in that it breaks down molecules transported to it from the outside, and an autophagic vacuole in that it is involved in the turnover of the cell's own plasma membrane.

It has been pointed out (Locke and Collins 1968) that if microvesicles are involved in transport in only one direction then multivesicular bodies which incorporate vesicles smaller than about 1000 A diameter are predominantly autophagic. Even if the whole of the center of a 1000 A diameter vesicle is filled with foreign molecules it is still 40–50% membrane. This suggested that multivesicular bodies may often be concerned primarily in plasma membrane turnover. With this in mind several insect cell types were examined to see if the form of MVBs varied in a way to suggest primarily one function or the other. The fat body (Locke and Collins, 1968) and epidermal MVBs (Locke, 1969b; Locke and Krishnan, 1971) are known to be active in protein uptake and have a dense matrix with few microvesicles. The apical MVBs of the gut (Sedlak, 1968) and malpighian tubules (McMahon, 1971) on the other hand contain little matrix and are predominantly composed of vesicles. If these MVBs are concerned in apical membrane turnover and not in protein uptake then we should expect that pinocytosis and MVB formation would take place in the absence of luminal protein. This has now been demonstrated in the primary cells of the malpighian tubules of *Calpodes* (McMahon, 1971). The apical face of these cells is normally exposed to a luminal filtrate containing no detectable protein. In this natural condition MVBs appear to be fed by pinocytosis from the apical face. When peroxidase is introduced as a tracer it passes to the MVBs by way of the pinocytosis vesicles, showing that these MVBs do indeed arise from the apical face. Since the lumen of these malpighian tubules is normally free of protein and the MVBs contain predominately membrane, we may presume that these apical MVBs are primarily autophagic. This hypothesis is summarized in Fig. 24.

The function of MVBs as organelles for the destruction of membranes is more plausible if the observations on the lipolytic enzymes of lysosomal granules are considered. Fowler and DeDuve (1959) reported that lysosomes isolated from rat liver contained enzymes capable of degrading a variety of lipids characteristic of cell membranes. Although these enzymes have not been demonstrated in specific components of the lysosomal system, it is

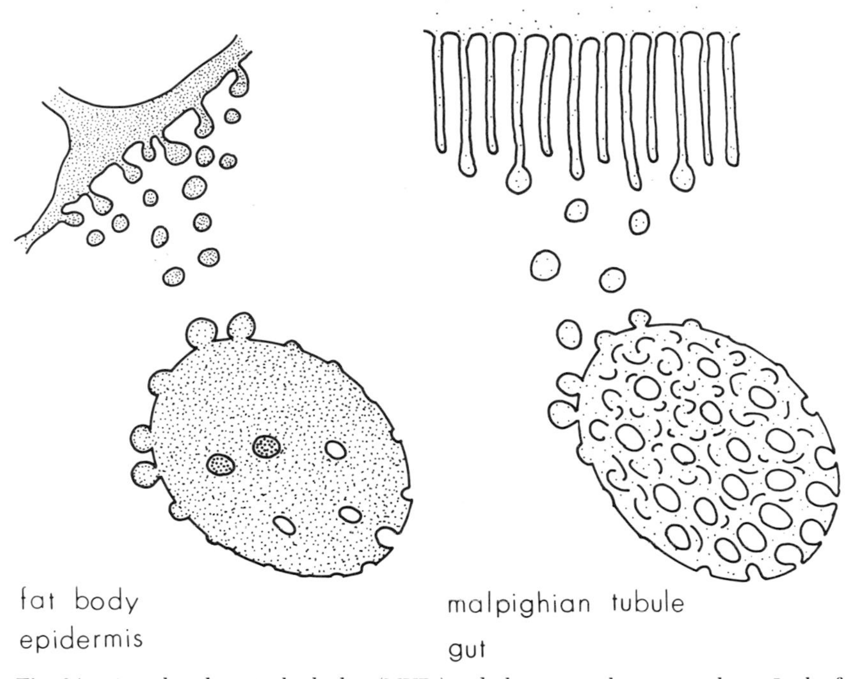


**Fig. 24.** Apical multivesicular bodies (MVBs) and plasma membrane autophagy. In the fat body and epidermis the MVBs are primarily concerned in the breakdown of pinocytosed macromolecules. In the gut and malpighian tubules, the MVBs are made up of empty vesicles and are presumed to be primarily concerned in plasma membrane turnover.

reasonable to expect their occurrence in autophagic vacuoles such as the apical multivesicular bodies considered here.

Plasma membrane turnover has been studied more recently (1975) in a variety of cell types. Secretion by exocytosis results in a significant transient increase in plasma membrane area and it is generally agreed that excess membrane is retrieved from the surface by endocytosis, although observations differ on the details of the process.

The extent to which cells recapture specifically that membrane contributed by secretory vesicles is receiving particular attention. De Camilli *et al.* (1976) have used freeze-fracture to study retrieval in the parotid gland, where membrane from secretory vesicles may be distinguished from luminal plasma membrane by the respective densities of their intramembrane particles. These authors concluded that mixing of granule and surface plasma membranes does not occur, and recapture is limited to the area of membrane corresponding to the secretory vacuoles. It has similarly been concluded that in cells of the adrenal medulla, membrane derived from secreted chromaffin granules (identified by their content of dopamine $\beta$-hydroxylase) is specifically recaptured (Winkler *et al.*, 1974). It has not been possible to identify with certainty membrane retrieved at synaptic terminals, but it seems likely that here also it is the limiting membrane of synaptic vesicles that is involved (Teichberg *et al.*, 1975; Ceccarelli *et al.*, 1973; Hurlbut and Ceccarelli, 1974; Heuser and Reese, 1973). Selective uptake of localized membrane regions is plausible in the light of observations that membranes, although fluid, do not show entirely unrestricted lateral movement of protein markers (Edelman, 1974; Edidin and Weiss, 1974; De Petris, 1974; Allison and Davies, 1974). Selective uptake by phagocytosis of surface receptors for concanavalin A and *Ricinis communis* agglutinin has been observed (Oliver *et al.*, 1974). It seems likely, that cells have the capacity to perform selective retrieval of membrane involved in massive exocytosis.

The mechanism of retrieval varies with cell type. In some tissues endocytosis of excess membrane involves formation of small vesicles which may be coated or smooth (Amsterdam *et al.*, 1969; Jamieson and Palade, 1971; Abrahams and Holtzman, 1973; Ceccarelli *et al.*, 1973; Heuser and Reese, 1973; Kramer and Geuze, 1974; Larsen, 1975; Nordmann *et al.*, 1974; Orci *et al.*, 1973; Pelletier, 1973). In other systems, larger structures—vacuoles or tubules continuous with the surface—have been implicated (Kramer and Geuze, 1974; De Camilli *et al.*, 1976; Nordmann and Morris, 1976). De Camilli *et al.* (1976) have observed both large indentations and apical vesicles in sequence in the parotid gland, and have suggested that they may represent phases of the same event.

The fate of the membrane taken up as microvesicles is inadequately understood. Although Pelletier (1973) observed reaction product from tracer Horseradish peroxidase in saccules of the Golgi complex, numerous studies have demonstrated the fusion of pinocytotic vesicles with multivesicular bodies (MVBs) following the uptake of macromolecules (Locke and Collins, 1968; Friend and Farquhar, 1967; Arstila *et al.*, 1971; Holtzman *et al.*, 1973; Pelletier, 1973; Kramer and Geuze, 1974). In *Calpodes* fat body, Larsen (1976) has quantified the volume of MVBs per cell during the period of membrane turnover at the pupal/adult metamorphosis. The volume of MVBs rises in the 24 hr prior to ecdysis and then falls in the succeeding 24 hr as would be expected if there is a turnover of the cell surface between pupa and adult. In *Calpodes* epidermis a new generation of plasma-membrane plaques (the regions of membrane specialized for certain sorts of cuticle secretion, often, but not always, at the tips of microvilli) arises at the larval to pupal molt. The old 5th-stage larval plaques are pinocytosed and transferred to the apical multivesicular bodies in microvesicles (Locke, 1976).

The function of MVBs as organelles for the destruction of membranes is more plausible if the observations on the lipolytic enzymes of lysosomal granules are considered. Fowler and de Duve (1969) reported that lysosomes isolated from rat liver contained enzymes capable of degrading a variety of lipids characteristic of cell membranes. We may therefore expect that lipases will be found particularly in apical multivesicular bodies.

## VI. Discussion and Conclusion

### A. *The Specificity of the Isolation Process*

These observations suggest very strongly that the destruction attendant upon cell remodeling at metamorphosis is an orderly process. There is a specific or at least hierarchical affinity for adhesion between isolation membranes and the organelles they envelop. Random movement in the fluid compartments of the cell may bring the vesicle antecedents of the isolation membranes sufficiently close to particular organelles for adhesion to take place, but at some degree of separation the vesicles either adhere or pass on to some other surface for adhesion. The final location of the isolation membranes is not random nor is the temporal change in that location. In order to understand the isolation of specific organelles, we must look for a mechanism by which the cell releases vesicles having membranes which adhere to a target. The interaction is probably not solely membrane to membrane since isolation membranes can cover RER coated with ribosomes. The specificity of the recognition process may not even reside in the isolation membrane. The demonstration of isolation in a temporal sequence of organelles could be achieved by changes in the organelle rather than the isolating membrane. It becomes particularly important to know whether the specificity is in the isolating membrane or in the organelle in relation to the mitochondria. Microbodies are completely lost from stage to stage, but mitochondria are only partly destroyed: part of the larval mitochondrial population survives through the pupa to repopulate the adult fat body cells (Larsen, 1970). Is the destruction of the larval mitochondria random until a certain low level is reached or is there a stem line resistant to destruction? Specificity in the organelle would favor the hypothesis of a stem line, specificity in the isolating membrane would favor random mitochondrial destruction. We can make little further progress until the vesicles have been prepared and their adhesion properties tested *in vitro*.

### B. *The Contents of the Isolating Envelope*

Isolation by envelopes solves the topological problems of how to externalize an organelle into a lytic environment. In addition to the osmium-staining reaction, the envelopes have two characteristics which suggest that they are more than empty compartments between membranes. The separation between inner and outer membranes is fairly constant at about 5–700 Å and the isolation bodies are all about the same size whether or not they contain discrete organelles or fragments of RER or smooth endoplasmic reticulum. It is difficult to account for the constancy of these dimensions without supposing that the envelope contains a structural component. It may be significant that the osmium-staining reaction occurs only between membranes separated by about 500–700 Å whether as vesicles, saccules, envelopes, or tubules. The smooth tubular endoplasmic reticulum of the oenocytes and wax gland are remarkably constant in their dimensions, as are the vesicular cristae of mitochondria, Golgi complex saccules and microvesicles and the nuclear envelope, all of which stain with the hot osmium reagent (Friend and Brassil, 1970). If the osmiophilia and constant dimensions can be correlated with a structural component it might help to explain the limited size range of the isolation bodies.

Since this chapter was written (1971) we have amplified some of the ideas based on the study of insect cells, and there have been important developments in the understanding of organelle turnover in other cell types, particularly in membrane turnover. We have also discovered the Golgi complex beads which seem to mark the exit gate for transition vesicles from the RER to the GC and which may have a role in regulating membrane flow (Locke and Huie 1975, 1976).

## References

Abrahams, S. J., and Holtzman, E. (1973). Secretion and endocytosis in insulin-stimulated rat adrenal medulla cells. *J. Cell Biol.* **56**, 540.

Allison, A. C., and Davies, P. (1974). Interactions of membranes, microfilaments and microtubules in endocytosis and exocytosis. *In* "Cytopharmacology of Secretion" (B. Ceccarelli, F. Clementi, and J. Meldolesi, eds.), pp. 237–248. Raven, New York.

Amsterdam, A., Ohad, I., and Schramm, M. (1969). Dynamic changes in the ultrastructure of the acinar cell of the rat parotid gland during the secretory cycle. *J. Cell Biol.* **41**, 753.

Arstila, A. U., and Trump, B. F. (1968). Studies on cellular autophagocytosis. The formation of autophagic vacuoles in the liver after glucagon administration. *Am. J. Pathol.* **53**, 687–733.

Arstila, A. U., and Trump, B. F. (1969). Autophagocytosis: Origin of membranes and hydrolytic enzymes. *Virchow's Arch. B* **2**, 85.

Arstila, A. U., H. O. Jauregui, J. Chang, and Trump, B. F. (1971). Studies on cellular autophagocytosis. Relationship between heterophagy and autophagy in HeLa cells. *Lab. Invest.* **24**, 162.

Ceccarelli, B., Hurlbut, W. P., and Mauro, A. (1973). Turnover of transmitter and synaptic vesicles of the frog neurotransmitter junction. *J. Cell Biol.* **57**, 499.

Cole, S., Matter, A., and Karnovsky, M. J. (1971). Autophagic vacuoles in experimental atrophy. *Exp. Mol. Pathol.* **14**, 158.

De Camilli, P., Peluchetti, D., and Meldolesi, J. (1976). Dynamic changes of the luminal plasmalemma in stimulated parotid acinar cells. A freeze-fracture study. *J. Cell Biol.* **70**, 59.

De Duve, C., and Wattiaux, R. (1966). Functions of lysosomes. *Annu. Rev. Physiol.* **28**, 435.

De Petris, S. (1974). Inhibition and reversal capping by cytochalasin B, vinblastine and colchicine. *Nature (London)* **250**, 54.

Edelman, G. M. (1974). Surface alterations and mitogenesis in lymphocytes. *In* "Control of Proliferation in Animal Cells" (B. Clarkson and R. Baserga, eds.), pp. 357–377. Cold Spring Harbour Laboratory, Cold Spring Harbour, New York.

Edidin, M., and Weiss, A. (1974). Restriction of antigen mobility in the plasma membranes of some cultures fibroblasts. *In* "Control of Proliferation in Animal Cells" (B. Clarkson and R. Baserga, eds.), pp. 213–219. Cold Spring Harbour Laboratory, Cold Spring Harbour, New York.

Fedorko, M. E., Hirsch, J. G., and Cohn, Z. A. (1968a). Autophagic vacuoles produced in vitro. I. Studies on cultured macrophages exposed to chloroquine. *J. Cell Biol.* **38**, 377.

Fedorko, M. E., Hirsch, J. G., and Cohn, Z. A. (1968b). Autophagic vacuoles produced in vitro. II. Studies on the mechanism of formation of autophagic vacuoles produced by chloroquine. *J. Cell Biol.* **38**, 392.

Fowler, S., and de Duve, C. (1969). Digestive activity of lysosomes III. The digestion of lipids by extracts of rat liver lysosomes. *J. Biol. Chem.* **244**, 471.

Friend, D. S. (1969). Cytochemical staining of multivesicular body and Golgi vesicles. *J. Cell Biol.* **41**, 269.

Friend, D. S., and Brassil, G. E. (1970). Osmium staining of endoplasmic reticulum and mitochondria in the rat adrenal cortex. *J. Cell Biol.* **46**, 252.

Friend, D. S., and Farquhar, M. G. (1967). Functions of coated vesicles during protein absorption in the rat vas deferens. *J. Cell Biol.* **35**, 357.

Heuser, J. E., and Reese, T. S. (1973). Evidence for recycling of synaptic vesicle membrane during transmitter release at the frog neuromuscular junction. *J. Cell Biol.* **57**, 315.

Hirsch, J. G., Fedorko, M. E., and Cohn, Z. A. (1968). Vesicle fusion and formation at the surface of pinocytic vacuoles in macrophages. *J. Cell Biol.* **38**, 629–632.

Holtzman, E., Teichberg, S., Abrahams, S. J., Atkowitz, E., Crain, S. M., Kawai, N., and Peterson, E. R. (1973). Notes on synaptic vesicles and related structures, endoplasmic reticulum, lysosomes and peroxisomes in nervous tissue and the adrenal medulla. *J. Histochem. Cytochem.* **21**, 349.

Hurlbut, W. P., and Ceccarelli, B. (1974). Transmitter release and recycling of synaptic vesicle membrane at the neuromuscular junction. *In* "Cytopharmacology of Secretion" (B. Ceccarelli, F. Clementi, and J. Meldolesi, eds.), pp. 141–154. Raven, New York.

Jamieson, J. D., and Palade, G. E. (1971). Synthesis, intracellular transport and discharge of secretory proteins in stimulated pancreatic acinar cells. *J. Cell Biol.* **50**, 135.

Kramer, M. E., and Geuze, J. J. (1974). Redundant cell membrane regulation in the exocrine pancreas cells after pilocarpine stimulation of secretion. *In* "Cytopharmacology of Secretion" (B. Ceccarelli, F. Clementi, and J. Meldolesi, eds.), pp. 87–97. Raven, New York.

Larsen, W. J. (1970). Genesis of mitochondria in insect fat body. *J. Cell Biol.* **47**, 373.

Larsen, W. J. (1971). Cell remodeling in the fat body at metamorphosis. Ph.D. thesis, Case Western Reserve University, June, 1971.

Larsen, W. J. (1976). Cell Remodeling in the Fat Body of an Insect. *Tissue & Cell* **8**(1).

Locke, M. (1969a). The ultrastructure of the oenocytes in the molt/intermolt cycle of an insect. *Tissue & Cell* **1**, 103.

Locke, M. (1969b). The structure of an epidermal cell during the development of the protein epicuticle and the uptake of molting fluid in an insect. *J. Morphol.* **127**, 7.

Locke, M. (1970a). The molt/intermolt cycle in the epidermis and other tissues of an insect *calpodes ethlius* (Lepidoptera, Hesperiidae). *Tissue & Cell* **2**, 197.

Locke, M. (1970b). The control of activities in insect epidermal cells. *Membr. Soc. Endocrinol.* **18**, 285.

Locke, M. (1976). The Role of Plasma Membrane Plaques and Golgi Complex Vesicles in Cuticle Deposition during the molt/intermolt cycle. *In* "The Insect Integument" (H. R. Hepburn, ed.), pp. 237–258. Elsevier, Amsterdam.

Locke, M., and Collins, J. V. (1965). The structure and formation of protein granules in the fat body of an insect. *J. Cell Biol.* **26**, 857–885.

Locke, M., and Collins, J. V. (1968). Protein uptake into multivesicular bodies and storage granules in the fat body of an insect. *J. Cell Biol.* **36**, 453.

Locke, M., and Huie, P. (1975). The Golgi complex/endoplasmic reticulum transition region has rings of beads. *Science* **188**, 1219.

Locke, M., and Huie, P. (1976). The Beads in the Golgi Complex/Endoplasmic Reticulum Region. *J. Cell Biol.* **70**, 384–394.

Locke, M., and Krishnan, N. (1971). Distribution of phenoloxidases and polyphenols during cuticle formation. *Tissue & Cell* **3**, 103.

Locke, M., and McMahon, J. T. (1971). The origin and fate of microbodies in the fat body of an insect. *J. Cell Biol.* **48**, 61.

Locke, M., and Sykes, A. K. (1975). The role of the Golgi complex in the isolation and digestion of organelles. *Tissue & Cell* **7**(1), 143.

McMahon, J. T. 1971. The structure and function of malpighian tubules in *calpodes.* Master's thesis, Case Western Reserve University, June, 1971.

Meldolesi, J. (1974). Secretory mechanisms in pancreatic acinar cells. Role of the cytoplasmic membranes. *In* "Cytopharmacology of Secretion" (B. Ceccarelli, F. Clementi, and J. Meldolesi, eds.), pp. 71–85. Raven, New York.

Nordmann, J. J., and Morris, J. F. (1976). Membrane retrieval at neurosecretory axon endings. *Nature (London)* **261**, 723.

Nordmann, J. J., Dreifuss, J. J., Baker, P. F., Ravazzola, M., Malaisse Lagae, F., and Orci, L. (1974). Secretion-dependent uptake of extracellular fluid by the rat neurohypophysis. *Nature (London)* **250**, 155.

Oliver, J. M., Ukena, T. E., and Berlin, R. D. (1974). Effect of phagocytosis and colchicine on the distribution of lectin binding sites on cell surfaces. *Proc. Nat. Acad. Sci. USA* **71**, 394.

Orci, L., Malaisse Lagae, F., Ravazzola, M., Amherdt, M., and Renold, A. E. (1973). Exocytosis-endocytosis coupling in the pancreatic $\beta$ cell. *Science* **181**, 561.

Pellettier, F. (1973). Secretion and uptake of peroxidase by rat adenohypophyseal cells. *J. Ultrastruct. Res.* **43**, 445–459.

Sedlak, B. J. (1968). The structure and development of the midgut. Master's thesis, Case Western Reserve University, September, 1968.

Teichberg, S., Holtzman, E., Crain, S. M., and Peterson, E. R. (1975). Circulation and turnover of synaptic vesicle membrane in cultured fetal mammalian spinal cord and neurons. *J. Cell Biol.* **67**, 215–230.

Winkler, H., Schneider, F. H., Rufener, C., Nakane, P. K., and Hortragl, H. (1974). Membranes of Adrenal-Medulla: Their Role in Exocytosis. *In* "Cytopharmacology of Secretion" (B. Ceccarelli, F. Clementi, and J. Meldolesi, eds.), pp. 127–139. Raven, New York.

# EDITORS' SUMMARY TO CHAPTER V

This chapter introduces the use of invertebrate models for studies of pathological reactions and certainly this is an area which needs much more recognition in terms of developing suitable systems for experimental and environmental pathology. Although such models have been used in order to solve problems in cellular and molecular biology (the squid axon, for instance, is extensively used in the study of ion fluxes across cell membranes), such models are badly needed in many other fields of membrane and molecular pathology. Probably in the near future many time-consuming and costly procedures in the study of carcinogenesis, mutagenesis, teratogenesis as well as studies of the toxic action of drugs to cell membranes can be supplemented by the use of these models.

The model presented in this chapter is not only important for the study of organelle turnover but also in the study of common mechanisms of tissue atrophy and possibly of aging. The great advantage of this model as discussed by the authors is the precise time patterns by which the metamorphosis takes place. So far these predictable sequences are not encountered in mammalian models.

Increasingly, autophagy (as discussed in this chapter) appears to be a very important if not the principal mechanism for organelle turnover. The author emphasizes the possible selectivity of the process for various organelles at various times. This is a point which needs much more work in order to see its general applicability, and also one that has problems with interpretation in that it has recently been shown in data from our laboratory that different organelles have different turnover times in the lysosomes and, therefore, it may be difficult to estimate selectivity from only the population of distribution of sequestered organelles (Glaumann and Trump, 1975), especially if the particles at the stage in question contain hydrolases. In the originally sequestered portions it may represent a true representation of selectivity. These points were also discussed by Pfeifer and Scheller (1975) in their studies on autophagocytosis.

A useful model to study organelle turnover that has been used in our laboratory is the intravenous injection of organelles into the rat (Glaumann *et al.*, 1975a,b). The uptake and digestion of these organelles then can be followed by both morphological and biochemical techniques to elucidate the sequence of changes involved in organelle turnover.

This has been further studied recently using the glucagon model with the conclusion that the stimulation of intracellular protein degradation in liver is a manifestation of deprivation-induced autophagy which results from a decrease in certain intracellular glucogenic amino acids, notably glutamine (Schwarter and Mortimer, 1979).

## References

Glaumann, H., and Trump, B. F. (1975). *Lab. Invest.* **33,** 262.

Glaumann, H., Berezesky, I. K., Ericsson, J. L. E., and Trump, B. F. (1975a). *Lab. Invest.* **33,** 239.

Glaumann, H., Berezesky, I. K., Ericsson, J. L. E., and Trump, B. F. (1975b). *Lab. Invest.* **33,** 252.

Pfeifer, U., and Scheller, H. J. (1975). *Cell Biol.* **64,** 608.

Schwarter, C. M., and Mortimer, G. E. (1979). *Proteins Nucleic Acid Synth.* **76,** 3169.

# CHAPTER VI

# REACTIONS OF LYSOSOMES TO CELL INJURY

**Hal K. Hawkins**



Lysosomes can react to cellular injury in several distinct ways. The relative simplicity of the structure and function of lysosomes allows these reactions to be divided into only a few patterns. Thus, all lysosomal alterations can be conceptualized as increases or decreases in: (a) quantity of specific lysosomal contents, (b) rate of specific membrane fusion events, or (c) permeability of lysosomal membranes to specific substances. Each type of lysosomal alteration has different consequences in terms of the whole cell and its survival, and observable effects of each type of lysosomal response can often be predicted.

Unfortunately, most of the methods used in the study of lysosomes give only indirect evidence of the nature of their alterations in abnormal conditions. The multiplicity of forms of cell injury and of organisms studied is such that only rather few lysosomal reactions to cell injury have been well characterized in terms of mechanisms at the organelle level. The situation is complicated slightly by the fact that lysosomal changes of one type may masquerade as a different type.



PATHOBIOLOGY OF CELL MEMBRANES, VOL. II

ISBN 0-12-701502-7

The purpose of this chapter is to selectively review the literature on lysosomal changes in cellular injury, and to classify the alterations in lysosomes into distinct patterns of response. Probable examples of each type of lysosomal aberration will be discussed, and consideration will be given to promising new methods for further improving our knowledge and understanding of this area of basic cellular pathology.

## I. Introduction and Definitions

Originally lysosomes were defined as a particulate fraction of homogenized rat liver cells comprising certain hydrolytic enzyme activities which were maximal only after exposure to physical or chemical treatments known to disrupt membranes. These particles sedimented in an ultracentrifuge between the mitochondrial and microsomal fractions, and their enzymes cleaved simple chemical groups from carbohydrates most rapidly in acid media, between pH 4 and 6 (de Duve *et al.*, 1955; de Duve and Wattiaux, 1966; Strauss, 1956; Claude, 1946). When a histochemical method was shown to identify sites of one of these enzymes, β-glycerophosphatase, and was adapted for electron microscopy, it became possible to identify the particles of the lysosomal fraction with a class of spherical organelles of intact rat liver cells composed of a homogeneous matrix surrounded by a single trilaminar "unit" membrane (Novikoff, 1960; Strauss, 1964). By general agreement all single membrane limited organelles which contain acid phosphatase activity have subsequently been considered to fit the cytologic definition of lysosomes (de Duve and Wattiaux, 1966). These particles were also shown to originate by budding from the "maturing face" of the Golgi apparatus in the same way as secretory droplets (Novikoff *et al.*, 1964). Subsequent work with histochemistry and ultracentrifugation of diverse cell types has demonstrated the presence of particles which fit the definition of lysosomes in every variety of cell studied except for mature erythrocytes. Many other enzyme activities have been associated with lysosomes, which catalyze hydrolysis of proteins, complex carbohydrates, and ester linkages of lipids. These enzymes are of low specificity, and with few exceptions have acid pH optima (Tappel, 1969). The prediction that lysosomes would be able to catalyze hydrolysis of all compounds synthesized by the host organism, to the extent that the products could freely cross the normal lysosomal membrane, is compatible with the evidence available at present.

Under proper circumstances, lysosomes fuse freely with certain other vacuoles and organelles of normal cells with confluence of the two membranes and mixing of the contents of the two membrane-bound spaces. If the structure fusing with a lysosome contains only material which can be degraded by lysosomal enzymes, the resulting hybrid structure or "secondary lysosome," after digestion is complete, is structurally identical with the original lysosome except for an excess of membrane material. Such excess membrane apparently can also be digested following a process of budding into the interior of the lysosome (Arstila *et al.*, 1971). All the structures which can fuse with lysosomes in viable cells comprise a special part of the vacuolar or cytocavitary system of the cell, which is of great importance in the homeostatic and specialized functions of both normal and injured cells.

## II. Injury to Lysosomes

Conditions which are injurious to cells may affect lysosomes directly, more often indirectly, or often in both ways. Only rarely has it been possible to reach firm conclusions as to whether direct lysosomal injury has occurred or the cellular response to some other effect has secondarily altered lysosomes, or as to whether changes in other organelles were causally related to abnormalities of lysosomes. However, based on the information available, certain principles can be delineated.

### A. *Primary Injury to Lysosomes*

Two types of cell injury are known in which experimental data as well as expectations strongly suggest that lethal cell injury is initiated by rupture of lysosomes. In both cases a potentially toxic agent is first accumulated by secondary lysosomes, the membrane of the secondary lysosome is damaged and the contents of the lysosomes are released to the cell sap, followed by cell death.

The first example is phagocytosis of microscopic crystals which interact destructively with membranes, e.g., silica ($SiO_2$) and sodium urate (Comolli, 1967; Allison, 1968a; Weissmann and Rita, 1972; Weissmann, 1974). In this system the time lag is apparently due to a necessity for macromolecules which are ingested with the crystals to be digested and eliminated from the secondary lysosomes, together with any excess water, before the crystalline particles are apposed to the lysosomal membrane tightly enough to interfere with its integrity (Weissmann *et al.*, 1971). The evidence that lysosomes are damaged before other components of the cell, in this model, is that crystals are seen in the cell sap without any enclosing membrane early in the course of cell injury, and that lysosomal enzymes appear in the extracellular medium at approximately the same time as do the enzymes of the cell sap (Weissman and Rita, 1972; Schumacher and Phelps, 1971). Of course, these results do not fully exclude the possibility that direct damage by crystals to other parts of the cell may be responsible for the death of the cells, but this possibility seems unlikely.

The second well-defined model of lethal cell injury due to primary damage to lysosomes is Allison's photosensitization experiment, in which a dye is concentrated by tissue culture cells in their lysosomes (Allison *et al.*, 1966; Allison, 1964, 1968b). Such dyes are known to damage membranes only when exposed to sufficient light of the appropriate wavelength, in the presence of oxygen (Blum, 1941). Thus cells which have been allowed to accumulate dye within their lysosomes in the dark are washed and irradiated with light, and the resulting cellular degeneration is considered to be a result of the activity of lysosomal enzymes. This interpretation is supported not only by cytochemical data and by the results of enzyme assays performed on extracellular media, but also by the ultrastructural observation that particles contained in secondary lysosomes appear diffusely in the cell sap following this type of experimental cell injury (Hawkins *et al.*, 1972; Brunk and Ericsson, 1972). Reduced temperature (0°–4°C), which inhibits enzymatic degradation much more than light-stimulated membrane damage (Blum, 1941), slows but does not prevent cellular degeneration in this model system (Brunk and Ericsson, 1972). The only objection to the conclusion that lysosomal enzyme release is responsible for the lethal cell injury in this experiment is that the photoperoxidation reaction which damages the lysosomal membrane is a self-perpetuating chain of free radical reactions which might well continue, after causing lysosomes to burst, to damage the cell surface membrane, mitochondria, or other cell components.

Other examples of cell injury which have been postulated to affect lysosomes primarily include antibody directed against leukocyte granules (Quie and Hirsch, 1964; Persellin, 1969) and several naturally occurring toxins including streptolysin O and other bacterial toxins, mellitin, phospholipase, brown recluse spider venom, and physical injuries including ultraviolet and ionizing radiation and high concentrations of oxygen (Kingdon and Sword, 1970; Weissman *et al.*, 1969; Beaufay and de Duve, 1959; Smith and Micks, 1970; Weissmann and Dingle, 1962; Desai *et al.*, 1964). The precise cellular effects of these agents have been less well defined than in the two injuries already discussed, and the evidence is discussed in the references cited.

### *B. Changes in Lysosomes Secondary to Other Cell Injuries*

In a number of types of cell injury, lysosomes are affected indirectly, as a secondary effect of some other alteration of cellular metabolism or structure. The direct effects of such injuries may be limited to the cell surface membrane, specific paths of cellular metabolism, specific stimuli such as hormonal influences which induce cells to undergo specialized functions, or alterations in the chemical or physical environment of cells such as changes in temperature, pH, or supply of oxygen.

The assumption is usually made, in interpreting these experiments, that only one identifiable cellular process is affected by the injury. Support for this assumption is very desirable but very often lacking. Two models of cell injury in which the primary action of the injurious agent is likely to be limited to the cell surface membrane have been studied in detail. These are the action of *p*-chloromercuribenzene sulfonate (PCMBS) on the flounder renal tubule and several studies of complement-mediated immune cytolysis (Sahaphong and Trump, 1971; Vansteveninck *et al.*, 1965; Goldberg and Green, 1959; Green *et al.*, 1959; Glick *et al.*, 1970). In both of these models, a rather sudden increase in permeability of the cell surface membrane leads to rapid changes in the composition of the cell sap, with subsequent alterations in lysosomes. Changes in lysosomes seem to occur well after cell death in these models. A few examples of cell injury due primarily to inhibition of specific steps in cellular energy metabolism have been studied (Trump and Bulger, 1968; Croker *et al.*, 1970; Saladino and Trump, 1969; Ginn *et al.*, 1968; Hawkins *et al.*, 1972).

Changes in the cytocavitary system have also been described by electron microscopic and biochemical techniques following tissue ischemia, a complex mode of cell injury. The observation by DeDuve in 1959, that ischemia of rat liver *in vivo* produced a rather consistent pattern of elevation of unsedimentable activities of lysosomal enzymes 2–4 hr after the onset of injury, stimulated great interest in the concept that lysosomes could kill cells (de Duve and Beaufay, 1959). These studies have not been followed by ultrastructural studies of the same model, however, so that the structural state of lysosomes and other organelles before homogenization is not known. Other work of de Duve has shown that the acidic cellular environment which is characteristic of ischemia causes lysosomal enzymes to be released *in vitro* from a particulate to an unsedimentable form. The changes in lysosomes in ischemic liver can be considered to be secondary to acidosis and possibly other factors. Studies by Baker on transient ischemia of rat liver showed that most cells became necrotic following 30–45 min of ischemia, so less than 1 hr of ischemia certainly represents irreversible injury in this model (Baker, 1956). The changes in lysosomes described by de Duve thus took place in dead cells. Histochemical study of the same model by Kerr indicated that lysosomes retained particulate staining for 3 hr, but diffusion of reaction product was seen after 6 hr (Kerr, 1965). Studies of autolysis *in vitro* of mouse liver slices at 37°C, a very similar condition, indicated that lysosomes did not change their histochemical staining pattern and showed intact membranes by electron microscopy of aldehyde-fixed tissue after 2 hr of ischemia *in vitro* (Trump *et al.*, 1962, 1965). Lability of lysosomes was indicated in the same studies, however, by the early appearance of breaks in their membranes as an osmium fixation artifact at 1 hr, and by detecting increased unsedimentable activity of acid phosphatase (Griffin *et al.*, 1965). Thus, in mouse or rat liver, deprivation of blood flow seems to affect lysosomes only after changes in other organelles and death of the cell, and thus secondarily.

More recently, a number of studies have demonstrated biochemical or histochemical alterations in cardiac lysosomes after coronary artery occlusion in several species (Wildenthal, 1975; Hoffstein *et al.*, 1975; Ricciutti, 1972). Generally these observations have been made after one or several hours of ischemia, whereas 40 min of ischemia due to circumflex coronary artery occlusion in the dog has been shown to be uniformly lethal to a significant mass of myocardial cells (Jennings *et al.*, 1960, 1969b). Severe abnormalities in mitochondrial structure and function occur after 1 hr of severe myocardial ischemia, and cell volume control is lost, but myocardial lysosomes have shown no abnormalities in ultrastructure in this early phase of irreversible injury (Jennings *et al.*, 1969a; Kloner *et al.*, 1974).

Thus at present it seems likely, but not certain, that lysosomes release their enzymes in ischemic myocardium after failure of mitochondrial function, failure of cell volume control, and cell death have occurred, and therefore, secondarily to some other event.

Of course, many varieties of cell injury, both experimental and natural, must affect the lysosomal system by a combination of direct influence on these organelles, indirect effects mediated by other cellular organelles, and effects on the chemical environment of lysosomes. In most types of cell injury, the extent of lysosomal reactions to injury is unknown. Even when alterations in lysosomes have been demonstrated, the nature of the cellular alterations which led to lysosomal damage has often remained unknown.

### *C. Pharmacologic Alterations of Lysosomes*

Cellular controls of organelle movements and of fusion events involving lysosomes have been greatly clarified by recent studies on inflammatory cells using chemical compounds which affect cyclic nucleotide metabolism or the structure and function of microfilaments and microtubules. These agents are capable of reproducing many reactions of lysosomes to cell injury, and can modify lysosomal responses to other cellular stresses, thus providing a valuable source of information on the mechanisms of response of the lysosome system. Another group of agents has been characterized as modifying lysosomal membrane permeability, or the response of the lysosomal membrane to stress. Moreover, a large and growing list of agents has been shown to accumulate in secondary lysosomes, and some have been designed to cause primary changes in lysosomes. These effects of drugs are important in the context of cell injury both as means of producing well-understood alterations in cells, and as tools for characterizing the mechanisms of lysosomal response.

### 1. *Labilizers and Stabilizers*

An important category of compounds which alter lysosomes includes those which have been classified as lysosomal "labilizers" and "stabilizers" (Weissmann, 1964; Ignarro, 1971). These are drugs or other chemicals which have the property of altering the response of a suspension of more or less purified lysosomes to a well-controlled stress such as hypotonicity or incubation at 37°C. Labilizers are compounds which increase the rate or extent of release of lysosomal enzymes in response to controlled stress *in vitro,* while stabilizers decrease enzyme release as compared to untreated control suspensions exposed to the same stress. It is thought that a related effect has been observed *in vivo* by treating intact animals, cells or tissues with chemical compounds, subsequently homogenizing experimental and control tissues in the same manner, and measuring the enzyme released (Brown and Schwartz, 1969). Although some of the labilizers do cause immediate release of enzymes from lysosomes, many such agents merely cause an increased sensitivity to stimuli which also cause the breakdown of untreated lysosomes. The labilization phenomenon also depends strongly on the concentration of the compound in the medium surrounding the lysosomes as well as on its chemical structure, since a number of compounds, particularly surface-active agents, have been observed to exert a labilizing influence on lysosomes at one concentration, and a stabilizing influence at another (Seeman, 1966, 1968). Of course, it is often difficult to predict the concentration which a drug will attain in the cell sap when a known quantity is present in the extracellular medium. Finally, many of the compounds which labilize lysosomes *in vitro* also have significant effects on other parts of intact cells. Thus the effect of a given drug administered to an intact animal may be very different from that of the same drug on isolated lysosomes, or those which might be predicted by considering only its influence on lysosome stability *in vitro* (Weissmann, 1968; Freidman *et al.*, 1969).

### 2. *Agents Which Affect Motility*

The drugs which alter motility of organelles and control the fusion events which lysosomes can undergo are providing new insight and control for these processes. Much of this work has been ably reviewed recently (Ignarro, 1975). Some function of microfilaments seems to be required for phagocytosis, since cytochalasin B inhibits phagocytosis, reducing the rate of this event of reverse membrane fusion (Allison *et al.*, 1971; Skosey *et al.*, 1973). Inhibition of microtubule function by colchicine does not inhibit phagocytosis, although it does prevent normal cell motility and inhibits degranulation (Malawista and Bodel, 1967; Allison *et al.*, 1971). Interestingly, inhibition of microfilament function causes release of lysosomal enzymes from leukocytes and macrophages, and enhances the rate of release induced by vitamin A (Davies *et al.*, 1973). Thus the processes of phagocytosis and fusion of lysosomes seem to involve distinctly separate mechanisms.

Fusion of lysosomes with phagosomes and the plasma membrane is under complex cellular control. With few exceptions, pharmacologic agents have the same effects on this process of fusion and degranulation whether it occurs during phagocytosis or during contact with immune complexes, which may be soluble or bound to surfaces ("frustrated phagocytosis"). Inhibition of lysosomal fusion and secretion result when cells are treated with epinephrine, theophylline, prostaglandin $E_1$, aspirin, or indomethacin (Ignarro, 1974; Ignarro *et al.*, 1974; Hawkins, 1974; Weissman *et al.*, 1971). These are all agents which directly or indirectly increase cellular content of cyclic AMP. In addition, cyclic AMP analogues themselves, colchicine, glucocorticoids, and inhibitors of glycolysis inhibit lysosomal enzyme release. A product of *Mycobacterium* tuberculosis, interestingly, also has the property of inhibiting fusion between lysosomes and phagosomes (Goren *et al.*, 1976). Release of lysosomal enzymes is stimulated by cyclic GMP and its analogues, by acetylcholine and prostaglandin $F2\alpha$, which increase cyclic GMP levels, and by calcium in the presence of the cation iontophore A23187 (Smith and Ignarro, 1975; Ignarro, 1974). This highly specialized process of increased rate of fusion of lysosomes thus occurs in response to a series of events which may include binding of a stimulant to a cell surface receptor, synthesis of cyclic GMP, alteration of microtubules and limited influx of calcium. These events are clearly subject to complex cellular regulation (Rossi *et al.*, 1976).

### 3. *Lysosomotropic Agents*

Another important group of drugs affecting lysosomes is those which reach their highest concentration in lysosomes because of selective accumulation in the cytocavitary system. These were elegantly classified and discussed by de Duve and his colleagues in 1974, in their article on lysosomotropism. Agents of this type may enter lysosomes by way of selective accumulation in the acidic interior of lysosomes, as do many cationic dyes and drugs including acridine orange, or by way of endocytic uptake. Considerable current research has been stimulated by the hope that drug vehicles can be designed which would allow correction of the enzyme deficiencies in lysosomal storage diseases, thus curing the patients (Desnick *et al.*, 1976). "Piggyback endocytosis" of enzymes or drugs attached to liposomes, erytbrocytes, antibodies, DNA, or other readily phagocytosed materials which also protect the agent from degradation or recognition as an antigen, is the subject of active research, and development of these techniques offers considerable promise.

## III. Responses of Lysosomes

Whatever the initial event which injures or stimulates a cell, a variety of distinct responses are possible on the part of the cell's vacuolar system. The fundamental changes which may occur can be classified as increases or decreases in general or specific lysosomal contents, increases or decreases in specific or generalized membrane fusion activities, or alterations of the permeability of the vacuolar membrane to various substances. Depending on the context or extent of change, each of these alterations may be an appropriate adaptation on the part of the cell, a conservative response in which the cell is protected from further injury, or an inappropriate response leading to cellular degeneration.

### A. *Alteration in Quantity of Lysosomal Contents*

#### 1. *Increased Enzyme Activity*

A number of different stimuli have been clearly shown to lead to increased activity of lysosomal enzymes. Although increases per organelle are possible, based either on increased volume of lysosomes or changes in relative quantities of their contents, increases in total lysosomal enzyme activity per cell or in whole tissues have been much more commonly described. The clearest example is the rapid new synthesis of lysosomal enzymes which occurs in response to phagocytosis in peritoneal macrophages (Axline and Cohn, 1970; Gordon and Cohn, 1973). This increase clearly serves the purpose of adaptation to an increased supply of substrates to be digested. A similar large burst of enzyme synthesis follows the stimulation of isolated lymphocytes by phytohemagglutinin (Hirschhorn *et al.*, 1968; Biberfeld, 1971). An increase in lysosomal enzyme activity may be presumed to occur *in vivo* whenever the mononuclear cell system responds to a new antigen by the activation and differentiation of a group of activated lymphocytes and mononuclear phagocytes. The extensive differentiation of macrophages during the formation of granulomata represents an extreme response of cells in which a very large number of lysosomes are produced by new synthesis (Cohn and Benson, 1965; Sutton and Weiss, 1966). The mechanisms of cellular control of these changes remain unclear. Other examples of increases in activity of lysosomal enzymes are those which occur during the differentiation of insect tissues, the involutional responses of endocrine organs to decreases in demand for their specific synthetic products, and regression of cardiac hypertrophy (Locke, 1969; Lockshin, 1969; Radford and Misch, 1971; Helminen and Ericsson, 1968; Helminen *et al.*, 1968; Cutilleta *et al.*, 1976). In all these cases increased lysosomal enzyme activity adaptively prepares the cell to eliminate macromolecules which are no longer needed and to recycle the products of digestion in new synthesis of useful cell products.

### 2. *Increased Nonenzymatic Contents*

Permanent engorgement of lysosomes with undigested material is seen in several situations, the most familiar being the storage of carbon, silica, asbestos, or other inhaled dusts in pulmonary macrophages and lymph nodes. Storage of hemosiderin pigment near sites of hemorrhage and of melanin granules near sites of destruction of skin, both in mononuclear phagocytes, are also familiar. Several varieties of ceroid, lipofuscin, or aging pigment are stored in lysosomes of many human and animal tissues (Streghler, 1964; Desai *et al.*, 1975; Gosden *et al.*, 1978). These pigments tend to ccumulate with increasing age in cells with low turnover rates and little capacity for exocytosis. Accumulation of similar material is associated with a class of hereditary degenerative diseases (Zeman, 1974). The increase in lysosomal accumulation of lipofuscin pigment which occurs in animals fed diets deficient in the antioxidants vitamin E and selenium suggests that the pigments may accumulate as a result of free-radical peroxidative reactions which convert natural tissue lipids to an indigestible form (Tappel, 1975).

The accumulation of lipid in lysosomes of smooth muscle cells which occurs in atherosclerotic plaques has been the subject of much recent scientific interest (de Duve, 1975; Peters, 1975). The excess lipid is predominantly cholesterol esters. The lipid-filled lysosomes are being well characterized by biochemical, cell fractionation, histochemical, and electron microscopic methods (Goldfischer *et al.*, 1975; Shio *et al.*, 1974; Peters and de Duve, 1974; Peters *et al.*, 1972; Wolinsky *et al.*, 1975). These studies have demonstrated increased specific activities of many lysosomal enzymes, with the remarkable exception of cholesterol esterase, in atherosclerotic lesions as compared to normal arterial wall (Goldstein *et al.*, 1975a,b). This increase in enzyme activity could be considered as an adaptation to an increased supply of lipid to the lysosomes of these smooth muscle cells. It is not yet clear to what extent the initiation of an atherosclerotic plaque is due to presentation of an excessive quantity of lipoproteins and other factors to smooth muscle cells which normally respond by phagocytosis and proliferation (Garfield *et al.*, 1975; Coltoff-Schiller *et al.*, 1976), or alternatively, to a relative deficiency in the ability of some smooth muscle cells to degrade lipid. The results of this current work will be of great biological as well as medical interest.

### 3. *Pseudohypertrophy*

Two different types of responses of cells to stimuli can be easily misinterpreted as primary increases in lysosomal enzyme activity per cell, when this is not the case. The first includes all the lysosomal storage diseases, in which a single lysosomal enzyme is inactive on the basis of a genetic defect. The situation is similar when lysosomal enzymes are inhibited. The predictable response to a deficiency of any enzyme activity in lysosomes is engorgement of the lysosomes with undigested material, leading to greatly increased size and visibility of lysosomes. Thus, on the basis of histochemical studies which indicated an increased number of enlarged lysosomes, several of the lysosomal deficiency diseases were first interpreted as examples of primary hypertrophy of lysosomes. In fact, detailed studies of lysosomes in several deficiency states have indicated that enzyme activities other than the ones which are deficient are often increased, presumably as a compensatory response (McKusick, 1970).

In a second situation, in the field of morphogenesis, an apparent increase in lysosomal enzyme activity of tissue led to the suggestion that hypertrophy of lysosomes was occurring in cells under precise control, but later work provided a different explanation. Beginning with Weber's early work on involution of the tail of the tadpole during metamorphosis, several observers have noted increases of total lysosomal enzyme activity in tissues which, in the development of the organism, were beginning to undergo involution (Weber, 1969). On the basis of such observations, a hypothesis has been developed in which "programmed cell death," a general phenomenon in embryonic morphogenesis which occurs, for example, during the separation of the digits of the hands, might be the result of accumulation within the doomed cells of large numbers of lysosomes, which would be triggered to kill their host cells at the appropriate time. In a few more detailed studies of embryonic morphogenesis, however, it has been found that increases in lysosomal enzyme activity did not occur in cells destined to disappear, but in exogenous macrophages which entered the area and phagocytosed degenerating cells (Dawd and Hinchcliffe, 1971).

### 4. *Decreased Enzyme Activity*

Another important lysosomal response is a decrease in enzyme activity, which may be specific or generalized. The genetic deficiencies of lysosomal enzymes, and their corresponding diseases, have been thoroughly described at clinical, tissue, subcellular, and chemical levels. Each enzyme deficiency predictably leads to massive intralysosomal accumulation of the product which is the usual substrate for the missing enzyme. These diseases have been very well reviewed (Hers and VanHoof, 1973), and specific ultrastructural appearances of the accumulated products have been documented for many of these diseases (Wallace *et al.*, 1966). One of the simplest means by which lysosomal enzyme activity may be inhibited is by increasing the pH of the extracellular medium. This alone has been shown to cause significant accumulation of material within lysosomes in diploid tissue culture cells (Lie, 1973). Such accumulation might be predicted from the assumption that degradation of parts of cells occurs continuously in cultured cells, and from the facts that: (a) most lysosomal enzyme activities are maximal around pH 5, and (b) lysosomes appear to maintain a pH gradient of only about 1.5 units between their exterior and interior surfaces (de Duve and Beaufay, 1959; Reijngoud and Tager, 1973). Likewise, those cationic dyes and drugs which accumulate in lysosomes (acridine orange, chloroquine) may also alter the internal pH of lysosomes, since these compounds become protonated in an acid environment, removing hydrogen ions from solution (Homewood *et al.*, 1972). It has been observed that several of the cationic compounds which do accumulate within lysosomes apparently cause the gradual engorgement of lysosomes in tissue culture cells with lamellar inclusions (Robbins *et al.*, 1964; Fedorko *et al.*, 1968). Inhibitors of lysosomal enzymes have also been described, the most specific being an antibody against cathepsin D (Dingle *et al.*, 1971; Dingle, 1973) and the inhibitors of lysosomal elastase (Janoff, 1972).

### *B. Alteration in Rate of Membrane Fusion Events*

A change in the rate of fusion between organelles is another important response of cells to certain forms of injury. The event of fusion between lysosomes and any other space in the vacuolar system can be considered separately and certain of these can be produced specifically, suggesting that fusion events within the vacuolar system are under rather precise cellular control (Ericsson, 1968b).

#### *1. Fusion between Plasmalemma, Phagosomes, and Lysosomes*

An increased rate of heterophagy, and subsequent fusion between newly formed phagocytic vacuoles and primary or secondary lysosomes, follows promptly after exposure of phagocytic cells to particulate foreign materials. The rate of phagosome–lysosome fusion, of course, would also be increased by agents which induce the differentiation of primitive cells into those capable of undergoing phagocytosis, e.g., phytohemagglutinin. A related fusion event, cellular regurgitation or "defecation," i.e., the fusion of a phagosome or secondary lysosome with the cell surface membrane, is rarely observed in normal cells, but has been observed following treatment with drugs (Abraham and Hendy, 1970; Kerr, 1970). This process is analogous to the discharge of secretory products and prevents engorgement of cells with debris. Neoplastic cells in tissue culture slowly release lysosomal enzymes without any evidence of release of enzymes associated with the cell sap, so lysosome–plasmalemma fusion must occur in them continuously. In polymorphonuclear leukocytes and macrophages, the process of phagocytosis is uniformly associated with discharge of lysosomal enzymes to the extracellualr space (Zucker-Franklin and Hirsch, 1964; Davies and Allison, 1976). This appears to occur by means of premature fusion of primary lysosomes with phagocytic vacuoles which have invaginated, but not yet completely separated from the cell surface membrane. The best established agent which leads to regurgitation of lysosomal enzymes is vitamin A, a result which was first reported in chick limb rudiments in tissue culture (Fell, 1969; Danes and Bearn, 1966; Dingle, 1968). Discharge of lysosomal contents to the bile canaliculi and sinusoids has been described in ultrastructural studies of rat liver (Kerr, 1970; Abraham and Hendy, 1970).

Extracellular release of lysosomal enzymes is a useful adaptation which serves a wide variety of specialized cellular functions. Resorption of bone is apparently accomplished by means of controlled regurgitation of lysosomal enzymes (Vaes, 1968, 1969) and it appears that the formation of the keratinized layer of the skin may well depend upon the regurgitation of lysosomal materials by degenerating epidermal cells (Goldstein *et al.*, 1975). Development of the inflammatory response is mediated in part by the discharge of lysosomal contents which then generate chemotactic factors, compounds which induce increased vascular permeability, and pyrogenic factors (Baggiolini *et al.*, 1969; Crowder *et al.*, 1969; Greenbaum and Kim, 1967; Movat *et al.*, 1971a,b; Chayen and Bitensky, 1971; Janoff and Zeligs, 1968; Janoff, 1972; Ranadive and Cochrane, 1968). Bradykinin and several other highly vasoactive polypeptide kinins are generated in the body by enzymatic cleavage of larger polypeptides, and it appears likely that lysosomal enzymes can participate in the generation of these vasoactive compounds as well as in their destruction (Melmon and Cline, 1967; Carvalho and Diniz, 1966). Finally, the tendency of malignant cells to separate from each other as single cells or clumps, which are then free to metastasize, seems likely to reflect a modification of the surface coat of malignant cells which may be brought about in part by the activity of lysosomal enzymes (Weiss and Holyoke, 1969). Invasion of normal tissue and of blood vessels by neoplastic tissue may be mediated by lysosomal enzymes; some tumors have been associated with increased local concentrations of lysosomal enzymes in the extracellular space (Sylven, 1968; Poole, 1973; Carter *et al.*, 1971). Of course, the presence of lysosomal enzymes in the extracellular space can be the result either of regurgitation from viable cells, or of diffusion of enzymes from necrotic cells, after their lysosomes rupture. Regurgitation of lysosomal enzymes by tumor cells would be a useful adaptation from the point of view of the tumor, although lethal to the host.

### 2. *Crinophagy*

An interesting phenomenon of fusion between lysosomes and secretory vacuoles of endocrine cells occurs under hormonal control. This process, known as crinophagy, has been elegantly described in the rat pituitary gland, in which an excess of circulating thyroid hormone leads to fusion between primary lysosomes and specific secretory granules of thyrotrophic cells, with subsequent degradation of the specific secretory product of these cells, thyroid-stimulating hormone (Farquhar, 1969). Crinophagy is a successful adaptation to efficiently recycle excess cellular synthetic products, and illustrates the precise control which can be exerted by the organism upon cellular fusion events involving lysosomes.

### 3. *Autophagy*

Fusion between autophagic vacuoles and lysosomes occurs in cells exposed to a long and growing list of stresses (Ericsson, 1969). Since the demonstration of this process requires fixation for ultrastructural histochemistry, it is often difficult to determine whether an increase in the number of observed autophagic lysosomes is entirely due to an increase in the rate of fusion events, or partly due to a decreased rate of degradation of material taken in by autophagy. The factors which lead to increases in the rate of autophagosome–lysosome fusion are becoming clarified, however, and patterns are emerging. Stimulation of cellular autophagy has been observed in response to deprivation of essential amino acids, protein starvation, and involution of the mammary gland (Hruban *et al.*, 1963; Lane and Novikoff, 1965; Mitchener *et al.*, 1976; Helminen *et al.*, 1968). In these situations it is reasonable to suppose that autophagic self-digestion is useful in providing protein, amino acids, and other metabolic necessities for the organism, and in removing unnecessary tissue. Many other stimuli have been shown to increase the number of autophagic vacuoles in cells, including hyperoxia, glucagon, vinblastine, and cyclic AMP analogues (Terry *et al.*, 1970; Shelburne *et al.*, 1973a,b; Arstila *et al.*, 1974). Autophagocytosis may well be a normal mechanism for turnover of aged or defective organelles, since the half-life of mitochondria has been found to be far less than that of their host cells in several organs (Beattie *et al.*, 1967; Dean, 1975; Fletcher and Sanadi, 1961; Segal, 1975).

### 4. *Decreased Rate of Fusion*

Decreased evidence of fusion of lysosomes with other organelles has been described in a naturally occurring situation only once, in the leukocytes of the chronic granulomatous disease of children, but this finding could not be confirmed in later investigations (Quie *et al.*, 1967; Baehner *et al.*, 1969). As already noted, many drugs have been shown to decrease the rate of fusion of lysosomes with the plasma membrane or other parts of the cytocavitary system, by increasing cellular cyclic AMP levels or by inhibiting the cell's metabolic or structural machinery.

### C. *Alteration in Permeability of the Lysosomal Membrane*

The final type of significant alteration in lysosomes which might be expected following cell injury is an alteration in the permeability of the lysosomal membrane. A great many studies of cell injury have provided evidence suggestive of an increase in permeability or rupture of lysosomal membranes. With the possible exception of the lysosome stabilizers, however, a decrease in permeability of lysosomes has not been described.

#### 1. *Estimation of Membrane Permeability*

The most detailed studies of lysosomal permeability in intact cells have been carried out in mouse peritoneal macrophages (Cohn and Ehrenreich, 1969; Ehrenreich and Cohn, 1969). It was shown that indigestible dipeptides, disaccharides, or larger compounds were taken up into these cells with prominent swelling and engorgement of phagolysosomes. Monosaccharides and amino acids did not lead to vacuolization, although they were taken up by the cells. In fact, digestion of the larger compounds within swollen lysosomes, which could be induced by administration of the appropriate enzymes, was accompanied by a decrease in lysosomal size, strongly suggesting that many disaccharides and dipeptides were retained by the lysosomal membrane, but that their products of hydrolysis were able to cross this membrane freely. This observation agrees with the suggestion that lysosomal membrane permeability is more or less perfectly matched to the normal function of the lysosome, in that materials taken into lysosomes are retained by its membrane until they have been degraded to the level of single sugar or amino acid units, which are ready for reutilization in other parts of the cell.

Observations on isolated lysosomes have allowed detailed examination of the ability of various compounds, in the medium surrounding lysosomes, to protect them from osmotic lysis (Lloyd, 1969). Osmotic protection apparently depends on limited transfer of a compound across the lysosomal membrane. These experiments indicate that even certain inorganic salts offer some osmotic protection, acetate being for example more effective than chloride (Berthet *et al.*, 1951). Sucrose, as would be expected, is a more effective osmotic protective agent than glucose (Lloyd, 1969). The membranes of rat liver lysosomes appear to be permeable to some dipeptides, disaccharides, and most smaller molecules (Lloyd, 1971). Observations on binding of dyes and drugs in isolated lysosomes strongly indicate that even *in vitro*, lysosomes maintain an effective internal pH approximately 1 to 1.5 units below that of the medium, even in the absence of any supply of metabolic energy (Reijngoud and Tager, 1973, 1976; Henning, 1975). That is, they maintain a transmembrane gradient of hydrogen ion concentration, approximately 10- to 30-fold greater inside lysosomes, and this gradient appears to require an intact lysosomal membrane. As previously noted, many of the substrates used to measure lysosomal enzyme activity are capable of crossing the normal lysosomal membrane only very slowly if at all, so that they are "latent," and some damage to lysosomal membranes is necessary for measurement of total enzyme activity *in vitro* (de Duve *et al.*, 1955).

### 2. *Selective Increases in Membrane Permeability*

It is possible that a limited injurious stimulus to cells might cause an increase in the permeability of lysosomal membranes to some compounds but not to all. A few experimental results suggest that this does occur. Anything which increases the rate of flow of water and salt solutions across the lysosomal membrane would be expected to lead to a greater sensitivity to disruption by hypotonic solutions, and this effect is observed following treatment by several lysosomal labilizers. In intact tissue culture cells, acridine orange dye, previously localized within lysosomes, has been observed to disperse throughout the cell at a time when macromolecules were retained, following treatment with metabolic inhibitors (Hawkins *et al.*, 1972). Since the accumulation of this dye probably depends on the pH gradient normally maintained between the lysosome and the cell sap, the release of dye from lysosomes may well reflect a loss of this pH gradient.

An increase in lysosomal permeability to the substrates used to assay enzyme activity has been observed in cell homogenates following mild injury, as an increase in free activity. Histochemical studies of acid phosphatase localization in minimally injured, unfixed tissue culture cells have suggested that the same process, called first-stage activation, occurs as an early stage of cell injury. Normal cells incubated briefly in the medium for histochemical demonstration of acid phosphatase produce no reaction product, presumably because their lysosomes, or the cell surface membrane, or both, are not permeable to the $\beta$-glycerophosphate used as a substrate. The appearance of reaction product after brief incubation of cells, localized entirely to lysosomes, has been described as a response to injury. The appearance of reaction product diffusely throughout the cell is a much later stage of reaction to lethal cell injury in this system (Chayen and Bitensky, 1968). The histochemical substrate solution is itself toxic to cells, however, so that some alterations in lysosomal or cell surface membranes may occur during incubation in this experimental system.

It has been assumed that release of lysosomal enzymes from the particulate phase into solution indicated rupture or complete dissolution of the lysosomal membrane, and that all lysosomal contents would be released from the particulate phase simultaneously. However, a few recent observations suggest that diffusion of enzymes from lysosomes is possible when lysosomal enzymes are morphologically intact, and retain tracer materials such as Thorotrast (Brunk and Ericsson, 1972; Decker *et al.*, 1977). These observations suggest the possibility that in a certain stage of their response to injury, lysosomal membranes may exhibit selective permeability even in the size range of protein macromolecules.

### 3. *Rupture of Lysosomes*

Disintegration of the membranes of lysosomes would be expected to release all hydrolytic enzymes and other contents of lysosomes into the cell sap. Cellular self-digestion would be expected to follow this event, as an effect of the neutral proteinase present in many lysosomes as well as the activity of many other hydrolases including RNase and DNase, which retain significant activity at neutral pH. At the acid pH present in ischemia or autolysis, lysosomal enzymes should be capable of digesting essentially all cellular components. It has been shown by direct experiments that the interaction of lysosomal enzymes with mitochondria causes rapid deterioration of oxidative phosphorylation (Mellors *et al.*, 1967), and it would be expected that abnormal plasmalemmal permeability would result from the attack of enzymes or products of digestion of lipids on the inside surface of the plasmalemma (Weglicki *et al.*, 1974). It has been suggested that intracellular release of lysosomal enzymes might damage the DNA of surviving cells, causing formation of abnormal chromosomes and possibly neoplasia itself. Limitations of methodology, unfortunately, are such that intracellular rupture of the lysosomal membrane cannot be demonstrated conclusively by any one of the usual biochemical, histochemical, or electron microscopic techniques, so that any isolated report based on a single technique must be received with some skepticism.

*a. Observations Suggesting Rupture of Lysosomes.* In the 1960s many observations accumulated which suggested that rupture of lysosomes occurred relatively early in many different types of cell injury. The work of de Duve and Beaufay on ischemia has already been discussed. Histochemical study of the effects of cytotoxic viruses in cell culture showed that release of reaction product from lysosomes accompanied cell lysis (Allison and Mallucci, 1965; Wolff and Bubel, 1964; Fine *et al.*, 1970; Koschel *et al.*, 1974). In the case of poliovirus infection of cell culture monolayers, nearly simultaneous release of both lysosomal and cytoplasmic enzymes to the medium was demonstrated, which is consistent with the concept that rupture of lysosomes preceded cell lysis, and altered sedimentation properties of lysosomes were observed as an early cytotoxic event in this model system (Blackman and Bubel, 1969). Histochemical studies showed that ingestion of silica caused redistribution of lysosomal enzymes in cultured macrophages (Comolli, 1967). An antiserum against leukocyte lysosomes was reported to kill leukocytes, and it was presumed that rupture of lysosomes mediated this effect (Quie and Hirsch, 1964). Several workers observed an increase in lysosomal enzyme activity in serum of animals exposed to hemorrhagic shock, suggesting that rupture of lysosomes was occurring in this condition, although the other changes in the cells which released these enzymes were not known (Reich *et al.*, 1965; Janoff, 1964).

*b. Cellular Consequences.* The most significant consequence of lysosomal enzyme release which could be anticipated as a response to cell injury is the death of the host cells. This "suicide bag" hypothesis has stimulated much experimental work on lysosomes in injured cells. The suggestion that lysosomes could be responsible for cell death after certain forms of tissue injury was made by de Duve, on the basis of theoretical considerations and the results of an experiment reported in 1959 on the lysosomal and other enzyme activities of ischemic rat liver (de Duve and Beaufay, 1959). It was observed that during the first 4 hr of total ischemia of a lobe of the liver, the proportion of enzyme activity which could be removed by centrifugation (sedimentable activity) steadily decreased, while the proportion of enzyme activity which was soluble increased to approximately 40% of the total. Although it was clearly pointed out that this finding could reflect either spontaneous intracellular rupture of lysosomes, increased sensitivity to the homogenization procedure, or the effect of some toxic substance released during homogenization, the first interpretation was favored. This evidence, suggesting that intracellular rupture of lysosomes occurred "early" in the course of cellular autolysis in ischemic tissues, led to the reasonable suggestion that, just as autolysis is prevented during the life of the cell by the lysosomal membrane, rupture of this membrane might lead to cellular self-digestion early in the course of lethal cell injury. Thus, like a spy taken into captivity, an injured cell might sacrifice itself to the greater good by suicidal disruption of its lysosomes. The hypothesis that rupture of lysosomes could itself be responsible for the death of injured cells was boldly proposed by de Duve for further study.

*c. Methods for Demonstrating Intracellular Rupture of Lysosomes and Its Cellular Effects.* Simply to demonstrate unequivocally that lysosomes have released their enzymes within an intact cell is a difficult problem, since each of the methods which may be applied to the study of cell injury induces its own artifacts. For example, studies of tissue homogenates, which can provide good documentation of changes in the distribution of lysosomal enzymes, do not give exact information about the state of lysosomes prior to homogenization. In a strict sense these methods indicate only the responses of lysosomes to the stress of mechanical disruption of tissue. Simple enlargement of lysosomes, whether due to heterophagy or autophagy, or fusion or swelling of primary lysosomes, leads to changes in their sedimentation properties, and larger vacuoles are known to be more sensitive to disruption in tissue homogenizers, simply because larger structures experience greater shearing forces (Deter and de Duve, 1967). Thus an increase in unsedimentable lysosomal enzyme activity in a tissue homogenate can simply reflect enlargement of lysosomes rather than their disruption *in vivo*.

Histochemical techniques, while they do allow identification of intracellular sites of lysosomal enzyme activity, suffer from well-documented artifacts of selective enzyme inactivation by fixative agents, membrane alterations which occur as a result of fixation or in spite of it, and toxic effects of the histochemical incubation mixtures themselves (Holt, 1959; Chayen and Bitensky, 1968; Lake and Ellis, 1976; Hündgen, 1977).

Ultrastructural studies of well-fixed tissues are also severely limited in their ability to demonstrate slight changes in lysosomes. Lysosomal membranes may occasionally appear to be broken even with fresh tissue and optimal fixation. On the other hand, the thin-sectioning technique allows observation of only a single plane through each lysosome, so that even large breaks in lysosomal membranes which might occur outside the plane of section would not be observed. Under certain special conditions, however, ultrastructural study of thin sections can provide a sensitive method of testing for the presence of gaps in lysosomal membranes. Since cells with active phagocytic capibilities can be induced to take up large numbers of small particles from the extracellular space, electron-dense particles such as ferritin, thorium dioxide, saccharated iron oxide, or colloidal gold can be administered to such cells and used as tracers to mark the interior of secondary lysosomes (Ericsson, 1968a). Since these phagocytized particles would be expected to be released along with other hetero-lysosomal contents through any breaks in the membranes of these organelles which might occur in injured cells, this method can be used to assay for lysosome rupture in any type of cell injury. Although the method is strictly limited to secondary lysosomes, and to rather large gaps in lysosomal membranes, it does provide a sensitive test for such gaps, inasmuch as single tracer particles can be identified and localized by electron microscopy.

Demonstration of lysosome rupture as the cause of cell death in any experimental situation would require clear evidence that lysosome rupture occurred prior to cell death. This would thus require fairly precise identification of the times of lysosome rupture and of cell death. In addition, it would require demonstration that intracellular release of lysosomal enzymes would in fact cause fatal injury to the host cell, which in turn would depend upon a method for rupturing lysosomes without damaging other cell components. As mentioned above, the photosensitization mechanism seems to be such a method. Cells which accumulate acridine orange within their lysosomes and are subsequently exposed to ultraviolet light do appear to undergo rapid rupture of their lysosomes, and the evidence suggests that release of lysosomal enzymes by this process is sufficient to cause death of the host cells, although other interpretations of the data are possible.

Circumstantial evidence that lysosomes may participate in causing irreversible changes in injured cells can be obtained in a variety of ways. For example, it could be shown that chemical treatments which inhibit the release of enzymes from damaged lysosomes increase the survival rate of injured cells, or vice versa. Such evidence has been obtained on the effect of lysosomal labilizers in increasing the severity of carbon tetrachloride induced liver necrosis, and in a few other studies of acute lethal cell injury (Iturriaga *et al.*, 1969; Weissmann, 1968). An effort has even been made to take advantage of this theoretical effect of lysosomal labilizers in the development of cancer chemotherapy (Brandes *et al.*, 1966). Dexamethasone and cortisol have been shown to decrease the release of creatine phosphokinase from ischemic myocardium, presumably by decreasing infarct size, and because the best-known subcellular effect of these drugs is stabilization of lysosomes, it has been suggested that prevention of enzyme release prevents cell death after ischemia (Spath and Lefer, 1975). However, it has been found that dexamethasone or medylprednisolone administered to cats was more concentrated in the plasma membrane fraction than in lysosomal or other subcellular fractions of the heart, and seemed to protect the plasma membrane from ischemic injury (Okuda *et al.*, 1976).

*d. Examples of Problems of Interpretation.* Probably because of these difficulties in identifying the exact cellular events which are responsible for cell death, lysosome rupture has been thought to be responsible for cellular degeneration in several experimental situations in which this "suicide bag" hypothesis is no longer tenable. The effect of streptolysin O on leukocytes, once thought to be mediated by rapid rupture of lysosomes on the basis of phase contrast light microscopy, was subsequently shown by ultrastructural methods to be associated with breaks in the cell surface membrane and degenerative changes in other organelles, prior to the time when changes in lysosomes could be observed (Zucker-Franklin and Hirsch, 1964; Zucker-Franklin, 1965). The observation that vitamin A excess led to degradation of cartilage matrix and release of lysosomal enzymes to the medium in organ cultures of embryonic limb rudiments was first interpreted as an indication that vitamin A, a lysosome labilizer, had killed cells by causing their lysosomes to rupture intracellularly (Lucy *et al.*, 1961; Dingle, 1961). This hypothesis was abandoned when it was observed that vitamin A caused increased regurgitation of lysosomal enzymes by fusion of lysosomes with the cell surface membranes (Fell, 1969). These examples are presented to show that caution is necessary in interpreting a lysosomal change in injured cells as evidence that rupture of lysosomes was responsible for cell death.

*e. Tracer Studies in Injured Cells.* An example of the use of ultrastructural tracer techniques to study the integrity of lysosomes in cell injury was a study by the author of complement-mediated immune cytolysis in tissue culture cells. For these studies, cells were incubated overnight in medium containing ferritin (Pentex, cadmium-free, 15–25 mg/ml) which led to heavy labeling of phagosomes and secondary lysosomes with ferritin molecules, the electron-dense iron cores of which are easily identifiable in electron micrographs of thin sections (Fig. 1). Monolayers were first exposed briefly to antiserum against whole cells, (in which complement had been inactivated by heating) and then washed, leaving antibody attached to the cell surface. The cells were subsequently exposed to complement, which led to swelling and degeneration of the majority of cells within approximately 30 min. Changes in cells were observed by electron microscopy at intervals, with and without previous incubation with ferritin. As shown in Figs. 2 and 3, not only was there no evidence of lysosome rupture during the early stages of cellular degeneration, disruption of labeled secondary lysosomes occurred surprisingly late in the course of the changes of cellular necrosis which were observed. Evidence of lysosome rupture was frequently observed only 4 hr or longer after the initial cell injury. These findings suggest that rapid, severe damage to the cell surface membrane, which causes rapid cell swelling and alterations in many organelles, leads to rupture of the membranes of secondary lysosomes only after major, irreversible cellular degeneration has already taken place. This impression was strengthened by the observation of living cells by fluorescence microscopy which had been incubated for 15 min in $10^{-5}$ gm/ml acridine orange dye. This vital stain probably accumulates in both primary and secondary lysosomes as a consequence of their relatively high internal concentration of hydrogen ions, and it identifies lysosomes as brilliant red granules. By this technique rapid, extensive cell swelling could be observed after adding complement to sensitized cells, followed by loss of cell motility and onset of Brownian motion. No loss of red fluorescence from lysosomes occurred until more than 2 hr after this severe injury (Fig. 4). Our results suggest that lysosomes are relatively resistant to cell injury mediated by disruption of the plasmalemma, and that lysosomes can persist without observable release of their contents for some time after cell death.

The technique of loading secondary lysosomes with electron-dense tracer substances is especially useful as a sensitive method of providing proof that lysosomal membranes have broken after experimental injury. The value of the technique in this situation was shown by illustrating release of ferritin from lysosomes in cells which were loaded with ferritin and acridine orange, then irradiated with ultraviolet light for 20 min. Dispersal of ferritin throughout the cell sap and nucleoplasm demonstrated that this photosensitization injury rapidly disrupted lysosomal membranes (Fig. 5).

*f. Cell Injuries Known to Cause Rupture of Lysosomes.* An increase in lysosomal membrane permeability, with escape of all contents from lysosomes, occurs after cell death as part of the process of necrosis; cell death is probably always followed by degradation by lysosomal enzymes *in vivo*. For many years it has been known that sterile incubation of isolated tissue is accompanied by autolysis, the extensive degradation of cellular protein and other macromolecules by endogenous hydrolases, which continues for days or weeks (Opie, 1922; Wilstätter and Rohdewald, 1932; Bradley, 1938; de Duve and Beaufay, 1959).

Much more interesting than this universal post-mortem disruption of lysosomes would be modes of cell injury in which loss of lysosomal contents could be clearly shown to precede cell death or to coincide exactly with the onset of irreversibility. Convincing evidence that disruption of lysosomes occurs at or before the point of cell death and probably causes fatal self-digestion has been presented for three forms of cell injury. Several indepen-

**Fig. 1.** Uninjured control Chang cell grown in monolayer culture in minimal essential medium with 10% fetal calf serum, and incubated overnight with cadmium-free ferritin, ca. 20 mg/ml. Many secondary lysosomes and phagosomes, most of which have the structure of multivesicular bodies, are filled with electron-dense ferritin particles. No ferritin is seen in the cell sap. Cells were fixed in a freshly made, buffered mixture containing 4% glutaraldehyde and 1% osmium tetroxide, and stained *en bloc* and as a thin section with uranium acetate. The same techniques were used in the following electron micrographs. ×32,000.

**Fig. 2.** Ferritin-loaded Chang cell exposed to heat-inactivated rabbit antiserum against whole Chang cells for 10 min, washed twice in fresh medium for 20 min, then exposed to 10% guinea pig serum as a source of complement, and fixed two hours after exposure to complement. Mitochondria are swollen and contain numerous amorphous matrical densities, rough endoplasmic reticulum is dilated, and the cell is considered to be dead. Ferritin is still restricted to secondary lysosomes and phagosomes with intact membranes. ×39,000.

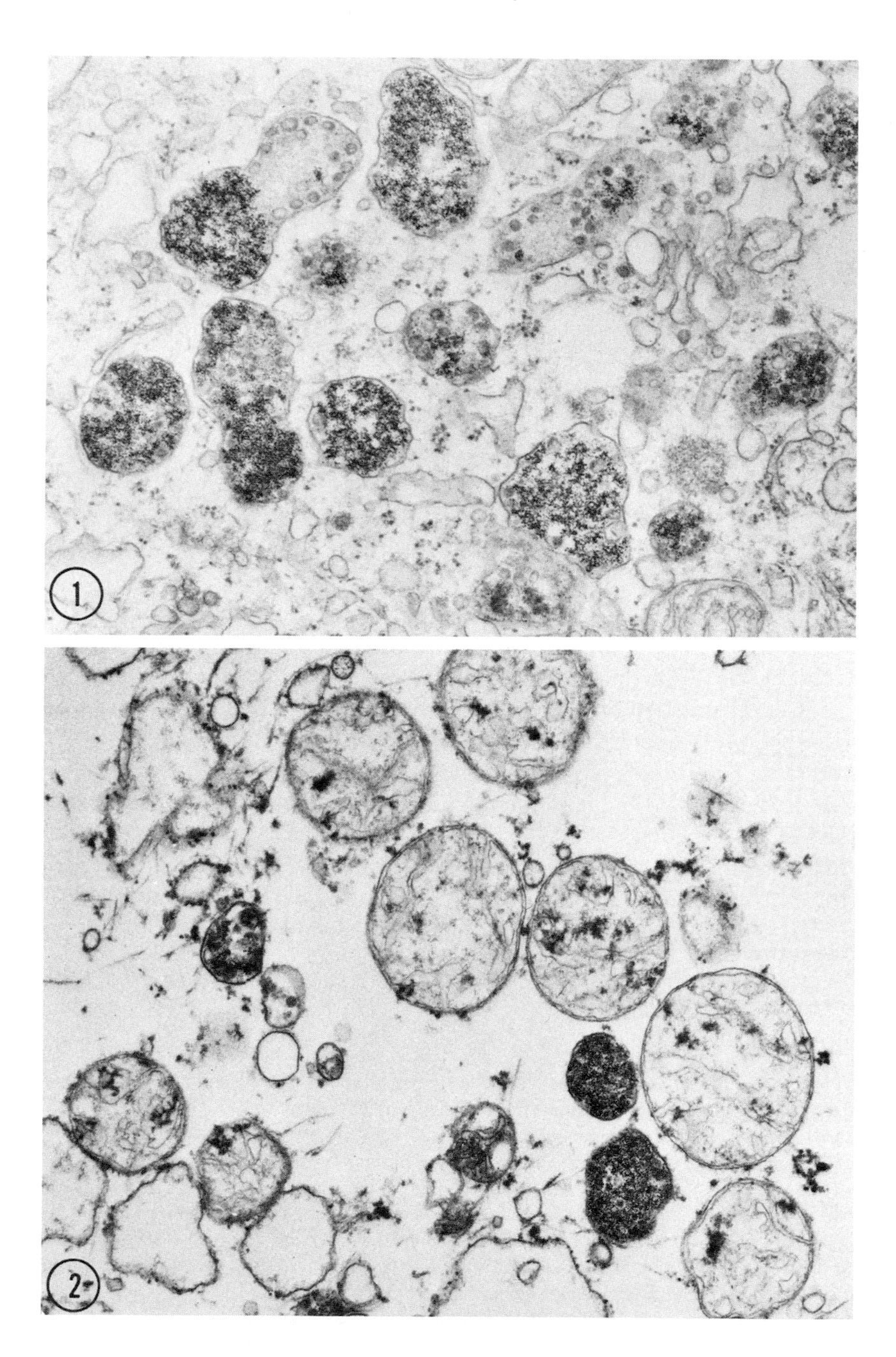
1
2

dent kinds of evidence support the "suicide bag" hypothesis in these cases, and they all fit a pattern of primary injury to lysosomal membranes.

Phagocytosis of particles of silica by cultivated macrophages has been shown to be followed within 24 hr by cell death, loss of lysosomal enzymes to the medium, loss of vital staining of lysosomes by acridine orange, loss of particulate staining of acid phosphatase, and, by electron microscopy, release of crystals from secondary lysosomes into the cell sap. Phagocytosis of diamond dust, or of silica particles coated by polyvinylpyrrolidone *N*-oxide or aluminum dust, did not damage lysosómes or cause cell death (Allison *et al.*, 1967). A possible toxic effect of silica on plasma membranes was not ruled out in these studies, but would have been expected to injure cells without a time lag, as occurs during hemolysis by silica particles. One consequence of the toxicity of silica to macrophages is release of a factor which stimulates fibrosis (Aalto *et al.*, 1976).

Similarly, phagocytosis of sodium urate crystals by leukocytes is followed by cell injury in which rupture of lysosomes is apparently the initiating event (Weissmann and Rita, 1972). In this model the evidence of lysosomal injury includes enzyme histochemistry, release of lysosomal enzymes to the medium simultaneously with cytoplasmic enzymes, and electron microscopy showing urate crystals in the cell sap space without a limiting membrane (Hoffstein and Weissmann, 1975; Schumacher and Phelps, 1971; Riddle *et al.*, 1967). Prevention of phagocytosis prevented cell injury, effectively ruling out the possibility of primary injury to the plasma membrane by crystals. Interesting similarities have been found between the surface structures of these crystals which induce inflammation by damaging membranes, as opposed to those which do not (Mandel, 1976).

Many compounds are selectively accumulated in lysosomes of cells in culture (Allison, 1964). When these cells are transferred to medium free of

---

**Fig. 3.** An antibody sensitized cell after nine hours of exposure to complement. Apparent rupture of this multivesicular body is releasing ferritin into the cell sap. ×42,000.

**Fig. 4.** Fluorescence micrograph of Chang cells exposed to acridine orange dye, $10^{-5}$ gm/ml, for 15 min, then returned to fresh culture medium. These cells were also treated with antiserum and complement as above and underwent swelling and lysis more than 2 hr before the picture was taken. Lysosomes fluoresced red-orange, and nucleoli green. ×600. (Reproduced from Hawkins *et al.*, 1972, with permission of *Am. J. Pathol.*)

**Fig. 5.** Electron micrograph of a Chang cell which was loaded with ferritin overnight, then labeled with acridine orange to resemble the cells shown above, then exposed to ultraviolet light (300–500 nm wavelength) for 30 min. Ferritin molecules are not restricted to lysosomes, but are scattered throughout the cell sap and nucleoplasm. Mitochondria appear abnormal but gross cell swelling is not apparent. ×32,000. (Reproduced from Hawkins *et al.*, 1972, with permission of *Am. J. Pathol.*)

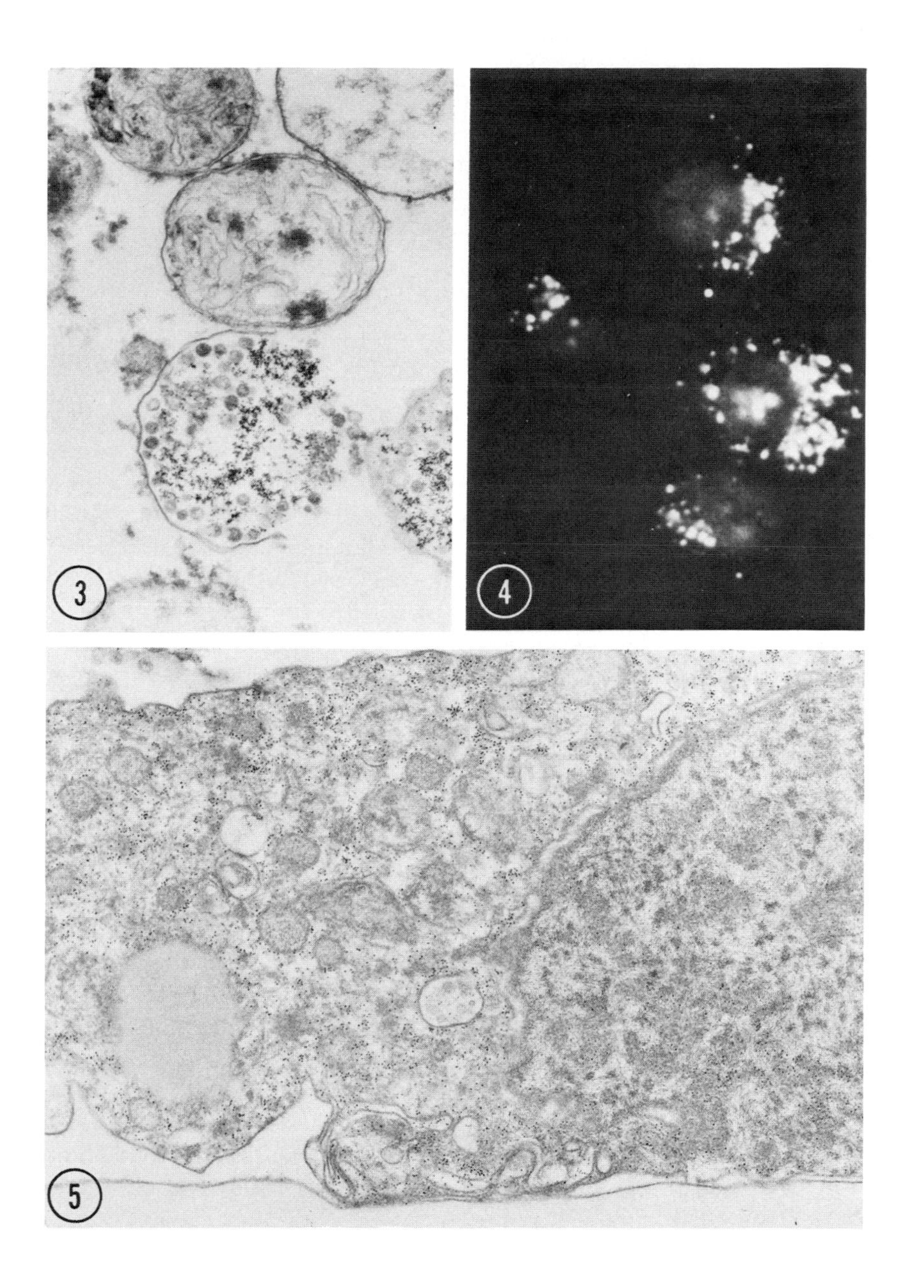
3
4
5

dye and irradiated with light absorbed by the dye in the presence of oxygen, lethal cell injury apparently occurs. This type of cell injury, using acridine orange as the sensitizing dye, has been shown by histochemical methods to cause rapid redistribution of acid phosphatase activity to the cell sap and release of lysosomal enzymes to the medium (Allison *et al.*, 1966; Allison, 1968b; Brunk and Ericsson, 1972). As shown in Fig. 5, electron-dense particles which were restricted to secondary lysosomes and phagosomes were rapidly released to the cell sap space after this kind of photosensitization injury (Hawkins *et al.*, 1972).

Only in these examples of primary injury to lysosomes has convincing evidence been presented to indicate that disruption of lysosomes preceded cell death and was the primary cause of lethal or irreversible cell injury.

*g. Early Rupture of Lysosomes in Ischemic Cell Injury.* In recent years several investigators have implicated intracellular disruption of lysosomes as an early event in ischemic injury in the heart. The problem of the mechanism of cell death in myocardial infarction is very important, since this disease is the leading cause of death in the United States. Active research is being directed at finding agents, either empirically or by rational selection, which can reduce the mass of tissue irreversibly injured by ischemia. The lysosomes of myocardium have only recently been characterized by modern techniques of cell fractionation; they have a relatively low equilibrium density and a high content of cathepsins as compared with liver lysosomes, and several distinct populations of lysosomes are present in heart muscle (Smith and Bird, 1975, 1976; Welman and Peters, 1976). It is critically important in studying the role of lysosomes in the pathogenesis of ischemic cell injury to consider the relationship of the onset of cell death to the observed changes in lysosomes. Lysosomal rupture is to be expected in dead cells as an autolytic change in cells undergoing necrosis, and becomes significant as a possible factor in causing cell death only if it occurs early. Fortunately, the time of onset of cell death is well defined as beginning in the papillary muscle 20 to 30 min following circumflex coronary artery occlusion in the dog (Jennings *et al.*, 1960), and the changes in ultrastructure, release of cytoplasmic enzymes such as creatine phosphokinase, and loss of volume control which are associated with cell death have been characterized (Jennings *et al.*, 1969a,b, 1975; Hawkins *et al.*, 1977). Several recent studies of myocardial ischemia have included substantial efforts to relate changes in lysosomes to the time of onset of cell death, so further clarification of this issue is to be expected (Wildenthal, 1975).

Biochemical studies of lysosomes as early as 1965 showed redistribution of lysosomal enzymes from sedimentable to unsedimentable form several hours after coronary artery occlusion (Brachfeld and Gemba, 1965). Ricciutti (1972) showed a decrease in lysosome stability 1 hr after circumflex coronary artery occlusion and suggested that this might have contributed to cell death. He noted depletion of tissue glycogen and maintenance of normal tissue potassium in the same injured tissue; however, these changes are consistent with irreversible cell injury (Jennings *et al.*, 1964). Weglicki and co-workers clearly demonstrated redistribution of several lysosomal enzymes from a sedimentable to a soluble form after 1 hr of occlusion of the left anterior coronary artery in dogs, but electron microscopy showed changes interpreted as early irreversible cell injury (Gottwick *et al.*, 1975). In an isolated perfused guinea pig heart preparation, deprivation of oxygen and glucose were shown to cause significant redistribution of several lysosomal enzymes in homogenates after only 30 to 60 min, but massive release to the perfusate of creatine phosphokinase and other intracellular enzymes was observed after 30 min, probably reflecting irreversible injury of many cells (Welman and Peters, 1977). Finally, asphyxia was reported to increase the proportion of free activity of lysosomal enzymes after only 6 to 10 min in dog hearts (Leighty *et al.*, 1967). Homogenization of myocardial tissue is difficult and tends to give a poor yield of sedimentable lysosomal enzymes, with little latency, making studies such as these technically difficult. Improved methods, e.g., homogenizing in hypertonic KCl, may reduce these problems.

A few histochemical studies have shown loss of the usual pattern of staining for acid phosphatase and arylsulfatase by electron microscopy after occlusion of the left anterior coronary artery in dogs for 3 1/2 hr in one study, and for 1 hr in another (Hoffstein *et al.*, 1975; Gottwick *et al.*, 1975). The ultrastructure of the cells was characteristic of irreversible injury in both cases, but further histochemical studies near the time of onset of cell death will be of considerable interest.

In another recent study, the new technique of measurement of lysosomal cathepsin D in cardiac lymph showed an increase in its specific activity after 2 hr of coronary artery occlusion in the dog (Araki and Takenaka, 1976). Further studies using this method should prove valuable.

It has recently become possible to identify lysosomal enzymes by immunohistochemical means using antibodies against lysosomal enzymes (Tulkens *et al.*, 1970; Dingle *et al.*, 1971). This promising new method allowed demonstration of redistribution of cathepsin D, suggesting release from lysosomes within myocardial cells, only 30–45 min after occlusion of the circumflex coronary artery in rabbits (Decker *et al.*, 1977). Electron microscopy showed changes consistent with sublethal or reversible cell injury at this time, so that the data in this single report strongly suggest that release of lysosomal cathepsin D preceded cell death.

Considered separately, these observations are at most suggestive that release of enzymes from lysosomes may occur near the time of cell death in myocardial ischemia, and several only show post-mortem changes. Taken together, however, they certainly indicate that the challenging possibility still exists, though not yet proven, that disintegration of lysosomes might be one of the significant alterations leading to cell death in ischemic injury of the heart.

## IV. Summary

This chapter represents an attempt to organize and discuss current knowledge of the pathology of lysosomes, and to organize the best established observations into a logical classification, emphasizing particularly the challenges of detecting changes in lysosomal permeability *in situ* and relating the changes to the causes of events leading to cell death. Clearly, all abnormalities of lysosomes, whether due to hereditary deficiency, alteration in response to disease, or laboratory manipulation, have to fall into one of these six categories: lysosomes can be abnormal only by increasing or decreasing either the quantity of some or all of their contents, the rate of their fusion with other parts of the cell, or the permeability of their limiting membranes. A worthwhile goal of experimentation in this area would be to define as accurately as possible the nature of the primary abnormality occurring in lysosomes. This is often complicated by superimposed secondary events. For example, deficiency of a specific lysosomal enzyme is often followed by not only engorgement of lysosomes with undigested substrate, but also an apparent adaptation on the part of lysosomes with an increase in their enzyme activity generally. Similarly, in the Chediak–Higashi syndrome, an abnormal rate or specificity of fusion of lysosomes is accompanied by extreme lysosomal enlargement, and it has proved difficult to discover which event causes the other (Oliver and Essner, 1975; Root *et al.*, 1972). Nevertheless, it would be a good working hypothesis to assume that one of the six types of lysosomal abnormality is the primary alteration in any study of lysosomal pathology. Further improvement of our knowledge of the responses of lysosomes to various kinds of stress should reveal patterns of lysosomal response, leading to improved hypotheses. Reliable knowledge in this area is very limited at the present time. Some of the changes in lysosomes discussed in this chapter can be shown to occur by straightforward application of existing methods, while others are more difficult to document since existing methods tend to give ambiguous results. Of course, it is relatively easy to document increases or decreases of lysosomal size, shape, or number, by direct observation of living tissue with a light microscope, or observation of fixed tissue by light or electron microscopy. In some cases stereological methods have been successfully applied to provide direct measurements of the volume fraction of lysosomes within cells or of mean lysosome diameter. Histochemical techniques often allow tentative identification of increases in specific lysosomal contents, as was done in the early studies on glycogen storage disease. Biochemical analysis of enzymes has of course been impressively productive in identifying specific lysosomal enzyme deficiencies, each with its associated congenital disease.

Detection of changes in fusion rates in lysosomes and other cell components, however, is much more difficult, and is usually inferred from indirect observations. An increased number of autophagic vacuoles per cell, such as we have detected after protein and amino acid starvation in tissue culture (Mitchener *et al.*, 1976), indicates the strong likelihood that organelles are being enclosed in autophagic vacuoles at an increased rate, which then fuse with lysosomes at an increased rate. An equally plausible alternative explanation of this observation, however, would be that organelles are being degraded more slowly once they are enclosed within autophagic vacuoles. Clear demonstration of the rate of formation of autophagic lysosomes would probably best be achieved by a combination of biochemical evidence of increased organelle turnover or increased catabolism, combined with morphological evidence that the process of autophagy is observed more frequently than in controls. As the effects of drugs on lysosomal functions are clarified, more sophisticated pharmacologic means will be found to produce predictable changes in rates of lysosomal fusion.

Changes in permeability of the lysosomal membrane are probably the most difficult type of abnormality to demonstrate reliably. In a preparation of isolated lysosomes, membrane permeability can be tested by measuring the distribution space of various radioactive tracers, by studying the effects of different compounds in protecting against osmotic lysis, or by measuring changes in light scattering as a result of osmotic effects of various compounds. However, when an abnormality of lysosomes is suspected, particularly when the structure and chemical state of the tissue is drastically abnormal, such as after ischemia, the process of isolation itself causes serious problems: the fraction of lysosomes which is selected may not be representative of the tissue as a whole, or lysosomes may be destroyed or rendered abnormally permeable during the isolation process. Changes in yield, sedimentation properties, or permeability or other abnormalities in lysosomes after isolation from an injured tissue certainly demonstrate that some abnormality was present before isolation, but do little to clarify the nature of that alteration. The direct demonstration of abnormal permeability of lysosomes within intact cells poses an interesting challenge. Introduction of tracer materials from outside the cell, which by definition gives information only about secondary lysosomes and phagosomes, has been useful in the study of this problem. Elegant studies of changes in size of macrophage lysosomes in response to exogenous sugars and peptides have helped to clarify the normal permeability of lysosomal membranes *in situ* (Ehrenreich and Cohn, 1969). Our work with acridine orange and ferritin as exogenous tracers made it possible to study the responses of lysosomes to experimental cell injury, and similar techniques were of value in providing direct demonstration of the rupture of lysosomes as a consequence of their interaction with sodium urate crystals, and as a result of ingestion of silica. Improvement in techniques of this sort should make it possible to gain additional information about lysosome rupture or more subtle changes in permeability within intact cells, and changes in primary lysosomes can be subjected to study as well, using powerful immunohistochemical methods.

Abnormalities in lysosomes have been observed in a wide variety of disease states and tissue reactions to injury. Application of modern techniques of cell biology should make it possible to greatly clarify the nature of these lysosomal abnormalities. Specific changes in quantity of lysosomal contents, rates of fusion, or membrane permeability, should be documented with increasing frequency in the near future, making it possible to devise a more rational scheme of pathology of lysosomes. As a result of the present active research on lysosomotropic drugs and on the mechanisms of inflammation, pharmacologic control of properties of the lysosomal system should become much more sophisticated. It may become possible to effectively correct abnormalities of lysosomes, whether these abnormalities represent the primary causes of disease, consequences of tissue injury leading to premature necrosis, mediators of an inflammatory response, or byproducts of disease. Improved understanding of lysosomal responses to cellular injury may thus be expected not only to provide increased clarity in this segment of the study of cell injury, but in some cases to lead to more rational therapy of disease.

## References

Aalto, M., Potila, M., and Kulonen, E. (1976). *Exp. Cell Res.* **96,** 193.

Abraham, R., and Hendy, R. (1970). *Exp. Mol. Pathol.* **12,** 148.

Allison, A. C. (1964). *Life Sci.* **3,** 1407.

Allison, A. C. (1968a). *In* "Scientific Basis of Medicine Annual Reviews," p. 18. Oxford Univ. Press (Athlone), London and New York.

Allison, A. C. (1968b). *In* "The Interaction of Drugs and Subcellular Components in Animal Cells," p. 218. Churchill, London.

Allison, A. C., and Mallucci, L. (1965). *J. Exp. Med.* **121,** 463.

Allison, A. C., Magnus, J. A., and Young, M. R. (1966). *Nature (London)* **209,** 874.

Allison, A. C., Harington, J. S., and Birbeck, M. (1967). *J. Exp. Med.* **124,** 141.

Allison, A. C., Davies, P., and dePetris, S. (1971). *Nature (London), New Biol.* **232,** 153.

Araki, H., and Takenaka, F. (1976). *Life Sci.* **17**, 613.
Arstila, A. U., Jauregui, H. O., Chang, J., and Trump, B. F. (1971). *Lab. Invest.* **24**, 162.
Arstila, A. U., Nuuja, I. J. M., and Trump, B. F. (1974). *Exp. Cell Res.* **87**, 249.
Axline, S. G., and Cohn, Z. A. (1970). *J. Exp. Med.* **131**, 1239.
Baehner, R. L., Karnovsky, M. J., and Karnovsky, M. L. (1969). *J. Clin. Invest.* **48**, 187.
Baggiolini, M., Hirsch, J. G., and de Duve, C. (1969). *J. Cell Biol.* **40**, 529.
Baker, H. deC. (1956). *J. Pathol. Bacteriol.* **71**, 135.
Beattie, D. S., Basford, R. E., and Doritz, S. B. (1967). *Biol. Chem.* **242**, 4584.
Beaufay, H., and de Duve, C. (1959). *Biochem. J.* **73**, 604.
Berthet, J., Berthet, L., Appelmans, F., and de Duve, C. (1951). *Biochem. J.* **50**, 182.
Biberfeld, P. (1971). *Exp. Cell Res.* **66**, 433.
Blackman, K. E., and Bubel, H. C. (1969). *J. Virol.* **4**, 203.
Blum, H. F. (1941). "Photodynamic Action and Diseases Caused by Light." Van Nostrand-Reinhold, Princeton, New Jersey.
Brachet, J., Decroly-Briers, M., and Hoyez, J. (1958). *Bull. Soc. Chem. Biol.* **40**, 2037.
Brachfeld, N., and Gemba, T. (1965). *J. Clin. Invest.* **44**, 1030.
Bradley, H. C. (1938). *Physiol. Revs.* **18**, 173.
Brandes, D., Anton, E., Schofield, B., and Barnard, S. (1966). *Cancer Chemother. Rep.* **50**, 47.
Brown, J. H., and Schwartz, N. L. (1969). *Proc. Soc. Exp. Biol. Med.* **131**, 614.
Brunk, U. T., and Ericsson, J. L. E. (1972). *Histochem. J.* **4**, 479.
Carter, R. L., Birbeck, M. S. C., and Stock, J. A. (1971). *Int. J. Cancer* **7**, 34.
Carvalho, I. F., and Diniz, C. R. (1966). *Biochim. Biophys. Acta* **128**, 136.
Chayen, J., and Bitensky, L. (1968). *In* "The Biological Basis of Medicine" (E. E. Bittar and N. Bittar, eds.), pp. 337–368. Academic Press, New York.
Chayen, J., and Bitensky, L. (1971). *Ann. Rheum. Dis.* **30**, 522.
Claude, A. (1946). *J. Exp. Med.* **84**, 61.
Cohn, Z. A., and Benson, B. (1965). *J. Exp. Med.* **121**, 153.
Cohn, Z. A., and Ehrenreich, B. A. (1969). *J. Exp. Med.* **129**, 201.
Coltoff-Schiller, B., Goldfischer, S., Adamany, A. M., and Wolinsky, H. (1976). *Am. J. Pathol.* **83**, 45.
Comolli, R. (1967). *J. Pathol. Bacteriol.* **93**, 241.
Croker, B. P., Saladino, A. J., and Trump, B. F. (1970). *Am. J. Pathol.* **59**, 247.
Crowder, J. G., Martin, R. R., and White, A. (1969). *Am. J. Pathol.* **74**, 436.
Cutilleta, A. F., Reddy, M. K., Dowell, R. T., Zak, R., and Rabinowitz, M. (1976). *Recent Adv. Stud. Card. Struct. Metab.* **7**, 111.
Danes, B. S., and Bearn, A. G. (1966). *J. Exp. Med.* **124**, 1181.
Davies, P., and Allison, A. C. (1976). *Agents Actions* **6**, 60.
Davies, P., Allison, A. C., and Haswell, A. D. (1973). *Biochem. J.* **134**, 33.
Dawd, D. S., and Hinchcliffe, J. R. (1971). *J. Embryol. Exp. Morphol.* **26**, 401.
Dean, R. T. (1975). *Nature (London)* **257**, 414.
Decker, R. S., Poole, A. R., Griffin, E. E., Dingle, J. T., and Wildenthal, K. (1977). *J. Clin. Invest.* **59**, 911.
De Duve, C. (1975). *Circulation* **51** (Supp. II), 2.
De Duve, C., and Beaufay, H. (1959). *Biochem. J.* **73**, 610.
De Duve, C., and Wattiaux, R. (1966). *Am. Rev. Physiol.* **28**, 435.
De Duve, C., Pressman, B. C., Gianetto, R., Wattiaux, R., and Appelmans, F. (1955). *Biochem. J.* **60**, 604.
De Duve, C., Wattiaux, R., and Wilson, M. (1962). *Biochem. Pharmacol.* **9**, 97.
De Duve, C., DeBarsy, T., Poole, B., Trouet, A., Tulkens, P., and VanHoff, F. (1974). *Biochem. Pharmacol.* **23**, 2495.

Desai, I. D., Sawant, P. L., and Tappel, A. L. (1964). *Biochim. Biophys. Acta* **86**, 277.
Desai, I. D., Fletcher, B. L., and Tappel, A. L. (1975). *Lipids* **10**, 307.
Desnick, R. J., Thorpe, S. R., and Fiddler, M. B. (1976). *Physiol. Rev.* **56**, 57.
Deter, R. L., and de Duve, C. (1967). *J. Cell Biol.* **33**, 437.
Dingle, J. T. (1961). *Biochem. J.* **79**, 509.
Dingle, J. T. (1968). *Br. Med. Bull.* **24**, 141.
Dingle, J. T. (1973). *J. Bone and J. Surg.* **55B**, 87.
Dingle, J. T., Barrett, A. S., and Weston, P. D. (1971). *Biochem J.* **123**, 1.
Ehrenreich, B. A., and Cohn, Z. A. (1969). *J. Exp. Med.* **129**, 227.
Ericsson, J. L. E. (1968a). *Acta Pathol. Microbiol. Scand.* **72**, 451.
Ericsson, J. L. E. (1968b). *Ann. Acad. Sci. Fenn.* **4**, 128.
Ericsson, J. L. E. (1969). *In* "Lysosomes in Biology and Pathology" (J. T. Dingle and H. B. Fell, eds.), Vol. 2, pp. 345–394. North-Holland Publ., Amsterdam.
Farquhar, M. G. (1969). *In* "Lysosomes in Biology and Pathology" (J. T. Dingle and H. B. Fell, eds.), Vol. 2, p. 462. North-Holland Publ., Amsterdam.
Fedorko, M. E., Hirsch, J. G., and Cohn, Z. A. (1968). *J. Cell Biol.* **38**, 377, 392.
Fell, H. B. (1969). *In* "The Fat-Soluble Vitamins" (H. F. DeLuca and J. W. Suttie, eds.), p. 186. Univ. of Wisconsin Press, Madison.
Fine, D. L., Lake, R. S., and Ludwig, E. H. (1970). *J. Virol.* **5**, 226.
Fletcher, M. J., and Sanadi, D. R. (1961). *Biochim. Biophys. Acta* **51**, 356.
Friedman, I., Laufer, A., and Davies, A. M. (1969). *Experimentia* **25**, 1092.
Garfield, R. E., Chacko, S., and Blose, S. (1975). *Lab. Invest.* **33**, 418.
Ginn, F. L., Shelburne, J. D., and Trump, B. F. (1968). *Am. J. Pathol.* **53**, 1050.
Glick, A. D., Horn, R. G., Collins, R. D., and Bryant, R. E. (1970). *Exp. Mol. Pathol.* **12**, 275.
Goldberg, B., and Green, H. (1959). *J. Exp. Med.* **109**, 505.
Goldfischer, S., Schiller, B., and Wolinsky, H. (1975). *Am. J. Pathol.* **78**, 497.
Goldstein, J. L., Brunschede, G. Y., and Brown, M. S. (1975a). *J. Biol. Chem.* **250**, 7854.
Goldstein, J. L., Dana, S. E., Faust, J. R., Beaudet, A. L., and Brown, M. S. (1975b). *J. Biol Chem.* **250**, 8487.
Gordon, S., and Cohn, Z. A. (1973). *Int. Rev. Cytol.* **36**, 171.
Goren, M. B., Hart, P. D., Young, M. R., and Armstrong, J.A. (1976). *Proc. Nat. Acad. Sci. USA* **73**, 2510.
Gosden, R. G., Hawkins, H. K., and Gosden, C. A. (1978). *Am. J. Pathol.* **91**, 155.
Gottwik, M. G., Kirk, E. S., Hoffstein, S., and Weglicki, W. B. (1975). *J. Clin. Invest.* **56**, 914.
Green, H., Fleischer, R. A., and Barrow, P., and Goldberg, B. (1959). *J. Exp. Med.* **109**, 511.
Greenbaum, L. M., and Kim, K. S. (1967). *Br. J. Pharmacol. Chemother.* **29**, 238.
Griffin, C., Waravdekar, V. S., Trump, B. F., Goldblatt, P. J., and Stowell, R. E. (1965). *Am. J. Pathol.* **47**, 833.
Hawkins, D. (1974). *Clin. Immunol. Immunopath.* **2**, 141.
Hawkins, D., and Peters, S. (1971). *Lab. Invest.* **24**, 483.
Hawkins, H. K., Ericsson, J. L. E., Biberfeld, P., and Trump, B. F. (1972). *Am. J. Pathol.* **68**, 255.
Hawkins, H. K., Hill, M. L., Klotman, S., and Jennings, R. B. (1977). *Am. J. Pathol.* **89**, 45a.
Helminen, H. J., and Ericsson, J. L. E. (1968). *J. Ultrastruct. Res.* **25**, 214.
Helminen, H. J., Ericsson, J. L. E., and Orrenius, S. (1968). *J. Ultrastruct. Res.* **25**, 240.
Henning, R. (1975). *Biochim. Biophys. Acta* **401**, 307.
Henning, R., Kaulen, H. D., and Stoffel, W. (1970). *Hoppe-Seyler's Z. Physiol. Chem.* **351**, 1191.
Hers, H. G., and VanHoof, F., eds. (1973). "Lysosomes and Storage Diseases." Academic Press, New York.

Hirschhorn, R., Brittinger, G., Hirschhorn, K., and Weissman, G. (1968). *J. Cell Biol.* **37**, 412.
Hoffstein, S., and Weissmann, G. (1975). *Arch. Rheum.* **18**, 153.
Hoffstein, S., Gennaro, D. E., Weissmann, G., Hirsch, J., Streuli, F., and Fox, A. C. (1975). *Am. J. Pathol.* **79**, 193.
Holt, S. S. (1959). *Exp. Cell Res.* Suppl. **7**, 1.
Homewood, C. A., Warhurst, D. C., Peters, W., and Baggaley, V. C. (1972). *Nature (London)* **235**, 50.
Hruban, Z., Spargo, B., Swift, H., Wissler, R. W., and Kleinfeld, R. G. (1963). *Am. J. Pathol.* **42**, 657.
Hündgen, M. (1977). *Int. Rev. Cytol.* **48**, 281.
Ignarro, L. J. (1971). *Biochem. Pharmacol.* **20**, 2847.
Ignarro, L. J. (1974). *Agents Actions* **4**, 241.
Ignarro, L. J. (1975). *In* "Lysosomes in Biology and Pathology" (J. T. Dingle and H. B. Fell, eds.), Vol. 4, p. 481. Academic Press, New York.
Ignarro, L. J., Lint, T. F., and George, W. J. (1974). *J. Exp. Med.* **139**, 1395.
Iturriaga, H., Posalaki, I., and Rubin, E. (1969). *Exp. Mol. Pathol.* **10**, 231.
Janoff, A. (1964). *In* "Shock" (S. G. Hershey, ed.), p. 93. Little, Brown, Boston, Massachusètts.
Janoff, A. (1972). *Annu. Rev. Med.* **23**, 177.
Janoff, A., and Zeligs, J. D. (1968). *Science* **161**, 702.
Jennings, R. B., and Ganote, C. E. (1974). *Circ. Res.* **34–35** *Suppl.* **III**, 156.
Jennings, R. B., Sommers, H. M., Smyth, G. A., Flack, H. A., and Linn, H. (1960). *Arch. Pathol.* **70**, 68.
Jennings, R. B., Sommers, H. M., Kaltenbach, S. P., and West, J. J. (1964). *Circ. Res.* **14**, 260.
Jennings, R. B., Herdson, P. B., and Sommers, H. M. (1969a). *Lab. Invest.* **20**, 548.
Jennings, R. B., Sommers, H. M., Herdson, P. B., and Kaltenbach, J. P. (1969b). *Annu. N.Y. Acad. Sci.* **156**, 61.
Jennings, R. B., Ganote, C. E., Kloner, R. A., Whalen, D. A., and Hamilton, D. G. (1975). *In* "Recent Advances in Studies on Cardiac Structure and Metabolism" (A. Fleckenstein and G. Rona, eds.), Vol. 6, p. 405. University Park Press, Baltimore, Maryland.
Kerr, J. F. R. (1965). *J. Pathol. Bacteriol.* **90**, 419.
Kerr, J. F. R. (1970). *J. Pathol.* **100**, 99.
Kingdon, G. C., and Sword, C. P. (1970). *Infect. Immun.* **1**, 356.
Kloner, R. A., Ganote, C. E., Whalen, D. A., and Jennings, R. B. (1974). *Am. J. Pathol.* **74**, 399.
Koschel, K., Aus, H. M., and Meulen, V. T. (1974). *J. Gen. Virol.* **25**, 359.
Lake, B. D., and Ellis, R. B. (1976). *Histochem. J.* **8**, 357.
Lane, N. S., and Novikoff, A. B. (1965). *J. Cell Biol.* **27**, 603.
Leighty, E. G., Stoner, C. D., Ressallat, M. M., Passananti, T., and Sirak, H. D. (1967). *Circ. Res.* **21**, 59.
Lie, S. O., Schofield, B. H., Taylor, H. A., and Doty, S. B. (1973). *Pediatr. Res.* **7**, 13.
Lloyd, J. B. (1969). *Biochem. J.* **115**, 703.
Lloyd, J. B. (1971). *Biochem. J.* **121**, 245.
Locke, M. (1969). *Tissue Cell* **1**, 103.
Lockshin, R. (1969). *In* "Lysosomes in Biology and Pathology" (J. T. Dingle and H. B. Fell, eds.), Vol. 1, p. 363. North-Holland Publ., Amsterdam.
Lucy, J. A., Dingle, J. T., and Fell, H. B. (1961). *Biochem. J.* **79**, 500.
McKusick, V. (1970). *Annu. Rev. Genet.* **4**, 1.
Malawista, S. E., and Bodel, P. T. (1967). *J. Clin. Invest.* **46**, 786.
Mandel, N. S. (1976). *Arch. Rheum.* **19**, 439.
Margulis, L. (1973). *Int. Rev. Cytol.* **34**, 333.

Mellors, A., Tappel, A. L., Sawant, P. L., and Desai, I. D. (1967). *Biochim. Biophys. Acta* **143**, 299.
Melmon, K. L., and Cline, M. J. (1967). *Nature (London)* **213**, 90.
Mitchener, J. S., Shelburne, J. D., Bradford, W. D., and Hawkins, H. K. (1976). *Am. J. Pathol.* **83**, 485.
Movat, H. Z., Macmorine, D. R. L., and Takeuchi, Y. (1971a). *Int. Arch. Allergy Appl. Immunol.* **40**, 218.
Movat, H. Z., Uriuhara, T., Takeuchi, Y., and Macmorine, D. R. L. (1971b). *Int. Arch. Allergy Appl. Immunol.* **40**, 197.
Novikoff, A. B. (1960). *In* "Developing Cell Systems and Their Control" (D. Rudnick, ed.), pp. 167–203. Ronald Press, New York.
Novikoff, A. B., Essner, E., and Quintana, N. (1964). *Fed. Proc. Fed. Am. Soc. Exp. Biol.* **23**, 1010.
Okuda, M., Young, K. R., and Lefer, A. M. (1976). *Circ. Res.* **39**, 640.
Oliver, C., and Essner, E. (1975). *Lab. Invest.* **32**, 17.
Opie, E. L. (1922). *Physiol. Rev.* **2**, 552.
Persellin, R. H. (1969). *J. Immunol.* **103**, 39.
Peters, T. A. (1975). *In* "Lysosomes in Biology and Pathology" (J. T. Dingle and R. T. Dean, eds.), Vol. 4, p. 47. Academic Press, New York.
Peters, T. J., and de Duve, C. (1974). *Exp. Mol. Pathol.* **20**, 228.
Peters, T. J., Müller, M., and de Duve, C. (1972). *J. Exp. Med.* **136**, 1117.
Poole, A. R. (1973). *In* "Lysosomes in Biology and Pathology" (J. T. Dingle, ed.), Vol. 3, p. 303. North-Holland Publ., Amsterdam.
Quie, P. G., and Hirsch, J. G. (1964). *J. Exp. Med.* **120**, 149.
Quie, P. G., White, J. G., Holmes, B., and Good, R. A. (1967). *J. Clin. Invest.* **46**, 668.
Radford, S. V., and Misch, D. W. (1971). *J. Cell Biol.* **49**, 702.
Ranadive, N. S., and Cochrane, C. G. (1968). *J. Exp. Med.* **128**, 605.
Reich, T., Dierolf, B. M., and Reynolds, B. M. (1965). *J. Surg. Res.* **5**, 116.
Reijngoud, D. J., and Tager, J. M. (1973). *Biochim. Biophys. Acta* **297**, 174.
Reijngoud, D. T., and Tager, J. M. (1976). FEBS Letters **64**, 231.
Ricciutti, M. A. (1972). *Am. J. Cardiol.* **30**, 492, 498.
Riddle, J. M., Bluhm, G. B., and Barnhart, M. I. (1967). *Am. Rheum. Dis.* **26**, 389.
Robbins, E., Marcus, P. I., and Gonatas, N. K. (1964). *J. Cell Biol.* **21**, 49.
Root, R. K., Rosenthal, A. S., and Balestra, D. J. (1972). *J. Clin. Invest.* **51**, 649.
Rossi, F., Romeo, D., and Patriarca, P. (1976). *Agents Actions* **6**, 50.
Sahaphong, S., and Trump, B. F. (1971). *Am. J. Pathol.* **63**, 277.
Saladino, A. J., and Trump, B. F. (1969). *Am. J. Pathol.* **52**, 737.
Schumacher, H. R., and Phelps, P. (1971). *Arthritis Rheum.* **14**, 513.
Seeman, P. (1966). *Biochem. Pharmacol.* **15**, 1632, 1767.
Seeman, P. (1968). "Symposium on The Interaction of Drugs and Subcellular Components in Animal Cells," p. 212. Churchill, London.
Segal, H. L. (1975). *In* "Lysosomes in Biology and Pathology" (J. T. Dingle and R. T. Dean, eds.), Vol. 4, pp. 295–302. Amer. Elsevier, New York.
Shelburne, J. D., Arstila, A. U., and Trump, B. F. (1973a). *Am. J. Pathol.* **72**, 521.
Shelburne, J. D., Arstila, A. U., and Trump, B. F. (1973b). *Am. J. Pathol.* **73**, 641.
Shimoyama, M., Niitani, H., Taniguchi, T., Inagaki, J., and Kimura, K. (1969). *Gann* **60**, 33.
Shio, H., Farquhar, M. G., and de Duve, C. (1974). *Am. J. Pathol.* **76**, 1.
Skosey, J. L., Chow, D., Damgaard, E., and Sorensen, L. B. (1973). *J. Cell Biol.* **57**, 237.
Slater, T. F., and Greenbaum, A. L. (1975). *Biochem. J.* **96**, 484.
Smith, A. L., and Bird, J. W. C. (1975). *J. Mol. Cell Cardiol.* **7**, 39.

Smith, A. L., and Bird, J. W. C. (1976). *Recent Adv. Stud. Card. Struct. Metab.* **7,** 41.
Smith, R. J., and Ignarro, L. J. (1975). *Proc. Nat. Acad. Sci. USA* **72,** 108.
Spath, J. A., and Lefer, A. M. (1975). *Am. Heart J.* **90,** 50.
Straus, W. (1956). *J. Biophys. Biochem. Cytol.* **2,** 513.
Straus, W. (1964). *J. Cell Biol.* **20,** 497.
Strehler, B. L. (1964). *Adv. Geront. Res.* **1,** 343.
Sutton, J. S., and Weiss, L. (1966). *J. Cell Biol.* **28,** 303.
Suzuki, H., and Kurosumi, K. (1972). *J. Electron Microsc.* **21,** 285.
Sylven, B. (1968). *Int. J. Cancer* **4,** 463.
Tappel, A. L. (1969). *In* "Lysosomes in Biology and Pathology" (J. T. Dingle and H. B. Fell, eds.), Vol. 1, pp. 207–244. North-Holland Publ., Amsterdam.
Tappel, A. L. (1975). *In* "Pathobiology of Cell Membranes" (B. F. Trump and A. U. Arstila, eds.), Vol. 1, pp. 145–172. Academic Press, New York.
Terry, R. D., Wisniewski, H., and Johnson, A. B. (1970). *J. Neuropathol. Exp. Neurol.* **29,** 142.
Trump, B. F., and Bulger, R. G. (1968). *Lab. Invest.* **18,** 721.
Trump, B. F., Goldblatt, P. J., and Stowell, R. E. (1962). *Lab. Invest.* **11,** 986.
Trump, B. F., Goldblatt, P. J., and Stowell, R. E. (1965). *Lab. Invest.* **14,** 1946.
Tulkens, P., Trouet, A., and VanHoof, F. (1970). *Nature (London)* **228,** 1282.
Vaes, G. (1968). *J. Cell Biol.* **39,** 676.
Vaes, G. (1969). *In* "Lysosomes in Biology and Pathology" (J. T. Dingle and H. B. Fell, eds.), Vol. 1, p. 217. North-Holland Publ., Amsterdam.
Vansteveninck, J., Weed, R. F., and Rothstein, A. (1965). *J. Gen. Physiol.* **48,** 618.
Wallace, B. J., Volk, B. W., Schueck, L., and Kaplan, H. (1966). *J. Neuropathol. Exp. Neurol.* **25,** 76.
Weber, R. (1969). *Gen. Comp. Endocrinol. Suppl.* **2,** 408.
Weglicki, W. B., Owens, K., Ruth, R. C., and Somnenblick, E. H. (1974). *Cardiovasc. Res.* **8,** 237.
Weiss, L., and Holyoke, E. D. (1969). *J. Nat. Cancer Inst.* **43,** 1045.
Weissmann, G. (1964). *Fed. Proc. Fed. Am. Soc. Exp. Biol.* **23,** 1038.
Weissmann, G. (1968). "Symposium on The Interaction of Drugs and Subcellular Components in Animal Cells," p. 203. Churchill, London.
Weissmann, G. (1974). *Adv. Int. Med.* **19,** 239.
Weissmann, G., and Dingle, J. T. (1962). *Exp. Cell Res.* **25,** 207.
Weissmann, G., and Rita, G. A. (1972). *Nature (London), New Biol.* **240,** 167.
Weissmann, G., Hirschhorn, R., and Krakauer, K. (1969). *Biochem. Pharmacol.* **18,** 1771.
Weissmann, G., Dukor, P., and Zurier, R. B. (1971). *Nature (London), New Biol.* **231,** 131.
Welman, E., and Peters, T. J. (1976). *J. Mol. Cell. Cardiol.* **8,** 443.
Welman, E., and Peters, T. J. (1977). *J. Mol. Cell. Cardiol.* **9,** 101.
Wildenthal, K. (1975). *In* "Lysosomes in Biology and Pathology" (J. T. Dingle and R. T. Dean, eds.), Vol. 4, pp. 167–190. Academic Press, New York.
Wilstätter, R., and Rohdewald, M. (1932). *Z. Physiol. Chem.* **208,** 258.
Wolinsky, H., Goldfischer, S., Daly, M. M., Kajak, L. E., and Coltoff-Schiller, B. (1975). *Circ. Res.* **36,** 553.
Zeman, W. (1974). *J. Neuropathol. Exp. Neurol.* **33,** 1.
Zucker-Franklin, D. (1965). *Am. J. Pathol.* **47,** 419.
Zucker-Franklin, D., and Hirsch, J. G. (1964). *J. Exp. Med.* **120,** 569.
Wolff, D. A., and Bubel, H. C. (1964). *Virology* **24,** 502.

# EDITORS' SUMMARY TO CHAPTER VI

On a theoretical basis lysosomes could be involved with damage to cells in the following three ways.

## I. Release of Enzymes into the Cell Itself

This is the "suicide bag" hypothesis reviewed here by Dr. Hawkins and also discussed in this series of treatises by Drs. Ericsson and Brunk (Volume I, Chapter IV). This hypothesis is a very difficult one to confirm. So far, there are not any clear-cut experiments that unequivocally support the release of lysosomal enzymes into the cell prior to cell death. On the other hand, the release of lysosomal enzymes in necrotic cells clearly seems to occur and is no doubt responsible for the final breakdown of cellular components in dead cells facilitating turnover and scavenging of necrotic cells products. This is probably a very important function of the lysosomes *in vivo* as many developmental and repair processes seem to involve this effect. The best example of a type of injury that might involve lysosomally induced cell suicide is the urate crystal studies of Weissman (Zurrier *et al.*, 1973). However, alternate interpretations of that are still possible, namely, that perforation of the lysosome occurs while the vesicle or vacuole is still in contact with the cell exterior.

Recently, Weissman's group (Hoffstein *et al.*, 1976) has been studying lysosomal changes in myocardial ischemia. It has been found that the status of lysosomes in muscle, including the heart, is still a bit of a problem in the sense that in most cells lysosomal enzyme activity, as demonstrated by conventional cytochemical procedures, is found in residual or lipofuscin bodies in the perinuclear region. Only a few acid phosphatase containing lysosomes are seen in more peripheral parts of the cell. Cytochemical studies of canine myocardial cells, however, reveal acid phosphatase and aryl sulfatase activities in elements of the sarcoplasmic reticulum and the authors note that within 2 hr after ischemia, evidence of more diffuse enzyme reaction production is seen suggesting release. Such escape of lyososomal enzymes could have a number of effects including hydrolysis of lipids and proteins and formation of toxic products such as lysolecithin resulting from phospholipase A2 action. These myocardial studies, of course, do not prove or disprove a primary role for lysosomal enzymes in provoking the tissue injury in myocardial infarction but they do provide clear evidence that they are among the organelles that are injured fairly early in the course of infarction. Furthermore, the authors provide biochemical data that show an increase in unsedimentable activity of acid phosphatase and $\beta$-glucuronidase. This is of course subject to various interpretations including increased mechanical fragility. They also show that early labilization in this sense is prevented by treatment with methylprednisolone given 30 min after initiation of injury. It is important to keep in mind, however, that most drugs including methylprednisolone which stabilize lysosomes *in vitro* can also have similar stabilizing effects on the plasma membrane so this alone cannot be taken as evidence of a primary role for the lysosomes in cell killing.

## II. Release of Lysosomal Enzymes to the Extracellular Space

This is another theoretical way in which lysosomal enzymes can exert cell damage not to the cell in question but to neighboring cells. Lysosomal enzymes can be released to the extracellular space in several ways. First, during normal cell eating or phagocytosis, lysosomal enzyme leakage commonly occurs because the phagosome retains continuity with the cell surface even after fusion with secondary lysosomes or primary lysosomes occurs. Second, is abnormal phagocytosis or frustrated phagocytosis when the process is completely or partially blocked as with cytochalasin B where fusion of lysosomes occurs directly with the cell surface rather than with phagosomes. Third, is where cell death and necrosis occur when leak of enzymes trom the necrotic cell into the extracellular space clearly occurs and could be damaging to other cells. Evidence of damage includes direct attack on macromolecules by proteases or lipases, changes in the immune or inflammatory response due to changes in C3- and C5-cleaving enzymes which were found in the lysosomes by Ward and his group (1972), and changes in number or concentration of antibacterial substances found in lysosomes. There is a role for potential decrease in this process. A theoretical possibility in normal tissue activity remains to be seen.

## III. Altered Turnover of Contents

Alterations in turnover or digestion of lysosomal contents is a very large category and many changes in this occur in these processes.

### A. *Increased Digestion*

Many reactions to sublethal injury are accompanied by increased numbers of autophagic vacuoles and increased organelle turnover. These include the effects of glucagon on the liver cell or parathormone on the proximal tubular epithelium. These hormones result in increased autophagocytosis and increased protein catabolism along with other effects such as increased gluconeogenesis. This problem is discussed in more detail by Drs. Locke and Collins in Chapter V in this volume and is also discussed by Drs. Trump and Arstila in Volume I, Chapter I.

Examples of harmful events resulting from interactions within the lysosomes appear to be increasing. These are somewhat analogous to the Peters' concept of "lethal" digestion. According to this concept as applied to the lysosomes, lysosomotropic materials, often particulate, are taken up by phagocytosis and stored in the lysosomes. Interactions including digestion in the lysosome then release or activate the deleterious materials. These may include procarcinogens such as benzo(a)pyrene which are inhaled into the lung usually bound to airborne particulates, e.g., asbestos or $Fe_2O_3$. These particulates are phagocytized by pulmonary macrophages and the procarcinogen metabolized by the marcophage with formation of mutagens and ultimate carcinogens.

### B. *Decreased Turnover*

This is a very large category and can result from either intrinsic or extrinsic causes.

#### 1. *Intrinsic*

Intrinsic deficiencies in turnover result from congenital enzyme deficiency seen in the large series of lysosomal storage diseases in which there is overloading of the lysosomes with normal components. This results in enormous enlargement of the lysosomal compartment in affected cells and in some cases in organ failure, e.g., in the myocardium or the nervous system. The mechanism of this organ failure is not clear although there is the possibility that the enormous overloading of the lysosomal compartment has some indirect effects on cell functions such as myocardial contractility. Overloading can also occur because even the normal enzyme components are inadequate in quality or quantity to digest a particular material. Some examples of this type of overloading are: heavy metals such as iron, mercury, and lead which are concentrated in lysosomes and polysaccharides such as mannitol or sucrose that are poorly digested, resulting in inclusions or vacuoles in the cell. The mannitol or sucrose inclusions are often reversible and may exert only minor interference with cell function. Intrinsic deficiencies of antibacterial substances seem to represent an important category in diseases such as granulomatous disease and NADPH oxidase deficiency.

2. *Extrinsic*

There are relatively few clear-cut examples of extrinsic materials interfering with lysosomal digestion. One example may be trypan blue which has been shown to result in congenital anomalies when given to chick embryos and has also been shown to inhibit some lysosomal enzyme activities. Since remodeling involving autophagy and cell death plays an important part in metamorphosis in various organ systems, it is conceivable that this inhibition is somehow related to the production of these congenital anomalies.

## References

Hoffstein, S., Weissman, G., and Fox, A. C. (1976). *Circulation* **53**, *Suppl. I*, 1.
Ward, P. A., and Hill, J. H. (1972). *J. Immunol.* **108**, 137.
Zurrier, R. B., Hoffstein, S., and Weissman, G. (1973). *J. Cell Biol.* **58**, 97.

# CHAPTER VII

# PATHOLOGY OF SKELETAL MUSCLE MEMBRANES

Stephen R. Max, Kenneth R. Wagner, and David H. Rifenberick



## I. Introduction

Because it occupies approximately 40% of an adult's body weight and provides the primary force for moving the skeletal system, skeletal muscle has been studied for centuries (Needham, 1971). Research on muscle disease also has a long, complex history. However, in spite of intensive efforts, the etiology of most muscle disorders remains unknown (cf. Walton, 1974; Adams, 1975). This chapter presents research findings on membrane sys-



PATHOBIOLOGY OF CELL MEMBRANES, VOL. II

ISBN 0-12-701502-7

tems in muscle diseases. The literature on muscle and muscle disease is prodigious, and a complete review is not attempted; rather, we describe recent trends. Where appropriate, we cite comprehensive reviews. In addition, the structure and function of normal muscle is not covered extensively since this is the subject of several recent publications (Pellegrino and Franzini-Armstrong, 1969; Bourne, 1972; Fuchs, 1974; Gergely, 1974; Morel and Pinset-Harström, 1975; Squire, 1975;Ebashi, 1976; Mannherz and Goody, 1976; Trentham, 1977; Homsher & Kean, 1978; Caputo, 1978). Biochemical aspects of muscle disease have been reviewed by Pennington (1971).

### A. *Ultrastructure of Normal Muscle*

The extrafusal skeletal muscle fiber in the adult mammal is a multinucleated cell which is surrounded by plasma and basement membranes and contains numerous myofibrils which represent its contractile machinery.

The basic contractile unit within the skeletal muscle fiber is the sarcomere, the area between two Z disks (Fig. 1). The sarcomere is composed of a light (isotropic) zone, the I band, and a dark (anisotropic) zone, the A band, bisected by the M line and H zone. The dark and light regions represent the thick (myosin) and thin (actin) filaments, respectively (Huxley, 1972). In the I band, only thin filaments are present, whereas both thick and thin filaments are present in the A band. The sliding filament hypothesis is the generally accepted theory of muscle contraction. According to this theory, contraction results from the interdigitation of filaments by a process involving cross-bridges which extend from the thick filaments to the adjacent thin filaments (Huxley, 1972) (Fig. 1). Contraction is triggered by calcium which is released from the sarcoplasmic reticulum (for review see Ebashi, 1976).

### B. *Fiber Types*

The various systems of fiber type nomenclature were summarized by Close (1972) and discussed by Guth and Yellin (1971). Guathier (1969, 1970) and Padykula and Guathier (1970) differentiated three types of skeletal muscle fibers based on ultrastructure. These types were called red, white, and intermediate. At least three fiber types can also be demonstrated histochemically (Barnard *et al.*, 1971). The various fiber types differ in mitochondrial content and distribution as well as form, fiber diameter, width of Z lines, sarcoplasmic reticulum, and muscle color (Table I). Furthermore, muscles composed primarily of intermediate fibers are slow-twitch muscles with low-myosin ATPase, while muscles composed primarily of white or of red fibers are fast-twitch muscles with higher ATPase. Barnard *et al.* (1971)

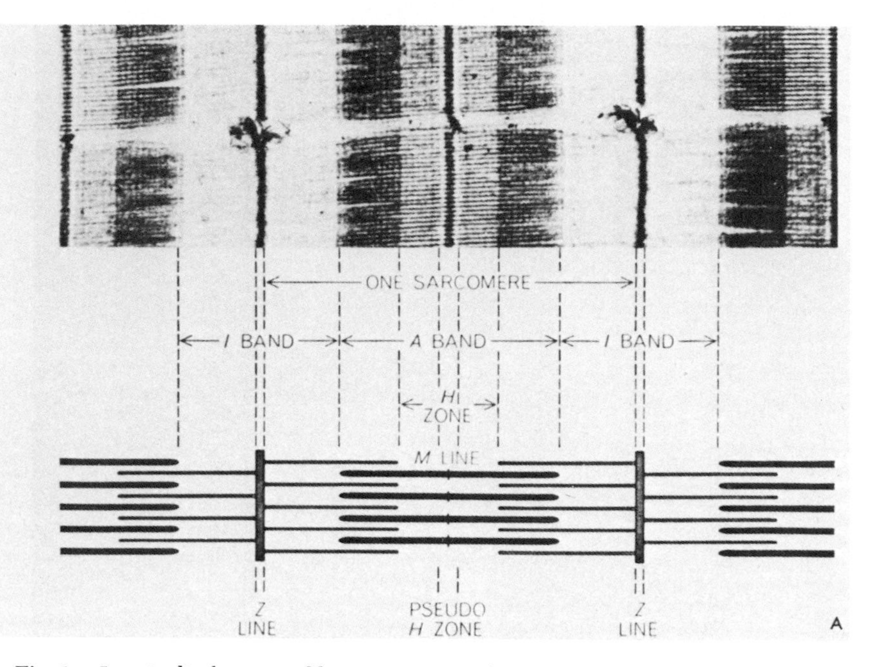


**Fig. 1.** Longitudinal section of frog sartorius muscle (top) with diagram showing overlap of filaments giving rise to the band pattern. The A band is most dense in the lateral zones where thick and thin filaments overlap. The central zone of the A band (H zone) is less dense, since it contains thick filaments only. I bands are even less dense because they contain only thin filaments. (Reproduced from Huxley, 1972, with permission of the author and Academic Press.)

recommended the terms fast-twitch red, fast-twitch white, and slow-twitch intermediate for the three types. Subsequent biochemical studies by the same investigators showed the red fibers to be high in oxidative and glycolytic activity (fast-oxidative glycolytic, FOG), the intermediate fibers to be rich in oxidative enzymes (slow-oxidative, SO) and the white fibers to be rich in glycolytic activity (fast-glycolytic, FG) (Peter *et al.*, 1972a; cf. Romanul, 1964).

Within the muscle, fibers are organized into functional units under the direction of the anterior horn cell forming an individual motor unit (the lower motor neuron, its axon, and the muscle fibers it innervates). Differences between these units within a given muscle are based upon physiological, histochemical, and morphological characteristics. Repetitive stimulation of a single fiber within the ventral root results in glycogen depletion of the muscle fibers innervated by the stimulated motor neuron (Kugelberg and Edström, 1968; Edström and Kugelberg, 1968). Such studies revealed that the muscle fibers of individual motor units are not grouped

**TABLE I**

*Some Characteristics of Three Muscle Fiber Types*[a]

| *Muscular fibers* | *Neuromuscular junction* | *Color* | *Myosin ATPase* | *Speed of contraction* | *Mitochondrial content* | *Fiber diameter* | *Z line width* | *Oxidative activity* | *Glycolytic activity* |
|---|---|---|---|---|---|---|---|---|---|
| Fast oxidative glycolytic (red) | Small, simple | Red | High | Fast | High | Small | Large | High | Intermediate |
| Fast glycolytic (white) | Large, complex | White | High | Fast | Low | Large | Small | Low | High |
| Slow oxidative (intermediate) | Intermediate | Red | Low | Slow | Intermediate | Intermediate | Intermediate | Intermediate | Low |

[a] The nomenclature used here is that of Peter *et al.* (1972a). The contents of the table have been compiled from Barnard *et al.* (1971), Peter *et al.* (1972a), and Close (1972).

together but are scattered throughout the muscle (Edström and Kugelberg, 1968; Brandstater and Lambert, 1969; Doyle and Mayer, 1969).

Histochemical profiles and morphological characteristics of muscle fibers of the cat triceps surae have been correlated with physiological properties of single motor units (Burke *et al.*, 1971, 1973). Physiological study revealed three types of motor units in the cat gastrocnemius: (a) fast-contracting fatigable (FF); (b) fast-contracting fatigue-resistant (FR); and (c) slow-contracting fatigue-resistant (S). Both FF and FR units have relatively fast-contraction speeds but differ markedly in fatigability. The S group has a slow contraction speed and a high resistance to fatigue. The S units of cat soleus differ physiologically (no posttentanic potentiation), histochemically and morphologically from gastrocnemius S units (Burke *et al.*, 1973).

According to Padykula and Guathier (1970), neuromuscular junctions of red, white, and intermediate fibers can be distinguished by shape and size of axonal endings, synaptic vesicle numbers, distribution and spacing of secondary synaptic clefts, and appearance of axoplasmic and sarcoplasmic mitochondria (Table I). In contrast, Santa and Engel (1973) found no significant differences between fiber types of soleus and gastrocnemius in terms of terminal size, nerve terminal mitochondrial content, or in postsynaptic folds and clefts. Higher (1–4 times) synaptic vesicle concentration and smaller vesicle diameter were found in junctions of red fibers than in those of white fibers. The mean postsynaptic to presynaptic membrane length ratio was larger for white than for intermediate and red type junctions (Santa and Engel, 1973).

The fiber-type profile of a given muscle can change with variation in muscle usage (Guth and Yellin, 1971). Thus, it was suggested that a new type of nomenclature be devised to consider the dynamic nature of muscle fiber types (Guth and Yellin, 1971).

### C. *Organelles*

The normal skeletal muscle fiber contains the same basic organelles as other cells (namely, plasma membrane, mitochondria, Golgi apparatus, lysosomes, ribosomes, and nucleus) with some additional specializations: the sarcoplasmic reticulum, the transverse tubules, and the motor endplate.

#### 1. *Plasma Membrane*

The plasma membrane (sarcolemma) of skeletal muscle is in many respects like that of other cell membranes (Singer and Rothfield, 1973). Its uniqueness is due to the neuromuscular junction, i.e., the synapse that muscle forms with nerve. The postsynaptic portion of this junction is the motor end plate (Fig. 2). A basement membrane, containing a substantial amount of polysaccharide, surrounds the sarcolemma. This external lamina may function as a microskeleton to help maintain muscle cell shape and structure.

### 2. *Nucleus*

The skeletal muscle nucleus of the mature muscle fiber is present immediately beneath the sarcolemma. Nuclei are surrounded by the nuclear envelope, which consists of two membranes separated by a space and perfo-

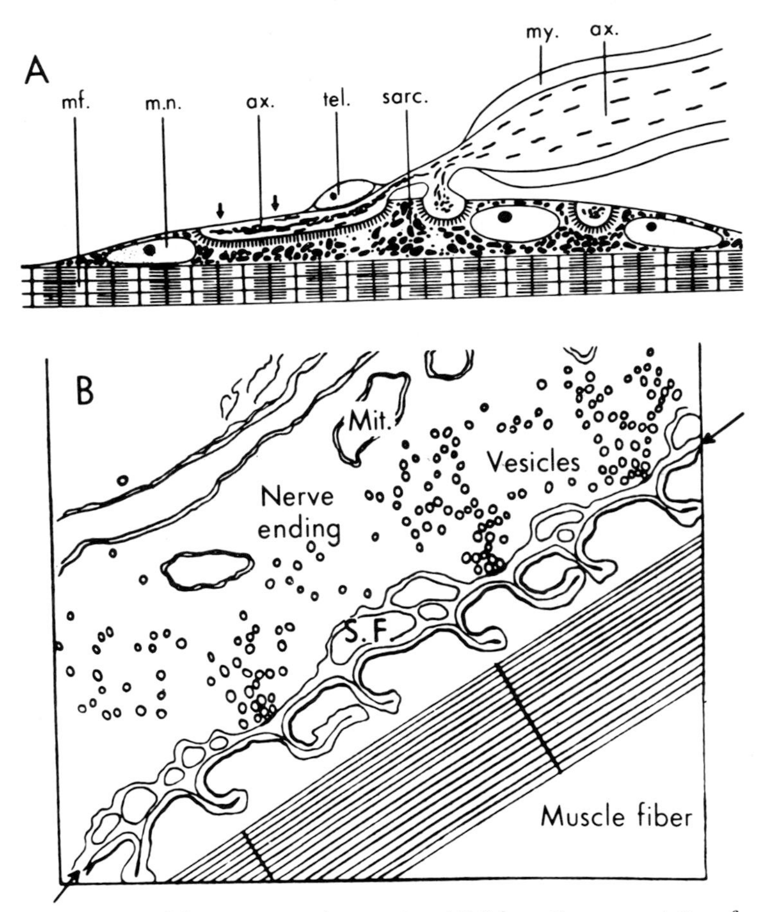


**Fig. 2.** Structure of the neuromuscular junction. (A) Schematic representation of a motor endplate. Axoplasm with mitochondria, ax.; myelin sheath, my.; teloglia (terminal Schwann cells), tel.; sarcoplasm of muscle fiber with mitochondria, sarc.; muscle nuclei, m.n.; muscle fiber, mf. The terminal nerve branches lie in troughs (from Couteaux, 1958). (B) Drawing of electron micrograph of part of frog sartorius neuromuscular junction (from Birks *et al.*, 1960). Mitochondrion, mit. (Reproduced from Eccles, 1973, with permission of the authors and McGraw-Hill.)

rated at intervals by "pores" which may be important sites for the exchange of materials between the intra- and extranuclear compartments. The pores in the nuclear membranes are not true perforations but are covered by a thin diaphragm.

### 3. *Mitochondria*

Mitochondria in skeletal muscle serve the same physiological functions as in other cells, namely, they provide ATP for endergonic processes.

Mitochondria are enclosed by two membranes: a smooth outer membrane and a more complex inner membrane which invaginates into the interior of the mitochondria. The conformation of these invaginations, or cristae, (Lehninger, 1964) is altered by changes in the energy status of the mitochondria (Hackenbrock, 1966).

Distinct localization of enzymes in the inner and outer membranes and in the matrix space has been observed in liver mitochondria (Schnaitman and Greenwalt, 1968). Similar studies have not been done on muscle mitochondria. Mitochondria from the various fiber types differ in their enzymatic components (Kark *et al.*, 1971). For example, mitochondria from red muscle oxidize $\beta$-hydroxybutyrate and pyruvate at higher rates and contain a higher specific activity of $\beta$-hydroxybutyrate dehydrogenase than mitochondria from white muscle. However, both muscle types catalyze succinate and palmitate oxidation at the same rate. Furthermore, mitochondria in the two red muscle types (red and intermediate fibers) are not equivalent. Ashmore *et al.* (1972) reported the specific activity of mitochondrial succinate dehydrogenase to be higher in fast-twitch red fibers than in slow-twitch intermediate (called $\alpha$R or $\beta$R by Ashmore *et al.*, 1972). Neither red nor intermediate ($\alpha$R or $\beta$R) fibers oxidized $\alpha$-glycerophosphate in this study, while fast-twitch white ($\alpha$W) fibers did. Only slow-twitch intermediate fibers oxidized $\beta$-hydroxybutyrate.

### 4. *Sarcotubular System*

The network which includes the plasma membrane, transverse tubules (T tubules) and sarcoplasmic reticulum has been extensively reviewed (Porter and Franzini-Armstrong, 1965; Hoyle, 1970; Ebashi, 1976). The transverse tubular system arises as invaginations of the plasma membrane, penetrates the cell in a radial fashion, and enwraps each fibril at the junction of the A and I bands (in mammalian muscle). The longitudinal vesicles of the sarcoplasmic reticulum surround each fibril and run in both directions along the fibril. At their termination at the A–I junctions, the vesicles form swellings, called terminal cisternae, which are in contact with the transverse tubules. This structure is the triad, so-called because it consists of two terminal cisternae on either side of one T tubule (Fig. 3). The T-tubular system connects directly with the extracellular space via small openings (caveolae) (Huxley and Taylor, 1958; Huxley, 1971; Zampighi *et al.*, 1974). A muscle action potential is conducted inward from the plasma membrane to the interior of the myofiber via the T tubules. Depolarization of the T-system

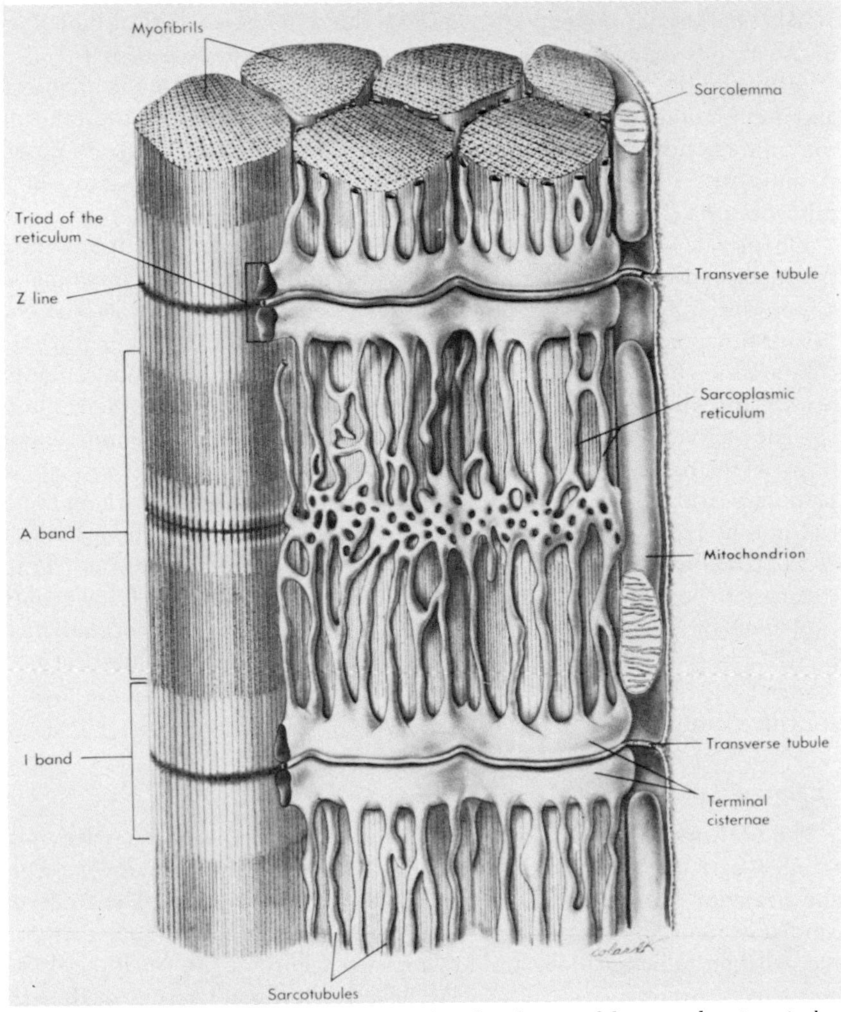


**Fig. 3.** Frog skeletal muscle structure, including distribution of the sarcoplasmic reticulum around the myofibrils. In frog muscle the triads are at the Z line. In mammalian muscle there are two to each sarcomere, located at A–I junctions (modified from the original drawing of L. Peachy; cf. B. A. Curtis, 1972). (Reproduced with permission of the author and the Rockefeller University Press.)

membrane causes release of calcium from the terminal cisternae of the sarcoplasmic reticulum initiating muscular contraction. As a cycle of depolarization ends, calcium is pumped back into the sarcotubular system by a calcium-sensitive ATPase and is thereby removed from the contractile elements. This uptake of calcium results in muscle relaxation.

Cyclic AMP is localized in the membranes of the lateral sacs of the sarcoplasmic reticulum at points of contact with transverse tubules. The level of cyclic AMP in muscle is enhanced by epinephrine (Scholze *et al.*, 1972).

### 5. *Neuromuscular Junction*

The structure and function of the neuromuscular junction was reviewed in a monograph by Zacks (1974). This synaptic apparatus transfers impulses from motor nerve endings to muscle cells, thereby initiating contraction. The postsynaptic portion of the myoneural junction, the motor endplate, is a highly specialized structure shown schematically in Figs. 2 and 4. The nerve terminal is apposed to the postsynaptic membrane and lies in the synaptic "gutter" which contains many junctional folds (secondary clefts) rich in acetylcholinesterase (Barnard *et al.*, 1975).

The acetylcholine receptor has been localized by electron microscope autoradiography using $I^{125}$-labeled $\alpha$-bungarotoxin (Barnard *et al.*, 1975). It appears to be situated at the crests of the junctional folds, opposite to areas in the nerve terminal membrane thought to be sites of interaction of synaptic vesicles (Barnard *et al.*, 1975; Fig. 4).

### 6. *Lysosomes*

The activities of lysosomal hydrolases in skeletal muscle are low compared with other tissues, such as liver. Although they are infrequently seen in electron microscope studies, lysosomes have been isolated from normal skeletal muscle (Stagni and DeBernard, 1968; Canonico and Bird, 1970). Peter *et al.* (1972b) demonstrated differences in distribution of acid hydrolases in the different fiber types. The specific activities of arylsulfatase A, $\beta$-*N*-acetylglucosaminidase, $\beta$-galactosidase, cathepsin D and acid phosphatase were highest in slow-twitch intermediate, medium in fast-twitch red, and lowest in fast-twitch white fibers.

### 7. *Additional Organelles*

Skeletal muscle contains, as do other cells, a Golgi apparatus, glycogen granules, ribosomes, triglyceride droplets, and lipofuscin. Also, satellite cells, found between the basement membrane and plasma membrane, may be of great importance in muscle regeneration (Murray, 1972; Bischoff, 1975; Mauro, 1979).

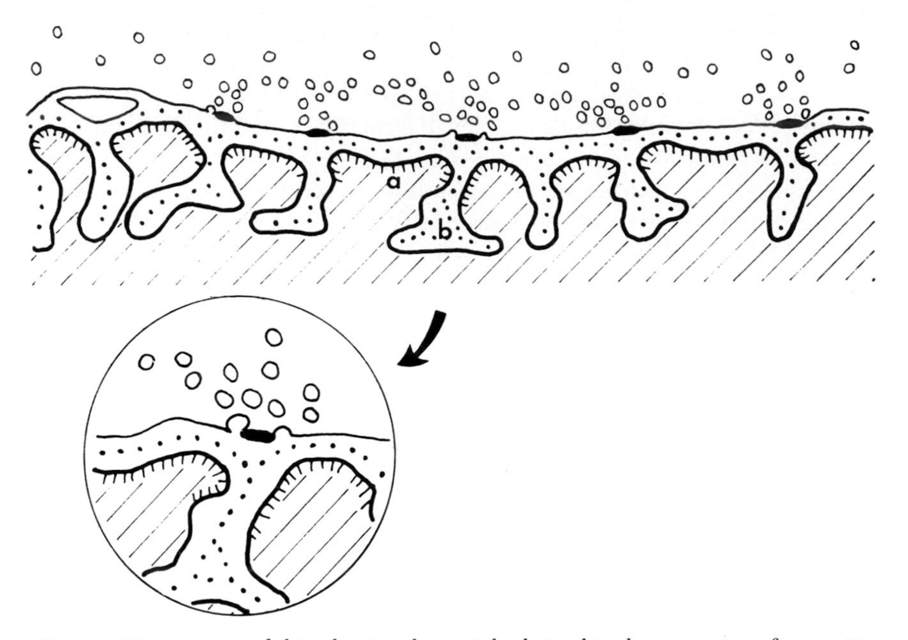


**Fig. 4.** Frog motor endplate showing the spatial relationships between sites of transmitter release and concentrations of acetylcholine receptors (AChR) and acetylcholinesterase. The vesicles characteristically stack opposite the mouths of the postjunctional folds. Dense bars on the presynaptic membrane opposite the fold mouths represent ridges seen in freeze-etch preparations. The striations along the postjunctional membrane at the fold crest correspond to the receptor-rich region (a) of the subneural apparatus. Dotted lines show the extent of visible cleft substance and, it is presumed, of the acetylcholinesterase. The inset shows, in detail, the fusion of vesicles with the presynaptic membrane as determined by Couteaux and Pecot-Dechavassine (1970) and Heuser *et al.* (1974). The dense presynaptic ridge prevents the vesicles from releasing transmitter directly into the receptor-poor-acetylcholinesterase-rich region of the fold depths (b). Instead, the ACh is released from vesicles at specific sites on either side of the presynaptic ridge (The "zones actives" of Couteaux) so that it is expelled onto the receptor-rich crests of the folds. It is proposed that upon release from receptors on the fold crests, the ACh diffuses into the fold depths where it is hydrolyzed by the acetylcholinesterase located there. (Reproduced from Barnard *et al.*, 1975, with permission of the authors and Academic Press.)

## II. Pathological and Clinical Studies of Muscle Diseases

In this section, certain features of some neuromuscular disorders are considered. Specific membrane systems are discussed in a later section. For an excellent presentation of the clinical aspects of these diseases, the reader is referred to Walton (1974) and Adams (1975).

### A. *Progressive Muscular Dystrophy*

The nature of the etiological lesions in the muscular dystrophies has so far escaped detection. Current theories of pathogenesis consider the cause to be myogenic, neurogenic, and vascular, possibly as part of a generalized membrane defect (Rowland, 1976). Ultrastructural studies from different laboratories have yielded contradictory findings, leading to numerous suggestions about the primary lesion in muscular dystrophy. These changes include: early vacuolization of the sarcoplasmic reticulum (Van Breeman, 1960), mitochondrial alterations (Fisher *et al.*, 1966), abnormal Z disks (Santa, 1969), and sarcolemmal defects (Platzen and Chase, 1964; Mokhri and Engel, 1975). Evidence that sarcolemmal changes may be of primary importance will be considered further in a later section.

Ultrastructural studies on muscle biopsy specimens from mothers of Duchenne progressive muscular dystrophy patients revealed mitochondrial hypertrophy and vacuolization and fragmentation of myofibrils (Fisher *et al.*, 1972). Many of these changes are similar to those of their clinically affected offspring. Another study of muscle from carriers of Duchenne dystrophy revealed focal lesions including nuclear centralization, sarcomeres poor in organelles and lacking fibrillar elements, and occasionally, aggregates of mitochondria. Fine structural lesions observed in this study were thought to be sufficient for identification of subclinical muscular dystrophy (Ionescu *et al.*, 1975). Although clinical symptoms of muscular dystrophy may not be present until early childhood, Toop and Emery (1974) concluded that there may be distinctive histological abnormalities in muscle from fetuses at risk for Duchenne dystrophy. The earliest histological signs observed *in utero* were a 50% increase in the variability of muscle fiber size and some hyaline degeneration of fibers. In a morphometric study of the microvasculature of muscle in Duchenne dystrophy, no evidence was found to suggest that Duchenne dystrophy is caused by a primary abnormality of muscle microcirculation (Jerusalem *et al.*, 1974a). Similarly, Jerusalem *et al.* (1974b) found no morphological support for the neurogenic hypothesis (McComas *et al.*, 1971). Mastaglia *et al.* (1970) concluded that the dystrophic fiber is capable of regeneration, although the resulting development is often abnormal and incomplete. The changes observed included regenerating fibers containing polysomes and large nuclei with dispersed chromatin and prominent nucleoli.

In studies of the enhanced protein catabolism in dystrophic muscle Kar and Pearson (1976) found a three-fold elevation of a $Ca^{2+}$-activated neutral protease. Use of protease inhibitors leupeptin and pepstatin *in vivo* delayed degeneration of muscle in dystrophic chickens (Stracher *et al.*, 1978).

Leupeptin has also been reported to decrease protein degradation in muscle *in vitro* in normal mice and in those with hereditary muscular dystrophy (Libby and Goldberg, 1978).

### *B. Myotonic Dystrophy*

Although it is frequently considered with other forms of muscular dystrophy, myotonic dystrophy is distinguished by the presence of myotonia and by the fact that distal muscles are more severely affected than proximal muscles. According to Schroder and Adams (1968), nearly all components of the muscle cell are involved in a dystrophic or atrophic process, including the myofilaments, Z disks, triads of the T system, nuclei, mitochondria, and sarcolemma. The characteristic feature of myotonic dystrophy is cytoplasmic masses filled with disoriented myofilaments and other sarcoplasmic components. Similar findings were reported by Wechsler and Hager (1961), Aleu and Afifi (1964), Milhaud *et al.* (1964), Fisher *et al.* (1966), and Schotland et *al.* (1966). In addition, Schroder and Adams (1968) described lipofuscin and lipid bodies, and pathology unique to myotonic dystrophy including sarcoplasmic masses and myofilamentous disorientation with hypernucleation and enlarged fibers. Schroder and Adams (1968) suggested that the dystrophic process in myotonic dystrophy results from a defect different from that seen in other muscular dystrophies. Other pathology has been reported in myotonic dystrophy including T-system alterations consisting of well-ordered intermyofibrillar networks of anastomosing tubules (Schotland, 1968) and centrally placed subsarcolemmal nuclei containing complex vacuolar areas with myofibrillar material (Johnson and Woolf, 1969).

An infantile variety of myotonic dystrophy has been studied by Karpati *et al.* (1973). The clinical picture of this disease differs from that of the adult form. Although myotonia is rare, patients display hypotonia, facial diparesis, dysphagia, and abnormal motor development. Histologically, a number of aberrations are described including the presence of myofilaments between nuclei and sarcolemma, with lysosomal dense bodies in the same region.

A severe neonatal form of myotonic dystrophy, considered to be a distinct clinical entity (Sarnat *et al.*, 1976), is characterized by generalized weakness, hypotonia, multiple contractures, arthogryposis multiplex congenita, pharyngeal weakness, and early death. There is an apparent arrest in fetal muscle maturation attributed by Sarnat and Silbert (1976) to "unresponsiveness of an abnormal sarcolemma to trophic influences of normal innervation."

Histological evaluation of muscle and cutaneous nerves in patients with myotonic dystrophy revealed no evidence of morphological abnormality of peripheral nerves (Pollack and Dyck, 1976). Swash and Fox (1975) described an abnormality in intrafusal muscle fibers in myotonic dystrophy.

### C. *Myasthenia Gravis and the Myasthenic Syndrome*

Myasthenia gravis is a neuromuscular disease characterized by muscle weakness and fatigability (for recent reviews see Grob, 1976; Elias and Appel, 1976; Drachman, 1978a,b; Lennon, 1978; Heilbronn and Stalberg, 1978; Fenichel, 1978).

It is generally accepted that the neuromuscular junction is the site of involvement. However, whether the presynaptic nerve terminal (Elmqvist *et al.*, 1964; Desmedt, 1973) or the postsynaptic muscule membrane (Grob and Johns, 1961) is the primary site of pathology has been controversial. Indeed, involvement of both sites has been suggested (Simpson, 1971). However, recent electrophysiological (Grob and Namba, 1976; Albuquerque *et al.*, 1976; Engel *et al.*, 1976), immunochemical (Almon *et al.*, 1974; Toyka *et al.*, 1975), and biochemical (Fambrough *et al.*, 1973; Drachman *et al.*, 1976) studies revealed abnormalities of motor endplate acetylcholine receptors (AChR) in both human and experimental myasthenia, supporting a postjunctional defect.

Ultrastructural investigation of this disease was done by Santa *et al.* (1972a; Fig. 5), Engel *et al.* (1977), and Albuquerque *et al.* (1976). In myasthenic endplates, the postsynaptic region appeared less complex than normal. Secondary clefts were sparse, shallow, abnormally wide, or absent. In regions of poor secondary cleft development, the primary cleft was widened. In other endplate regions, abundant junctional sarcoplasm was present under simplified clefts and synaptic folds. Some nerve terminals were of reduced size, and some multiple small nerve terminals were applied against an extended length of the simplified postsynaptic membrane. Although histometric analysis revealed that there was a decrease in the nerve terminal area, the mean vesicle count per unit area, the mean synaptic vesicle diameter, and the mean mitochondrial area per nerve terminal were not different from controls. Postsynaptically, the mean area of folds and clefts per nerve terminal and the mean postsynaptic membrane profile were significantly reduced. Postsynaptic regions without nerve terminals were also observed. There appeared to be no mitochondrial alterations in nerve terminals or in junctional sarcoplasm from those seen in control material. Therefore, a distinct deviation of the average endplate in myasthenia gravis from control endplates was observed (Fig. 5). No relation could be demonstrated between symptom duration and quantitative endplate abnormalities. Bergman

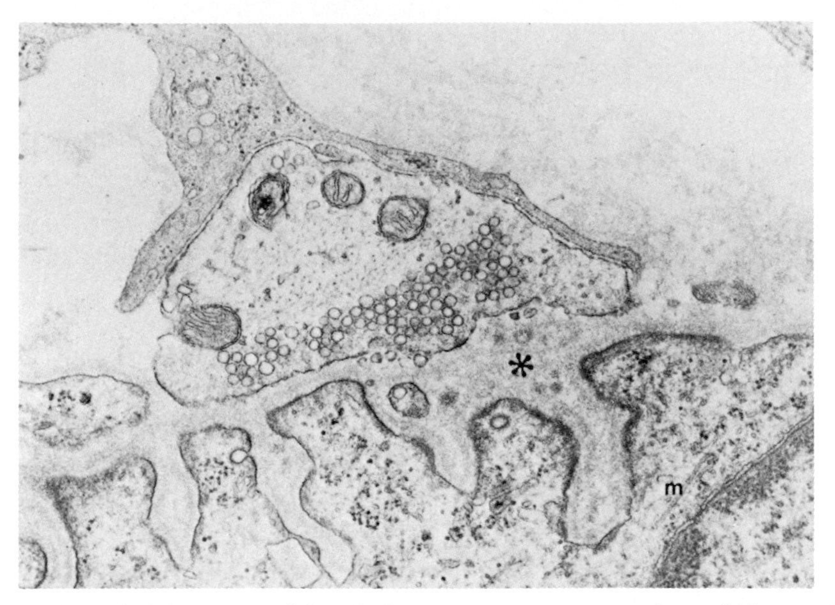


**Fig. 5.** Myasthenia gravis endplate. In the postsynaptic region sarcoplasmic folds are wide and secondary synaptic clefts sparse. Loosely arranged junctional sarcoplasm contains microtubules and ribosomes. Asterisk indicates widening of primary synaptic cleft. Microtubule, m. ×26,000. (Reproduced from Santa *et al.*, 1972a, with permission of the authors. Copyright New York Times Media Company. Micrograph provided by A. G. Engel.)

*et al.* (1971) reported that myasthenia is similar ultrastructurally to denervation atrophy. About 1% of the fibers seemed to be completely denervated, while 80 to 90% appeared to be partially denervated. Although observations of abnormal motor endplates suggested that the receptor is the locus of the primary defect, some changes may have been due to therapy with anticholinesterases or steroids (cf. Fenichel *et al.*, 1972).

The major electrophysiological abnormality in myasthenia gravis is reduction in size of miniature endplate potentials (Mepps) (Elmqvist, 1964). In addition, progressively reduced muscle action potentials occur in response to repetitive nerve stimulation. This deficiency improves upon anticholinesterase drug treatment (Grob, 1967). Although these abnormalities can be explained by a presynaptic deficiency, no reduction in size or number of ACh vesicles occurs in the disease (Elmqvist, 1964). Furthermore, the data of Grob and Namba (1976) and Ito *et al.* (1976) did not support the theory that decreased numbers of ACh molecules per quantum released are responsible for the abnormalities.

Studies of biopsy specimens from myasthenic patients revealed an 80% reduction in binding of α-bungarotoxin, a specific blocker of the AChR (Fambrough *et al.*, 1973). Furthermore, animal studies suggested that decreased availability of AChR may be responsible for the pathophysiology of myasthenia gravis (Satyamurti *et al.*, 1975). Injection of cobra α-toxin, which blocks AChR, caused electrophysiological and pharmacological changes similar to those of myasthenia gravis. Cobra α-toxin also produced a reduction in amplitude of miniature endplate potentials in the rat diaphragm (Chang and Lee, 1966). Using peroxidase-labeled bungarotoxin, Engel *et al.* (1977) also found decreased AChR at myasthenic endplates. The decrease in receptors was proportional to the decrease in mepp amplitude. Thus, a decrement of available AChR can account for the physiological defects of transmission in myasthenia gravis (Drachman *et al.*, 1976).

An autoimmune pathogenesis was proposed for myasthenia gravis by Smithers (1959), Simpson (1960), and Nastuk *et al.* (1960) on the basis of indirect evidence then available. This idea is further supported by observations of myasthenia patients with thymic hyperplasia (Castleman and Norris, 1949), thymona (Simpson, 1958), and of their improvement after thymectomy (Osserman and Genkins, 1971). Other immunological dysfunctions are (a) presence of antibodies to muscle structural proteins (Strauss, 1968) and muscle AChR (Almon *et al.*, 1974); (b) evidence for a genetically determined breakdown of immunological tolerance which results in immunological damage to motor endplates (Simpson *et al.*, 1976); (c) beneficial effects of repeated lymphocyte drainage (Bergstrom *et al.*, 1973); (d) presence of lymphocytes toxic to muscle *in vitro* (Armstrong *et al.*, 1973); and (e) association with certain histocompatibility antigens (Pirskanen, 1976).

Recently, Albuquerque *et al.* (1976) concluded that an immune response to the motor endplate produces a physical alteration in synaptic folds, causing a decrease in ACh receptor density. In ultrastructural studies of myasthenic endplates Rash *et al.* (1976), found a material which they suggested to be immunoglobin (IgG) ("fuzzy coat"). More recently both IgG and $C_3$ have been localized at myasthenic endplates on the postsynaptic membranes and on degenerating material in the synaptic space (Engel *et al.*, 1977).

Evidence for circulating serum factors has provided support for the immunological theory of myasthenia gravis. In some patients studied by Almon *et al.* (1974), a serum factor from a serum globulin fraction caused inhibition of α-bungarotoxin binding to AChR. Similarly, in animal studies, sera from rats immunized with AChR caused reduction in extrajunctional ACh sensitivity in denervated muscle (Bevan *et al.*, 1976). Radioimmunoassay has shown 87% of patients with myasthenia to have receptor-binding antibodies (Lindstrom *et al.*, 1976). The antibody titers corresponded closely to the clinical severity of the disease. Further evidence for a circulating serum factor came from passive transfer experiments which reproduced many features of the disease in mice after injection of immunoglobulin fraction from myasthenic serum (Toyka *et al.*, 1975). IgG alone produced the reduction in mepp and acetylcholine receptors. The effect was enhanced by complement, in particular $C_1$ to $C_3$ (Toyka *et al.*, 1977).

Immunization of animals with purified AChR produces an autoimmune response to AChR protein (Patrick and Lindstrom, 1973; Lennon *et al.*, 1976; Tarrab-Hazdai *et al.*, 1975; Lindstrom *et al.*, 1976).

Experimental autoimmune myasthenia gravis (EAMG) has proven to be an excellent animal model for human myasthenia gravis (Lindstrom *et al.*, 1976; Lennon, 1978). Chronic EAMG in rats resembles myasthenia gravis in every criterion examined, including clinical weakness, improvement with anticholinesterases and electrophysiological and ultrastructural properties (Engel *et al.*, 1976; Lennon, 1978). Rats inoculated once with purified AChR developed immunity to skeletal muscle AChR (Lennon *et al.*, 1976). There is evidence for both cell-mediated and antibody-mediated immunity to AChR in EAMG. Cell-mediated immunity, a property of thymus-derived (T) lymphocytes (Miller and Osaba, 1967), was suggested by delayed-type hypersensitivity reactions. Contact between an immune T cell and its antigen resulted in massive accumulation of predominantly mononuclear cells at the neuromuscular junction eight days after inoculation (Engel *et al.*, 1976; Lennon *et al.*, 1976).

The mechanisms of impairment of neuromuscular transmission by antibody to the acetylcholine receptor have been reputed to include: blockage of the active sites of the receptor, accelerated internalization and intracellular destruction, and destruction of the receptor-containing segments of the postsynaptic membrane by antibody-dependent, complement-mediated lysis (Engel, 1978; Drachman, 1978a). Sera from human myasthenic patients and from rats with EAMG caused a reduction in the acetylcholine sensitivity, an increased rate of degradation of the acetylcholine receptor, lowered density of $\alpha$-bungarotoxin binding sites on myotube membranes and direct impairment of ion channel function of AChR molecules in muscle cells in culture (Bevan *et al.*, 1976; Anwyl *et al.*, 1977; Bevan *et al.*, 1977; Kao and Drachman, 1977; Drachman *et al.*, 1977; Drachman, 1978a).

Anti-AChR antibodies of many different specificities have been described in sera from patients with myasthenia gravis. Some react more with the AChR of one heterologous species than with another (Garlepp and Dawkins, 1977). Some react preferentially with the extrajunctional receptors (Almon and Appel, 1975), and some inhibit binding of concanavalin A to the receptors suggesting that the antigenic determinant may exist at or near a carbohydrate moiety of the AChR (Mittag *et al.*, 1976; Lennon, 1978). Myasthenic patients reportedly have higher serum titers against extrajunctional than junctional receptors (Weinberg and Hall, 1979). Competition experiment with purified rat AChR demonstrated two classes of determinants, those common to both junctional and extrajunctional receptors and those present or exposed only on extrajunctional receptors. These results suggest molecular differences between the two types of receptors (Weinberg and Hall, 1979).

The importance of complement in the myasthenic process is supported by studies on rats decomplemented by cobra venom factor and then injected with anti-AChR antibodies (Lennon *et al.*, 1978; Lennon, 1978). The results showed that antibody was bound to the AChR without impairing neuromuscular transmission, that the amount of receptor was in the normal range after 72 hr even though 60% was complexed with antibody, and that no inflammatory cells accumulated at the neuromuscular junction. It was concluded that the activation of complement by binding of antibody to the receptor is required to initiate acute EAMG. This event appears both to provide a chemotactic signal for mononuclear inflammatory cells and to promote phagocytosis of the postsynaptic membrane. Thus, a causal relationship is suggested between the binding of antibody to the AChR, the reduction in muscle AChR, and the impairment of neurotransmission (Lennon, 1978).

Electron microscopic examination of neuromuscular junctions has also been made in the myasthenic syndrome (Santa *et al.*, 1972b; Fukuhara *et al.*, 1972). In this condition, unlike myasthenia gravis, the mean area of postsynaptic junctional folds and secondary synaptic clefts per nerve terminal and the relative amounts of presynaptic membrane are increased. It was suggested that the release of acetylcholine is defective.

### D. *Polymyositis*

Polymyositis is an inflammatory myopathy, the etiology of which is unknown (Adams, 1973, 1975; Bohan and Peter, 1975; Walton, 1978). The disease forms part of a group of disorders, including dermatomyositis, in which dermatitis, arthritis, and occult carcinoma may be present. The prognosis of dermatomyositis is worse than that of the potentially treatable polymyositis. Patients have proximal muscle weakness which may be acute, subacute or chronic (Pearson, 1966). As a result of the destructive muscle lesion, serum CPK and aldolase are elevated. In addition, characteristic EMG findings are present (Pearson, 1966; Adams, 1975; Bohan and Peter, 1975).

The essential lesion in polymyositis is an inflammatory necrosis which is either focal or involves many sarcomeres (Adams, 1975; Hughes and Esiri, 1975). Muscle biopsy studies revealed perivenous and interstitial infiltration of lymphocytes, macrophages, plasma cells, and polymorphonuclear leukocytes (Adams, 1975; Hughes and Esiri, 1975). Cell migration is an early response and may play a role in initiating fiber degeneration (Adams, 1975). Ultrastructural changes include redundant folding of the sarcolemma (Hughes and Esiri, 1975), degeneration or sarcolemmal nuclei and myofibrils, disorganization of sarcomeres with "streaming" of the Z bands, dilatation of the sarcoplasmic reticulum, and increased numbers of autophagic vacuoles and cytoplasmic bodies (Adams, 1975; Hughes and Esiri, 1975; Shafig *et al.*, 1967).

Prominent features in muscle biopsies of polymyositic patients are the regenerative changes present in segments which have undergone necrosis (Adams, 1975; Mastaglia and Kakulas, 1970). This regenerative response resembles myogenesis and can restore normal muscle structure and function if the inflammatory process is arrested. However, in chronic polymyositis, the destructive process can eventually exhaust the regenerative capacity of the muscle, resulting in replacement of muscle fibers by connective tissue and fat (Adams, 1975). In the late stages of chronic polymyositis, the pathology cannot be distinguished from that of progressive muscular dystrophy and differs from denervation only in having less group atrophy (Adams, 1973).

Based upon clinical, pathological, and immunological findings, polymyositis is considered to be an autoimmune disorder. This hypothesis forms the basis for current treatment with corticosteroids and/or immunosuppressive agents (Dawkins and Mastaglia, 1973; Haas, 1973; Metzger, 1974). Clinically, the disease may occur with connective tissue disorders, e.g., polyarteritis and Sjogrens syndrome (Bloch, 1974) or with autoimmune diseases, e.g., systemic lupus erythematosus and rheumatoid arthritis (Pearson, 1972a). An association with visceral malignancy exists, with an incidence of about 15% in cases of dermatomyositis (cf. references in Bohan and Peter, 1975).

Immunological studies of patients with polymyositis and experimentally induced myositis in laboratory animals suggest that humoral, cellular, or even viral factors may be important (Bohan and Peter, 1975). Significant abnormalities of leukocyte migration and secretion of migration inhibition factor in the presence of muscle antigens occur in a significant number of patients with polymyositis (Goust *et al.*, 1974). In addition, lymphocytes from polymyositis patients are cytotoxic to cultured muscle cells, an effect prevented by antilymphocyte antiserum (Currie *et al.*, 1971). Animal studies disclosed that lymph node cells obtained from rats sensitized to muscle destroyed muscle fibers *in vitro* (Kakulas, 1966). Mastaglia *et al.* (1970) described lymphocyte adherence to muscle cell membranes which was considered to be an integral part of the mechanism of cell-mediated cytoxicity.

Numerous experimental models have been employed to study the immune response in polymyositis. A generalized myositis with lymphocytic infiltrates was produced in guinea pigs injected with heterologous muscle homogenates in Freund's complete adjuvant (FCA) (Dawkins, 1965). Transfer of experimental allergic myositis to normal syngeneic animals was produced by injections of lymphoid cells from the spleens of rats given two or more immunizing injections of heterologous muscle in FCA (Esire and Maclennan, 1974, 1975). Guinea pigs injected with a myofibrillar fraction developed lesions typical of those seen in animals injected with whole muscle homogenates. The antibody, as detected by immunofluorescence, was directed against myosin in the A band of the myofibril (Manghani *et al.*, 1974).

A mechanism of vascular damage mediated through immunoglobulin and $C_3$, possibly in the form of immune complexes, has been suggested as the cause of muscle damage in childhood dermatomyositis (Whittaker and Engel, 1972). Examination of muscle capillaries in polymyositis revealed hypertrophy of endothelial cells and pericytes and replication of the basal lamina around capillaries in 24–74% of cases of inflammatory muscle disease, which may indicate repeated cycles of capillary degeneration and regeneration (Jerusalem *et al.*, 1974c).

### *E. Denervation Atrophy*

An important ultrastructural study of denervation atrophy of the rat gastrocnemius, soleus, and plantaris muscles was made by Pellegrino and Franzini (1963). They found the diameter of all fibers of denervated muscles to be markedly decreased but at different times and to a different degree. After the first two weeks following denervation, they noted the presence of small lysosomes between myofibrils. During this period there was reduction in the number of myofibrils and appearance of areas of disorganization of contractile material. Alterations appeared first at the Z disks, which lost their straight line configuration, became bent, and sometimes reached the H band. Because the Z-line alterations, the filaments lost their parallel arrangement. While there were no early changes in the sarcoplasmic reticulum, this structure later became disrupted and gave rise to isolated vesicles. After two weeks large peripheral areas containing only glycogen, some mitochondria and remnants of sarcoplasmic reticulum were present. These remnants consisted of isolated vesicles and tubules seemingly derived from intermediate elements of the triads.

After one month, most fibers were greatly reduced in diameter, and the number of fibrils was decreased. Fibril shape was irregular and the diameter variable. The sarcoplasmic reticulum was disordered and overdeveloped.

Nuclei were often found in long rows, occasionally centrally located in the fibers. Mitochondria were still present between the fibers, cristae were not so closely packed as in normal mitochondria, and the overall population of mitochondria was reduced.

After three months, there was overproduction of sarcoplasmic reticulum and T tubules and reduction in mitochondrial content. Mitochondria appeared to be only secondarily involved. Similar results were obtained for the denervated rat diaphragm (Miledi and Slater, 1968, 1969). Proliferation of the sarcoplasmic reticulum and transverse tubular system were described by Schrodt and Walker (1966).

Denervation atrophy of the frog semitendinosus muscle was characterized by a prominence of sarcoplasmic reticulum and the presence of numerous small mitochondria (Muscatello *et al.*, 1965). A transient increase in amount of sarcoplasmic reticulum and mitochondria was seen during the early stages after denervation. Denervated muscles showed a progressive decrease in myofibrillar diameter after denervation and an increase in the number of polyribosomes and membrane-bound ribosomes.

### *F. Disuse Atrophy*

Disuse atrophy can be produced by a number of techniques, including spinal cord and/or dorsal root section (Tower, 1937), tenotomy (Vrbova, 1963), surgical pinning (Solandt *et al.*, 1943; Fischbach and Robbins, 1969), and plaster casts (Tabary *et al.*, 1972; Booth and Kelso, 1973).

In an ultrastructural study (Klinkerfuss and Haugh, 1970) in which the Tower technique was employed, alterations were largely limited to myofibrillar material and to collections of glycogen granules while mitochondria appeared normal. Normal histochemical-staining intensity was maintained up to one month in both gastrocnemius and plantaris muscles. Fibers richest in oxidative activity showed the greatest atrophy, while the size of phosphorylase-rich fibers did not change up to 30 days. Lysosomes appeared at the margin of sarcoplasmic masses in phosphorylase-rich fibers which contained remnants of the T–system and Z disks.

Tenotomy of rat soleus muscle was studied by Shafiq *et al.* (1969). Two striking changes, both of which developed within 7–10 days following tenotomy, were observed, i.e., central core fibers and nemaline rods. Tenotomy appeared to cause lesions in rat soleus which are characteristic of the hereditary central core and nemaline myopathies.

Disuse atrophy of human skeletal muscles produced by plaster casting or by traction was studied histochemically by Patel (1969). There was no difference between immobilized and control muscles in succinate dehydrogenase activity. No discernible change in the structure of muscle fibers was observed, with the exception of a relative increase of nuclei in atrophied myofibers.

### *G. Hereditary Muscular Dystrophies in Animals*

Murine dystrophy is a commonly used model of the human dystrophies with which it shares several features (Pinckney *et al.*, 1963). In an ultrastructural study of this disease, Bray and Banker (1970) observed large muscle fibers with a forked or branched appearance. In necrotic fibers, mitochondria were either aggregated or scattered, and mitochondria displayed irregular dense inclusions, or vacuoles and disintegration of the membranes or vesiculation of cristae. The sarcolemma was absent from necrotic fibers. No lysosomes were seen except in macrophages. The changes in sarcolemma, nuclei, mitochondria, and sarcoplasmic reticulum were reminiscent of changes in necrotic muscle produced by experimental procedures including ischemia (Stenger *et al.*, 1962; Moore *et al.*, 1956), crush (Allbrook, 1962), cold (Price *et al.*, 1964), heat (Shafiq and Gorycki, 1965), vitamin E deficiency (Van Fleet *et al.*, 1968), plasmocid (Price *et al.*, 1962), and cortisone (D'Agostino and Chiga, 1966).

Although many investigators have considered mouse dystrophy to be a primary muscle disorder (Michelson *et al.*, 1955; Pinckney *et al.*, 1963; Pearce and Walton, 1963; West *et al.*, 1966; Blaxter, 1969), numerous workers have suggested that this disease may be neural in origin (Baker *et al.*, 1960; McComas and Mossawy, 1965; Harris and Wilson, 1971; Salafsky, 1971). Previously, the neurogenic theory received a great deal of support (Askenas and Hee, 1974; Bray and Aguayo, 1975; Stirling, 1975; Weinberg *et al.*, 1975; Hironaka and Miyata, 1975; Wood and Boegman, 1975; Komiya and Austin, 1974; Tang *et al.*, 1974; Law and Caccia, 1975; Jablecki and Brumijoin, 1974; Panayiotopoulas and Scarpalezos, 1976). However, it has received little support from the most recent work (Bradley and Jenkison, 1975; Neerunjun and Dubowitz, 1974a,b; Douglas, 1975; Ashmore and Doerr, 1976; Harris and Marshall, 1973; Paul and Powell, 1974; Montgomery, 1975; Johnson and Montgomery, 1975; Law *et al.*, 1976; Drachman and Fambrough, 1976; Howe *et al.*, 1976).

Pachter *et al.* (1974) studied the dystrophic mouse (gene dy 25) as a model for human myotonia and found fiber splitting as a prominent feature. Alterations in motor endplates were seen in extraocular muscles from dystrophic mice (Davidowitz *et al.*, 1976; Pachter *et al.*, 1976). The findings of this study were consistent with motor endplate disruption as a primary factor in mouse dystrophy with the myotonic gene.

## III. Pathology of Specific Membrane Systems

### A. *Plasma Membrane*

That muscle plasma membranes might be damaged in disease processes was demonstrated by Sibley and Lehninger (1949) and confirmed by Schapira *et al.* (1953). Several subsequent investigators (e.g., Jacob and Neuhaus, 1954) reported that serum concentrations of certain muscle enzymes (aldolase, glutamate oxalacetate transaminase, glutamate pyruvate transaminase, and creatine phosphokinase) become elevated during the course of progressive muscular dystrophy. These findings are significant diagnostically in that they help to differentiate muscular dystrophy from neurogenic atrophy in which serum enzyme concentrations are normal or less elevated (Shaw *et al.*, 1967). Hudson *et al.* (1969) reported increased serum enzymes in hyperkalemic periodic paralysis and slightly increased enzymes in myotonic dystrophy. Similarly, serum aldolase is increased in the dystrophic mouse (Schapira *et al.*, 1957) and in rats with vitamin E deficiency (Beckmann and Buddecke, 1958).

Serum enzyme elevation is apparently due to increased muscle permeability to the enzymes (Zierler and Rogus, 1957). Muscle permeability to aldolase is increased by depolarization and by metabolic inhibitors (Zierler and Rogus, 1958), as well as by tenotomy and ischemia (Pellegrino and Bibbiani, 1963). Some investigators (e.g., Platzen and Chase, 1964) considered the plasma membrane to represent the site of the primary defect in muscular dystrophy, a finding strongly supported at present (Rowland, 1976). Because increased intracellular $Ca^{2+}$ can cause muscle cell damage, Duncan (1978) suggested that a genetic lesion of the sarcolemma $Ca^{2+}$ channels exists in dystrophy.

Mendell *et al.* (1972) proposed that the increase in plasma enzyme concentrations in Duchenne dystrophy may be a consequence of functional ischemia. They developed a rat model employing abdominal aortal ligation below the renal arteries which causes functional ischemia after injection of serotonin. In addition to producing lesions similar to those observed in Duchenne muscular dystrophy (Mendell *et al.*, 1971), this model of ischemia causes fluctuation in serum enzyme patterns. Mendell *et al.* (1972) proposed that enzyme fluctuations in muscular dystrophy represent short episodes of functional ischemia resulting in focal areas of muscle necrosis. However, little experimental support has been adduced for the importance of defective microvasculature in dystrophy (Bradley *et al.*, 1975).

Further abnormalities of plasma membranes in dystrophic muscle have been reported. For example, the resting membrane potential of muscle fibers from dystrophic mice is lower than that of healthy fibers (Conrad and

Glaser, 1961; Harris, 1971), possibly accounting for their increased excitability. However, the release of acetylcholine and the responsiveness of the postsynaptic membrane to acetylcholine are intact in Duchenne muscular dystrophy (Sakakibara, 1977).

In dystrophy, skeletal muscle membranes have markedly elevated $Na^+$, $K^+$-activated ATPase (Sulakhe *et al.*, 1971). The concept of a cellular membrane defect was further supported by demonstration of abnormalities in phospholipid metabolism in nerve and muscle (Hughes, 1972; Kwok *et al.*, 1976). Diverse ATPase abnormalities were also demonstrated (Dhalla *et al.*, 1973, 1975) in muscles from dystrophic humans, hamsters, and vitamin E deficient rats (cf. Wrogemann *et al.*, 1974).

Freeze-fracture studies of skeletal muscle from Duchenne patients showed nonuniform distribution and depletion of particles on both protoplasmic and extracellular faces of sarcolemma (Schotland *et al.*, 1977).

Fine structural studies of dystrophic mouse muscle also revealed abnormalities which are consistent with a theory of defective membranes. Embryonic myogenic cells apparently were not affected by dystrophy, whereas adult myogenic cells lost some developmental capacities when compared with normal muscle (Platzen and Powell, 1975). Other reported abnormalities in dystrophic mouse muscle include a decreased permeability to potassium (Lipicky and Hess, 1974) and a reduced activity of adenylate kinase (Kitchen and Watts, 1974).

Substantial evidence supports the theory that the basic defect in myotonia is in the muscle plasma membrane. Abnormalities include hyperirritability of curare-resistant bursts of repetitive depolarization after stimulation (Peter and Fiehn, 1973). In myotonia congenita, reduction of chloride conductance may account for most of the electrical abnormalities (Rudel and Senger, 1972). This chloride hypothesis was recently well studied and confirmed by Barchi (1975). Also, inclusion of 3% potassium iodide in the drinking water of immature chickens caused myotinia due to a block of chloride conductance (Morgan *et al.*, 1975).

McComas and Mrozek (1968) showed that the resting membrane potential in myotonic dystrophy was abnormally low. Since the resting membrane potential in myotonia congenita was normal, this deficit was considered to result from the dystrophic process. However, Lipicky and Bryant (1966) reported that the resting membrane resistance in the myotonic goat was four to five times normal.

An animal model for myotonia can be produced by injection of 20,25-diazacholesterol, an inhibitor of cholesterol synthesis, into rats (Kuhn *et al.*, 1968; Eberstein and Goodgold, 1972). This experimental myotonia is similar to the impaired relaxation after muscular contraction observed in human myotonia congenita and myotonic dystrophy. In addition to myotonia, diazacholesterol caused cataracts in treated rats (Peter *et al.*, 1973). Because it alters membrane ATPase activity, diazacholesterol was considered to alter membrane ion transport (Brown *et al.*, 1968). Seiler (1971) also found diminished ATPase activity in the sarcolemma of rats after diazacholesterol treatment. Recent studies of the biochemical effects of diazacholesterol (Peter and Fiehn, 1973; Peter *et al.*, 1975a) indicated that desmosterol accumulates in palsma, red blood cells and, to a lesser extent in muscle during diazacholesterol-induced myotonia. Electron spin resonance (ESR) experiments on erythrocyte membranes from rats with diazacholesterol-induced myotonia suggested increased surface membrane fluidity (Butterfield and Watson, 1977). Similar findings are seen in human congenital myotonia (Butterfield *et al.*, 1976). The abnormal steroid composition of membranes may have caused the decreased chloride conductance and other electrical abnormalities observed in myotonic muscle (Peter and Fiehn, 1973). Diazacholesterol myotonia was prevented in animals by feeding them a high-cholesterol diet (Peter *et al.*, 1975b).

That a generalized defect in plasma membranes may represent the etiological lesion in muscular dystrophies has received considerable support. In accordance with the expectation that a genetic defect in muscle membranes would be expressed in other membrane systems as well, numerous investigators have studied erythrocyte ghosts. However, recent experiments have resulted in contradictions among various investigators and variability in results among Duchenne patients. Indeed, experiments showing an induction of an abnormal ouabain response by the ATPase of normal erythrocytes are highly suggestive of the absence of a primary defect in the red cell membrane (Siddigui and Pennington, 1977). These authors suggested that factor(s) leaking into the blood from the diseased muscle may be responsible for the altered response. Scanning electron microscopy of erythrocyte ghosts from dystrophic mice disclosed surface alterations (Morse and Howland, 1973). Abnormal red cells were subsequently seen in Duchenne, limb-girdle, facio-seapulo-humeral dystrophies (Matheson and Howland, 1974). However, other workers have been unable to reproduce these findings. These erythrocyte alterations were not present in all patients. Wide ranges of values with considerable overlap between patients and controls were seen, and the appearance of stomatocytes rather than echinocytes has been reported (Maile *et al.*, 1975; Roses and Appel, 1974; Lumb and Emery, 1976). More recently Matheson *et al.* (1976) in a study of erythrocytes from Duchenne patients and age-matched controls found no abnormalities of erythrocytes. On the contrary, erythrocyte morphology was found to be very sensitive to various cell treatments. It was concluded that erythrocyte morphology should not be considered an established diagnostic test from Duchenne patients or carriers (Matheson *et al.*, 1976). Adornato *et al.*

(1977) investigated hemolysis in Duchenne patients and found no increase in red cell turnover by indirect measurements or turnover of $^{51}$Cr-labeled erythrocytes. They concluded that their results failed to support a significant erythrocyte membrane defect. In contrast, Wakayama *et al.* (1979) in a freeze-fracture study of erythrocyte plasma membranes from Duchenne patients found a marked depletion of intramembranous particles as compared to controls. They concluded that the internal molecules architecture of the erythrocyte membrane is abnormal in Duchenne dystrophy.

Using ESR spectrometry, Butterfield *et al.* (1974) demonstrated that the myotonic erythrocyte membrane is more fluid and less polar than the normal or Duchenne red cell membrane. Changes were considered to be more marked near the surface than in the interior of the membrane. Erythrocyte membrane alterations were not specific but were seen in myotonic dystrophy, myotonia congenita, and oculopharyngeal muscular dystrophy (Roses *et al.*, 1975a). ESR examination of erythrocyte membranes from patients with Duchenne and myotonic dystrophy suggest alterations in membrane protein conformation and/or organization (Butterfield, 1977). Roses *et al.* (1975b) studied phosphorylation of erythrocyte ghost proteins from Duchenne dystrophic patients and showed that protein phosphorylation of bands II and III increased in myotonic dystrophy patients. They concluded that Duchenne and myotonic dystrophies are inherited membrane disorders with widespread tissue involvement. This conclusion was supported by Percy and Miller (1975) who demonstrated reduced deformability of erythrocyte membranes from Duchenne patients. However, their technique has been challenged in that it does not specifically determine membrane properties (Somer *et al.*, 1979). Rather, Somer *et al.* (1979), using microsieving or flow channel measurements and taking the effect of a smaller mean corpuscular volume of Duchenne erythrocytes into account, found no significant differences from controls in membrane deformability. Although they found an increased osmotic fragility, individual values overlapped with controls and only 50% of Duchenne erythrocyte values were abnormal.

Phenytoin, a drug with a mechanism of action involving membranes, was shown to normalize "fluidity" differences in ESR spectra of myotonic red cells with no effect on normal membranes (Roses *et al.*, 1975b). Treatment of dystrophic chicks with diphenylhydantoin improved their righting ability and reduced acetylcholinesterase activity to normal values (Entrikin *et al.*, 1977). Red cell, muscle, and liver membranes in Suchenne dystrophy have enhanced microviscocity (Lumb and Emery, 1975; Sha'afi *et al.*, 1975).

Further alterations in red cell surfaces were demonstrated by Bosmann *et al.* (1976) who found increased electrophoretic mobility of "dystrophic" (Duchenne, myotonic, mouse, chicken) erythrocytes. These results are consistent with the concept of dystrophy as a systemic membrane disease not limited to muscle. Miller *et al.* (1976), in a scanning electron microscope study, found cup-shaped red cells to be common in myotonic and Duchenne patients and in Duchenne carriers.

In addition, ion transport appears to be abnormal in red blood cells from patients with myotonic dystrophy (Hull and Roses, 1976). While the normal sodium "pump" expels three sodium for two potassium ions per ATP hydrolyzed, the pump in myotonic dystrophy red cells exchanged sodium and potassium in a ratio of two for two per ATP. Thus, there was a decrease in the active extrusion of sodium by red cells from patients with myotonic dystrophy.

Hull and Roses (1976) suggested that the altered pump activity was related to decreased membrane protein phosphorylation. Howland (1974) found greatly increased potassium conductance in red cell membranes from Duchenne patients.

The $Na^+$, $K^+$-activated ATPase activity of Duchenne erythrocytes and the effects of ouabain on the enzyme activity have been a subject of controversy in various laboratories (Siddiqui and Pennington, 1977; Souweine *et al.* 1978). The behavior of a variety of sensitive hydrophobic fluorescent probes in isolated erythrocyte membrane ghosts from myotonic dystrophy patients showed that intrinsic changes in membrane physical properties, if present, must be small or localized within lateral microdomains. Alternatively, the abnormalities previously reported in whole erythrocytes may reflect specific interactions between the membrane and the intracellular environment. The concept of a generalized membrane defect in myotonic dystrophy must be interpreted with caution (Chalikian and Barchi, 1979).

Roses *et al.* (1976) attempted to utilize alteration in endogenous protein phosphorylation as a marker of Duchenne dystrophy carriers. There was a significant increase in protein phosphorylation in mothers of children with Duchenne dystrophy, including those of children whose dystrophy was seemingly due to a new mutation. However, due to the large range of the data, individual carriers could not be detected. Roses *et al.* (1976) suggested that mutations are uncommon and that cases so classified are a result of failure of current methods to detect the carrier state.

The reduction in membrane protein phosphorylation in myotonia was attributable to a membrane glycoprotein (component a) in erythrocytes. This is of interest for myotonia since Bretscher (1973) has proposed an ion channel role for component a. It thus seems possible that altered protein phosphorylation in membranes is related to the electrical changes in myotonic cells.

Max *et al.* (1970) showed that denervation results in increased synthesis of a specific ganglioside ($GM_3$, or hematoside) which probably is a membrane component. This observation was confirmed by Cotrufo and Appel (1971), who also reported an increase in the content of protein-bound sialic acid in denervated gastrocnemius (but not soleus) muscles (Cotrufo and Appel, 1973). Dystrophic mouse muscle, on the other hand, showed ganglioside abnormalities different from those of denervation (Max and Brady, 1971). Other lipid studies showed increased incorporation of $^{32}P_i$ into all glycerophosphatides in denervated frog muscle membranes (Bunch *et al.*, 1970).

Significant alterations are present in neutral lipids and phospholipids in isolated sarcolemma from dystrophic mice (DeKretser and Livett, 1977).

### *B. The Sarcotubular System*

The role of the sarcotubular system in excitation–contraction coupling and relaxation has stimulated research interest regarding possible function of these structures in muscle disease of both humans and experimental animals. For example, Samaha and Gergely (1969) studied calcium uptake and ATPase activity of the sarcoplasmic reticulum in progressive muscular dystrophy and in myotonic dystrophy. In progressive muscular dystrophy, low-calcium uptake was found with low ATPase activity and normal transport efficiency. These changes had no apparent relationship to the stage of the disease. In myotonic dystrophy, however, sarcoplasmic reticulum preparations exhibited a high initial and normal total calcium uptake, and normal or greater than normal efficiency. Samaha and Gergely (1969) suggested that there is a primary involvement of the sarcoplasmic reticulum in red fibers in myotonic dystrophy. Dhalla and Sulakhe (1973) found no alteration in calcium transport in sarcoplasmic reticulum preparations from genetically dystrophic hamsters, except in older animals. These results suggested that observed changes were secondary to other processes. Similarly, in myotonic dystrophy, calcium transport by sarcotubular vesicles was found to be normal (Peter and Worsfold, 1969). Biochemical alterations have been reported in the fragmented sarcoplasmic reticulum from dystrophic mouse, chicken, and human muscle (Samaha and Gergeley, 1969; Takagi *et al.*, 1973; Martinosi, 1968; Hanna and Baskin, 1977; Scales *et al.*, 1977) including alterations in calcium transport, ATP hydrolysis and phosphoenzyme formation (Hanna and Baskin, 1977; Mrak and Baskin, 1978; Hanna and Baskin, 1978). Samaha and Congedo (1977) found two distinctly abnormal patterns in the distribution of six proteins of sarcoplasmic reticulum membranes. Although the groups of patients could not be distinguished on clinical grounds these authors suggested the possibility of two distinct biochemical diseases in these patients.

In diazacholesterol-induced myotonia, Seiler *et al.* (1970) found that calcium storage capacity and initial rate of calcium uptake was increased, while Briggs and Kuhn (1968) found no change in the rate of calcium uptake. Seiler *et al.* (1970) suggested that sarcotubular vesicles from myotonic animals suppressed myofibrillar ATPase less effectively.

Chemical studies (Seiler and Kuhn, 1969) demonstrated esterified diazacholesterol in sarcoplasmic reticulum vesicles. Such changes were not, however, substantiated by Peter and Fiehn (1973). They reported an unchanged rate of calcium uptake by the sarcoplasmic reticulum in diazacholesterol-induced myotonia. In six weeks, values were equal to or less than normal, while ATPase activity was enhanced during the same period. Calcium uptake in the absence of oxalate and initial rates in the presence or absence of oxalate were decreased. The rate or release of calcium from preloaded sarcoplasmic reticulum vesicles was increased. Essential fatty acid deficiency was found to cause impaired function of the calcium pump (Seiler and Hasselbach, 1971). Baskin (1970) found no change in calcium uptake and ATPase activity in dystrophic chick preparations.

In periodic paralyses (hypokalemic), it appears that paralysis is not caused by hyperpolarization of the plasma membrane since the resting membrane potential is not raised during an attack (McArdle, 1969; Pearson, 1972b). Electron microscopy revealed characteristic vacuoles, thought to arise from dilatation of the sarcoplasmic reticulum. It is not known whether the vacuoles are the cause of periodic paralysis.

In Gamstorp's disease (adynamic episodica hereditaria, hyperkalemic periodic paralysis) (Kissel *et al.*, 1969), there is swelling of mitochondria, dilatation of the sarcoplasmic reticulum and an increase in the number of glycogen particles in muscle. Between crises, muscle chloride and sodium were elevated, while muscle potassium was reduced. Crises accentuated these abnormalities. The cell membrane polarization decreased during crisis and administration of potassium induced a crisis. Kissel *et al.* (1969) suggested that the paralytic episodes are due to elevated muscle fiber chloride and a polarization corresponding to chloride accumulation which extends throughout the membrane. The disorders of potassium and chloride are apparently interdependent. The origin of the disturbance in Gamstorp's disease is not clear at this time.

Other studies have shown that sarcoplasmic reticulum from denervated muscle binds more calcium than normal (Thorpe and Seeman, 1971). Permeability to sodium in denervated muscle (Creese *et al.*, 1968) was twice that of normal muscle.

### C. *Lysosomes*

Since the discovery of lysosomes by de Duve and co-workers (Wattiaux, 1969), these particles have been the object of numerous pathological investigations. Their content of hydrolases active at acid pH has suggested a role in degradation of cellular materials. Indeed, a large number of diseases have been identified in which deficiencies of lysosomal hydrolases represent the primary defect (Brady, 1972; Van Hoof and Hers, 1968; Tallman *et al.*, 1971). The topic of lysosomes in muscle pathology has been reviewed by Weinstock and Iodice (1969). Hudgson and Pierce (1969) considered the role of lysosomes in pathogenesis of muscle disease and concluded that "in most instances, they have only a secondary role in damaging a muscle fiber and act as intracellular garbage collectors." An exception to this concept may be Pompe's disease (glycogenosis type II), which is characterized by deficiency of $\alpha$-glucosidase (or acid maltase) and appears to be a primary lysosomal disorder (Hers, 1963).

Acid hydrolase activities are increased substantially in various pathological conditions of muscle including hereditary muscular dystrophy (Tappel *et al.*, 1962; Iodice and Weinstock, 1965; Pennington and Robinson, 1968; Wrogemann and Blanchaer, 1971; Kar and Pearson, 1972a; Abdullah and Pennington, 1968; Iodice *et al.*, 1972); vitamin E deficiency dystrophy (Zalkin *et al.*, 1962; Bond and Bird, 1966; Bunyan *et al.*, 1967; Desai, 1966); denervation atrophy (Weinstock and Lukacs, 1965; Syrovy *et al.*, 1966; Hajek *et al.*, 1964; Pollack and Bird, 1968; Pellegrino *et al.*, 1957); tenotomy (Pollack and Bird, 1968); limb immobilization (Max *et al.*, 1971); selenium deficiency dystrophy (Whanger *et al.*, 1969, 1970); and ischemia (Archangeli *et al.*, 1973). Kar and Pearson (1972a) reported that acid phosphatase and $\beta$-glucosaminidase are enhanced only in severely damaged dystrophic muscles. Cathepsin D and DNase II and other acid hydrolase activities were found to be increased in muscles from young dystrophic chicks (Tappel *et al.*, 1963) before histological evidence of inflammatory reaction was present. These studies suggested an important role for lysosomes in muscle pathology. However, such a role remains to be demonstrated.

An important unanswered questions concerns the origin of lysosomes in muscle. Canonico and Bird (1970) concluded from zonal centrifugation studies that there are two populations of lysosomes, one from muscle and the other possibly from macrophages. Since denervation causes proliferation of lysosomes (Pellegrino and Franzini, 1963) and an increase in acid hydrolase activities, Schiaffino and Hanzlikova (1972) addressed themselves to these questions by studying the effects of denervation on developing muscle. Most lysosomes were seen near the Golgi apparatus. It was considered that the sarcoplasmic reticulum might be involved in the formation of lysosomes or that lysosomes may exist as part of the sarcoplasmic reticulum in skeletal muscle. In ultrastructural studies, Schiaffino and Hanzlikova (1972) concluded that myofibril breakdown is not initiated by lysosomes since disintegration of myofibrils took place in a part of the cell without lysosomes and since autophagic vacuoles did not contain myofilaments. Lysosome formation appeared to be secondary to myofibrillar changes.

In a biochemical study of atrophy caused by limb immobilization, Max *et al.* (1971) found an elevation of a number of acid hydrolases at different times in atrophy. β-Glucosidase was elevated early in disuse, whereas other enzymes, β-galactosidase, β-*N*-acetylhexosaminidase, arylsulfatase A, and acid phosphatase were elevated later. Subsequent studies (J. Pargament and S. R. Max, unpublished) showed cathepsin D to be elevated at day seven following the production of disuse atrophy. This observation, along with the studies alluded to above, militates against lysosomes as causative agents of muscle wasting.

Supporting evidence for the idea that lysosomes are important in degradation of materials in the cell (e.g., mitochondria) was provided by Kadenbach (1969) who correlated lysosomal enzyme activity of various tissues with rate of turnover of cytochrome *c*. Tissues with high-lysosomal activity, such as liver, had a shorter half-life of cytochrome *c* than tissues with low-lysosomal activity, such as muscle.

Other investigators suggested that proteolytic enzymes not of lysosomal origin may be more important in muscle diseases than lysosomal enzymes. Thus, Pennington and Robinson (1968) showed that in human dystrophy there are increases in the activities of neutral and alkaline cathepsins as well as acid cathepsin. Kohn and co-workers (Kohn, 1966, 1969; Pearlstein and Kohn, 1966; Pater and Kohn, 1967) studied autolysis in denervated muscle and concluded that three components of the muscle homogenate are necessary for autolysis, namely, the residue fraction, the mitochondrial fraction, and the soluble microsomal fraction; Pearlstein and Kohn (1966) suggested that accleration of catabolism, rather than repressed synthesis of muscle proteins, is important in denervation atrophy. Alkaline phosphatase was found to be increased only in severe stages of muscle diseases, and, therefore, is secondary (Kar and Pearson, 1972b).

Further support for a role of nonlysosomal hydrolases in degradation of skeletal muscle was given in a report that a ribonuclease inhibitor was abnormal in dystrophic mouse muscle. This was accompanied by elevated RNase II (alkaline RNase) activity (Meyeer and Little, 1970).

Proteolytic activity during growth of hypertrophic and atrophic muscles of genetically dystrophic chickens was also studied. Cathepsins A, B, C, and D were assayed, and at day one, all activities were elevated in dystrophic and normal muscle. The activities declined in normal muscles but remained elevated in dystrophic muscles. Cathepsin D was considered to be related to recurrent growth and degeneration of atrophic muscle in chicken dystrophy (Peterson *et al.*, 1971).

### *D. Mitochondria*

Because of the nature of muscular function, the need for efficient energy-transducing mechanisms in muscle tissue is obvious. Consequently, energy metabolism has been widely studied in muscle in both normal and pathological states. Biochemical, ultrastructural, and histochemical defects of mitochondria have been reported in muscle in many different and diverse muscle disorders (for a review see DiMauro *et al.*, 1974). Although mitochondria are altered in many disease states of muscle, in only a few, the so-called "mitochondrial myopathies", can mitochondrial abnormalities be said to represent a primary lesion. In most cases, it appears that mitochondrial damage is secondary to other pathological alterations in muscle fibers. Nevertheless, even if mitochondrial lesions are not primary, they may contribute significantly to the pathology of a given condition. Several reports suggested that impairment of energy metabolism may be an important factor in the initiation and progression of atrophy secondary to denervation (Carafoli *et al.*, 1964).

Work from the authors' laboratory on atrophy secondary to disuse induced by limb immobilization (Max, 1972) has yielded results which are in general agreement with Carafoli *et al.* (1964). Mitochondria isolated from disused rat gastrocnemius muscles by a simple, reproducible technique (Max *et al.*, 1972) had a decreased respiratory control index (Max, 1973). Subsequent experiments (Rifenberick *et al.*, 1973) revealed that atrophic muscles contained fewer mitochondria than controls. Furthermore, isolated mitochondria displayed a selective loss of a matrix enzyme (malate dehydrogenase) with sparing of inner and outer mitochondrial membrane enzymes (Rifenberick *et al.*, 1973). Further studies revealed abnormal substrate utilization by disused muscles (Rifenberick and Max, 1974a) and differences in recovery from immobilization of fast and slow muscles (Rifenberick and Max, 1974b). Margreth *et al.* (1972) also demonstrated loss of malate dehydrogenase from mitochondria isolated from denervated extensor and soleus muscles. Decreased activities of mitochondrial enzymes following denervation have been reported (Nachmias and Padykula, 1958; Humoller *et al.*, 1951; Schmidt, 1952; Hearn, 1959; Humoller *et al.*, 1952; Romanul and Hogan, 1965; Hogan *et al.*, 1965). Denervated rat diaphragm has impaired acetate utilization (Kouvelas and Manchester, 1968) and decreased oxidation of other substrates of oxidative metabolism (Koski and Max, 1974). Several authors reported alterations of levels of high energy compounds in denervated muscle (cf. Gutmann, 1962) and various diseased muscles. For example, Ronzoni *et al.* (1958) reported that creatine and creatine phosphate levels were decreased in biopsy material from patients with muscular dystrophy, and Dhalla *et al.* (1972) demonstrated that ATP and total adenine nucleotide levels in muscle from genetically dystrophic hamsters were significantly reduced.

Several investigators (Lochner and Brink, 1967; Jacobson *et al.*, 1970; Wrogemann and Blanchaer, 1968; Wrogemann *et al.*, 1969, 1970) studied oxidative phosphorylation in the dystrophic hamster. Oxidative phosphorylation was found to be defective in mitochondria isolated only from those animals of advanced age. Similarly, abnormalities in oxidative phosphorylation were shown in muscle from dystrophic mouse (Wrogemann and Blanchaer, 1967) and chicken (Ashmore and Doerr, 1970). Peter *et al.* (1970) reported normal respiratory control and ADP to O ratios for mitochondria isolated from patients with progressive muscular dystrophy, limb girdle dystrophy, facioscapulohumeral dystrophy, myotonic dystrophy, polymyositis, and neurogenic atrophy. Abnormal mitochondria could be isolated only in later stages in many of these diseases. In these cases, the mitochondrial damage was considered secondary to muscle cell necrosis and probably not of significance in the etiology of the disease. The results of a study by Max and Mayer (1972) agreed with this conclusion. In addition, Olson *et al.* (1968) reported that oxidative phosphorylation was normal in muscle in human muscular dystrophy.

A great number of mitochondrial myopathies have been reported since the description of the first case (Luft *et al.*, 1962), the majority of which are based upon ultrastructural abnormalities (for references see Spiro *et al.*, 1970a,b and DiMauro *et al.*, 1974; and recent papers, Julien *et al.*, 1974; Tamura *et al.*, 1974; McLeod *et al.*, 1975; Bender and Engel, 1975; Okamura *et al.*, 1976). Mitochondrial alterations such as increased number and size, packed cristae, and paracrystalline inclusions were also present in a number of neuromuscular disorders; e.g., neurogenic atrophy, muscular dystrophy, and polymyositis (Pearce, 1966; Chou, 1969, Shafiq *et al.*, 1968). It was suggested that diagnosis of a mitochondrial myopathy be based upon biochemical as well as ultrastructural and clinical criteria (DiMauro *et al.*, 1974).

The first *bona fide* "mitochondrial myopathy" was described by Ernster *et al.* (1959) and by Luft *et al.* (1962). This disorder selectively involved skeletal muscle mitochondria and caused a nonthyroidal hypermetabolism. Mitochondria isolated from biopsied muscle revealed absence of respiratory control, and respiration which was insensitive to oligomycin and only slightly enhanced by 2,4-dinitrophenol. Electron microscopic examination of a muscle biopsy specimen disclosed increased numbers of mitochondria (Fig. 6) contained large numbers of densely packed cristae, cristae replaced by tubular structures, and inclusions. Some cylindrical mitochondria were surrounded by concentric sheaths. The high respiratory rate of the patient's mitochondria, "loosely coupled" in state 3 or state 4, accounted for the inability to adapt respiratory function to energy requirements with the result

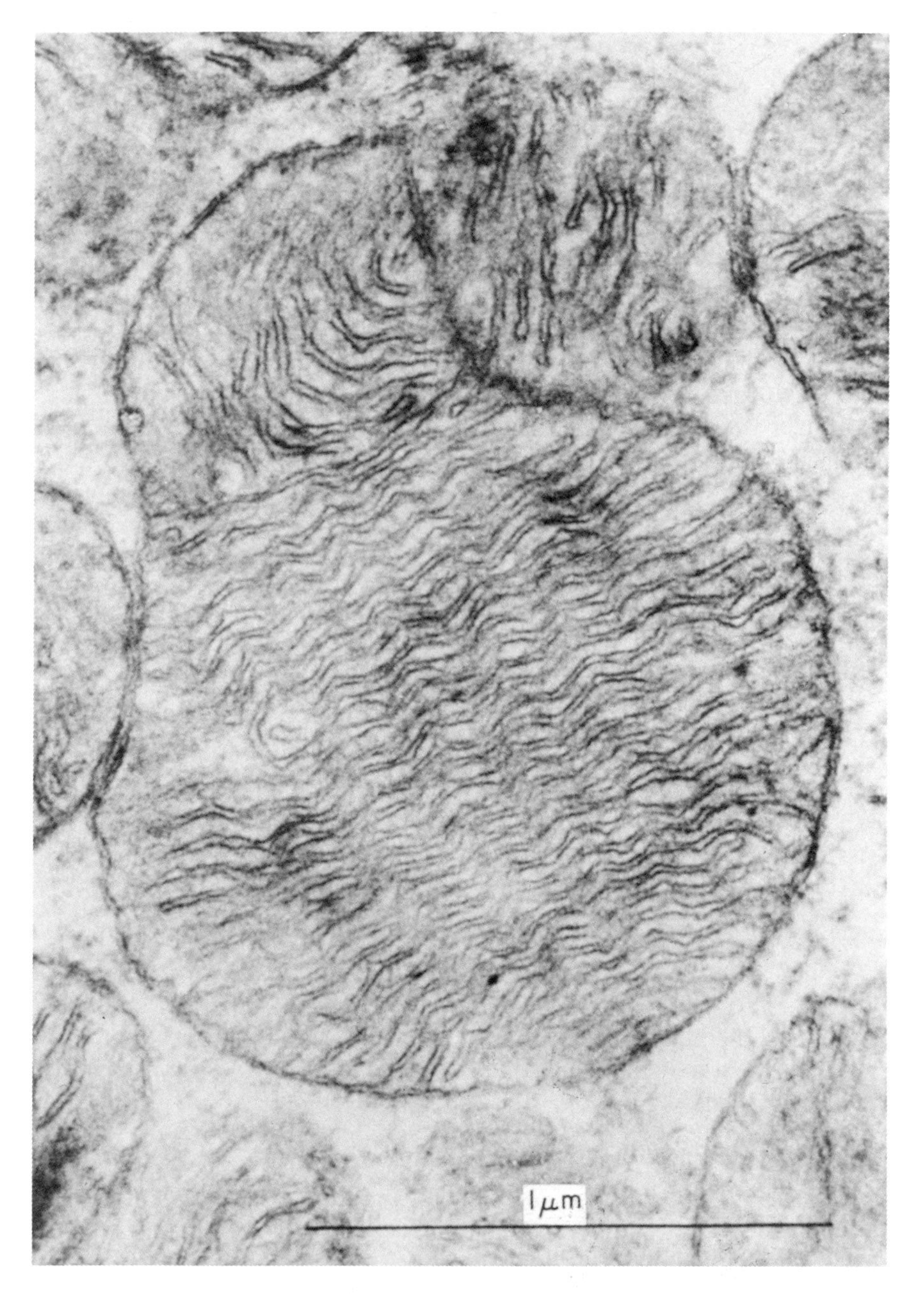


**Fig. 6.** Mitochondria from hypermetabolic patient. Note the zig-zag-like arrangement of the densely packed cristae. The mitochondria contain no electron-dense particles. ×54,000. (Reproduced from Luft *et al.*, 1962, with permission of the authors and *The Journal of Clinical Investigation.*)

that energy was wasted as heat. The hypermetabolism and hyperthermia served to explain the remainder of the patient's clinical findings, namely, increased perspiration, thirst, polyphagia, and thinness. A second case of "Luft's Disease" was reported by Haydar *et al.* (1971) and Afifi *et al.* (1972). The defect was attributed to large overgrowth of skeletal muscle mitochondria which were anatomically bizarre and functionally characterized an excessively high respiratory rate and uncoupled oxidative phosphorylation. Since treatment with chloramphenicol (an inhibitor of mitochondrial protein synthesis) caused clinical improvement, defective mitochondrial protein may have caused the disorder. Chloramphenicol also retarded synthesis of abnormal mitochondria. Further biochemical studies on mitochondria from this patient have been reported by DiMauro *et al.* (1974, 1976). An abnormality of mitochondrial cation transport was suggested by the inhibitory effect of magnesium on ATPase activity, increased ATPase activity after freezing and thawing, inhibition of respiration by calcium, and the presence of an electron-dense material, thought to be calcium within mitochondria. Furthermore, a number of abnormalities of calcium transport were uncovered. It was suggested that the inability of the mitochondria to retain calcium (and possibly magnesium) may have resulted in recycling of either or both cations. The ability of mitochondria to transport ions is an energy-requiring process dependent on respiration (Lehninger, 1964). Since energy for ion transport is removed at a step prior to ADP phosphorylation, calcium transport may "steal" energy for phosphorylation (DiMauro *et al.*, 1974). Although a plausible explanation for clinical findings in Luft's disease can be based upon loose coupling of respiration, the true molecular abnormality remains undefined. A recent electron cytochemical study showed the presence of two populations of mitochondria in Luft's disease with an impairment of succinate and $\alpha$-glycerophosphate oxidation in large mitochondria with abnormal cristae (Bonilla *et al.*, 1977).

In contrast to clinical findings in patients with Luft's disease, other cases of muscle weakness without hypermetabolism and mitochondrial abnormalities have been reported. Biochemical study of muscle mitochondria from a patient with progressive ophthalmoplegia and facial and proximal limb weakness revealed a lack of respiratory control with $\alpha$-glycerophosphate as a substrate (DiMauro *et al.*, 1973). In a trichrome stain of a biopsy specimen, there was evidence of "ragged-red" fibers, and a carbohydrate-specific stain showed excessive glycogen accumulation. Mitochondria contained paracrystalline inclusions in addition to large deposits of glycogen. Association of glycogen accumulation and abnormal muscle mitochondria was previously reported in progressive ophthalmoplegia (Sluga and Maser, 1970) and other muscle disorders (French *et al.*, 1972; Jerusalem *et al.*, 1973). The glycogen storage in these cases may be the result of a mitochondrial abnormality.

Schotland *et al.* (1976) reported a patient with a slowly progressive congenital neuromuscular disorder, whose muscle biopsy disclosed widespread crystalline inclusions. Isolated mitochondria had a decreased respiratory rate and respiratory control index with both NAD- and FAD-linked substrates. Basal-, magnesium-, and 2,4-dinitrophenol-stimulated mitochondrial ATPase were greatly reduced. It was concluded that a defect of respiratory chain-linked energy transfer common to all three energy coupling sites was present and the result of a loss of functional mitochondrial inner membrane which was replaced by crystalline inclusions.

A number of studies of defective muscle mitochondrial utilization of lipids have been reported (for a review see Angelini, 1976). For example, a skeletal muscle disorder associated with intermittent symptoms and a possible defect of long-chain fatty acid metabolism was described by Engel *et al.* (1970). The patients were 18-year-old twin girls who had intermittent aching, muscle "cramps" without weakness, and myoglobinuria. DiMauro and Melis-DiMauro (1973) studied a patient who had symptoms similar to those reported by Engel *et al.* (1970). Muscle carnitine concentration was moderately increased, while muscle carnitine palmityltransferase activity was low. Long-chain fatty acyl-CoA synthetase activity was normal, while acetylcarnitine transferase activity was decreased by 40%. It was suggested that the myoglobinuria in this patient may have been due to a genetic defect of lipid metabolism in skeletal muscle.

Bank *et al.* (1975) studied two brothers with recurrent myoglobinuria without apparent muscle weakness. Because of elevated plasma triglycerides and free fatty acids with increased and plasma pre-$\beta$-lipoproteins, a defect in peripheral utilization of free fatty acids was considered. Examination of carnitine palmityltransferase activity revealed a virtual absence of the enzyme in crude muscle extracts and mitochondrial fractions. It was suggested that this metabolic block, i.e., an inability to utilize fatty acids for energy production, may have resulted in a deficiency of muscle ATP content. The ATP deficiency may then have resulted in an inability of the muscle to maintain sarcolemma integrity causing the myoglobinuria. Three other reports of muscle palmityltransferase deficiency have been reported (Cumming *et al.*, 1976; Herman and Nedler, 1977; Di Donato *et al.*, 1978).

Other disorders in muscle lipid metabolism were considered to result from a muscle carnitine deficiency. Engel and Siekert (1972) described clinical, histochemical, and ultrastructural observations of a woman with lipid storage myopathy. Muscle homogenates from this patient oxidized fatty acids at a slow rate which returned to control levels upon addition of carnitine (Engel and Angelini, 1973). The mean carnitine level was less than 20% of control. Since carnitine is not formed by muscle, a defect in the required active transport, lack of a hypothetical sarcoplasmic receptor substance, or impairment of hepatic carnitine synthesis would result in the carnitine deficit. The patient reported by Boudin *et al.* (1976) with a carnitine deficiency was suggested to have a defect in carnitine biosynthesis. Other reports of carnitine deficiency with lipid storage myopathy have been reported (Issacs *et al.*, 1976; Bradley *et al.*, 1978; Cornelio *et al.*, 1977).

Karpati *et al.* (1975) described an 11-year-old boy with recurrent episodes of hepatic and cerebral dysfunction and underdeveloped musculature. Excessive lipid was found in type I fibers. Hypertrophic smooth endoplasmic reticulum and excessive microbodies were present in liver. A marked carnitine deficiency occurred in skeletal muscle, plasma, and liver. Although oral replacement therapy restored plasma carnitine levels to normal, it did not affect liver or muscle carnitine levels. The cause of the deficiency is unknown.

Vandyke *et al.* (1975) described an 8-year-old boy who had a progressive muscle weakness and a predominant vacuolization of type I muscle fibers. Although serum carnitine was normal, muscle carnitine was markedly reduced. Prednisone treatment resulted in clinical improvement but no change in muscle histology. It was suggested that carnitine entry into muscle was impaired in this autosomal recessive disorder. Carnitine treatment was successful in a case in which the muscle biopsy showed a majority of type I fibers to be vacuolated and to contain excess lipid (Angelini *et al.*, 1976).

Jerusalem *et al.* (1975) suggested that lipid storage myopathies may represent a syndrome rather than a unique disease entity. They reported a patient with a congenital nonprogressive proximal muscle weakness whose muscle biopsy revealed "myriads" of neutral lipid in type I and type II fibers. Serum and muscle carnitine levels and muscle carnitine palmityltransferase activity were normal. No major mitochondrial or sarcotubular abnormalities were found. Thus, biochemical defects other than carnitine or palmityltransferase deficiency can occur in lipid storage myopathy. Yamaguchi *et al.* (1978) reported a case of von Gierke's disease (glycogen storage disease type I) with lipid storage myopathy.

A model of experimental mitochondrial myopathy has been reported by Melmed *et al.* (1975). Two uncouplers of mitochondrial oxidative phosphorylation (2,4-dinitrophenol and carbonyl cyanide-*m*-chlorophenylhydrazone) were infused into a branch of the lower abdominal aorta of rats, causing a severe hypermetabolic state, systemic lactic acidosis, stiffness of the lower extremities, and numerous "ragged red" fibers in plantaris muscles. Interestingly, infusion of DNP plus chloramphenicol, an inhibitor of mitochondrial protein synthesis, prevented "ragged red" fiber formation but not the formation of mitochondrial inclusions. Thus uncoupling of mitochondria by an exogenous substance could lead to a myopathic state within the muscle cell. A motor neuropathy with muscle mitochondrial involvement has been produced in rats on a thiamin-deficient diet (Kark *et al.*, 1975). The only changes considered to be myopathic in man (Engel, 1965) were fibers resembling "ragged-red" fibers (Olson *et al.*, 1972).

### *E. Protein Synthesis*

Because proteins are important constituents of cellular membranes, alterations in protein synthesis may be responsible for some aspects of disease processes. This seems especially likely in the case of skeletal muscle disorders involving atrophy and degeneration in which loss of protein is observed. Consequently, many investigators have addressed themselves to this problem.

It is now apparent that the protein-synthesizing activity of muscle is subject to neural influences. For example, the denervation-induced changes in acetylcholine sensitivity of skeletal muscle can be blocked by inhibitors of protein and RNA synthesis (Fambrough, 1970; Grampp *et al.*, 1972), suggesting that the motor nerve regulates gene expression in muscle cells. These changes may also be related to muscular activity, since atrophy is accompanied by decreased protein synthesis and increased degradation, while in hypertrophy the opposite is true (Goldberg, 1969a,b). These and a variety of other studies have been reviewed (Guth, 1971; Smith and Kreutzberg, 1976).

Recent work has also indicated that creatine, an end-product of muscle metabolism, is involved in the control of muscle protein synthesis (Ingwall *et al.*, 1972), since muscle cells were observed to form myosin heavy chain faster in the presence of creatine than in its absence. Another feature of muscle protein synthesis which must be taken into consideration in terms of assembly and degradation of contractile units is the recently documented view that myofibrillar proteins turn over at different rates rather than as a unit (Funabiki and Cassens, 1972; Low and Goldberg, 1973).

Alterations in protein synthesis have been implicated in muscle diseases by many investigators. Murine dystrophy has been studied extensively in this regard. For example, Srivastava and Berlinguet (1966) and Srivastava (1968) reported that dystrophic mouse muscle displays an increased rate of protein synthesis, and that microsomal preparations from dystrophic mouse muscle showed increased incorporation of labeled amino acids into proteins. More recently, Srivastava (1969) reported that ribosomes from dystrophic mouse muscle responded more actively to addition of poly(U) than control preparation. Watts and Reid (1969), who reviewed the subject of protein synthesis in dystrophic mouse muscle, have performed experiments which suggest that there is no absolute difference between the protein-synthesizing systems of normal and dystrophic mice. Differences observed by other workers were attributed to differences in concentration, resulting from changes in muscle volume (Watts and Reid, 1969). Protein synthesis was studied in chicken dystrophy by Battelle and Florini (1973), who showed that loss of protein from dystrophic muscles cannot be explained by inability of ribosomes to synthesize proteins efficiently. Battelle and Florini (1973) suggested that there may be changes in the types of soluble proteins synthesized in chicken dystrophy. Penn *et al.* (1972) found no differences between myosin from normal human muscle and from patients with either Duchenne, facio-scapulo-humeral, limb-girdle, or myotonic dystrophies.

## IV. Muscle Regeneration

It has been known for a long time that skeletal muscle is capable of regeneration following injury in a number of diseases and after transplantation (Studitsky and Stryanova, 1951; Carlson, 1972, 1973; Carlson and Gutmann, 1975; Mauro, 1979). Most studies of skeletal muscle regeneration have been morphological. Muscle regeneration appears to mimic enbryogenesis of muscle in that a population of myogenic cells gives rise to myoblasts, which fuse to form myotubes and subsequently become mature skeletal muscle fibers. Most workers consider the satellite cell of Mauro (1961), found between the sarcolemma and the basement membrane, to be the myogenic cell. Bischoff (1975) presented evidence that satellite cells can become myogenic cells in muscle regeneration *in vitro*. In some diseases, e.g., Meyer-Betz disease and polymyositis, muscle regeneration can be complete. In muscular dystrophies, on the other hand, regeneration appears to proceed normally, but regenerating fibers fail to mature and eventually succumb to the dystrophic process. It has been suggested that defective regeneration is a fundamental property of dystrophic muscle (Mastaglia and Kakulas, 1970). Others, however, feel that regeneration is normal in dystrophic muscle, but the regenerated fiber then degenerates as a result of disease (Hartman and Standish, 1974). Whether dystrophic muscle is deficient in regenerative capacity must remain a moot issue at this time. A number of experimental models have been used for the study of muscle regeneration. Among these, the minced muscle preparation, developed initially by Studitsky and studied extensively by Carlson (1972), has been employed to advantage. Recently, a myotoxic local anesthetic, Marcaine (bupivacaine, Winthrop), has been shown to cause widespread skeletal muscle degeneration followed by complete regeneration (Benoit and Belt, 1970). Combination of Marcaine with hyaluronidase causes destruction of essentially the entire muscle, which then regenerates in a synchronous fashion (Hall-Craggs, 1974). Carlson used Marcaine to enhance the survival of orthotopic-free grafts of rat extensor digitorum longus muscle (Wagner *et al.*, 1977). A number of biochemical studies of muscle regeneration have been initiated in the authors' laboratory. Using the Marcaine plus hyaluronidase and the Marcaine plus free grafting models, studies have been made of oxidative metabolism (Rifenberick *et al.*, 1974), glycolytic enzymes (Wagner *et al.*, 1976, 1977), lactate dehydrogenase and creatine kinase isoenzymes (Max *et al.*, 1975), enzymes involved in cholinergic transmission (acetylcholinesterase, choline acetyltransferase) (Max and Rifenberick, 1975), and glucose-6-phosphate dehydrogenase (Wagner *et al.*, 1978; Max and Wagner, 1979).

Other aspects of this topic have been reviewed by Pennington (1971).

## V. Concluding Remarks

It is clear from the foregoing that despite a great number of studies, the molecular basis of most muscle disorders is not understood. Many changes have been observed in diseased muscle. At this time, however, it is not possible to state with any certainty which, if any of these changes, represents a primary lesion. The solution to this problem must await further experimentation.

## Acknowledgments

We are grateful to Drs. L. D. Peachy, H. E. Huxley, R. Couteaux, B. Katz, J. C. Eccles, A. G. Engel, R. Luft, and E. X. Albuquerque for permission to include their material; to Drs. E. Nelson, C. L. Koski, and B. H. Sohmer for helpful suggestions; and to Ms. B. Pasko, Ms. K. Conway, and Ms. M. Doss for preparation of the typescript. Work from the authors' laboratory was supported by grants from the National Institutes of Health, the Muscular Dystrophy Association, and by The National Amyotrophic Lateral Sclerosic Foundation.

## References*

Abdullah, F., and Pennington, R. J. (1968). *Clin. Chim. Acta* **20**, 365.

Adams, R. D. (1973). *In* "The Striated Muscle" (C. M. Pearson and F. K. Mostofi, eds.). pp. 292–300. Williams & Wilkins, Baltimore, Maryland.

Adams, R. D. (1975). "Diseases of Muscle: A Study in Pathology," 3rd ed. Harper, New York.

Adornato, B. T., Corash, L., and Engel, W. K. (1977). *Neurology*, **27**, 1093.

Afifi, A. K., Ibrahim, M. Z., Bergman, R. A., Haydar, N. A., Mire, J., Bahuth, N., and Kaylani, F. (1972). *J. Neurol. Sci.* **15**, 271.

*The major literature review for this chapter was concluded in June, 1976. The chapter was revised and updated prior to publication in June, 1979.

Albuquerque, E. X., Rash, J. E., Mayer, R. F., and Satterfield, J. R. (1976). *Exp. Neurol.* **51**, 536.
Aleu, F. P., and Afifi, A. D. (1964). *Am. J. Pathol.* **45**, 221.
Allbrook, D. (1962). *J. Anat. (London)* **96**, 137.
Almon, R. R., and Appel, S. M. (1975). *Biochim. Biophy. Acta* **393**, 66.
Almon, R. R., Andrew, C. G., and Appel, S. H. (1974). *Science* **186**, 55.
Angelini, C., Lucke, S., and Cantarutti, F. (1976). *Neurology* **26**, 633.
Archangeli, P., Soldato, P. D., Digiesi, V., and Melani, F. (1973). *Life Sci.* **12**, 13.
Anwyl, R., Appel, S. M., and Narohashi, T. (1977). *Nature* **267**, 262.
Armstrong, R. M., Nowak, R. M., and Falk, R. E. (1973). *Neurology* **23**, 1078.
Ashmore, C. R., and Doerr, L. A. (1970). *Biochem. Med.* **4**, 246.
Ashmore, C. R., and Doerr, L. (1976). *Exp. Neurol.* **50**, 312.
Ashmore, C. R., Tompkins, G., and Doerr, L. (1972). *Exp. Neurol.* **35**, 413.
Askenas, V., and Hee, D. (1974). *J. Neuropathol. Exp. Neurol.* **33**, 541.
Baker, N., Wilson, L., Oldendorf, W., and Blahd, W. H. (1960). *Am. J. Physiol.* **198**, 926.
Bank, W. J., DiMauro, S., Bonilla, E., Capuzzi, D. M., and Rowland, L. P. (1975). *New Engl. J. Med.* **292**, 443.
Barchi, R. L. (1975). *Arch. Neurol. (Chicago)* **32**, 175.
Barnard, E. A., Dolly, J. O., Porter, C. W., and Albuquerque, E. X. (1975). *Exp. Neurol.* **48**, 1.
Barnard, R. J., Edgerton, V. R., Furukawa, T., and Peter, J. B. (1971). *Am. J. Physiol.* **220**, 410.
Baskin, R. J. (1970). *Lab. Invest.* **23**, 581.
Battelle, B. A., and Florini, J. R. (1973). *Biochemistry* **12**, 635.
Beckmann, R., and Buddecke, E. (1958). *Klin. Wochschr.* **34**, 818.
Bender, A. N., and Engel, W. K. (1975). *J. Neuropathol. Exp. Neurol.* **34**, 108 (Abs.).
Benoit, P. W., and Belt, W. D. (1970). *J. Anat.* **107**, 547.
Bergman, R. A., Johns, R. J., and Afifi, A. K. (1971). *Ann. N.Y. Acad. Sci.* **183**, 88.
Bergstrom, K., Franksson, C., Matell, G., and Von Reis, G. (1973). *Eur. Neurol.* **9**, 157.
Bevan, S., Heinemann, S., Lennon, V. A., and Lindstrom, J. (1976). *Nature (London)* **260**, 438.
Bevan, S., Kullberg, R. W., and Heinemann, S. F. (1977). *Nature (London)* **267**, 263.
Bischoff, R. (1975). *Anat. Rec.* **182**, 215.
Birks, R., Huxley, H. E., and Katz, B. (1960). *J. Physiol. (London)* **150**, 134.
Blaxter, K. L. (1969). *In* "Disorders of Voluntary Muscle" (J. N. Walton, ed.), 2nd ed., pp. 733–762. Little, Brown, Boston, Massachusetts.
Bloch, K. J. (1974). *Medicine* **44**, 187.
Bohan, A., and Peter, J. B. (1975). *New Engl. J. Med.* **292**, 344, 403.
Bond, J. S., and Bird, J. W. C. (1966). *Fed. Proc. Fed. Am. Soc. Exp. Biol.* **25**, 242 (Abs.).
Bonilla, E., Schotland, D. L., DiMauro, S., and Lee, C.-P. (1977). *J. Ultrastruct. Res.* **5**, 81.
Booth, F. W., and Kelso, J. R. (1973). *J. Appl. Physiol.* **34**, 404.
Bosmann, H. B., Gerstein, D. M., Griggs, R. C., Howland, J. L., Hudecki, M. S., Katyore, S., and McLaughlin, J. (1976). *Arch. Neurol. (Chicago)* **33**, 135.
Boudin, G., Mikol, J., Guillard, A., and Engel, A. G. (1976). *J. Neurol. Sci.* **30**, 313.
Bourne, G. H., ed. (1972). "The Structure and Function of Muscle," 2nd ed., Vol. I, Academic Press, New York.
Bradley, W. G., and Jenkison, M. (1975). *J. Neurol. Sci.* **25**, 249.
Bradley, W. G., O'Brien, M. D., Walder, D. N., Murchison, D., Johnson, M., and Newell, D. J. (1975). *Arch. Neurol. (Chicago)* **32**, 466.
Brady, R. O. (1972). *Semin. Hematol.* **9**, 273.
Brady, R. O. (1978). *Annu. Rev. Biochem.* **47**, 687.

Brady, W. G., Tomlinson, B. E., and Hardy, M. (1978). *J. Neurol. Sci.* **35,** 201.
Brandstater, M. E., and Lambert, H. E. (1969). *Bull. Am. Assoc. EMG Electrodiagn.* **82,** 15.
Bray, G. M., and Aguayo, A. J. (1975). *J. Neuropathol. Exp. Neurol.* **34,** 517.
Bray, G. M., and Banker, B. Q. (1970). *Acta Neuropathol. (Berlin)* **15,** 34.
Bretscher, M. D. (1973). *Science* **181,** 622.
Briggs, A. H., and Kuhn, E. (1968). *Proc. Soc. Exp. Biol. Med.* **128,** 677.
Brown, H. D., Chattopadhyay, S. K., and Patel, A. B. (1968). *Metab. Clin. Exp.* **17,** 555.
Bunch, W., Kallsen, G., Berry, J., and Edwards, C. (1970). *J. Neurochem.* **17,** 613.
Bunyan, J., Green, J., Diplock, A. T., and Robinson, D. (1967). *Br. J. Nutr.* **21,** 127.
Burke, R. E., Levine, D. N., Zajac, F. C., Tsairis, P., and Engel, W. K. (1971). *Science* **174,** 709.
Burke, R. E., Tsairis, P., Levine, D. N., Zajac, F. E., and Engel, W. K. (1973). *In* "New Developments in Electromyography and Clinical Neurophysiology" (J. E. Desmedt, ed.), Vol. 1, pp. 23–30. Karger, Basel.
Butterfield, D. A. (1977). *Biochim. Biophys. Acta* **470,** 1.
Butterfield, D. A., and Watson, W. E. (1977). *J. Membrane Biol.* **32,** 165.
Butterfield, D. A., Roses, A. D., Cooper, M. L., Appel, S. H., and Chestnut, D. B. (1974). *Biochemistry* **13,** 5078.
Butterfield, D. A., Chestnut, D. B., Roses, A. D., and Appel, S. M. (1976). *Nature (London)* **263,** 159.
Canonico, P. G., and Bird, J. W. C. (1970). *J. Cell Biol.* **45,** 321.
Carafoli, E., Margreth, A., and Buffa, P. (1964). *Exp. Mol. Pathol.* **3,** 171.
Carlson, B. M. (1972). *Monogr. Dev. Biol.* **4.**
Carlson, B. M. (1973). *Am. J. Anat.* **137,** 119.
Carlson, B. M., and Gutmann, E. (1975). *Anat. Rec.* **183,** 47.
Castleman, B., and Norris, E. H. (1949). *Medicine* **28,** 27.
Chang, C. C., and Lee, C. Y. (1966). *Br. J. Pharmacol. Chemother.* **28,** 172.
Chalikian, D. M., and Bardi, R. L. (1979). *Neurology* **29,** 557.
Chou, S. M. (1969). *Neurology* **17,** 309.
Close, R. I. (1972). *Physiol. Rev.* **52,** 129.
Conrad, J. T., and Glasser, G. H. (1961). *Arch. Neurol. (Chicago)* **5,** 46.
Cornelio, F., DiDonato, S., Peluchetti, D., Bizzi, A., Bertagnolio, B., D'Angelo, A., and Wiesmann, U. (1977). *J. Neurol. Neurosurg. Psychiat.* **40,** 170.
Cotrufo, R., and Appel, S. H. (1971). *Neurology* **21,** 416.
Cotrufo, R., and Appel, S. H. (1973). *Exp. Neurol.* **39,** 58.
Couteaux, R. (1958). *Exp. Cell Res. Suppl.* **5,** 294.
Couteaux, R., and Pecot-Dechavassine, M. (1970). *C. R. Acad. Sci. Ser. D* **271,** 2346.
Creese, R., El-Shafie, A. L., and Vrbova, G. (1968). *J. Physiol. (London)* **197,** 279.
Cumming, W. J. K., Hardy, M., Hudgson, P., and Walls, J. (1976). *J. Neurol. Sci.* **30,** 247.
Currie, S., Saunders, M., Knowles, M., and Brown, A. E. (1971). *Q. J. Med.* **40,** 63.
Curtis, B. (1972). *In* "An Introduction to the Neurosciences" (B. A. Curtis, S. Jacobson, and E. M. Marcus, eds.), pp. 104–119. Saunders, Philadelphia, Pennsylvania.
D'Agostino, A. N., and Chiga, M. (1966). *Neurology (Minneap.)* **16,** 257.
Davidowitz, J., Pachte, B. R., Phillips, G., and Breinin, G. M. (1976). *Am. J. Pathol.* **82,** 101.
Desmedt, J. E. (1973). *In* "New Developments in Electron Myography and Clinical Neurophysiology" (J. E. Desmedt, ed.), pp. 241–304. Karger, Basel.
Dawkins, R. L. (1965). *J. Pathol. Bacteriol.* **90,** 619.
Dawkins, R. L., and Mastaglia, T. L. (1973). *New Engl. J. Med.* **288,** 434.
DeKretser, T. A., and Livett, B. G. (1977). *Biochem. J.* **168,** 229.
Desai, I. D. (1966). *Nature (London)* **209,** 1349.

Dhalla, N. D., and Sulakhe, P. V. (1973). *Biochem. Med.* **7,** 159.
Dhalla, N. D., Fedelsova, M., and Toffler, I. (1971). *Can. J. Biochem.* **49,** 1201.
Dhalla, N. S., Fedelesova, M., and Toffler, I. (1972). *Can. J. Biochem.* **50,** 550.
Dhalla, N. D., McNamara, D. B., Balasubramanian, U., Greenlaw, R., and Tucker, F. R. (1973). *Res. Commun. Chem. Pathol. Pharmacol.* **6,** 643.
Dhalla, N. S., Singh, A., Lee, S. L., Anand, M. B., Berratsky, A. M., and Jasmin, G. (1975). *Clin. Sci. Mol. Med.* **49,** 359.
Di Donato, S., Cornelio, F., Pacini, L., Peluchetti, D., Rimoldi, M., and Spreafico, S. (1978). *Ann. Neurol.* **4,** 465.
DiMauro, S., and Melis-DiMauro, P. M. (1973). *Science* **182,** 929.
DiMauro, S., Schotland, D. L., Bonilla, E., Lee, C. P., Gambetti, P., and Rowland, L. P. (1973). *Arch. Neurol. (Chicago)* **29,** 170.
DiMauro, S., Schotland, D. L., Bonilla, E., Lee, C. P., Melis-DiMauro, P. M., and Scarpa, A. (1974). *In* "Exploratory Concepts in Muscular Dystrophy" (A. T. Milhorat, ed.), Vol. II, pp. 506–515. Excerpta Medica, Amsterdam.
DiMauro, S., Bonilla, E., Lee, C. P., Schotland, D. L., Scarpa, A., Conn, Jr., H., and Chance, B. (1976). *J. Neurol. Sci.* **27,** 217.
Douglas, W. B. (1975). *Exp. Neurol.* **36,** 647.
Doyle, A. M., and Mayer, R. F. (1969). *Bull. Sch. Med. Univ. Maryland* **54,** 11.
Drachman, D. B. (1978a). *New Engl. J. Med.* **298,** 136.
Drachman, D. B. (1978b). *New Engl. J. Med.* **298,** 186.
Drachman, D. B., and Fambrough, D. M. (1976). *Arch. Neurol. (Chicago)* **33,** 485.
Drachman, D. B., Kao, I., Pestronk, A., and Toyka, K. V. (1976). *Ann. N.Y. Acad. Sci.* **274,** 226.
Drachman, D. B., Kao, L., Angus, C. W., and Murphy, A. (1977). *Ann. Neurol.* **1,** 504.
Duncan, C. J. (1978). *Experientia* **34,** 1531.
Ebashi, S. (1976). *Annu. Rev. Physiol.* **38,** 293.
Eberstein, A., and Goodgold, J. (1972). *Exp. Neurol.* **34,** 183.
Eccles, J. C. (1973). "The Understanding of the Brain." McGraw-Hill, New York.
Edström, L., and Kugelberg, E. (1968). *J. Neurol. Neurosurg. Psychiatr.* **31,** 424.
Elias, S. B., and Appel, S. H. (1976). *Life Sci.* **18,** 1031.
Elmqvist, D., Hofmann, W. W., Kugelberg, J., and Quastel, D. M. J. (1964). *J. Physiol. (London)* **174,** 417.
Engel, A. G. (1978). *In* "Neurology" (W. A. den Hartog Jager, G. W. Bruyn, and A. P. J. Heijstee, eds.), pp. 162–172. Excerpta Medica, Amsterdam.
Engel, A. G., and Angelini, C. (1973). *Science* **179,** 899.
Engel, A. G., and Siekert, R. G. (1972). *Arch. Neurol. (Chicago)* **27,** 174.
Engel, A. G., and Tsujihata, M. (1976). *Ann. N.Y. Acad. Sci.* **274,** 60.
Engel, A. G., Tsujihata, M., Lambert, E. H., Lindstrom, J. M., and Lennon, V. A. (1976). **35,** 569.
Engel, A. G., Lindstrom, J. M., Lambert, E. H., and Lennon, V. A. (1977). *Neurology* **27,** 307.
Engel, W. K. (1965). *Clin. Orthop.* **39,** 2.
Engel, W. K., Vick, N. A., Glueck, C. J., and Levy, R. I. (1970). *New Engl. J. Med.* **282,** 697.
Entriken, R. K., Swanson, K. L., Weidoff, P. M., Patterson, G. T., and Wilson, B. W. (1977). *Science* **195,** 873.
Ernster, L., Ikkos, D., and Luft, R. (1959). *Nature (London)* **184,** 1851.
Esire, M. M., and Maclennan, I. C. M. (1975). *Clin. Exp. Immunol.* **19,** 513.
Fambrough, D. M. (1970). *Science* **168,** 372.
Fambrough, D. M., Drachman, D. B., and Satyamurti, S. (1973). *Science* **182,** 2'

Fenichel, G. M. (1978). *Arch. Neurol.* **35**, 97.
Fenichel, G. M., Kibler, W. B., Olson, W. H., and Dettbarn, W. D. (1972). *Neurology* **22**, 1026.
Fiehn, W., Kuhn, E., and Geldmacher, I. (1973). *FEBS Lett.* **34**, 163.
Fischbach, G. D., and Robbins, N. (1969). *J. Physiol. (London)* **201**, 305.
Fisher, E. R., Cohn, R. E., and Danowski, F. S. (1966). *Lab. Invest.* **15**, 778.
Fisher, E. R., Wissinger, H. A., Gerneth, J. A., and Danowski, T. S. (1972). *Arch. Pathol.* **44**, 456.
French, J. H., Sherard, E. S., Lubell, H., Brotz, M., and Moore, C. L. (1972). *Arch. Neurol. (Chicago)* **26**, 229.
Fuchs, F. (1974). *Annu. Rev. Physiol.* **36**, 461.
Fukuhara, N., Jakamori, M., Gutmann, L., and Chou, S.-D. (1972). *Arch. Neurol. (Chicago)* **27**, 67.
Funabiki, R., and Cassens, R. G. (1972). *Nature (London), New Biol.* **236**, 249.
Garlepp, M., and Dawkins, R. (1977). *In* "Workshop on Neuromuscular Diseases. Proceedings of the Third International Congress of Immunology" Australian Academy of Science, Sydney.
Gergely, J. (1974). *In* "Disorders of Voluntary Muscle" (J. N. Walton, ed.), 3rd ed., pp. 102–167. Livingstone, Edinburgh.
Goldberg, A. L. (1969a). *J. Biol. Chem.* **244**, 3217.
Goldberg, A. L. (1969b). *J. Biol. Chem.* **244**, 3223.
Goust, T. M., Castaigne, A., and Moulias, R. (1974). *Clin. Exp. Immunol.* **18**, 39.
Grampp, W., Harris, J. B., and Thesleff, S. (1972). *J. Physiol. (London)* **221**, 743.
Grob, D. (1967). *Physiol. Pharmacol.* **3**, 389.
Grob, D. (1976). *Acad. N.Y. Acad. Sci.* **271**, 1.
Grob, D., and Johns, R. J. (1961). *In* "Myastenia Gravis" (H. R. Viets, ed.), pp. 127–149. Thomas, Springfield, Illinois.
Grob, D., and Namba, T. (1976). *Ann. N.Y. Acad. Sci.* **274**, 143.
Guathier, G. F. (1969). *Z. Zellforsch. Mikrosk. Anat.* **95**, 362.
Guathier, G. F. (1970). *In* "The Physiology and Biochemistry of Muscle as a Food" (E. J. Briskey, N. G. Cassens, and B. B. Marsh, eds.), 2nd ed., pp. 103–130. Univ. of Wisconsin Press, Madison.
Guth, L. (1971). *In* "Contractility of Muscle Cells and Related Processes" (R. J. Podolsky, ed.), pp. 189–201. Prentice-Hall, Englewood Cliffs, New Jersey.
Guth, L., and Yellin, H. (1971). *Exp. Neurol.* **31**, 277.
Gutmann, E. (1962). "The Denervated Muscle." Czech. Acad. Sci., Prague.
Haas, D. (1973). *Neurology* **23**, 55.
Hackenbrock, C. R. (1966). *J. Cell Biol.* **30**, 269.
Hajek, I., Gutmann, E., and Syrovy, I. (1964). *Physiol. Bohemoslov.* **13**, 32.
Hall-Craggs, E. C. B. (1974). *Exp. Neurol.* **43**, 349.
Hanna, S. D., and Baskin, R. J. (1977). *Biochem. Med.* **17**, 300.
Hanna, S. D., and Baskin, R. J. (1978). *Biochim. Biophys. Acta* **540**, 144.
Harris, J. B. (1971). *J. Neurol. Sci.* **12**, 45.
Harris, J. B., and Marshall, M. W. (1973). *Exp. Neurol.* **41**, 331.
Harris, H., and Wilson, P. (1971). *Nature (London)* **229**, 61.
Hartman, K. S., and Standish, S. M. (1974). *Arch. Pathol.* **98**, 126.
Haydar, N. A., Conn, H. L., Jr., Afifi, A., Wakid, N., Ballas, S., and Fawaz, K. (1971). *Ann. Intern. Med.* **74**, 548.
Hearn, G. R. (1959). *Am. J. Physiol.* **196**, 465.
Hers, H. G. (1963). *Biochem. J.* **80**, 11.

Heilbronn, E., and Stalberg, E. (1978). *J. Neurochem.* **31**, 5.
Herman, G., and Nedler, H. L. (1977). *J. Pediatrics* **91**, 247.
Heuser, J. E., Reese, T. S., and Landis, D. M. D. (1974). *J. Neurocytol.* **3**, 109.
Hironaka, T., and Miyata, Y. (1975). *Exp. Neurol.* **47**, 1.
Hofmann, W. W., and DeNardo, G. L. (1968). *Am. J. Physiol.* **214**, 330.
Hogan, E. L., Dawson, D. M., and Romanul, F. C. A. (1965). *Arch. Neurol. (Chicago)* **13**, 274.
Homsher, E., and Kean, C. J. (1978). *Annu. Rev. Physiol.* **40**, 93.
Howe, P. R. C., Livett, B. G., and Austin, L. (1976). *Exp. Neurol.* **51**, 132.
Howland, J. C. (1974). *Nature (London)* **251**, 724.
Hoyle, G. (1970). *Sci. Am.* **222**, 85.
Hudgson, P., and Pearce, G. W. (1969). *In* "Disorders of Voluntary Muscle" (J. N. Walton, ed.), 2nd ed., pp. 277–317. Little, Brown, Boston, Massachusetts.
Hudson, A. J., Strickland, K. P., and Robert, J. (1969). *In* "Progress in Neurogenetics" (A. Barbeau and J.-R. Brunette, eds.), pp. 66–77. Excerpta Medica, Amsterdam.
Hughes, B. P. (1972). *J. Neurol. Neurosurg. Psychiat.* **35**, 658.
Hughes, J. T., and Esiri, M. M. (1975). *J. Neurol. Sci.* **25**, 347.
Hull, Jr., K. L., and Roses, A. D. (1976). *J. Physiol. (London)* **254**, 169.
Humoller, F. L., Griswold, B., and McIntyre, A. B. (1951). *Am. J. Physiol.* **164**, 742.
Humoller, F. L., Hatch, D., and McIntyre, A. R. (1952). *Am. J. Physiol.* **170**, 371.
Huxley, A. F. (1971). *Proc. R. Soc. London Ser. B* **178**, 1.
Huxley, A. F., and Taylor, R. E. (1958). *J. Physiol. (London)* **144**, 426.
Huxley, H. E. (1972). *In* "The Structure and Function of Muscle" (G. H. Bourne, ed.), 2nd ed., Vol. I, pp. 302–387. Academic Press, New York.
Ingwall, J. S., Morales, M. F., and Stockdale, F. E. (1972). *Proc. Nat. Acad. Sci. USA* **69**, 2250.
Iodice, A. A., and Weinstock, I. M. (1965). *Nature (London)* **207**, 1102.
Iodice, A. A., Chin, J., Perker, S., and Weinstock, I. M. (1972). *Arch. Biochem. Biophys.* **152**, 166.
Ionescu, V., Radu, H., and Nicolesin, P. (1975). *Arch. Pathol.* **99**, 436.
Isaacs, H., Heffron, J. J. A., Badenhorst, M., and Pickering, A. (1976). *J. Neurol. Neurosurg. Psychiatry* **39**, 1114.
Ito, Y., Miledi, R., Molenaar, P. C., and Vincent, A. (1976). *Proc. R. Soc. Lond.* **192**, 475.
Jablecki, C., and Brimijoin, S. (1974). *Nature (London)* **250**, 151.
Jacob, W., and Neuhaus, J. (1954). *Klin. Wochschr.* **32**, 923.
Jacobson, B. E., Blanchaer, M. C., and Wrogemann, K. (1970). *Can. J. Biochem.* **48**, 1037.
Jerusalem, F., Angelini, C., Engel, A. G., and Groover, R. V. (1973). *Arch. Neurol. (Chicago)* **29**, 162.
Jerusalem, F., Engel, A. G., and Gomez, M. R. (1974a). *Braini* **97**, 115.
Jerusalem, F., Engel, A. G., and Gomez, M. R. (1974b). *Brain* **97**, 123.
Jerusalem, F., Rakusa, M., Engel, A. G., and MacDonald, R. D. (1974c). *J. Neurol. Sci.* **23**, 391.
Jerusalem, F., Spiess, H., and Baumgartner, G. (1975). *J. Neurol. Sci.* **24**, 273.
Johnson, A. G., and Woolf, A. L. (1969). *Acta Neuropathol.* **12**, 183.
Johnson, M. A., and Montgomery, A. (1975). *J. Neurol. Sci.* **26**, 425.
Julien, J., Vital, C. L., Vallat, J. M., Vallat, M., Le Blanc, M. (1974). *J. Neurol. Sci.* **21**, 165.
Kadenbach, B. (1969). *Biochim. Biophys. Acta* **186**, 399.
Kakulas, B. A. (1966). *Nature (London)* **210**, 1115.
Kao, I., and Drachman, D. B. (1977). *Science* **195**, 74.
Kar, N. C., and Pearson, C. M. (1976). *Clin. Chim. Acta* **73**, 293.
Kar, N. C., and Pearson, C. M. (1972a). *Clin. Chim Acta* **40**, 341.
Kar, N. C., and Pearson, C. M. (1972b). *Proc. Soc. Exp. Biol. Med.* **141**, 4.

Kark, R. A. P., Blass, J. P., Avigan, J., and Engel, W. K. (1971). *J. Biol. Chem.* **246,** 4560.

Kark, P., Brown, W. J., Edgerton, V. R., Reynolds, S. F., Gibson, G. (1975). *Arch. Neurol. (Chicago)* **32,** 818.

Karpati, G., Carpenter, S., Watters, G. V., Eisen, A. A., and Andermann, F. (1973). *Neurology* **23,** 1066.

Karpati, G., Carpenter, S., Watters, G. V., Eisen, A. A., and Andermann, F. (1973). *Neurology* **23,** 1066.

Karpati, G., Carpenter, S., Engel, A. G., Watters, G., Allen, J., Rothman, S., Klassen, G., and Mamer, O. A. (1975). *Neurology* **25,** 16.

Kissel, P., Schmitt, J., Duc, M., and Duc, L. (1969). *In* "Progress in Neurogenetics" (A. Barbeau and J.-R. Brunette, eds.), pp. 194–198. Excerpta Medica, Amsterdam.

Kitchen, S. E., and Watts, D. C. (1974). *Biochem. Biophys. Acta* **364,** 272.

Klinkerfuss, G. H., and Haugh, M. J. (1970). *Arch. Neurol. (Chicago)* **22,** 309.

Kohn, R. R. (1966). *Am. J. Pathol.* **48,** 241.

Kohn, R. R. (1969). *Lab. Invest.* **20,** 202.

Komiya, Y., and Austin, L. (1974). *Exp. Neurol.* **43,** 1.

Koski, C. L., and Max, S. R. (1974). *Exp. Neurol.* **43,** 547.

Kouvelas, E. D., and Manchester, K. L. (1968). *Biochem. Biophys. Acta* **164,** 132.

Kugelberg, E., and Edström, L. (1968). *J. Neurol. Neurosurg. Psychiaty.* **31,** 415.

Kuhn, E., Doron, W., Kahlke, W., and Pfisterer, H. (1968). *Klin. Wochenschr.* **46,** 1043.

Kwok, C. T., Kuffer, A. D., Tang, B. Y., and Austin, L. (1976). *Exp. Neurol.* **50,** 362.

Law, P. K., and Caccia, M. R. (1975). *J. Neurol. Sci.* **24,** 251.

Law, P. K., Cosmos, E., Butter, J., and McComas, A. J. (1976). *Exp. Neurol.* **51,** 1.

Lehninger, A. L. (1964). "The Mitochondrion." Benjamin, New York.

Lennon, V. A. (1978). *Human Pathol.* **9,** 541.

Lennon, V. A., Lindstrom, J. M., and Seybold, M. E. (1976). *Ann. N.Y. Acad. Sci.* **274,** 283.

Lennon, V. A., Seybold, M. E., Lindstrom, J. M., Cochrane, C., and Ulevitch, R. (1978). *J. Exp. Med.* **147,** 973.

Libby, P., and Goldberg, A. L. (1978). *Science* **199,** 534.

Lindstrom, J. J., Lennon, V. A., Seybold, M. E., and Whittinghan, S. (1976a). *Ann. N.Y. Acad. Sci.* **274,** 254.

Lindstrom, J., Seybold, M. E., Lennon, V. A., Whittingham, S., and Duane, D. D. (1976b). *Neurology* **26,** 1054.

Lipicky, R. J., and Bryant, S. H. (1966). *J. Gen. Physiol.* **50,** 89.

Lipicky, R. J., and Hess, J. (1974). *Am. J. Pathol.* **226,** 592.

Lochner, A., and Brink, A. J. (1967). *Clin. Sci.* **33,** 409.

Low, R. B., and Goldberg, A. L. (1973). *J. Cell Biol.* **56,** 590.

Luft, R., Ikkos, D., Palmieri, G., Ernster, L., and Afzelios, B. (1962). *J. Clin. Invest.* **41,** 1776.

Lumb, E. M., and Emery, A. E. H. (1975). *Br. Med. J.* **3,** 467.

McArdle, B. (1969). *In* "Disorders of Voluntary Muscle" (J. N. Walton, ed.), 2nd ed., pp. 607–638. Little, Brown, Boston, Massachusetts.

McComas, A. J., and Mossawy, S. (1965). *In* "Proceedings, Third Symposium of The Muscular Dystrophy Group," p. 317. Pitman, London.

McComas, A. J., and Mrozek, K. (1968). *J. Neurol. Neurol. Neurosurg. Psychiat.* **31,** 441.

McComas, A. J., Sica, R. E. P., and Campbell, M. J. (1971). *Lancet* **1,** 321.

McLeod, J. G., Baker, W. De C., Shorty, C. D., and Kerr, C. B. (1975). *J. Neurol. Sci.* **24,** 39.

Manghani, D., Partridge, T. A., and Sloper, J. C. (1974). *J. Neurol. Sci.* **23,** 489.

Mannherz, H. G., and Goody, R. S. (1976). *Annu. Rev. Biochem.* **45,** 427.

Margreth, A., Salviati, G., DiMauro, D., and Turati, G. (1972). *Biochem. J.* **126,** 1099.

Martonosi, A. (1968). *Proc. Soc. Exp. Biol. Med.* **127,** 824.

Mastaglia, F. L., and Kakulas, B. A. (1970). *J. Neurol. Sci.* **10**, 471.
Mastaglia, F. L., Papadimitriou, J. M., and Kakulas, B. A. (1970). *J. Neurol. Sci.* **11**, 425.
Matheson, D. W., and Howland, J. L. (1974). *Science* **184**, 165.
Matheson, D. W., Engel, W. K., and Derrer, E. C. (1976). *Neurology* **26**, 1182.
Mauro, A. (1961). *Biochem. Cytol.* **9**, 493.
Mauro, A., ed. (1979). "Muscle Regeneration." Raven, New York.
Max, S. R. (1972). *Biochem. Biophys. Res. Commun.* **46**, 1394.
Max, S. R. (1973). *Biochem. Biophys. Res. Commun.* **52**, 1278.
Max, S. R., and Albuquerque, E. X. (1975). *Exp. Neurol.* **49**, 852.
Max, S. R., and Brady, R. O. (1971). *Nature (London), New Biol.* **233**, 55.
Max, S. R., and Mayer, R. F. (1972). *Trans. Am. Neurol. Ass.* **97**, 306.
Max, S. R., and Rifenberick, D. H. (1975). *J. Neurochem.* **24**, 771.
Max, S. R., and Wagner, K. R. (1979). *In* "Muscle Regeneration" (A. Mauro, ed.), pp. 475–483. Raven, New York.
Max, S. R., Nelson, P. G., and Brady, R. O. (1970). *J. Neurochem.* **17**, 1517.
Max, S. R., Mayer, R. F., and Vogelsang, L. (1971). *Arch. Biochem. Biophys.* **146**, 227.
Max, S. R., Garbus, J., and Wehman, H. J. (1972). *Anal. Biochem.* **46**, 576.
Max, S. R., Rifenberick, D. H., and Koski, C. L. (1975). *Neurology* **25**, 391. (Abs.).
Melmed, C., Karpati, G., and Carpenter, S. (1975). *J. Neurol. Sci.* **26**, 305.
Mendell, J. R., Engel, W. K., and Derrer, E. C. (1971). *Science* **172**, 1143.
Mendell, J. R., Engel, W. K., and Derrer, E. L. (1972). *Nature (London)* **239**, 522.
Metzger, A. L. (1974). *Ann. Intern. Med.* **81**, 182.
Meyeer, W. L., and Little, B. W. (1970). *Science* **170**, 747.
Miale, T. D., Frias, J. L., and Lawson, D. L. (1975). *Science* **187**, 453.
Michelson, A. M., Russell, E. S., and Harman, P. J. (1955). *Proc. Nat. Acad. Sci. USA* **41**, 1079.
Miledi, R., and Slater, C. R. (1968). *J. Cell Sci.* **3**, 49.
Miledi, R., and Slater, C. R. (1969). *Proc. Roy. Soc. London Ser. B* **174**, 253.
Milhaud, M., Fardeau, M., and La Presle, J. (1964). *C.R. Soc. Biol.* **158**, 2274.
Miller, J. F. A. P., and Osoba, D. (1967). *Physiol. Rev.* **47**, 437.
Miller, S. E., Roses, A. D., and Appel, S. H. (1976). *Arch. Neurol. (Chicago)* **33**, 172.
Mittag, T., Kornfeld, P., Tormay, A., and Woo, C. (1976). *New Engl. J. Med.* **294**, 691.
Mokhri, B., and Engel, A. G. (1975). *Neurology* **25**, 1111.
Montgomery, A. (1975). *J. Neurol. Sci.* **26**, 401.
Moore, D. H., Ruska, H., and Copenhaver, W. M. (1956). *J. Biochem. Biophys. Cytol.* **2**, 755.
Morel, J. E., and Pinset-Harström, (1975). *Biomedicine* **22**, 88, 186.
Morgan, K. G., Entrickin, R. K., and Bryant, S. H. (1975). *Am. J. Physiol.* **229**, 1155.
Morse, P. F., and Howland, J. L. (1973). *Nature (London)* **245**, 156.
Mrak, R. E., and Baskin, R. J. (1978). *Biochem. Med.* **19**, 47.
Murray, M. R. (1972). *In* "Structure and Function of Muscle" (G. H. Bourne, ed.), 2nd ed., Vol. I, pp. 237–299. Academic Press, New York.
Muscatello, U., Margreth, A., and Aloisi, M. (1965). *J. Cell Biol.* **27**, 1.
Nachmias, V. T., and Padykula, H. (1958). *J. Biophys. Biochem. Cytol.* **4**, 47.
Namba, T., Brown, S. B., and Grob, D. (1970). *Pediatrics* **45**, 488.
Nastuk, W. L., Plescia, O. J., and Osserman, K. E. (1960). *Proc. Soc. Exp. Biol. Med.* **105**, 177.
Needham, D. M. (1971). "Machina carnis." Cambridge Univ. Press, London and New York.
Neerunjun, J. S., and Dubowitz, V. (1974a). *J. Neurol. Sci.* **23**, 505.
Neerunjun, J. S., and Dubowitz, V. (1974b). *J. Neurol. Sci.* **23**, 521.
Okamura, K., Santa, T., Nagae, K., and Omae, T. (1976). *J. Neurol. Sci.* **27**, 79.
Olson, E., Vignos, P. J., Jr., Woodlock, J., and Perry, T. (1968). *J. Lab. Clin. Med.* **71**, 220.

Olson, W., Engel, W. K., Walsh, G. O., and Einaugler, R. (1972). *Arch. Neurol. (Chicago)* **26**, 193.
Osserman, K. E., and Genkins, G. (1971). *J. Mt. Sinai Hosp. N.Y.* **38**, 497.
Pachter, B. R., Davidowitz, J., Eberstein, A., and Breinin, G. M. (1974). *Exp. Neurol.* **45**, 462.
Pachter, B. R., Davidowitz, J., and Breinin, G. M. (1976). *Am. J. Pathol.* **82**, 111.
Padykula, H., and Gauthier, G. F. (1970). *J. Cell Biol.* **46**, 27.
Panayiotopoulas, C. P., and Scarpalezos, S. (1976). *J. Neurol. Sci.* **27**, 1.
Patel, N. N., Razzak, Z. A., and Dastur, D. K. (1969). *Arch. Neurol. (Chicago)* **20**, 413.
Pater, J. L., and Kohn, R. R. (1967). *Proc. Soc. Exp. Biol. Med.* **125**, 476.
Patrick, J., and Lindstrom, J. M. (1973). *Science* **180**, 871.
Paul, C. V., and Powell, J. A. (1974). *J. Neurol. Sci.* **21**, 365.
Pearlstein, R. A., and Kohn, R. R. (1966). *Am. J. Pathol.* **48**, 823.
Pearce, G. W. (1966). *Ann. N.Y. Acad. Sci.* **138**, 138.
Pearce, G. W., and Walton, J. N. (1963). *J. Pathol. Bacteriol.* **86**, 25.
Pearson, C. M. (1966). *Annu. Rev. Med.* **17**, 63.
Pearson, C. M. (1972a). *In* "Polymyositis and Dermatomyositis, Arthritis and Allied Conditions" (J. L. Hollander, D. J. McCarty, eds.), 8th ed., pp. 940–961. Lea & Febiger, Philadelphia, Pennsylvania.
Pearson, C. M. (1972b). *In* "The Metabolic Basis of Inherited Disease" (J. B. Stanbury, J. B. Wyngaarden, and D. S. Fredrickson, eds.), 3rd ed., pp. 1181–1203. McGraw-Hill, New York.
Pellegrino, C., and Bibiani, C. (1963). *Boll. Soc. Ital. Biol. Sper.* **39**, 1989.
Pellegrino, C., and Franzini, C. (1963). *J. Cell Biol.* **17**, 327.
Pellegrino, C., and Franzini-Armstrong, C. (1969). *Int. Rev. Exp. Pathol.*, **7**, 139.
Pellegrino, C., Villani, G., and Franzini, C. (1957). *Arch. Sci. Biol.* **41**, 339.
Penn, A. S., Cloak, R. A., and Rowland, L. P. (1972). *Arch. Neurol. (Chicago)* **27**, 159.
Pennington, R. J. (1971). *Adv. Clin. Chem.* **14**, 409.
Pennington, R. J., and Robinson, J. E. (1968). *Enzymol. Biol.* **9**, 175.
Percy, A. K., and Miller, M. E. (1975). *Nature (London)* **258**, 147.
Peter, J. B., and Fiehn, W. (1973). *Science* **179**, 910.
Peter, J. B., and Worsfold, M. (1969). *Biochem. Med.* **2**, 457.
Peter, J. B., Stempel, K., and Armstrong, J. (1970). *In* "Muscle Diseases, Proceedings of the International Congress" (J. N. Walton, ed.), pp. 228–235. Excerpta Medica, Amsterdam.
Peter, J. B., Barnard, R. J., Edgerton, V. R., Gillespie, K. E., and Stempel, K. E. (1972a). *Biochemistry* **11**, 2627.
Peter, J. B., Kar, N. C., Barnard, R. J., Pearson, C. M., and Edgerton, V. R. (1972b). *Biochem. Med.* **6**, 257.
Peter, J. B., Andiman, R. M., Bowman, R. C., and Nagatomo, T. (1973). *Exp. Neurol.* **41**, 738.
Peter, J. B., Stempel, K. E., Dromgoole, S. H., Campion, D. S., Bowman, R. C., Nagatomo, T., and Andiman, R. M. (1975a). *Exp. Neurol.* **49**, 429.
Peter, J. B., Dromgoole, S. H., Campion, D. S., Stempel, K. E., Bowman, R. L., Angiman, R. M., and Nagatomo, T. (1975b). *Exp. Neurol.* **49**, 115.
Peterson, D. W., Lilyblade, A. L., and Bond, D. C. (1972). *Proc. Soc. Exp. Biol. Med.* **141**, 1056.
Pinckney, J. H., Tassoni, J. P., Curtis, R. L., and Hollinshead, M. B. (1963). *In* "Muscular Dystrophy in Men and Animals" (G. H. Bourne and M. N. Golarz, eds.), pp. 407–449. Hafner, New York.
Pirskanen, R. (1976). *J. Neurol. Neurosurg. Psychiat.* **39**, 23.
Platzen, A. C., and Chase, W. H. (1964). *Am. J. Pathol.* **44**, 931.
Platzen, A. C., and Powell, J. A. (1975). *J. Neurol. Sci.* **24**, 109.

Pollack, M. D., and Bird, J. W. C. (1968). *Am. J. Physiol.* **215**, 716.
Pollock, M., and Dyck, P. J. (1976). *Arch. Neurol. (Chicago)* **33**, 33.
Porter, K. R., and Franzini-Armstrong, C. (1965). *Sci. Am.* **212**, 72.
Price, H. M., Pease, D. C., and Pearson, C. M. (1962). *Lab. Invest.* **11**, 549.
Price, H. M., Howes, E. L., Jr., and Blumberg, J. M. (1964). *Lab. Invest.* **13**, 1264.
Rash, J. E., Albuquerque, E. X., and Hudson, E. S. (1976). *Proc. Nat. Acad. Sci., USA* **73**, 4584.
Rifenberick, D. H., and Max, S. R. (1974a). *Am. J. Physiol.* **226**, 295.
Rifenberick, D. H., and Max, S. R. (1974b). *Am. J. Physiol.* **227**, 1025.
Rifenberick, D. H., Gamble, J. G., and Max, S. R. (1973). *Am. J. Physiol.* **225**, 1295.
Rifenberick, D. H., Koski, C. L., and Max, S. R. (1974). *Exp. Neurol.* **45**, 527.
Romanul, F. C. A. (1964). *Arch. Neurol. (Chicago)* **11**, 355.
Romanul, F. C. A., and Hogan, E. L. (1965). *Arch. Neurol. (Chicago)* **13**, 263.
Ronzoni, E., Wald, S., Berg, L., and Ramsey, R. (1958). *Neurology* **8**, 359.
Roses, A. D., and Appel, S. H. (1974). *Lancet* **2**, 1400.
Roses, A. D., Herbstreith, M. H., and Appel, S. H. (1975a). *Nature (London)* **254**, 350.
Roses, A. D., Butterfield, D. A., Appel, S. H., and Chestnut, D. B. (1975b). *Arch. Neurol. (Chicago)* **32**, 535.
Roses, A. D., Roses, M. J., Miller, S. E., Hull, Jr., K. L., and Appel, S. H. (1976). *New Engl. J. Med.* **294**, 193.
Rowland, L. P. (1976). *Arch. Neurol. (Chicago)* **33**, 315.
Rudel, R., and Senger, J. (1972). *Naunyn-Schmiedebergs Arch. Pharmakol.* **274**, 337.
Sakakibara, H., Engel, A. G., and Lambert, E. H. (1977). *Neurology* **27**, 741.
Salafsky, B. (1971). *Nature (London)* **229**, 270.
Samaha, F. J., and Congedo, C. Z. (1977). *Ann. Neurol.* **1**, 125.
Samaha, F. J., and Gergely, J. (1969). *New Engl. J. Med.* **280**, 184.
Santa, T. (1969). *Arch. Neurol. (Chicago)* **20**, 479.
Santa, T., and Engel, A. G. (1973). *In* "New Developments in Electromyography and Clinical Neurophysiology," (J. E. Desmedt, ed.) Vol. 1, pp. 41–54. Kager, Basel.
Santa, T., Engel, A. G., and Lambert, E. H. (1972a). *Neurology* **22**, 71.
Santa, T., Engel, A. G., and Lambert, E. H. (1972b). *Neurology* **22**, 370.
Sarnat, H. B., and Silbert, S. W. (1976). *Arch. Neurol. (Chicago)* **33**, 466.
Sarnat, H. B., O'Connor, T., and Byrne, P. A. (1976). *Arch. Neurol. (Chicago)* **33**, 459.
Satyamurti, S., Drachman, D. B., and Slone, F. (1975). *Science* **187**, 955.
Scales, D., Sabbadini, R., and Inesi, G. (1977). *Biochim. Biophys. Acta* **465**, 535.
Schapira, S. G., Dreyfuss, J. C., and Schapira, F. (1953). *Sem. Hosp.* **29**, 1917.
Shapira, F., Schapira, G., and Dreyfuss, J. C. (1957). *C.R. Soc. Biol.* **245**, 753.
Schiaffino, S., and Hanzlikova, V. (1972). *J. Ultrastruct. Res.* **39**, 1.
Schmidt, C. G. (1952). *Biochem. Z.* **323**, 266.
Schnaitman, C., and Greenwalt, J. W. (1968). *J. Cell Biol.* **38**, 158.
Scholze, W., Krause, E. G., and Wallenberger, A. (1972). *Adv. Cyclic Nucleotide Res.* **1**, 249.
Schotland, D. L. (1968). *J. Neuropathol. Exp. Neurol.* **27**, 109.
Schotland, D. L., Spiro, D., and Carmel, P. (1966). *J. Neuropathol. Exp. Neurol.* **25**, 431.
Schotland, D. L., Dimauro, S., Bonilla, E., Scarpa, A., Lee, and C.-Pu. (1976). *Arch. Neurol. (Chicago)* **33**, 475.
Schotland, D. L., Bonilla, E., and Van Meter, M. (1977). *Science* **196**, 1005.
Schroder, J. M., and Adams, R. D. (1968). *Acta Neuropathol.* **10**, 218.
Schrodt, G. R., and Walker, S. M. (1966). *Am. J. Pathol.* **49**, 33.
Seiler, D. (1971). *Experientia* **27**, 1170.

Seiler, D., and Hasselbach, W. (1971). *Eur. J. Biochem.* **21**, 385.
Seiler, D., and Kuhn, E. (1969). *Eur. J. Biochem.* **11**, 175.
Seiler, D., Kuhn, E., Fiehn, W., and Hasselbach, W. (1970). *Eur. J. Biochem.* **12**, 375.
Sha'afi, R. I., Rodan, S. B., Hintz, R. L., Fernandez, S. M., and Rodan, G. A. (1975). *Nature (London)* **254**, 525.
Shafiq, S. A., and Gorycki, M. A. (1965). *J. Pathol. Bacteriol.* **90**, 123.
Shafiq, S. A., Milhorat, A. T., and Gorycki, M. A. (1967). *J. Pathol. Bacteriol.* **94**, 139.
Shafiq, S. A., Milhorat, A. T., and Gorycki, M. M. (1968). *Neurology* **18**, 785.
Shafiq, S. A., and Gorycki, M. A., Asiedu, S. A., and Milhorat, A. T. (1969). *Arch. Neurol. (Chicago)* **20**, 625.
Shaw, R. F., Pearson, C. M., Chowdhury, S. R., and Dreyfus, F. E. (1967). *Arch. Neurol. (Chicago)* **16**, 115.
Sibley, J. A., and Lehninger, A. C. (1949). *J. Nat. Cancer Cancer Inst.* **9**, 303.
Siddiqui, P. Q. R., and Pennington, R. J. T. (1977). *J. Neurol. Sci.* **34**, 365.
Simpson, J. A. (1958). *Brain* **81**, 112.
Simpson, J. A. (1960). *Scott. Med. J.* **5**, 419.
Simpson, J. A. (1971). *Ann. N.Y. Acad. Sci.* **183**, 241.
Simpson, J. A., Behan, P. O., and Dick, H. M. (1976). *Ann. N.Y. Acad. Sci.* **274**, 382.
Singer, S. J., and Rothfield, L. I. (1973). *Neurosci. Res. Program. Bull.* **11**, 1.
Sluga, E., and Moser, K. (1970). *In* "Muscle Diseases" (J. N. Walton, N. Canal, and G. Scarlato, eds.), pp. 116–119. Excerpta Medica, Amsterdam.
Smith, B. H., and Kreutzberg, G. W. (1976). *Neurosci. Res. Program Bull.* **14**, 211.
Smithers, D. W. (1959). *J. Fac. Radiol. London* **10**, 3.
Solandt, D. Y., Partrige, R., and Hunter, J. (1943). *J. Neurophysiol.* **6**, 17.
Somer, H., Chien, S., Sung, L. A., and Thurn, A. (1979). *Neurology* **29**, 519.
Souweine, G., Bernard, J. C., Lasne, Y., and Lachanat, J. (1978). *J. Neurol.* **217**, 287.
Spiro, A. J., Moore, C. L., and Prineas, J. W. (1970a). *Arch. Neurol. (Chicago)* **23**, 103.
Spiro, A. J., Prineas, J. W., and Moore, C. L. (1970b). *Arch. Neurol. (Chicago)* **22**, 259.
Squire, J. M. (1975). *Ann. Rev. Biophys. Bioeng.* **4**, 137.
Srivastava, V. (1968). *Can. J. Biochem.* **46**, 35.
Srivastava, V. (1969). *Arch. Biochem. Biophys.* **135**, 236.
Srivastava, V., and Berlinguet, L. (1966). *Arch. Biochem. Biophys.* **114**, 320.
Stagni, N., and DeBernard, B. (1968). *Biochim. Biophys. Acta* **170**, 129.
Stenger, R. J., Spiro, D., Scolly, R. E., and Shannon, J. M. (1962). *Am. J. Pathol.* **40**, 1.
Stirling, S. (1975). *J. Anat.* **119**, 169.
Stracher, A., McGowan, E. B., and Shafiq, S. A. (1978). *Science* **200**, 50.
Strauss, A. J. (1968). *Adv. Intern. Med.* **14**, 241.
Strauss, A. J., Segal, B. C., Hsu, K. C., Burkholder, P. M., Nastur, W. L., and Osserman, K. E. (1960). *Proc. Soc. Exp. Biol. Med.* **105**, 184.
Studitsky, A. N., and Stryanova, A. R. (1951). *Dokl. Akad. Navk SSSR* **145**, 198. Cited from Carlson, B. M. (1973). *Am. J. Anat.* **137**, 119.
Sulakhe, P. V., Fedelesova, M., McNamara, D. B., and Dhalla, N. S. (1971). *Biochem. Biophys. Res. Commun.* **42**, 793.
Swash, M., and Fox, K. P. (1975). *J. Neurol. Neurosurg. Psychiatry* **38**, 91.
Syrovy, I., Hajek, I., and Gutmann, E. (1966). *Physiol. Bohemoslov.* **15**, 7.
Takagi, A., Schotland, D. L., and Rowland, L. P. (1973). *Arch. Neurol. (Chicago)* **28**, 380.
Tallman, J. F., Brady, R. O., and Suzuki, K. (1971). *J. Neurochem.* **18**, 1775.
Tamura, K., Santa, T., and Kuroiwa, Y. (1974). *Brain* **97**, 665.
Tang, B. Y., Komiya, Y., and Austin, L. (1974). *Exp. Neurol.* **43**, 13.

Tappel, A. L., Zalkin, H., Caldwell, K. A., Desai, I. D., and Shibko, S. (1962). *Arch. Biochem. Biophys.* **96**, 340.
Tappel, A. L., Sawant, P. C., and Shibko, S. (1963). *In* "CIBA Foundation Symposium on Lysosomes" pp. 78–108. Little, Brown, Boston, Massachusetts.
Tabary, J. C., Tabary, C., Tardieu, C., Tardieu, G., and Goldspink, G. (1972). *J. Physiol. (London)* **224**, 231.
Tarrab-Hazdai, R., Aharonov, A., Silman, I., and Fuchs, S. (1975). *Nature (London)* **256**, 128.
Thorpe, W. R., and Seeman, P. (1971). *Exp. Neurol.* **30**, 277.
Toop, J., and Emery, A. E. H. (1974). *Clin. Genet.* **5**, 230.
Tower, S. S. (1937). *J. Comp. Neurol.* **67**, 241.
Toyka, K. V., Drachman, D. B., Pestronk, A., and Kao, I. (1975). *Science* **190**, 397.
Toyka, K. V., Drachman, D. B., Griffin, D. E., Pestronk, A., Winkelstein, J. A., Fischback, K. H., and Kao, I. (1977). *New Engl. J. Med.* **296**, 125.
Trentham, D. R. (1977). *Biochem. Soc. Trans.* **5**, 5.
Van Breemen, V. L. (1960). *Am. J. Pathol.* **37**, 333.
Vandyke, D. H., Griggs, R. C., Markesbery, W., and DiMauro, S. (1975). *Neurology* **25**, 154.
Van Fleet, J. F., Hall, B. V., and Simon, J. (1968). *Am. J. Pathol.* **52**, 1067.
Van Hoof, F., and Hers, H. G. (1969). *Eur. J. Biochem.* **7**, 34.
Vrbova, G. (1963). *J. Physiol. (London)* **169**, 513.
Wagner, K. R., Max, S. R., Grollman, E. M., and Koski, C. L. (1976). *Exp. Neurol.* **52**, 40.
Wagner, K. R., Carlson, B. M., and Max, S. R. (1977). *J. Neurol. Sci.* **34**, 373.
Wakayama, Y., Hodson, A., Bonilla, E., Pleasure, D., and Schotland, D. L. (1979). *Neurology* **29**, 670.
Walton, J. N. (1974). "Disorders of Voluntary Muscle," 3rd ed. Livingstone, Edinburgh.
Wattiaux, R. (1969). *In* "Handbook of Molecular Cytology" (A. Lima-de Faria, ed.), pp. 1159–1178. North-Holland Publ., Amsterdam.
Watts, D. C., and Reid, J. D. (1969). *Biochem. J.* **115**, 377.
Weinberg, C. B., and Hall, Z. W. (1979). *Proc. Nat. Acad. Sci. USA* **76**, 504.
Weinberg, H. J., Spencer, P. S., Raine, C. S. (1975). *Brain Res.* **88**, 532.
Weinstock, I. M., and Iodice, A. A. (1969). *In* "Lysosomes in Biology and Pathology." (J. T. Dingle and H. B. Fell, eds.), pp. 450–468. North Holland Publ., Amsterdam.
Weinstock, I. M., and Lukacs, M. (1965). *Enzymol. Biol. Clin.* **5**, 89.
West, W. T., Meier, H., and Hoag, W. G. (1966). *Ann. N.Y. Acad. Sci.* **138**, 4.
Whanger, P., Weswig, P. H., Muth, O. H., and Oldfield, J. E. (1969). *J. Nutr.* **99**, 331.
Whanger, P., Weswig, P. H., Muth, O. H., and Oldfield, J. E. (1970). *J. Nutr.* **100**, 773.
Wagner, K. R., Kauffman, F. C., and Max, S. R. (1978). *Biochem. J.* **170**, 17.
Whittaker, J. N., and Engel, W. K. (1972). *New Engl. J. Med.* **286**, 333.
Wood, P. C., and Boegman, R. J. (1975). *Exp. Neurol.* **48**, 136.
Wrogemann, K., and Blanchaer, M. C. (1967). *Can. J. Biochem.* **45**, 1271.
Wrogemann, K., and Blanchaer, M. C. (1968). *Can. J. Biochem.* **46**, 323.
Wrogemann, K., and Blanchaer, M. C. (1971). *Enzyme* **12**, 322.
Wrogemann, K., Jacobson, B. E., and Blanchaer, M. C. (1969). *In* "Muscle Diseases, Proceedings of the International Congress" (J. N. Walton, ed.), pp. 290–293. Excerpta Medica, Amsterdam.
Wrogemann, K., Blanchaer, M. C., and Jacobson, B. E. (1970). *Can. J. Biochem.* **48**, 1332.
Wrogemann, K., Jacobson, B. E., and Blanchaer, M. C. (1974). *Can. J. Biochem.* **52**, 500.
Yamaguchi, K., Santa, T., Inoue, K., and Omae, T. (1978). *J. Neurol. Sci.* **38**, 195.
Zacks, S. I. (1974). "The Motor Endplate," 2nd ed. R. E. Krieger, Huntington.

Zalkin, H., Tappel, A. L., Caldwell, K. A., Shibko, S., Desai, I. D., and Holliday, J. A. (1962). *J. Biol. Chem.* **237**, 2678.
Zampighi, G., Vergara, J., Ramon, F. (1974). *J. Cell Biol.* **64**, 734.
Zierler, K. L., and Rogus, E. (1957). *Am. J. Pathol.* **190**, 201.
Zierler, K. L., and Rogus, E. (1958). *Am. J. Physiol.* **193**, 534.

# EDITORS' SUMMARY TO CHAPTER VII

From the point of human physiology, skeletal muscle tissue plays an important role as discussed by the authors. The skeletal muscle occupies about 40% of adult human body weight and has the primary role of providing force for moving the skeletal system. From the point of human pathology there are many other reasons to study this tissue; for instance, enough tissue can be obtained by biopsies both for morphological and biochemical purposes, and as compared to many other tissues the biopsies have little risk. Therefore, the sequence of pathological alterations might be followed by serial biopsies. Another advantage for pathological studies is the possibility to obtain electrophysiological information such as electromyographic data prior to the taking of the muscle biopsy. By this means the proper muscle, which is clinically but not too severely involved can be selected. On the other hand a complicating factor in the study of human skeletal muscle is the presence of two major fiber types. Since these fiber types have quite different metabolic properties it is difficult to interpret the biochemical results obtained from muscle tissue containing an unknown mixture of both fiber types. In the electron microscopic as well as histochemical studies this problem can be avoided since it is fairly easy to distinguish the fiber types by their histochemical properties as well as by the ultrastructure of their organelles such as Z bands and mitochondria.

Since skeletal muscle is one of the rare examples in which enough human tissue is readily available it is not surprising that there is a vast amount of knowledge of both biochemical and morphological alterations in many diseases. These studies have revealed interesting examples in which the "primary lesion" has been identified at the organelle level. Examples of these are many so-called mitochondrial myopathies, which as "experiments of the nature" have helped us to understand the normal molecular functions of mitochondria. Pompe's disease, as discussed by the authors, is another example of those so far rare diseases where the primary lesion can be identified at the organelle level. The damaging factor is the lack of one lysosomal enzyme. This leads to a sequence of secondary effects such as accumulation of large lysosomes in the muscle cells which occupy the spaces of muscle fibers and damage the function of both heart and skeletal muscle cells. However, in most muscle diseases the underlying molecular mechanisms are still unknown although many changes have been observed at the cell membranes both by biochemical and fine structural methods. In many instances it is not even possible to distinguish whether the disease process primarily affects the muscle cell or whether these changes are only secondary due to the damage of the innervating nerves and neuromuscular junctions.

The role of autophagocytosis in muscle atrophy is controversial. As discussed in this chapter there are several examples where atrophy is accompanied by increased levels of lysosomal enzymes as well as increase in the number of lysosomes. In some cases it appears that atrophy of muscle fibers is at least partly dependent upon nonlysosomal degradation. Another example of the possible combined effect of increased autophagocytosis and nonlysosomal degradation is the breakdown of the intersegmental muscles of silkmoths as described by Lockshin and Beaulaton (1974a,b). According to these authors the mitochondria appear to be digested within the autophagic vacuoles whereas the myofilaments are digested external to these organelles. Since it has been shown that the sarcoplasmic reticulum also contains acid hydrolase activities, it might be that the breakdown of myofilaments represents an effect of the release of acid hydrolases from the sarcoplasmic reticulum. The important role of mitochondria in human disease is also well illustrated in the various skeletal muscle diseases described in this chapter. The detailed study of these skeletal muscle diseases has helped us to understand the normal as well as the abnormal physiology of mitochondria.

Studies of skeletal muscle from the pathobiological point of view have been vastly neglected. This is a very important organ to study because of its size, as mentioned above, and equally important because of the access to biopsy techniques. Many more studies of the cell biology of muscle are needed in diverse diseases ranging from Raynaud's phenomenon to intermittent claudication. Moreover, the study of skeletal muscle could serve as an important model for delineation of problems in the myocardium. Although the myocardium has additional problems including intercellular conduction, etc. the basic muscle physiology may well be similar. Furthermore, the influence of training by exercise on either skeletal or cardiac muscle in terms of resistance to anoxia is not known although the effects of training on athletic performance are beginning to be well established. Questions include the distribution of red and white fibers as modified by training, the mechanism of changes in oxygen consumption, and the influence of training on microcirculation. All of these need much further basic investigation and are equally useful for studies on pharmacological interventions following myocardial infarction.

The role of lysosomes, through autophagy, is also suggested by studies on the rodent heart. Lysosomal enzymes appear to play an important role in catabolic processes including the accelerated degradation and atrophy of cardiac muscle (Hoffstein *et al.*, 1975). They appear to have a role in cardiac ischemia and may have an important injurious function in diabetes.

## References

Hoffstein, S., Gennaro, D. E., Weissmann, G., Hirsch, J., Streuli, F., and Fox, A. C. (1975). *Am. J. Pathol.* **79,** 193.

Lockshin, R. A., and Beaulaton, J. (1974a). *J. Ultrastruct. Res.* **46,** 43.

Lockshin, R. A., and Beaulaton, J. (1974b). *J. Ultrastruct. Res.* **46,** 63.

# CHAPTER VIII

# THE ALVEOLAR WALL IN THE SHOCK LUNG

**Norman B. Ratliff**



## I. Introduction

The lung as a target organ in shock is an entity upon which attention has been focused only in recent years. The picture is incomplete, but changes in structure and function can be correlated, and some insight into the pathogenesis of the shock lung has been achieved. To consider this entity in terms of injury to a functioning biological membrane, a brief review of the normal structure and function of the alveolus is necessary.



PATHOBIOLOGY OF CELL MEMBRANES, VOL. II

ISBN 0-12-701502-7

## II. The Alveolar Wall As a Biological Membrane

The alveolar wall functions as a biological membrane in a manner analagous to the renal glomerulus; the function of each being the exchange of solutes between the blood and the external environment. A regulated balance between perfusion of the alveolus with blood, ventilation, and diffusion of gases across the alveolar wall (blood–air interface) is essential for the normal exchange of oxygen and carbon dioxide.

## III. Maintenance of Alveolar Structure

Ventilation and perfusion are dependent on maintenance of alveolar structure. The connective tissue which supports the alveoli may be viewed conceptually as a three-dimensional set of guy wires embedded in an elastic honeycomb in which the principle structural wires pass through the areas where two or more alveoli join. The network of collagen, reticulum, and elastic together with blood vessels and the connective tissue of the bronchial tree are responsible not only for structural integrity but for the elasticity of the normal lung (Kilburn, 1970).

The surfactants, a mixture of saturated phospholipids with lesser amounts of lipo- and glycoproteins (Weibel and Gill, 1968) are important in preventing the alveoli from filling with fluid (Kilburn, 1970). The small radius and inward curvature of an alveolus creates a natural pressure gradient favoring the movement of fluid into the alveoli (Pattle, 1965). The low surface tension created by the surfactants negates this pressure and is also of importance in the opening of closed alveoli (Pattle, 1965).

The surfactants are metabolized at least in part by the granular pneumocytes (great alveolar cells, type II cells) (Bensch *et al.*, 1964; Kilburn, 1970) which normally occupy the niches formed by the junction of alveoli (Fig. 1). Less is known about the membranous pneumocytes (type I cells) which form the flat pavement epithelium lining alveoli and which, when viewed with the electron microscope, are very similar in appearance to the endothelial cells of capillaries (Bertalanffy, 1965; Fig. 2). They are joined to each other by zonulae occludentes (tight junctions) which are less permeable to tracers than those of endothelial cells (Schneeberger-Keely and Karnovsky, 1968); these tight junctions probably play a role in preventing the leakage of fluid and proteins into the alveoli.

## IV. Perfusion

Perfusion of the alveoli is controlled in large measure by the arterioles which, in contrast to the capillaries, are not exposed to alveolar pressure but

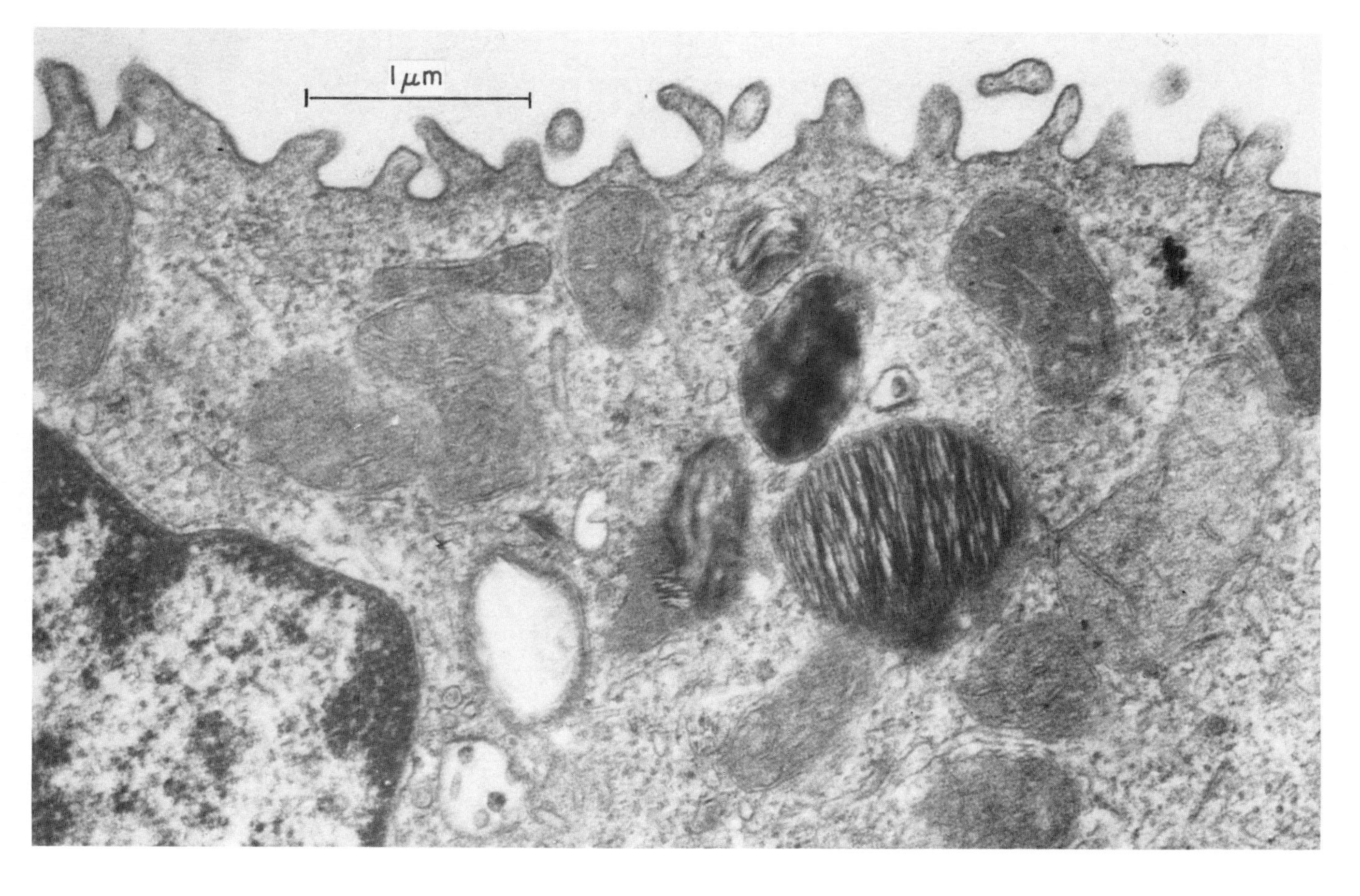


**Fig. 1.** Normal granular pneumocyte. The cytoplasm is filled with organelles. Numerous mitochondria exhibit a variety of shapes. A cluster of four laminated osmiophilic bodies in the cytoplasm are in different stages of development. These may be the source of surfactant. The cell surface facing the alveolus has numerous microvilli. Glutaraldehyde and osmium fixation. Lead citrate and uranyl acetate staining. ×30,000.

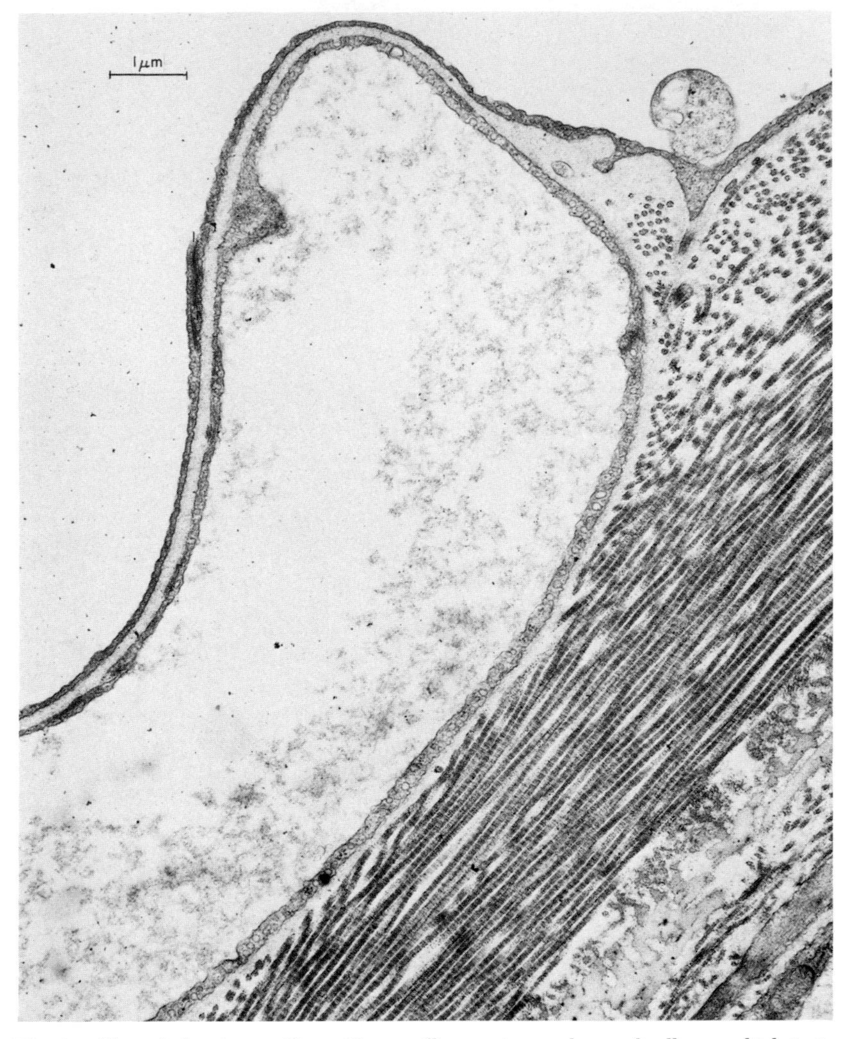


**Fig. 2.** Normal alveolar capillary. The capillary rests on a layer of collagen which in turn rests on a layer of elastin. Facing the alveolus the capillary wall is composed of membranous pneumocytes and continuous capillary endothelium joined by a common basement membrane. Both the membranous pneumocytes and the endothelium are of single cell thickness. Glutaraldehyde and osmium fixation. Lead citrate and uranyl acetate staining. ×10,000. (Reprinted with permission from the *Am. J. Pathol.*, Ratliff *et al.*, 1970a.)

are surrounded by the interstitium and are classed as extraalveolar or muscular vessels, as are the veins (Howell *et al.*, 1961; Permutt, 1965; West *et al.*, 1965; West, 1968;)(Figs. 3 and 4). The negative pressure of the interstitium is important in the maintenance of the patency of the extraalveolar vessels (Mackay *et al.*, 1969).

All of the above are important in normal alveolar function, and all are altered in the complex condition now known as the shock lung, or adult respiratory distress syndrome.

## V. The Syndrome

### A. *History and Clinical Course*

A syndrome of respiratory insufficiency following nonthoracic trauma and shock was recognized in World War II (Brewer *et al.*, 1946); and the pathological changes were described shortly thereafter (Moon, 1948). However, this syndrome received little attention until the Viet Nam conflict (Sealy, 1968). Now the syndrome has been clearly recognized as frequently following not only combinations of soft tissue injury, sepsis, and shock (Kamada and Smith, 1972) but also complicating cardiogenic shock, cardiopulmonary bypass (Ratliff *et al.*, 1973), and hemodialysis (Craddock *et al.*, 1977a). The recognition of this syndrome is due at least in part to the rapid evacuation of injured soldiers and civilians via helicopter with the result being that therapy is instituted in patients who in previous years would have died before reaching a treatment center. The rise in frequency is also related to increasing success in the management of the renal and hemodynamic problems of shock (Lucas *et al.*, 1968). A typical clinical course is stabilization of the acute hemodynamic problems followed by a quiescent period of hours or days, then the insidious onset of respiratory insufficiency (Bredenberg *et al.*, 1969). The patient may go on to die from respiratory insufficiency with the pulmonary insult having been worsened by therapeutic measures employed in the initial therapy (Bryant *et al.*, 1970). Implicit in these case studies is the notion that the pulmonary injury is rarely, if ever, a cause of irreversible shock (Bryant *et al.*, 1970; Sealy, 1968).

### B. *Functional Impairment*

The hallmarks of the syndrome are arterial hypoxemia which does not respond to the administration of oxygen, late ensuing hypercarbia, decreased compliance (increased stiffness), and radiological evidence of military atelectasis (Bredenberg *et al.*, 1969; Long *et al.*, 1968; Lucas *et al.*, 1968; Powers *et al.*, 1972).

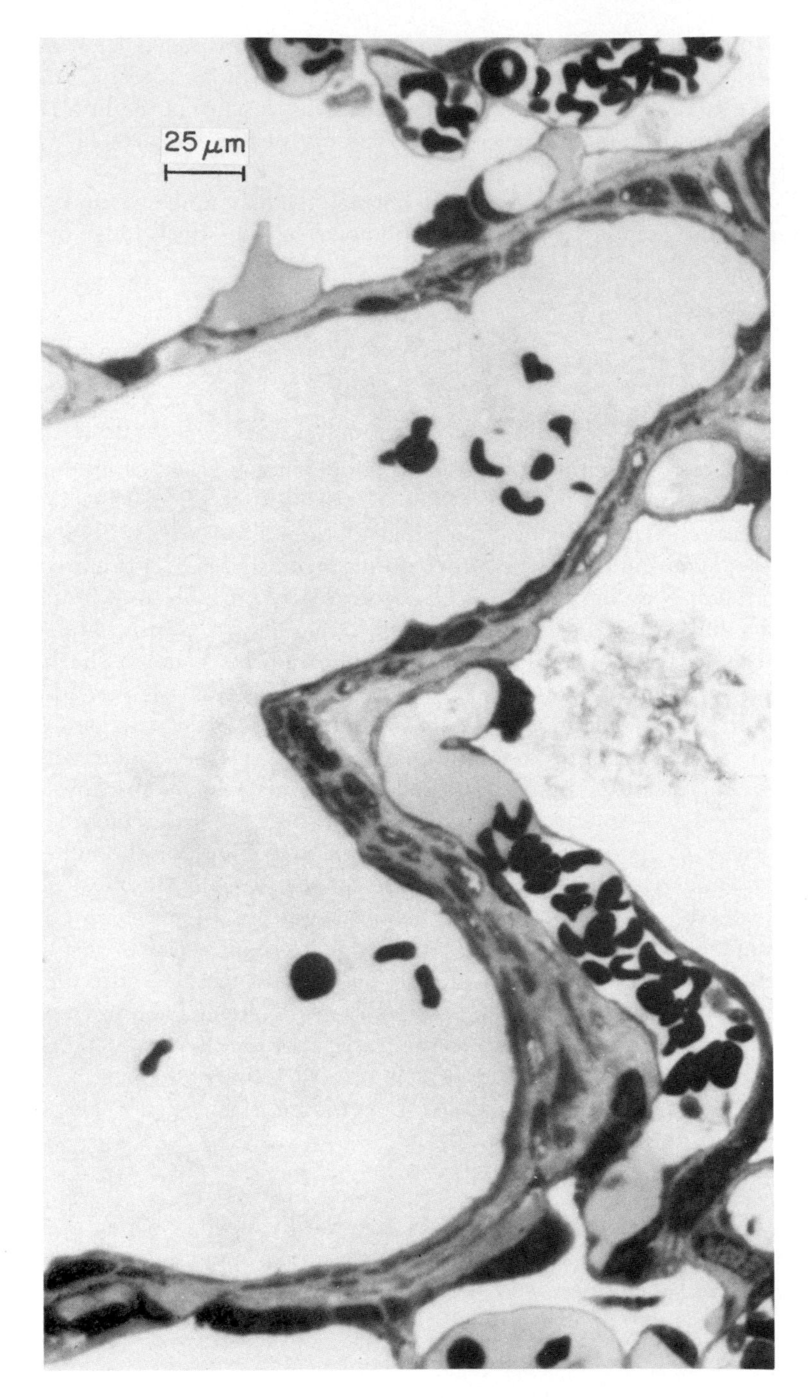
25 μm

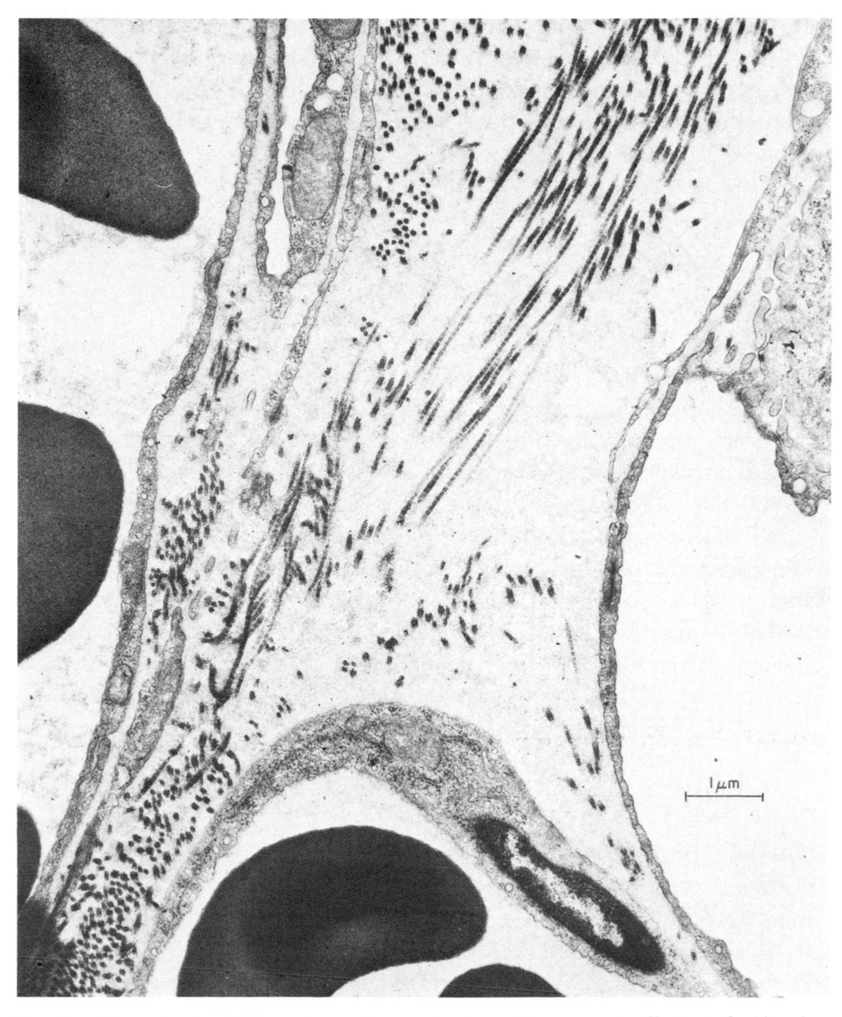


**Fig. 4.** Normal interstitium. A septal vascular branching point is illustrated. Alveolar air space is on the right. Capillaries are at the bottom and to the left. Note the fine granularity and electron density of the interstitial ground substance. Glutaraldehyde and osmium fixation. Lead citrate and uranyl acetate staining. ×10,000. (Reprinted with permission from the *Am. J. Pathol.*, Ratliff *et al.*, 1970a.)

---

**Fig. 3.** Normal lung, light micrograph. Intraalveolar vessels (capillaries) and an extraalveolar vessel (a vein) are present with a normal interstitium surrounding the vein. Glutaraldehyde and osmium fixation. Toluidine blue staining. ×400.

Arterial hypoxemia which does not respond to oxygen administration is clear evidence of ventilation–perfusion imbalance (shunting). Hypercarbia indicates that diffusion is also impaired, while loss of compliance and the development of atelectasis are indicative of injury to the structural system of the lung.

## VI. Morphology

### A. *Human*

Martin *et al.* (1968) have provided an extensive study of the lungs of combat casualties. Their report is noteworthy for the high incidence of hyaline membranes in alveoli and of pleural effusions. Both of these entities are absent in experimental studies of the shock lung (Clowes *et al.*, 1968; DePalma *et al.*, 1967; McKay *et al.*, 1966, 1967; Ratliff *et al.*, 1970a; Sugg *et al.*, 1968; Wilson *et al.*, 1970). This discrepancy may be due in part to the duration of survival of a number of the patients in the study of Martin and his colleagues. In experimental studies the animals have generally not been studied at comparably long intervals following shock. The discrepancy does add weight to the idea that the pulmonary insult sustained in shock may be made worse by therapeutic measures such as blood transfusion or oxygen (Bryant *et al.*, 1970; Veith *et al.*, 1968).

### B. *Experimental*

Experimental studies of the morphology of the shock lung are complicated by the fact that there is considerable variability from animal to animal as well as, in any given animal, from one area of the lung to the next (Wilson *et al.*, 1970). In spite of this, a clear picture has been obtained of the morphological alterations in several different experimental animals: interstitial edema progressing to intraalveolar edema; intense vascular congestion; lymphatic congestion; perivascular, peritubular, and intraalveolar hemorrhage; and multifocal (miliary) atelectasis (Cook and Webb, 1968; Lowery *et al.*, 1970; Wilson *et al.*, 1970; Wilwerth *et al.*, 1967; Fig. 5). With the exception of the discrepancies noted above, the changes are similar to those noted in human lungs. Ultrastructural studies have extended these observations and have documented injury to all structural elements of the alveolar wall. Within the edematous interstitium, the connective tissue elements, in sharp contrast to their normal proximity, are dispersed and separated by edema fluid, much of which is a transudate (Fig. 6). Surprisingly, this disruption is consistently located between the adventitial and muscular layers of extraalveolar vessels

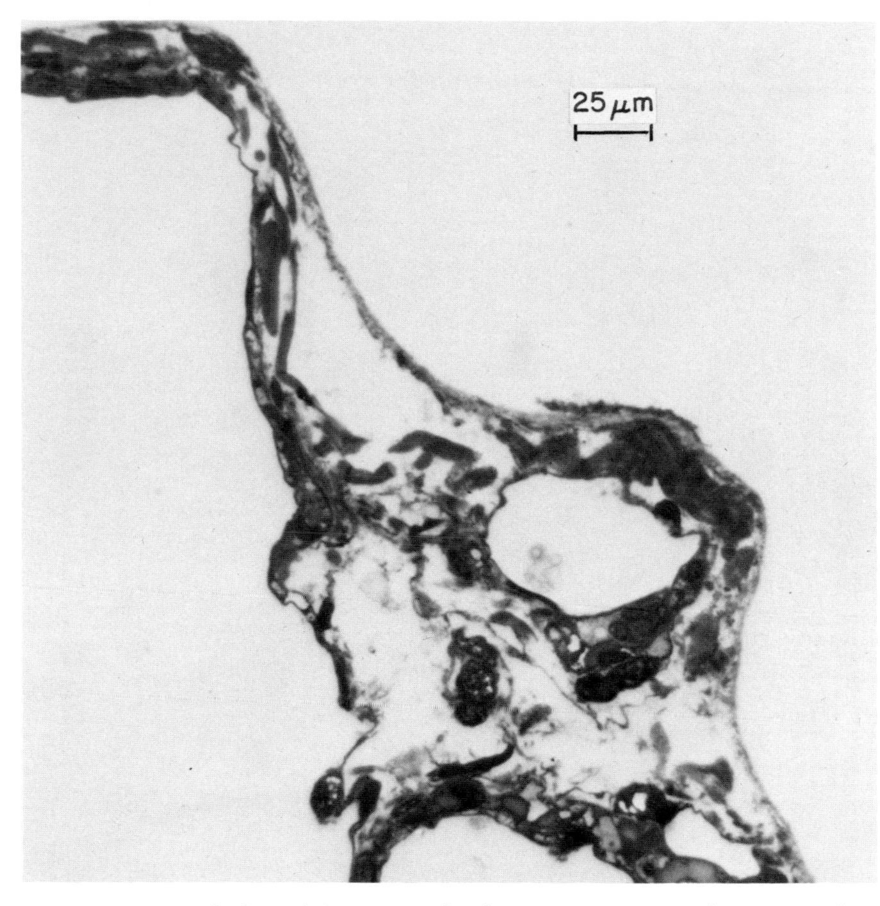


**Fig. 5.** Interstitial edema, light micrograph. The interstitium surrounding an arteriole is disrupted by edema at a septal branching point. The lumen of the arteriole is devoid of erythrocytes. Compare with Fig. 3. Glutaraldehyde and osmium fixation. Lead citrate and uranyl acetate staining. ×400. (Reprinted with permission from the *Am. J. Pathol.*, Ratliff *et al.*, 1970a.)

at the junction of alveolar septae and does not extend out into the thinnest portion of the alveolar wall, that part which consists only of a membranous pneumocytes and endothelial cells joined by a common basement membrane (Ratliff *et al.*, 1970a). Extensive swelling of membranous pneumocytes and endothelial cells is present, primarily involving the cell sap, with a variable degree of mitochondrial swelling (Figs. 7–12). Mitochondrial swelling is more frequent in endothelial cells than in membranous pneumocytes, and is

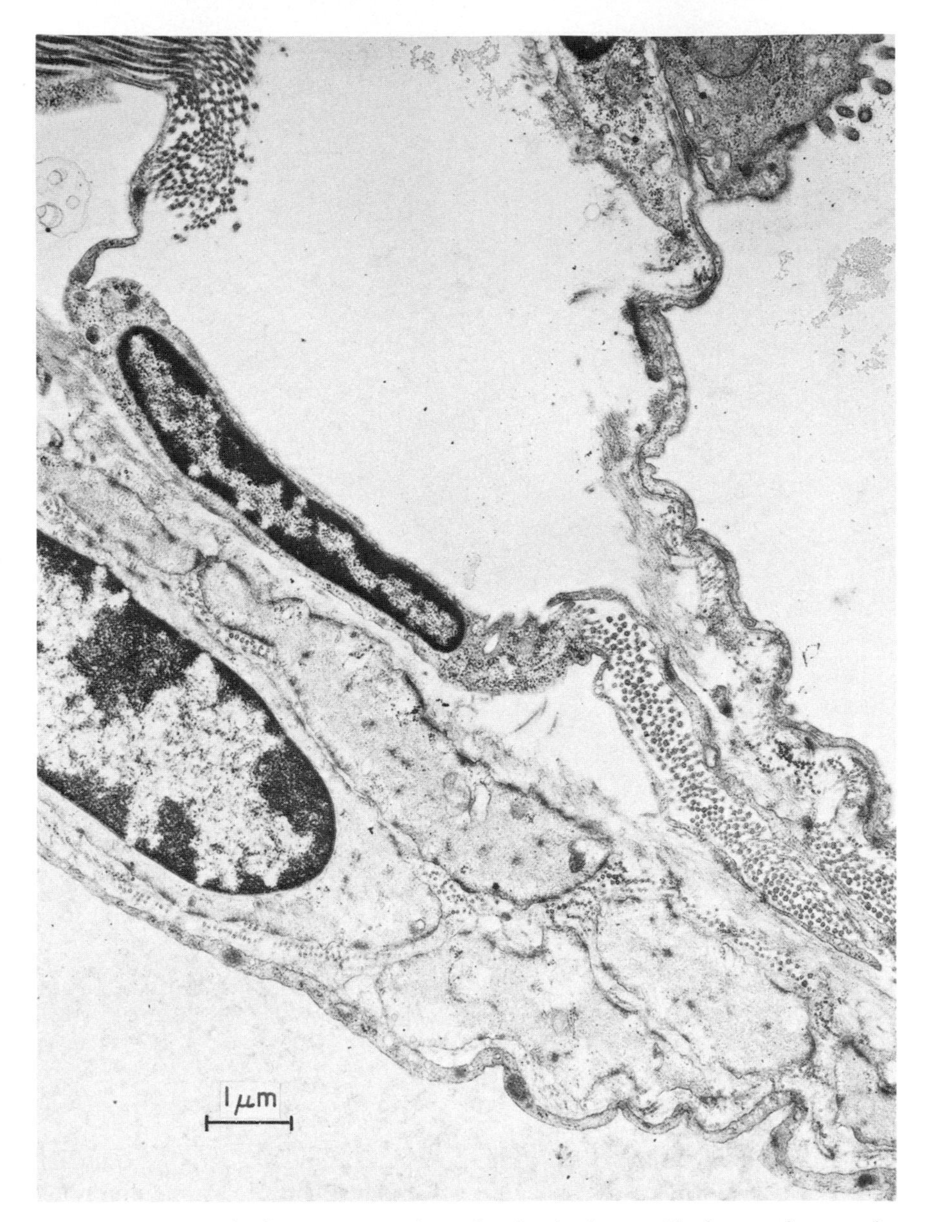


**Fig. 6.** Interstitial edema. At top right is the alveolar lumen. The lumen of a muscular vessel is at the bottom. The interstitium at the top is disrupted by edema, without evidence of protein deposition within the edematous interstitium. Note that the edema is between the muscular and adventitial layers of the wall of the muscular vessel. Glutaraldehyde and osmium fixation. Lead citrate and uranyl acetate staining. ×11,000. (Reprinted with permission from the *Am. J. Pathol.*, Ratliff *et al.*, 1970a.)

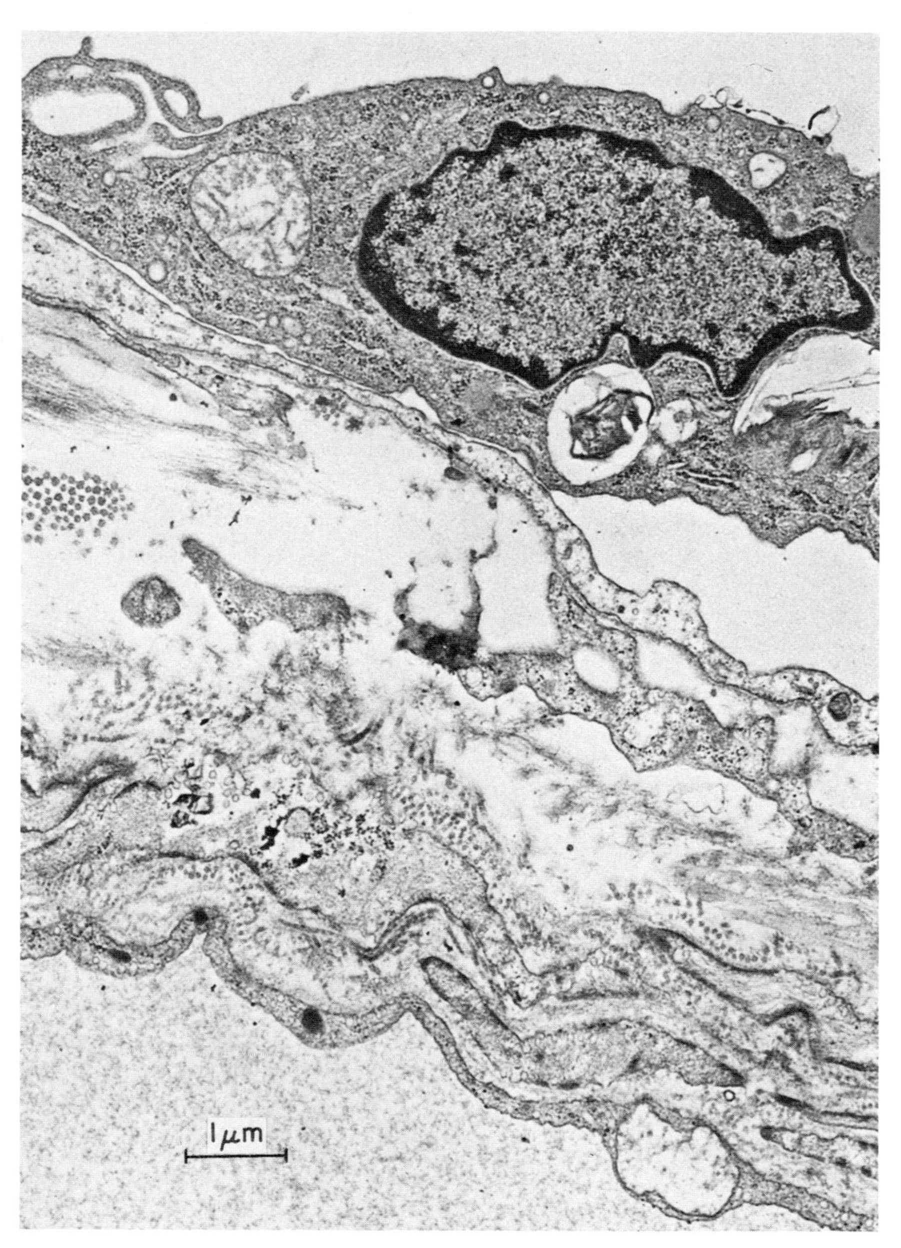


**Fig. 7.** The alveolar wall containing a muscular vessel. At the top is the alveolar lumen and a granular pneumocyte. Beneath this cell is a membranous pneumocyte and beneath this the interstitium and wall of a muscular vessel. At the bottom is the lumen of this vessel. Interstitial edema is present, as in Fig. 6. Mitochondria in the granular pneumocyte and in an endothelial cell are swollen, as is the cell sap of the membranous pneumocyte. Glutaraldehyde and osmium fixation. Lead citrate and uranyl acetate staining. ×13,000. (Reprinted with permission from the *Am. J. Pathol.*, Ratliff *et al.*, 1970a.)

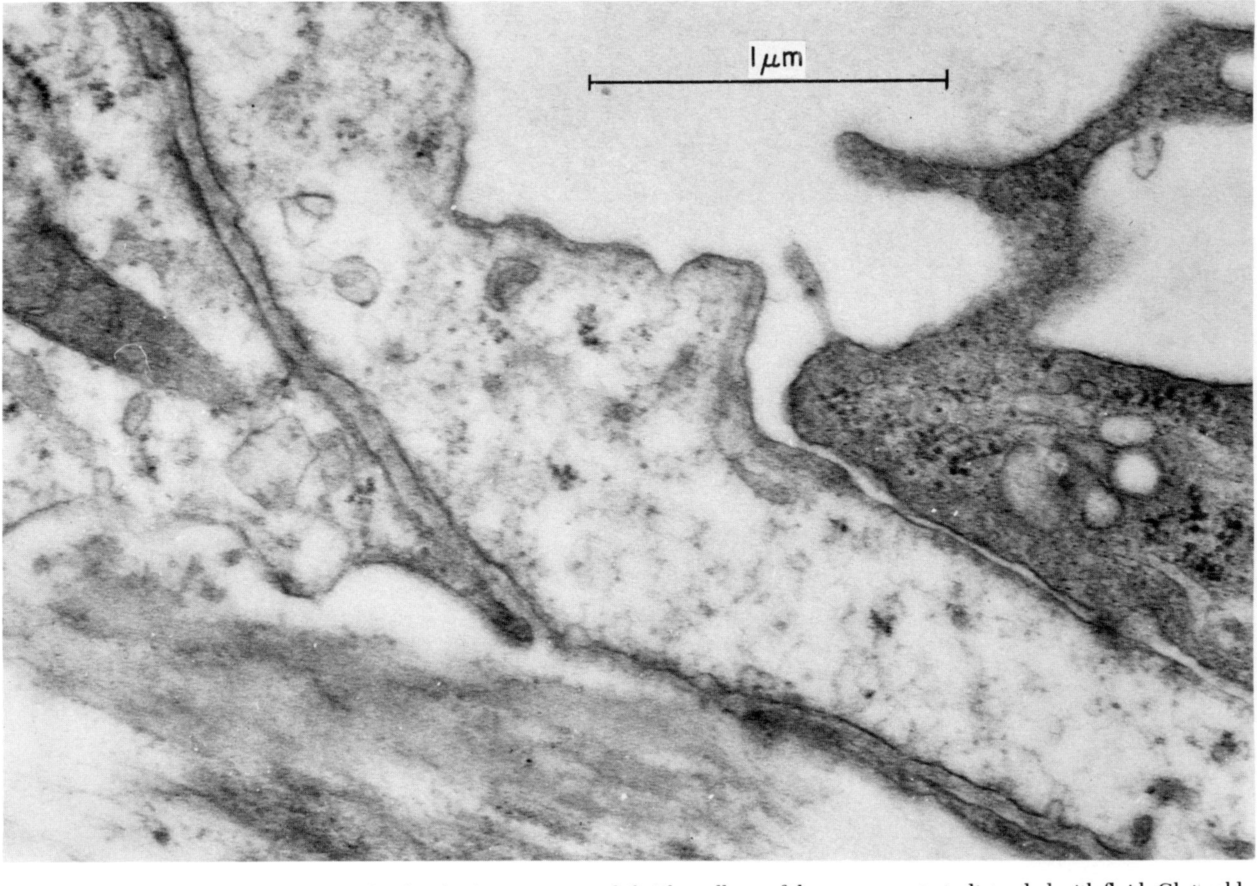


**Fig. 8.** Membranous pneumocyte. The alveolar lumen is at top left. The cell sap of the pneumocyte is distended with fluid. Glutaraldehyde and osmium fixation. Lead citrate and uranyl acetate staining. ×47,000. (Reprinted with permission from the *Am. J. Pathol.*, Ratliff

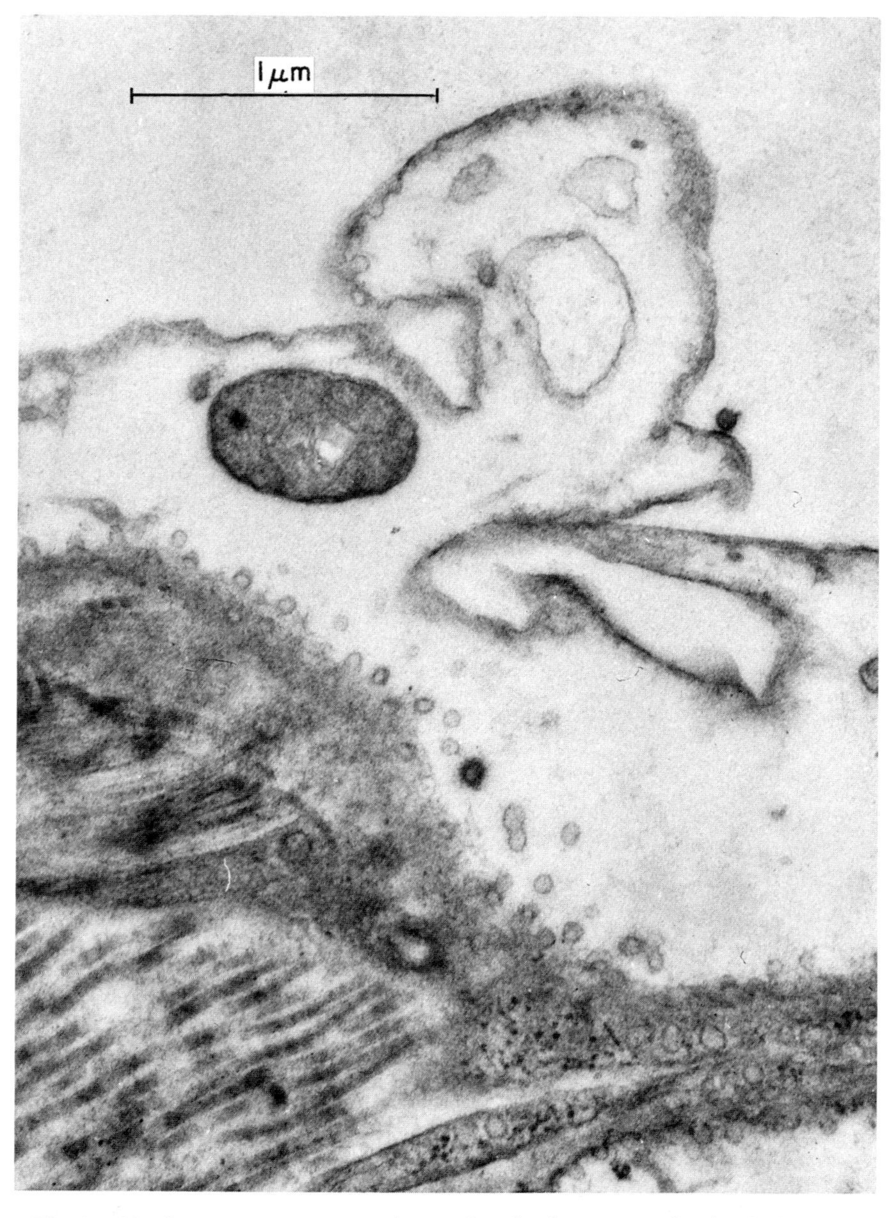


**Fig. 9.** Membranous pneumocyte at the tip of an alveolar septum. The alveolar lumen is at the top. The pneumocyte is severely swollen, but a mitochondrion within the cell is not swollen. Martin *et al.* (1968) reported that hyaline membranes were usually found overlying the tips of alveolar septae. Glutaraldehyde and osmium fiaxation. Lead citrate and uranyl acetate staining. ×40,000. Reprinted with permission from the *Am. J. Pathol.*, Ratliff *et al.*, 1970a.)

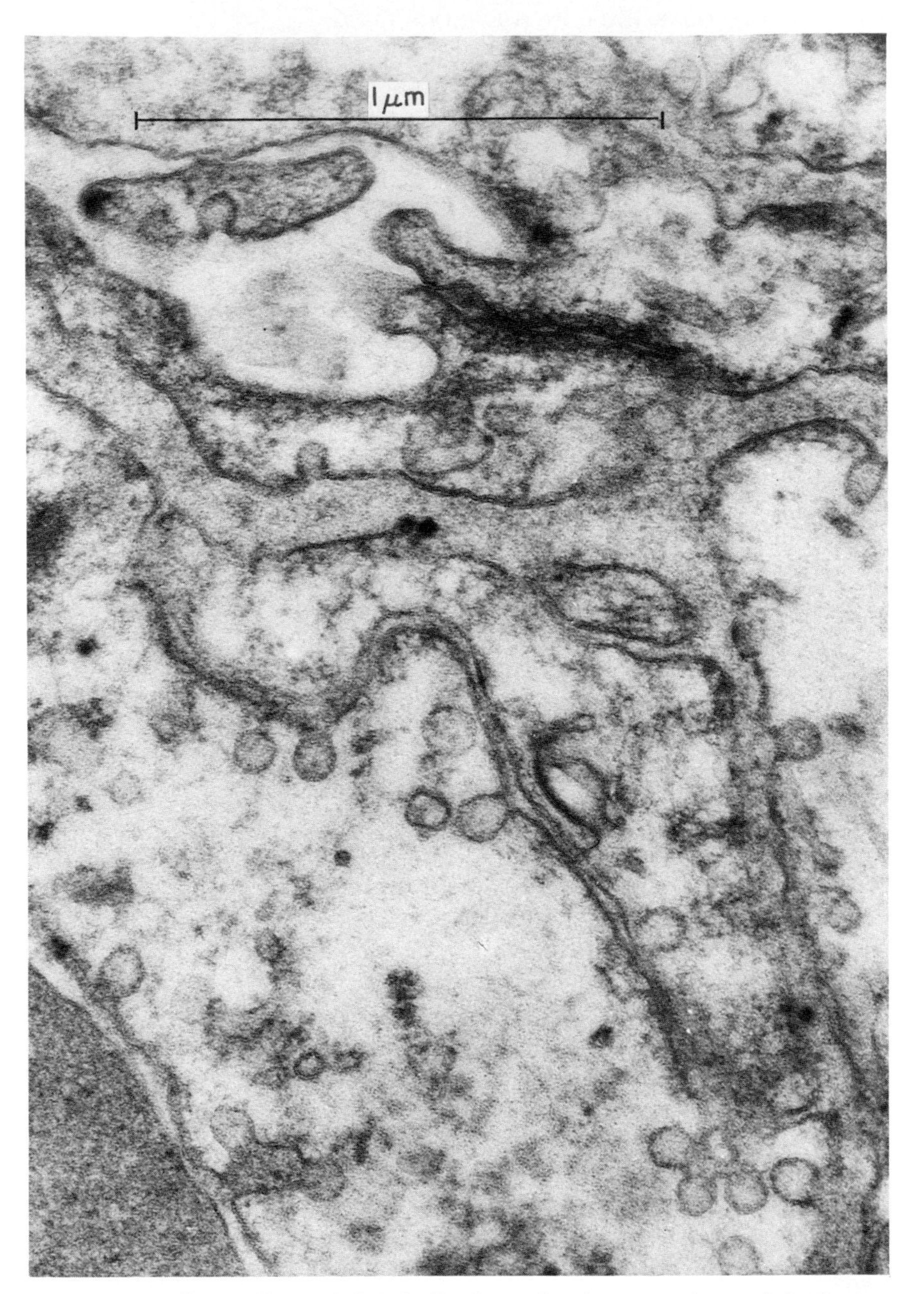


**Fig. 10.** Swollen capillary endothelial cell. The capillary lumen is at bottom left. The density of the cell sap is decreased and the plasmalemmal vesicles are dispersed by edema. The plasmalemma is intact. Glutaraldehyde and osmium fixation. Lead citrate and uranyl acetate staining. ×69,000. (Reprinted with permission from the *Am. J. Pathol.*, Ratliff *et al.*, 1970a.)

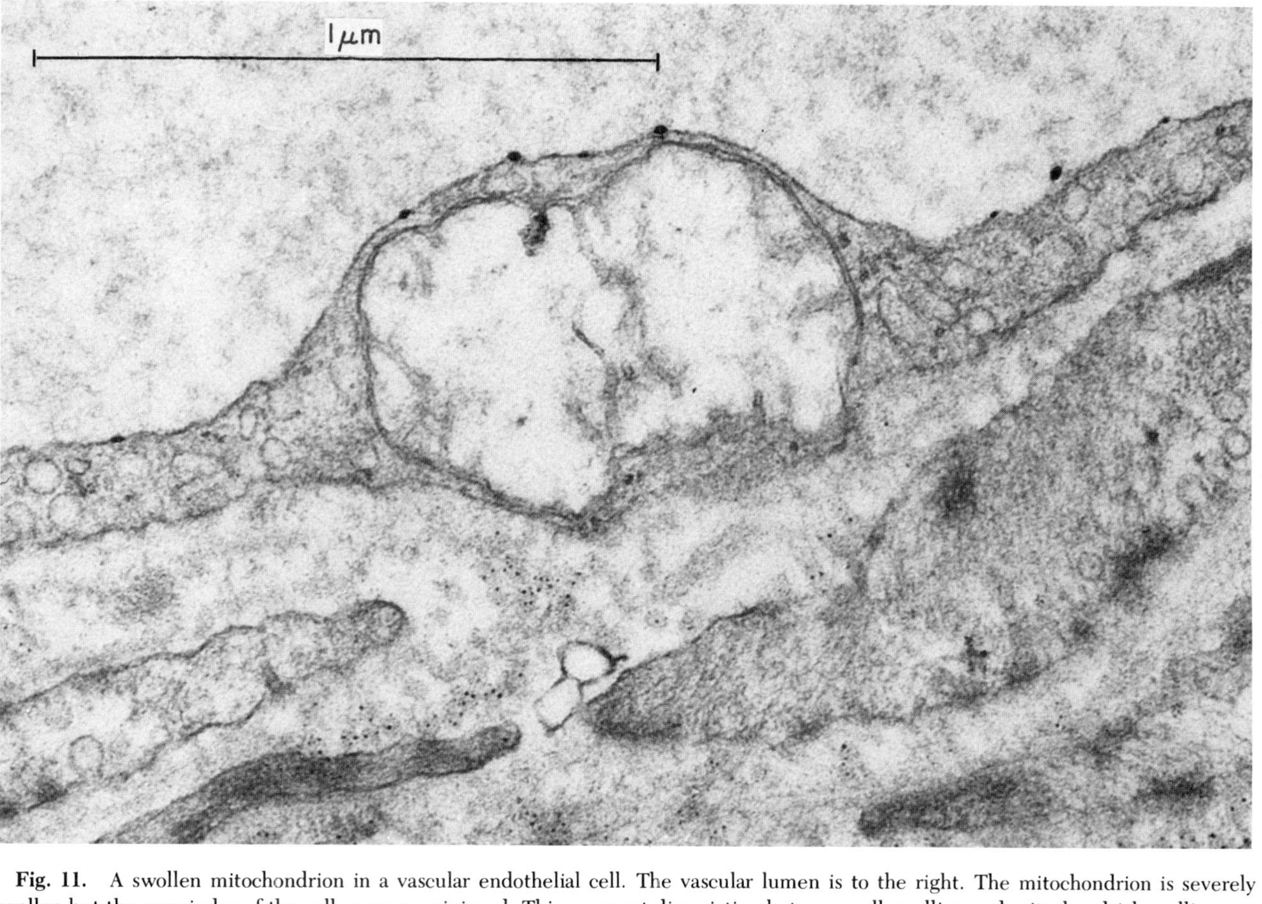


**Fig. 11.** A swollen mitochondrion in a vascular endothelial cell. The vascular lumen is to the right. The mitochondrion is severely swollen but the remainder of the cell appears uninjured. This apparent dissociation between cell swelling and mitochondrial swelling was often observed in membranous pneumocytes and vascular endothelial cells, but not in granular pneumocytes. Glutaraldehyde and osmium fixation. Lead citrate and uranyl acetate staining. ×82,000.

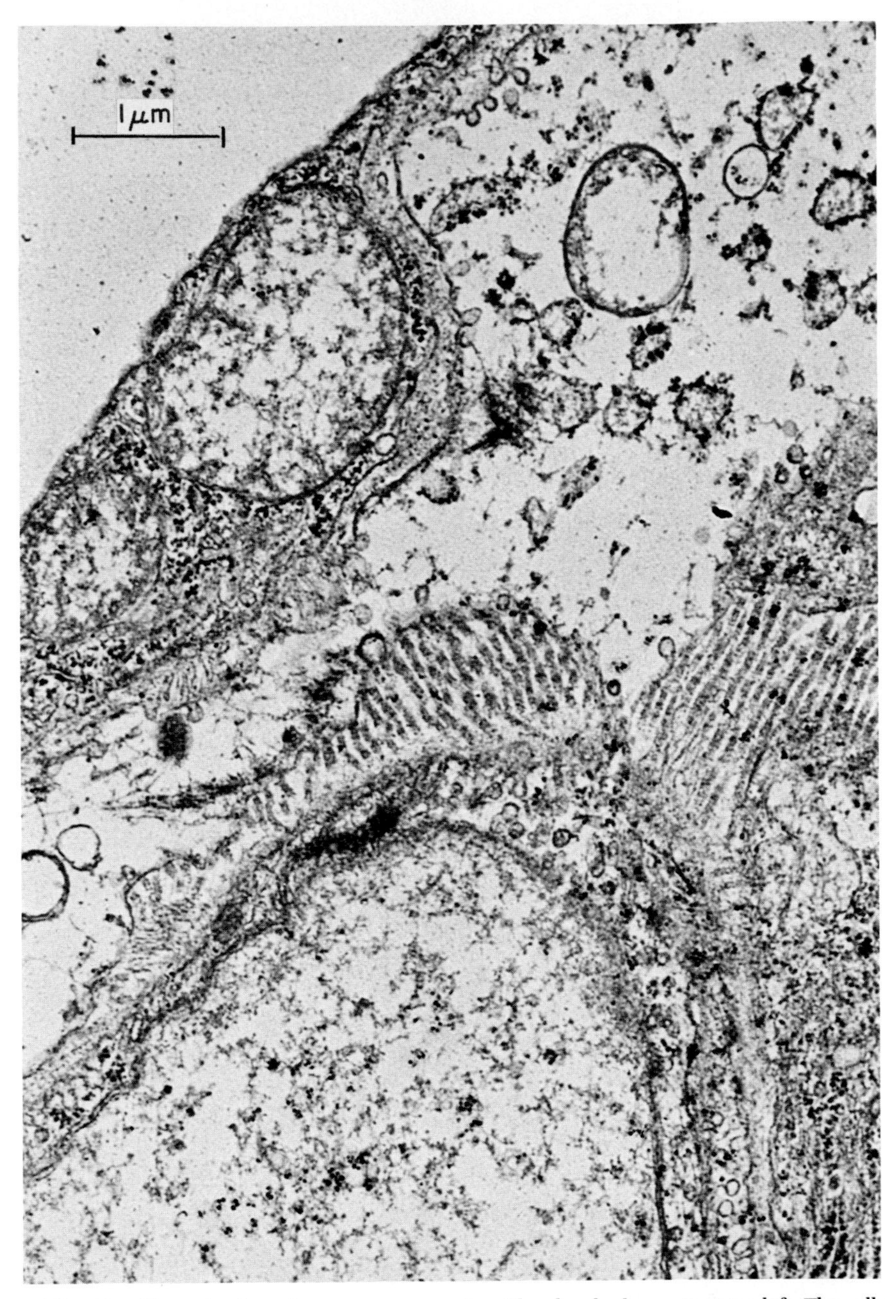


**Fig. 12.** Necrosis of membranous pneumocytes. The alveolar lumen is at top left. The cells are swollen and their cytoplasm contains granular debris. The only recognizable organelles are severely swollen mitochondria. Glutaraldehyde and osmium fixation. Lead citrate and uranyl acetate staining. ×20,000. (Reprinted with permission from the *Am. J. Pathol.*, Ratliff *et al.*, 1970a.)

less common in both than is swelling of the cell sap (Ratliff *et al.*, 1970a). In contrast, the matrix compartment of the mitochondria of granular pneumocytes is often swollen (Barkett *et al.*, 1968; Ratliff *et al.*, 1970a).

## VII. Blood Cells

### A. *Platelets*

Hardaway *et al.* (1962) have consistently put forth the idea that intravascular coagulation is of paramount importance in shock. There is an increasing body of evidence that intravascular microaggregates are screened out in the blood vessels of the lungs and that this process contributes to the development of the shock lung. Platelet aggregates are found in the pulmonary microcirculation in shock (Allardyce *et al.*, 1969; Lim *et al;.*, 1967; Ratliff *et al.*, 1970a; Robb, 1963; Figs. 13 and 14). Where these aggregates adhere to endothelium they may exhibit partial degranulation, but often the platelets adhere to endothelium which appears normal (Ashford and Freidman, 1968; Ratliff *et al.*, 1970a). Transient increases in platelet adhesiveness and aggregability occur after severe hemorrhage or soft tissue trauma, and these platelets aggregate in the lung (Ljungqvist, 1973). In addition to possible mechanical obstruction of vessels, platelets may release chemical vasoconstrictors which may be sufficient to cause an increase in pulmonary vascular resistance (Ljungqvist, 1973). Further, Margaretten and McKay (1969) have proposed that platelets may trigger intravascular coagulation in endotoxin shock. The role of platelets in shock and trauma has been comprehensively reviewed (Ljungqvist, 1973).

### B. *Erythrocytes and Flow Patterns*

Erythrocytes also seem to aggregate in hemorrhagic shock, and may contribute to stasis in the pulmonary circulation. Investigators who have observed the alveolar circulation *in vivo* have described the development in shock of a granular appearance of the blood followed by "rocking chair" or to-and-fro flow which progresses to the development of visible erythrocyte clumping and "sludging" of the blood (Murakami *et al.*, 1970; Wilson *et al.*, 1970). As stagnation progresses, large aggregates of blood cells develop in the venules, and these vessels may actually sacculate. At the same time a segmental appearance of arterioles and venules is noted (Wilson *et al.*, 1970). Reversals of flow in pulmonary vessels and bypassing of blood around capillary beds are visible correlates of the ventilation–perfusion imbalance which is characteristic of the shock lung (Murakami *et al.*, 1970; Wahrenbrook *et al.*, 1970; Wilson *et al.*, 1970). With re-infusion of shed blood, there

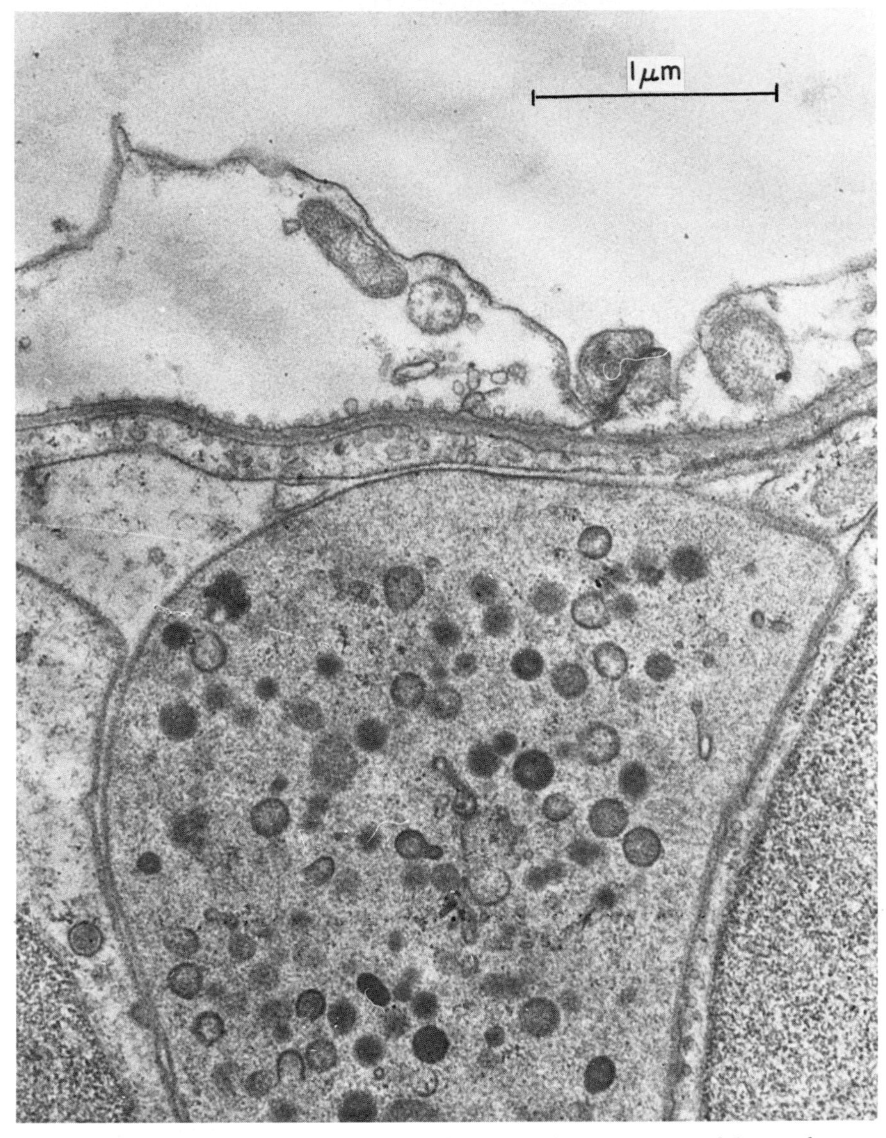


**Fig. 13.** Platelet in a pulmonary capillary. The alveolar lumen is at top and the membranous pneumocytes are swollen. The platelet is deformed by adjacent blood cells but it is not degranulated. The endothelium appears uninjured. Osmium fixation. Lead citrate and uranyl acetate staining. ×32,000.

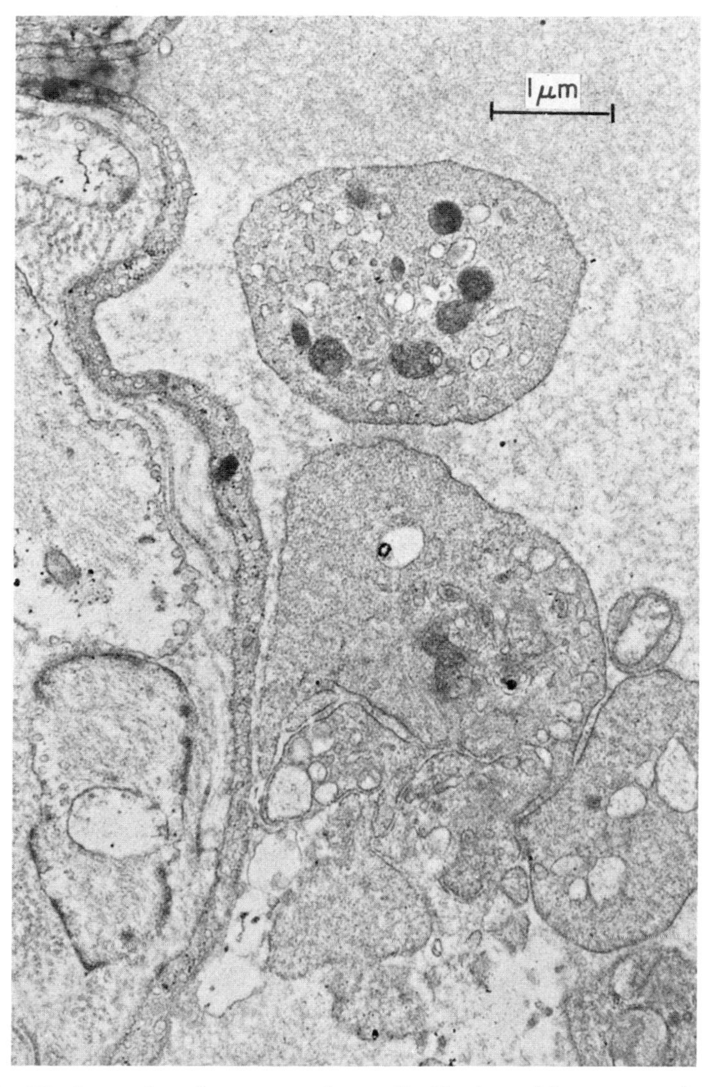


**Fig. 14.** Platelets in the pulmonary circulation. The blood vessel lumen is at right. Within the blood vessel is a mass of platelets which are adherent to the apparently uninjured vascular endothelium. Several of the platelets have undergone degranulation. Glutaraldehyde and osmium fixation. Lead citrate and uranyl acetate staining. ×16,000. (Reprinted with permission from the *Am. J. Pathol.*, Ratliff *et al.*, 1970a.)

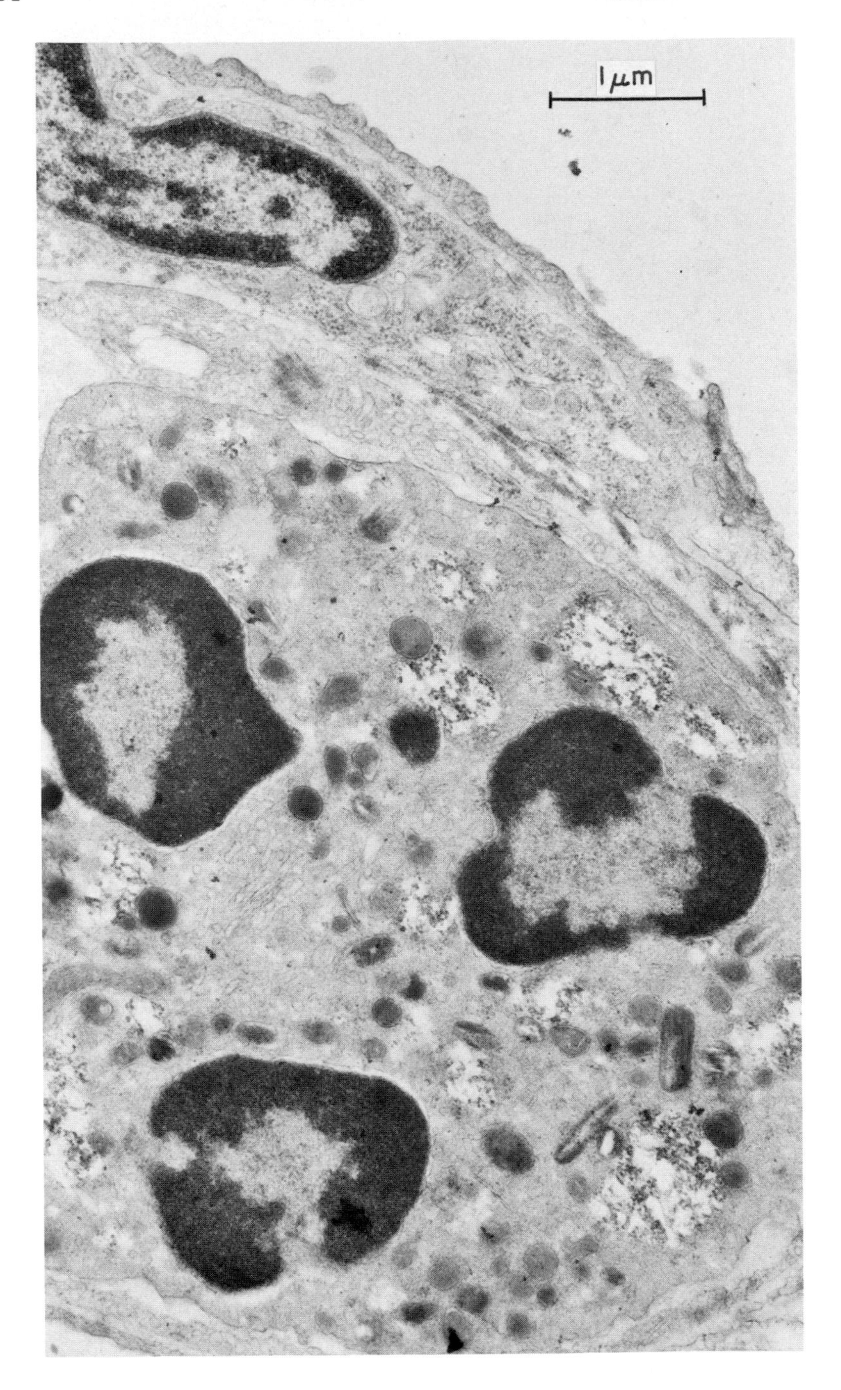
1 μm

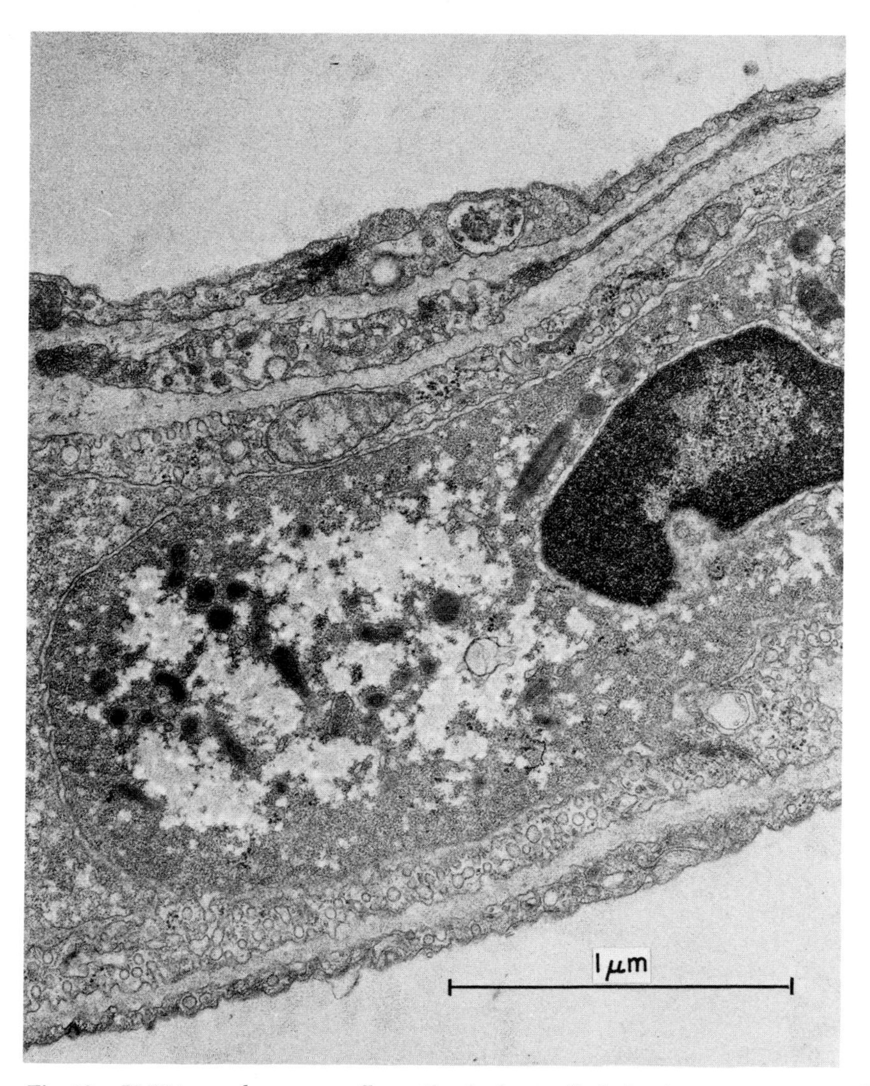


**Fig. 16.** PMN in a pulmonary capillary. Alveolar lumina flank this thin septum at top and bottom. The capillary is filled with a PMN. The changes in the cytoplasm of this PMN are similar but more severe than those in Fig. 15. Glutaraldehyde and osmium fixation. Lead citrate and uranyl acetate staining. ×45,000.

---

**Fig. 15.** Neutrophilic polymorphonuclear leukocyte (PMN) in a pulmonary capillary. The alveolar lumen is at the top. The capillary is filled by the PMN. Within the cytoplasm of the PMN are irregular electron lucent spaces which are not enclosed by membranes. These spaces contain variable amounts of stainable glycogen and they were noted to be larger and more numerous when the severity and duration of shock was prolonged. The variability in this amount of stainable glycogen may be the result of extraction during processing of the tissue. Glutaraldehyde and osmium fixation. Lead citrate and uranyl acetate staining. ×20,000.

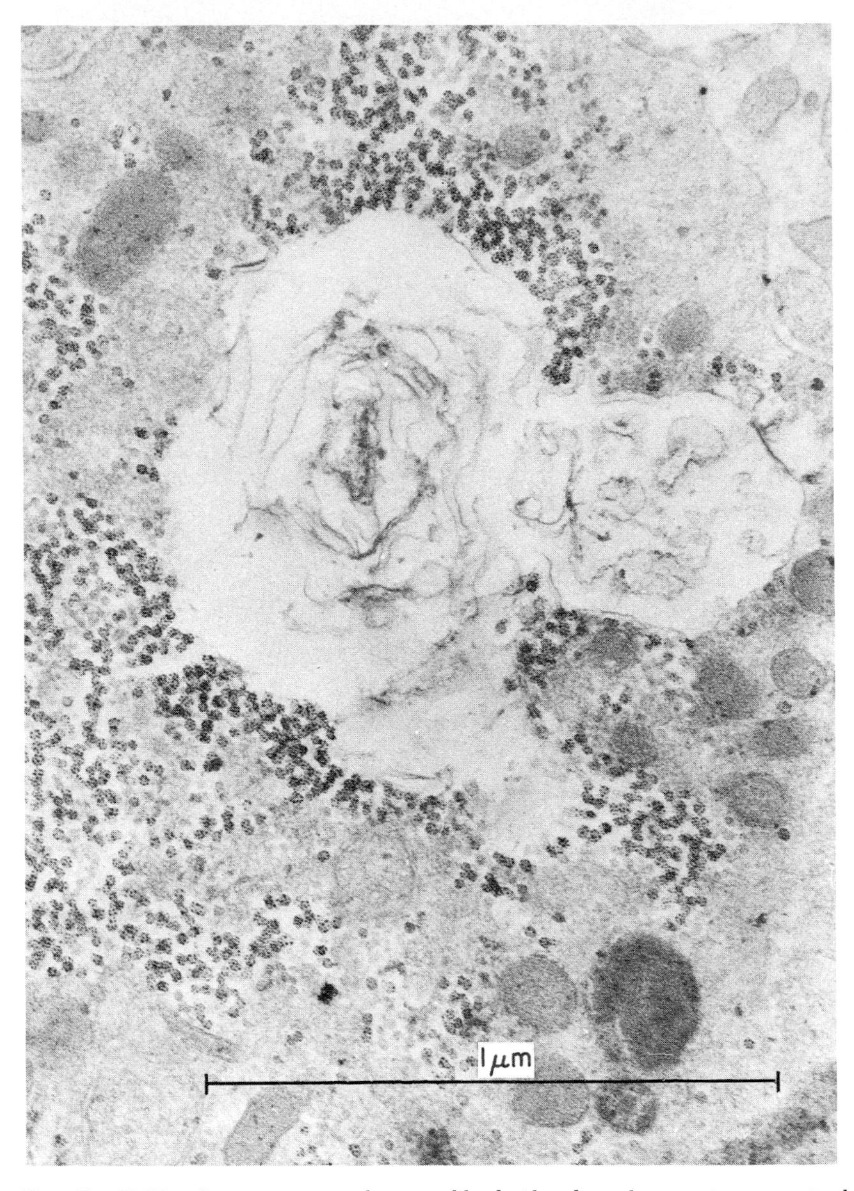


**Fig. 17.** PMN isolated from central venous blood. Abundant glycogen is present in the cytoplasm and surrounds an irregular electron lucent space containing disorganized membranes. Glutaraldehyde and osmium fixation. Lead citrate and uranyl acetate staining. ×75,000.

is either break up of aggregates, engorgement of vessels and resumption of flow (Murakami *et al.*, 1970); or the resumption of circulation is uneven, and stagnation and "sludging" persists in many vascular beds (Wilson *et al.*, 1970).

### C. *Leukocytes*

Recent evidence indicates that granulocytes may play a crucial role in the shock lung (Craddock *et al.*, 1977a; Craddock *et al.*, 1977b; Sacks *et al.*,

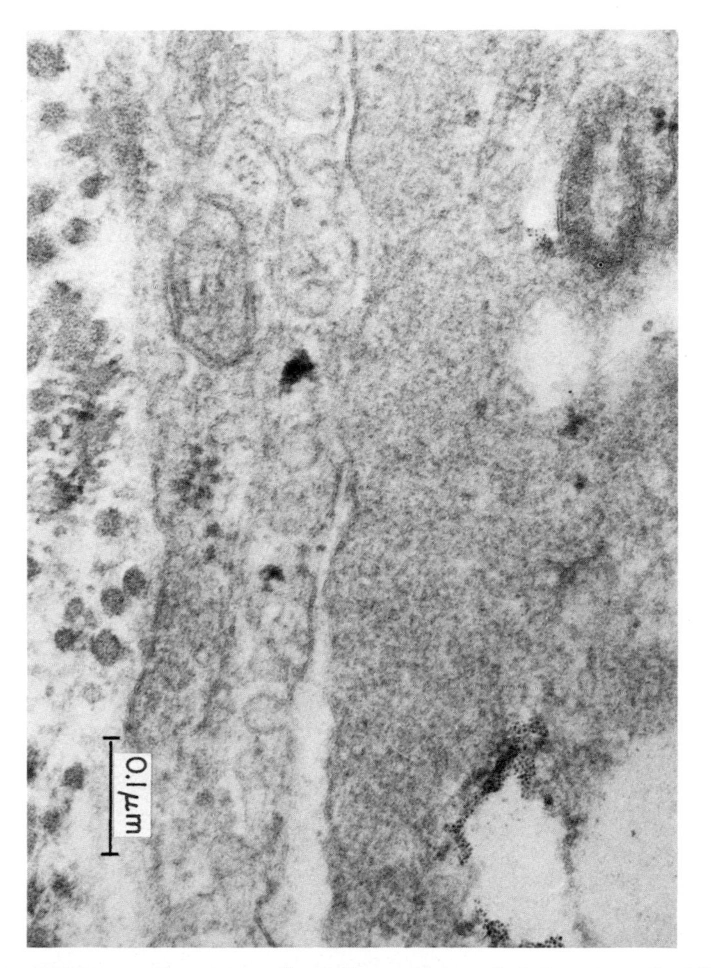


**Fig. 18.** PMN in a pulmonary capillary. The capillary endothelium is on the left, and a portion of the PMN is on the right. At bottom center there is a generous space between PMN and endothelium, but near the top this space becomes very small. Glutaraldehyde and osmium fixation. Lead citrate and uranyl acetate staining. ×150,000.

1978; Bowers *et al.*, 1980, Craddock *et al.*, 1979). Leukocytes begin to assume a spherical shape and stick to arteriole and capillary endothelium at the same time that aggregation of erythrocytes becomes prominent (Wilson *et al.*, 1970). Histological examination of the lungs after hemorrhagic shock revealed numerous granulocytes in capillaries and arterioles, and, in heparinized animals, in venules as well (Lim *et al.*, 1967; Lowery *et al.*, 1970; Ratliff *et al.*, 1971; Sugg et al., 1968). Studies *in vivo* suggest that the granulocytes had undergone morphological alterations prior to adherence, and these observations were verified by subsequent ultrastructural studies (Ratliff *et al.*, 1971; Wilson *et al.*, 1970; Fig. 15). The granulocytes contained large cytoplasmic spaces probably filled with glycogen (Harrison *et al.*, 1969; Simpson and Ross, 1971; Vye and Fischman, 1970; Figs. 16–20). There was loss of granule density, and the space between leukocyte and endothelial plasmalemma was frequently obliterated. The underlying endothelium often appeared normal, suggesting that the sticking of leukocytes did not depend on prior endothelial injury (Ratliff *et al.*, 1971). This phenomenon is not restricted to hemorrhagic shock. McKay and colleagues have called attention

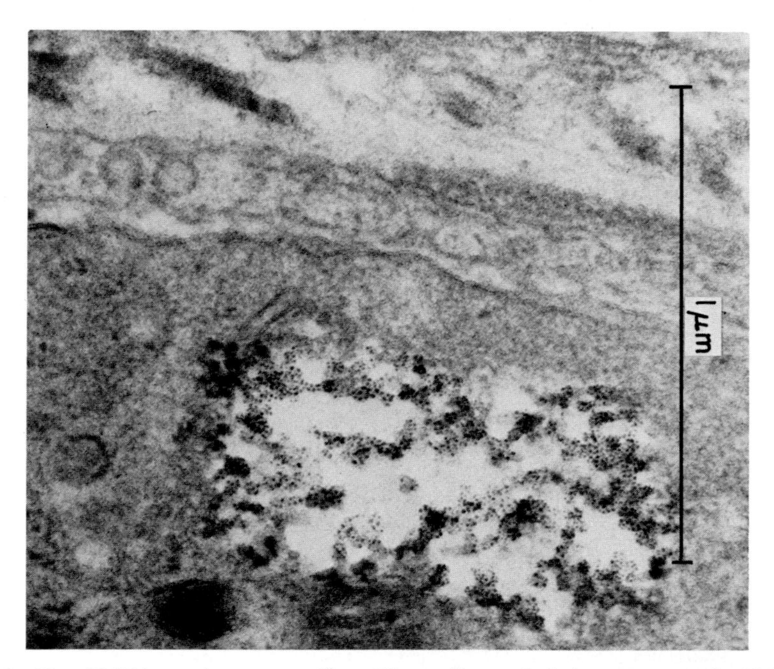


**Fig. 19.** PMN in a pulmonary capillary. The capillary endothelium is at top, the PMN is at bottom. Illustration of the proximity between PMN and capillary endothelium. In several places it is difficult to decide whether the plasmalemma is part of the PMN or of the endothelial cell. Glutaraldehyde and osmium fixation. Lead citrate and uranyl acetate staining. ×62,000.

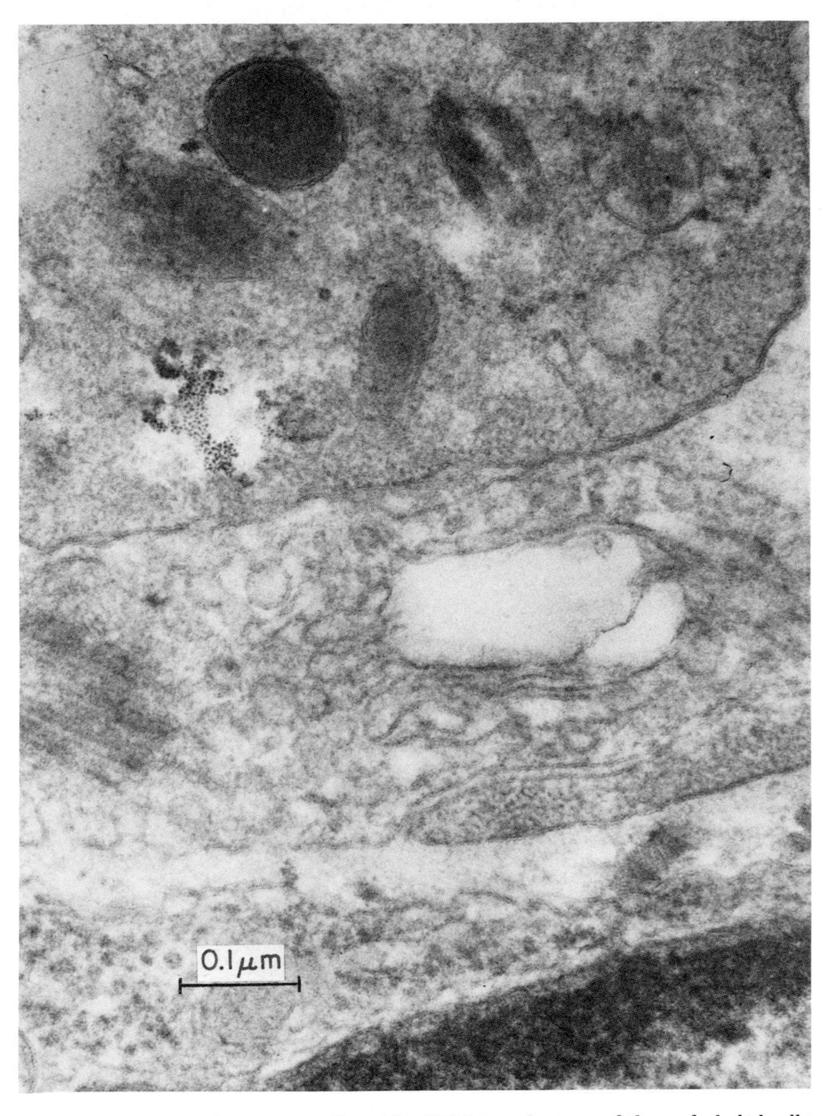


**Fig. 20.** PMN in a pulmonary capillary. The PMN is at the top and the endothelial cell is at the bottom. The proximity between endothelium and PMN at several points gives the appearance of spot welding. Glutaraldehyde and osmium fixation. Lead citrate and uranyl acetate staining. ×150,000.

to intravascular coagulation in endotoxin shock and have noted engorgement of pulmonary capillaries with clumps of granulocytes and platelets and disintegration of the granulocytes. This occurred without demonstrable endothelial damage and led McKay and colleagues to conclude that the trapping of granulocytes in the pulmonary circulation is responsible for the neutropenia that follows endotoxemia (McKay *et al.*, 1966, 1967). These observations have been confirmed. DePalma *et al.* (1967) noted similar trappings of granulocytes in the pulmonary circulation. Alterations in the cytoplasm and granules of granulocytes in pulmonary capillaries following endotoxin administration are similar to these of hemorrhagic shock (Harrison *et al.*, 1969). Furthermore, dogs subjected to hemorrhagic shock develop progressive neutropenia similar to that of endotoxemia (Ratliff *et al.*, 1971; Figs. 21 and 22). Similar trappings of platelets and leukocytes in the lung also occurs following regional shock (Lim *et al.*, 1967; Stallone, 1970). Elucidation of the pathogenesis of this leukocyte and platelet trapping may have been provided by studies of the role of complement activation in promoting granulocyte aggregation in the pulmonary microvasculature (Craddock *et al.*, 1977a;

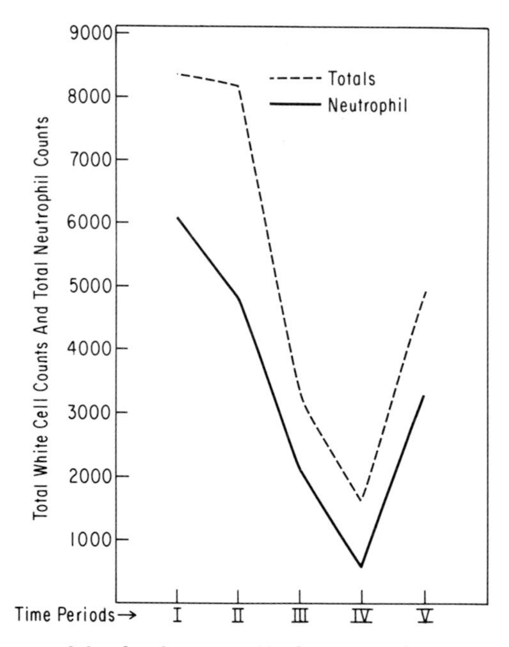


**Fig. 21.** Time course of the development of leukopenia and neutropenia in hemorrhagic shock in blood drawn from the inferior vena cava. On the ordinate time period I is control preanesthesia; II is control postanesthesia; III is 60 min. of shock; IV is 90 min. of shock; and V is after infusion of the shed blood. Cells per cubic millimeter of blood are on the abscissa. (Reprinted with permission from the *Am. J. Pathol.*, Ratliff *et al.*, 1971.)

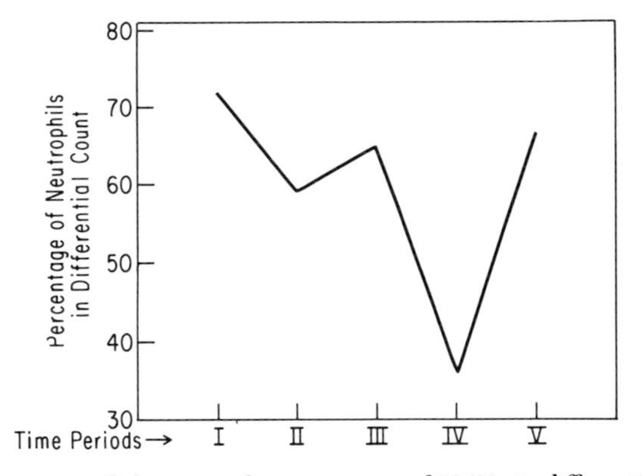


**Fig. 22.** Time course of change in the percentage of PMNs in differential count of white blood cells in blood samples drawn from the inferior vena cava in hemorrhagic shock. The percentage of PMNs is plotted on the abscissa. The time periods on the ordinate are the same as in Fig. 21. Note that in addition to the neutropenia and leukopenia illustrated in Fig. 21, there is a further selective diminution in PMNs which becomes severe after 90 min of shock. In contrast to the total leukocyte count, the differential count is restored to control values by infusion of the shed blood. (Reprinted with permission from the *Am. J. Pathol.*, Ratliff *et al.*, 1971.)

Craddock *et al.*, 1977b; Sacks *et al.*, 1978; Bowers *et al.*, 1980; Craddock *et al.*, 1979). Complement, primarily the C5a component, activated via the alternative pathway, induces pulmonary dysfunction and leukostasis (Craddock *et al.*, 1977a; Fehr and Jacob, 1977). This reaction is accelerated by zymosan activated complement (Bowers *et al.*, 1980), and superoxide anion released by aggregating granulocytes is a potent platelet aggregant.

## VIII. Sieving, a Uniform Response

It is clear from the foregoing that the lung responds in similar fashion to soft tissue injury, endotoxemia, and hemorrhage. This point has been noted by a number of authors (Clowes *et al.*, 1968; McLean *et al.*, 1968; Veith *et al.*, 1968). The explanation for this similar response may lie in a normal nonrespiratory function of the lung. The vasculature of the lung, with its tapering arterioles and right angle branches, is ideally suited to act as a filter. There is evidence that the lung is normally a regulator of circulating leukocyte levels, and perhaps of platelets and erythrocytes, and that this regulation is accomplished by filtration (Heineman and Fishman, 1969). Swank (1968) has developed sound evidence that there is increased adhesiveness between circulating blood cells in shock and that the degree of adhesiveness is related to the severity of shock, possibly through the activation of complement (Craddock *et al.*, 1977a; Fehr and Jacob, 1977). If the adhesiveness is prevented pharmacologically, there is a notable decrease in the severity of the pulmonary injury (Swank, 1968). Furthermore, if all of the venous blood is filtered by the liver, microaggregates do not reach the lung and the degree of lung injury is reduced (Stallone, 1970). In addition, blood stored for transfusion contains aggregates of leukocytes and platelets; it reasonably follows that the pulmonary injury of shock might be aggravated by transfusion (Swank, 1968). There seems to be no question that the filtration of microaggregates by the lung is increased in shock (Allardyce *et al.*, 1968). Furthermore, there is increasing evidence that the adhesiveness of granulocytes and the trapping of microaggregates in the pulmonary vasculature is related to complement activation, specifically the C5a fraction in shock, trauma, sepsis, hemodialysis, and cardiopulmonary bypass (Craddock *et al.*, 1977a, b; Fehr and Jacob, 1977; Sacks *et al.*, 1978; Bowers *et al.*, 1980; Craddock *et al.*, 1979).

There is also evidence that catecholamines can induce intravascular coagulation and produce severe lung injury (Whitaker *et al.*, 1969); that serotonin can induce microaggregation (Swank, 1968); and that antineutrophil serum can produce neutropenia and morphological changes in polymorphoneuclear leukocytes similar to those observed in hemorrhagic shock and endotoxemia (Simpson and Ross, 1971). Divalent cation concentrations are important in determining the adherence of leukocytes (Allison *et al.*, 1963), and the plasma concentrations of magnesium and calcium are elevated in hemorrhagic shock (Ratliff *et al.*, 1970b; Hackel *et al.*, 1972). However low-flow states alone do not increase leukocyte adhesiveness; the addition of tissue injury is apparently necessary to increase adhesiveness (Branemark, 1968). While not sufficient to permit the drawing of definite conclusions regarding the cause of the microaggregates, this information supports the thesis that complement activation is important in the formation of intravascular microaggregates and in the pathogenesis of the shock lung syndrome. Such microaggregates may obstruct the microcirculation and release potent vasoconstrictive agents; and polymorphonuclear leukocytes carry large quantities of lysosomal enzymes which are potentially capable of digesting all structural elements of the lung.

## IX. Correlation of Changes in Structure and Function

The development of interstitial edema contributes to the loss of compliance by disrupting the delicate balance between structural rigidity and elasticity (Powers *et al.*, 1970). It may also contribute to perfusion–

ventilation imbalance by its effect on the extraalveolar vessels. The patency of these vessels is dependent on the pulling force of the surrounding negative interstitial pressure (Howell *et al.*, 1961; Permutt, 1965; West *et al.*, 1965; West, 1968). An increase in this pressure would promote collapse of these vessels. Flow through the veins would probably cease prior to arterial closure and would promote pulmonary congestion. A more active role for the veins, active constriction, has been proposed but remains a somewhat controversial issue (Murakami *et al.*, 1970). A combination of vascular slowing and adequate ventilation would lead to a decrease in the gas partial pressure gradient between blood and alveolar lumen. If the gas in the blood approaches equilibrium with alveolar gas, then a condition of phase equilibrium may be approached. In this state random nucleation of gas in the interstitium might be expected to occur (Hills, 1966; Hills and Lemessurier, 1969). "Bubbles" with the optical characteristics of an air–liquid interface have been observed in the interstitium of the lung *in vivo* during shock and were noted to be preferential sites for subsequent interstitial hemorrhage (Wilson *et al.*, 1970).

Production of surfactants is diminished several hours after shock (Henry *et al.*, 1967) and there is evidence of injury to granular pneumocytes in shock (Barkett *et al.*, 1968; Ratliff *et al.*, 1970a). However, it is not clear whether or not loss of surfactants contributes to the development of atelectasis.

Swelling of endothelial cells and of membranous pneumocytes increases the diffusion path length of gases between alveoli and capillaries and may contribute to the ventilation–perfusion imbalance. Injury to these cells also disrupts the integrity of the alveolar wall and could permit the passage of proteins from blood into the alveoli (Schneeberger-Kelley and Karnovsky, 1968). The presence of hyaline membranes is evidence that such leakage does occur, and hyaline membranes would certainly inhibit diffusion of gases.

## X. Conclusions

In summary, all of the alveolar components are injured in shock. From the standpoint of patient care the most important factor may be the development of vascular shunting which creates perfusion–ventilation imbalance (Wahrenbrook *et al.*, 1970). It is not known whether the shunting is functional or whether there are actual anatomic arterial–venous connections. Shock, atelectasis, and acidosis all contribute to shunting. Thus disruption of the interstitium may be of signal import; the mechanism of this disruption is loss of vascular integrity with subsequent formation of interstitial edema (Powers *et al.*, 1970). Shunting is increased by adding shock and acidosis to atelectasis, implicating the altered characteristics of the circulating blood and the importance of microaggregates, especially granulocytes (Swank, 1968). Finally, inadequate diffusion becomes apparent after the onset of shunting, implicating swelling of the cellular components and loss of integrity of the alveolar wall (Lucas *et al.*, 1968). The result is the shock lung.

## Acknowledgment

Supported by Grant HE-05875 and research training fellowship 5 TI-GM-726 from the U.S. Public Health Service, by a grant from the Walker P. Inman Cardiovascular Research Fund at Duke University, and by a grant from the Upjohn Company.

## References

Allardyce, B., Hamit, H. F., Matsumoto, T., and Moseley, R. V. (1969). Pulmonary vascular changes in hypovolemic shock: Radiography of the pulmonary microcirculation and the possible role of platelet embolism in increasing vascular resistance. *J. Trauma* **9**, 403.

Allison, F., Lancaster, M. G., and Crosthwaite, J. L. (1963). Studies on the pathogenesis of acute inflammation. V. an assessment of factors that influence in vitro the phagocytic and adhesive properties of leukocytes obtained from rabbit peritoneal exudates. *Am. J. Pathol.* **43**, 775.

Ashford, T. P., and Freidman, D. G. (1968). Platelet aggregation at sites of minimal endothelial injury. *Am. J. Pathol.* **53**, 599.

Barkett, V. M., Coalson, J. J., and Greenfield, L. J. (1968). Protective effects of pulmonary denervation in hemorrhagic shock. *Surg. Forum* **19**, 538.

Bensch, K., Schaefer, K., and Avery, M. E. (1964). Granular pneumocytes: electron microscopic evidence of their exocrine function. *Science* **145**, 1318.

Bertalanaffy, F. D. (1965). On the nomenclature of the cellular elements in respiratory tissue. *Am. Rev. Respir. Dis.* **91**, 605.

Bowers, T. K., Ozolins, A. E., Ratliff, N. B., and Jacob, H. S. (1980). Hyperacute necrotizing pulmonary vasculitis produced in rabbits by activated complement infusion. *Am. J. Pathol.* (In press).

Branemark, P. I. (1968). *In* "Microcirculation as Related to Shock". (D. Shepro and G. P. Fulton, eds.), pp. 161–180. Academic Press, New York.

Bredenberg, C. E., James, P. M., Collins, J., Anderson, R. W., Martin, A. M. Jr., and Hardaway, R. M. III, (1969). Respiratory Failure in Shock. *Ann. Surg.* **169**, 392.

Brewer, L. A., Burbank, B., Samson, P. C., and Schiff, C. A. (1946). The "wet lung" in war casualties. *Ann. Surg.* **123**, 343.

Bryant, L. R., Trinkler, J. K., and Dubilier, L. (1970). Acute respiratory pathophysiology after hemorrhagic shock. *Surgery* **68**, 512.

Clowes, G. H. A., Jr., Zuschneid, W., Dragacevic, S., and Turner, M. (1968). The nonspecific pulmonary inflammatory reactions leading to respiratory failure after shock, gangrene and sepsis. *J. Trauma* **8**, 899.

Cook, W. A., and Webb, W. R. (1968). Pulmonary changes in hemorrhagic shock. *Surgery* **64**, 85.

Craddock, P. R., Fehr, J., Brigham, K. L., Kronenberg, R. S., and Jacob, H.S. (1977a). Complement and leukocyte mediated pulmonary dysfunction in hemodialysis. *New Engl. J. Med.* **296**, 769.

Craddock, P. R., Fehr, J., Dalmasso, A. P., Brigham, K. L., and Jacob, H. S. (1977b). Hemodialysis leukopenia: Pulmonary vascular leukostasis resulting from complement activation by dialyzer cellophane membranes. *J. Clin. Invest.* **59**, 879.

Craddock, P. R., Hammerschmidt, D. E., Moldow, C. F., Yamada, O., and Jacobs, H. S. (1979). Granulocyte aggregation as a manifestation of membrane interaction with complement: Possible role in leukocyte margination, microvascular occlusion, and endothelial damage. *Sem. Hematol.* **16**, 140.

DePalma, R. G., Coil, J., Davis, J. H., and Holden, W. D. (1967). Cellular and ultrastructural changes in endotoxemia: A light and electron microscopic study. *Surgery* **62**, 505.

Fehr, J., and Jacob, H. S. (1977). *In vitro* granulocyte adherence and *in vivo* margination: Two associated complement-dependent functions. *J. Exp. Med.* **146**, 641.

Hackel, D. B., Ratliff, N. B., Mikat, E. and Graham, T. (1972). *In* "Comparative Pathophysiology of Circulatory Disturbances: Advances in Experimental Medicine and Biology" (C. M. Bloor ed.), pp. 277–288. Plenum, New York.

Hardaway, R. M., Brune, W. H., Geever, E. F., Burns, J. W., and Mock, H. P. (1062). Studies on the role of intravascular coagulation in irreversible hemorrhagic shock. *Ann. Surg.* **155**, 241.

Harrison, L. H., Beller, J. J., Hinshaw, L. B., Coalson, J. J., and Greenfield, L. J. (1969). Effects of endotoxim on pulmonary capillary permeability, ultrastructure and surfactant. *Surg. Gynecol. Obstec.* **129**, 723.

Heineman, H. O., and Fishman, A. P. (1969). Nonrespiratory functions of mammalian lung. *Physiol. Rev.* **49**, 1.

Henry, J. N., McArdle, A. H., Bounows, G., Hampson, L. G., Scott, H. J., and Guard, F. N. (1967). The effect of experimental hemorrhagic shock on pulmonary alveolar surfactant. *J. Trauma* **7**, 691.

Hills, B. A. (1966). "A Thermodynamic and Kinetic Approach to Decompression Sickness." Libraries Board of South Australia, Adelaide.

Hills, B. A., and LeMessurier, D. H. (1969). Unsaturation in living tissue relative to the pressure and composition of inhaled gas and its significance in decompression theory. *Clin. Sci.* **36**, 185.

Howell, J. B. L., Permutt, S., Proctor, D. F., and Riley, R. L. (1961). Effects of inflation of the lung on different parts of pulmonary vascular bed. *J. Appl. Physiol.* **16**, 71.

Kamada, R. O. and Smith, J. R. (1972). The phenomenon of respiratory failure in shock: the genesis of "shock lung". *Am. Heart J.* **83**, 1.

Kilburn, K. H. (1970). Alveolar microenvironment. *Arch. Intern. Med.* **126**, 435.

Lim, R. C., Jr., Blaisdell, F. W., Goodman, J. R., Hall, A. D., and Thomas, A. N. (1967). Electron microscopic study of pulmonary microemboli in regional and systemic shock. *Surg. Forum* **18**, 25.

Ljungqvist, U. (1973). Current research review: platelets in shock and trauma. *J. Surg. Res.* **15**, 132.

Long, D. M., Kim, S. I., and Shoemaker, W. C. (1968). Vascular responses in the lung following trauma and shock. *J. Trauma* **8**, 715.

Lowery, B. D., Mulder, D. D., Joyal, E. M., and Palmer, W. H. (1970). Effect of hemorrhagic shock on the lung of the pig. *Surg. Forum* **21**, 21.

Lucas, C. E., Ross, M., and Wilson, R. F. (1968). Physiologic shunting in the lungs in shock or trauma. *Surg. Forum* **19**, 35.

MacKay, P. A., Burgess, J. H., Finlayson, M. H. and Hampson, L. G. (1969). Hypoxemia and atelectasis in experimental hemorrhagic shock: its decrease by periodic hyperinflation of the lungs. *Can. J. Surg.* **12**, 351.

McKay, D. G., Margaretten, W., and Csavossy, I. (1966). An electron microscopic study of the effects of bacterial endotoxin on the blood vascular system. *Lab. Invest.* **15**, 1815.

McKay, D. G., Margaretten, W., and Csavossy, I. (1967). An electron microscope study of endotoxin shock in rhesus monkeys. *Surg. Gyneol. Obstet.* **125**, 825.

McLean, A. P. H., Duff, J. H., and MacLean, L. D. (1968). Lung lesions associated with septic shock. *J. Trauma* **8**, 891.

Margaretten, W., and McKay, D. G. (1969). The role of the platelet in the generalized Schwartzman reaction. *J. Exp. Med.* **129**, 585.

Martin, A. M., Jr., Soloway, H. B., and Simmons, R. L. (1968). Pathologic anatomy of the lungs following shock and trauma. *J. Trauma* **8**, 687.

Moon, V. H. (1948). The pathology of secondary shock. *Am. J. Pathol.* **24**, 235.

Murakami, T., Stennis, D. W., and Webb, W. R. (1970). Pulmonary microcirculation in hemorrhagic shock. *Surg. Forum* **21**, 25.

Pattle, R. E. (1965). Surface lining of lung alveoli. *Physiol. Rev.* **45**, 48.

Permutt, S. (1965). Effect of interstitial pressure of the lung on pulmonary circulation. *Medicina Thoracalis* **22**, 118.

Powers, S. R., Burdge, R., Leuther, R., Monaco, V., Newell, J., Sardar, S., and Smith, E. J. (1972). Studies of pulmonary insufficiency in non-thoracic trauma. *J. Trauma* **12**, 1.

Powers, S. R., Jr., Gump, F. E., and Bendixen, H. H. (1970). Trauma workshop report: the lung. *J. Trauma* **10**, 1047.

Ratliff, N. B., Wilson, J. W., Hackel, D. B., and Martin, A. M., Jr. (1970a). The lung in hemorrhagic shock. II. observations on alveolar and vascular ultrastructure. *Am. J. Pathol.* **58**, 353.

Ratliff, N. B., Hackel, D. B., Cruz, P. J., and Wilson, J. W. (1970b). Myocardial and serum electrolyte alterations in dogs in hemorrhagic shock. *Fed. Proc. Fed. Am. Soc. Exp. Biol.* **29**, 422.

Ratliff, N. B., Wilson, J. W., Mikat, E., Hackel, D. B., and Graham, T. C. (1971). The lung in hemorrhagic shock. IV. the role of neutrophilic polymorphonuclear leukocytes. *Am. J. Pathol.* **65**, 325.

Ratliff, N. B., Young, W. G., Jr., Hackel, D. B., Mikat, E., and Wilson, J. W. (1973). Pulmonary injury secondary to extra-corporeal circulation: An ultrastructural study. *J. Thorac. Cardiovasc. Surg.* **65**, 425.

Robb, H. J. (1963) The role of micro-embolism in the production of irreversible shock. *Ann. Surg.* **158**, 685.

Sacks, T., Moldow, C. F., Craddock, P. R., Bowers, T. K., and Jacob, H. S. (1978). Oxygen radicals mediate endothelial cell damage by complement-stimulated granulocytes. *J. Clin . Invest.* **61**, 1161.

Schneeberger-Keeley, E. E., and Karnovsky, M. J. (1968). The ultrastructural basis of alveolar-capillary membrane permeability to peroxidase as a tracer. *J. Cell Biol.* **37**, 781.

Sealy, W. C. (1968). The lung in hemorrhagic shock. *J. Trauma* **8**, 774.

Simpson, D. M., and Ross, R. (1971). Effects of heterologous antineutrophil serum in guinea pigs: hematologic and ultrastructural observations. *Am. J. Pathol.* **65**, 79.

Stallone, R. J. (1970). Mechanism of cardiopulmonary failure in shock. *Rev. Surg.* **27**, 212.

Sugg, W. L., Webb, W. R., and Ecker, R. R. (1968). Prevention of lesions of the lung secondary to hemorrhagic shock. *Surg. Gynecol. Obstret.* **127**, 1005.

Swank, R. L. (1968). Platelet aggregation: its role and cause in surgical shock. *J. Trauma* **8**, 872.

Veith, F. J., Hagstrom, J. U. C., Panossian, A., Nehlsen, S. L., and Wilson, J. W. (1968). Pulmonary microcirculatory response to shock, transfusion and pump oxygenator problems: a unified mechanism underlying pulmonary damage. *Surgery* **64**, 95.

Vye, M. V., and Fischman, D. A. (1970). The morphological alteration of particulate glycogen by en bloc staining with uranyl acetate. *J. Ultrastruct. Res.* **33**, 278.

Wahrenbrock, E. A., Carrico, C. J., Amundsen, D. A., Trummer, M. J., and Severinghaus, J. W. (1970). Increased atelectatic pulmonary shunt during hemorrhagic shock in dogs. *J. Appl. Physiol.* **29**, 615.

Weibel, E. R., and Gil, J. (1968). Electron Microscopic demonstration of an extracellular duplex lining of alveoli. *Resp. Physiol.* **4**, 42.

West, J. B. (1968). *In* "Microcirculation as Related to Shock" (D. Shepro and G. P. Fulton, eds.), pp. 23–40. Academic Press, New York.

West, J. B., Dollery, C. T., and Heard, B. E. (1965). Increased pulmonary vascular resistance in the dependent zone of isolated dog lung caused by perivascular edema. *Circ. Res.* **17**, 191.

Whitaker, A. N., McKay, D. G., and Csavossy, I. (1969). Studies of catecholamine shock. I. disseminated intravascular coagulation. *Am. J. Pathol.* **56**, 153.

Wilson, J. W., Ratliff, N. B., and Hackel, D. B. (1970). The lung in hemorrhagic shock. I. in vivo observations of pulmonary microcirculation in cats. *Am. J. Pathol.* **58**, 337.

Wilwerth, B. M., Crawford, F. A., Young, W. G., Jr., and Sealy, W. C. (1967). The role of functional demand in the development of pulmonary lesions during hemorrhagic shock. *J. Thorac. Cardiovasc. Surg.* **54**, 658.

# EDITORS' SUMMARY TO CHAPTER VIII

In this chapter, Dr. Ratliff approaches the problem of what is often called the shock lung. This syndrome, which achieved prominence during the Vietnam War, has been known by many names in the literature. It has become manifest in recent years as the therapy of various types of shock has improved, resulting in patients who survive longer. The lung weight at autopsy has become increasingly greater. This increased weight is probably due to the accumulation of fluid predominantly in the interstitium and, therefore, the shock lung is also called the wet lung syndrome. The pathophysiology involved in the development of this syndrome has not been worked out. Among the current concepts is the notion that therapy itself, especially in the infusion of large amounts of colloid-free salt solutions may, in part at least, precipitate the occurrence of edema in lungs. Other factors might include anoxia and/or ischemia affecting the cells of the alveolar wall itself and the occurrence of many other predisposing factors that occur in the several types of shock such as platelet factors, leukocytic lysosomal enzymes, changes in pressure relationships, etc. all of which might modify the endothelial barrier including the endothelial junctions.

For example, Barrios *et al.* (1977), studying intercellular junctions in the shock lung in alveolar and capillary epithelium in dogs following hemorrhagic shock, showed that zonulae occludentes (tight junctions) in the epithelium revealed alterations that were not present in controls. These were similar to those reported in such junctions following exposure to osmotic gradients. The appearance of tight junctions in the capillary endothelium which were of a rather poorly organized leaky-type in controls were generally unaltered; however, disintegration and disappearance of junctional strands were occasionally focally observed. The authors suggest that these junctional changes were related to the presumed increase in permeability. It does, however, seem clear as discussed by Dr. Ratliff that changes in the pulmonary circulation, demonstrated within seconds or minutes following the development of experimental shock, are reflected pathologically by such alterations as margination of neutrophilic leukocytes along the pulmonary capillaries. This is accompanied simultaneously by the accumulation of fluid in the interstitium reflected by interstitial edema. This edema is ultimately replaced in the irreversible phase by the development of cell division and by mitosis-promoting factors possibly present in the lymph with fibroplasia and ultimate fibrosis. During this phase an increased DNA content is evident in the lung. Initial damage probably involves Type I alveolar lining cells which undergo degenerative changes with changes in volume and membrane surface area, detachment of cell junctions and extrusion of fluid and cells from the circulation into the interstitium. This is followed by proliferation of Type II alveolar cells which repopulate and, indeed, by changes in surfactant characteristics which have been described.

In later stages the cellular alterations and edema are followed by fibroblastic proliferation and interstitial fibrosis which further accentuate the changes. There have been reports that administration of glucocorticoids may ameliorate the process. The mechanism is unknown; however, according to Dr. Ratliff's interpretation it is reasonable to believe that the steroids may affect the leukocyte degranulation and endothelial and Type I epithelial degeneration more than any direct effect on the lysosomes per se.

One interesting point which also relates to Chapter X is that there is evidence that endothelial change in the lung following shock is correlated with the changes of the endothelial cells in skeletal muscle. This might suggest that by assessment the relatively noninvasive procedure of skeletal muscle may be able to estimate the degree of damage to the lung which in conjunction with modern therapy could result in improved treatment.

The role of membrane changes in microcirculatory alterations probably cannot be overestimated. Such changes doubtlessly involve parallel changes in the cytoskeleton and are also linked to changes in ion and water shifts (see summary to Chapter 1). Webb and Brunswick (in press) have recently emphasized the role of endothelial changes in the microcirculation especially in various types of shock, confirming or at least supporting the point of view of Dr. Ratliff in this chapter.

## References

Barrios, R., Inoue, S., and Hogg, J. C. (1977). *Lab. Invest.* **36,** 628.

Schlog, G., Voigt, W. H., Schnells, G., and Glatzel, A. (1977). *Anesthetic* **26,** 12.

Webb, W. R., and Brunswick, R. A. (in press). *In* "Pathophysiology of Shock, Anoxia and Ischemia" (R. A. Cowley and B. F. Trump, eds.). Williams & Wilkins, Baltimore, Maryland.

# CHAPTER IX

# PATHOLOGY OF JAUNDICE AND CHOLESTASIS AT THE ULTRASTRUCTURAL LEVEL

Kyuichi Tanikawa



Jaundice has been a classical symptom of hepatic disorders for clinicians, and also has been an important topic of research for investigators who are interested in the pathophysiology of hepatic diseases.

Recent advances in the physiology and biochemistry of bilirubin metabolism have lead to better understanding of constitutional hyperbilirubinemia, and our accumulated knowledge about bile excretion has helped to clarify the mechanism of cholestasis.

Electron microscopy has also contributed much to this field by relating morphologic aspects to biochemical and physiological alterations.

In this chapter, ultrastructural aspects of jaundice and cholestasis will be reviewed in conjunction with our experiences in this field.



PATHOBIOLOGY OF CELL MEMBRANES, VOL. II

ISBN 0-12-701502-7

## I. Bilirubin Metabolism, Bile Excretion, and the Related Fine Structure of the Liver

Bilirubin originating from hemoglobin of a senescent erythrocyte is taken up by a hepatocyte mainly along the plasma membrane facing the space of Disse. The hepatocyte seems to have a selective capacity to remove unconjugated bilirubin and other organic anions very rapidly from the plasma. At the plasma membrane, which is permeable to nonpolar molecules, the unconjugated bilirubin become separated from albumin. Specific intracellular protein, known as ligandin, and Z protein (Levi *et al.*, 1969) act as acceptors and facilitate the transfer of bilirubin and other organic anions into the hepatocyte across the plasma membrane by nonionic diffusion (Goresky and Back, 1970).

Thus,ligandin and Z protein appear to regulate hepatocytic uptake of bilirubin, as well as its subsequent biotransformation to conjugated derivatives and its excretion into the bile canaliculus.

Deficiency of ligandin and/or Z protein may be related to one variety of Gilbert's disease, said to be caused by disturbance in bilirubin uptake.

Morphologically the perisinusoidal space or space of Disse is in free communication with the sinusoidal lumen through numerous fenestrae in the endothelial cells (Fig. 1a). In normal human liver, only a few reticulum fibers are noted in the space and basement membrane formation is absent.

The sinusoidal surface of the plasma membrane extends numerous microvilli into the space, facilitating efficient exchange of materials at the plasma membrane between the hepatocyte and plasma (Fig. 1b). Thus, accumulation of fibrous materials or basement membrane formation in the space of Disse, or a decreased number of fenestrae in the endothelial cell would inhibit the exchange of materials including organic anions such as bilirubin between the hepatocyte and plasma. This exchange or uptake would also be inhibited by loss or stunting of the microvilli. These morphological findings would explain some of the unconjugated hyperbilirubinemia seen in Gilbert's disease or cirrhosis.

After its entry into the hepatocyte, bilirubin is transfered to the smooth endoplasmic reticulum (SER) where it is converted into polar conjugates. The major biotransformation of bilirubin in man involves conjugation with glucuronic acid, which is catalyzed by UDPglucuronyltransferase, located in the SER. Following administration of phenobarbital UDPglucuronyltransferase becomes markedly increased in association with predominant proliferation of the SER (Remmer and Merker, 1963). It seems likely that this change is at least partially responsible for the improvement of unconjugated hyperbilirubinemia in Crigler–Najjar syndrome or in one variety of Gilbert's disease in which glucuronyltransferase is partially deficient.

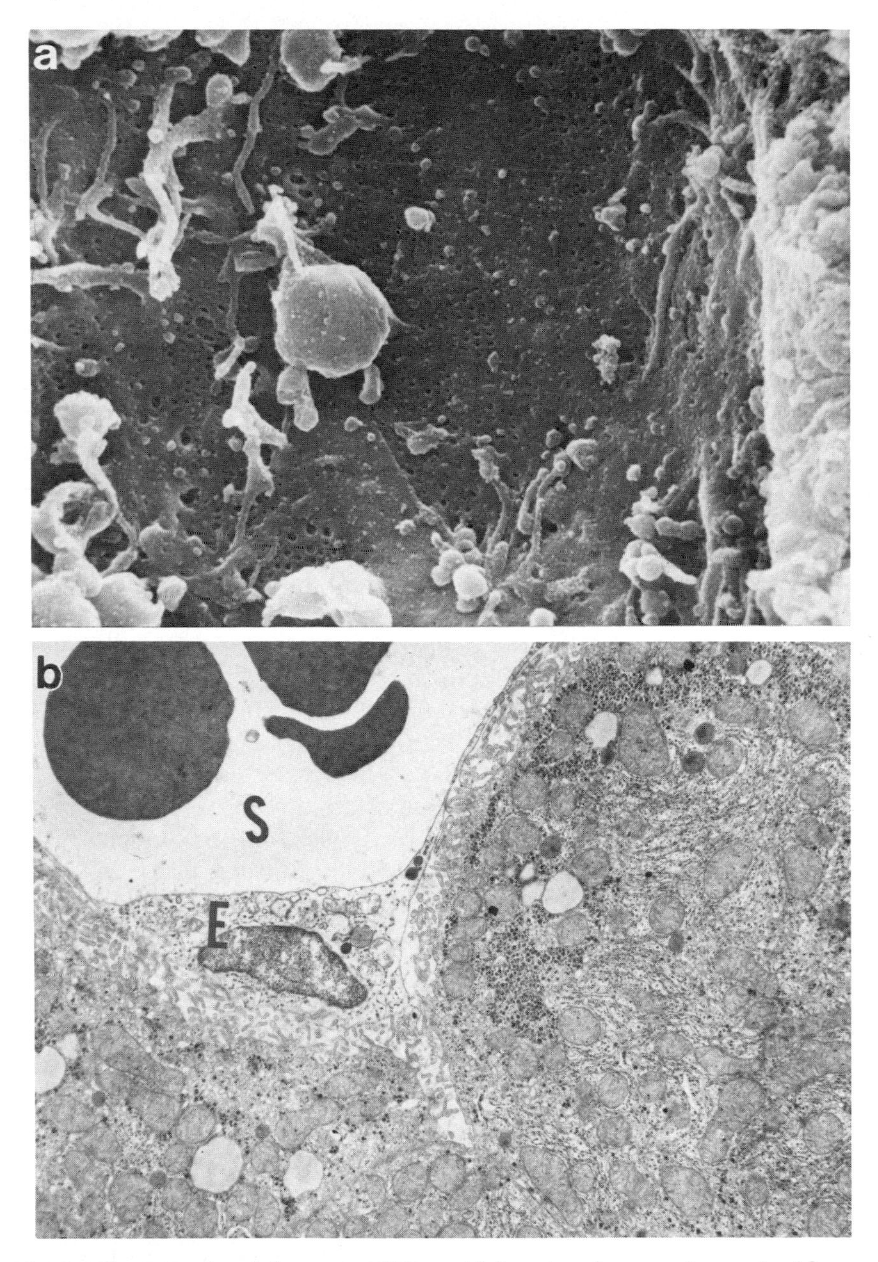


**Fig. 1.** The sinusoid and the space of Disse. (a) Scanning electron micrograph of human sinusoid. Numerous fenestrae measuring about 0.1 $\mu$m in diameter are observed on the wall of the endothelial cells, forming sieve plates. ×10,000. (b) Normal rat liver. Hepatocytes extend numerous microvilli into the space of Disse. Sinusoid, S; endothelial cell, E. ×7000.

A recent study indicates that two separate enzyme systems seem to be involved in the formation of bilirubin diglucuronide. The first step occurs in the SER which catalyzes the formation of bilirubin monoglucuronide. This bilirubin monoglucuronide either is excreted directly into the bile canaliculus or is converted in a second step to bilirubin diglucuronide by transferase located in the canalicular membrane before the excretion (Jansen *et al.*, 1977). The diglucuronide is a major portion of bilirubin in normal human bile, with the monoglucuronide representing a minor portion (Fevery *et al.*, 1972; Gordon *et al.*, 1976).

Both mono- and diglucuronide have been recently isolated in a pure form (Jacobson, 1969; Ostrow and Murphy, 1970). Thus, monoglucuronide is not a complex of unconjugated bilirubin and bilirubin glucuronide. Various nonglucuronide bilirubin conjugates such as sulfate (Noir *et al.*, 1970), glucose, or xylose (Fevery *et al.*, 1972) have been described. Although their functional significance remains uncertain, they seem to have little physiological importance in man since in disorders characterized by a UDPglucuronyltransferase deficiency they do not appear to be increased (Arias *et al.*, 1969).

Bilirubin glucuronide is rapidly excreted from the hepatocyte into the bile canaliculus. This process which takes place against a concentration gradient had been postulated to be carrier-mediated and energy-dependent. Bilirubin IX is not a linear tetrapyrrole as thought previously, having an involuted intramolecular hydrogen-bonded structure (Fog and Jellum, 1973; Bennett and Davis, 1976) and needing to be conjugated for biliary excretion. However, its isomers such as bilirubin IX$\beta$, IX$\gamma$, and IX$\delta$ seem to be excreted into the bile canaliculus in the unconjugated form (Blanckaert *et al.*, 1977). These isomers are more polar because such involuted intramolecular hydrogen-bonded structures are not possible. Biliverdin IX$\gamma$ also does not seem to require conjugation for excretion (Colleran and O'Carra, 1970). Thus, it seems likely that no specific carrier is needed in the canalicular membrane for bilirubin excretion, and bilirubin excretion across the canalicular membrane depends more on the molecular structure of bilirubin.

The biliary solids including bilirubin presumably pass the Golgi zone or directly from the SER to pericanalicular ectoplasm (Biava, 1964a), probably in vesicles. The Golgi apparatus probably plays a role in bile excretion by analogy with its function in other secretory cells. However, this function in the hepatocyte is still unclear, although morphometric studies suggest this possibility (Bianchi *et al.*, 1971).

Although bilirubin is conventionally considered the characteristic component of bile, the bile salts are the most important element in the regulation of bile excretion. Bile acids stimulate bile production. An apparently linear relation between bile acid excretion rates and bile flow has been demonstrated in man and animals. The main driving force for the movement of water and other solutes into the canalicular lumen is considered to be provided by the osmotic gradient created by the excretion of bile acids into the canalicular lumen. In addition, there are two types of a bile acid-independent bile flow. One is probably mediated through a Na+, K+-activated ATPase system along the canalicular membrane (Erlinger *et al.*, 1970). The other occurs in the bile ductules or duct chiefly in response to secretin.

The bile salts, formed mainly in the SER of the hepatocyte derive from cholesterol. Cholesterol is sterically transformed to trihydroxycholic acid by $7\alpha$ and $12\alpha$ hydroxylation of a hydroxy group to $3\gamma$ and to a lesser extent to dihydroxychenodeoxycholic acid by $7\alpha$ hydroxylation alone. These hydroxylations occur in the SER. Side chain oxidation to C-24 occurs in the mitochondria. When this side chain oxidation occurs first, ring hydroxylation is inhibited and monohydroxy bile acids are formed. If the side chain oxidation is completed after $7\alpha$ and $12\alpha$ hydroxylation, cholate is formed.

In normal human bile the bile salts are conjugated by microsomal enzymes containing taurine or glycine in a three to one ratio.

The bile salts exist in a polyanionic form as spherical molecular aggregates or micelles, rather than in a monomolecular form in the bile, and probably also in the cytoplasm of the hepatocyte. These micelles, whose site of formation is unknown represent mixtures of bile salts, phospholipid, and cholesterol.

The biliary tract begins with the bile canaliculus. The bile canaliculus lies between two or occasionally three hepatocytes and is lined by a specialized part of the cell membrane of the hepatocytes. The cell membrane facing the canaliculus extends fingerlike microvilli into the lumen (Fig. 2a). In the centrolobular area of normal human liver, the lumen appears about 1 $\mu$m in diameter (average area 1.77 $\mu m^2$) and its microvilli are about eight in number and approximately 0.37 $\mu$m in length on cross-section (Miyakoda, 1975). Freeze-fracture replica studies revealed that the fracture surface of the microvilli are diffusely studded with fine granules measuring about 60–100 Å in diameter (Miyai *et al.*, 1974).

This canalicular membrane represents approximately 13% of the total cell membrane area of the hepatocyte (Weibel *et al.*, 1969) and is estimated at about 10 $m^2$ for a whole human liver (Erlinger and Dhumeaux, 1974). The cell membranes of adjacent hepatocytes on each side of the canaliculus are firmly bound to each other by structures known as junctional complexes. Thus, in normal liver, the bile canaliculus does not communicate directly with the intercellular space. However, the intercellular space is in free communication with the space of Disse.

The canalicular membrane is covered with a relatively prominent fuzzy polysaccharide layer: the glycocalix. The functional significance of the

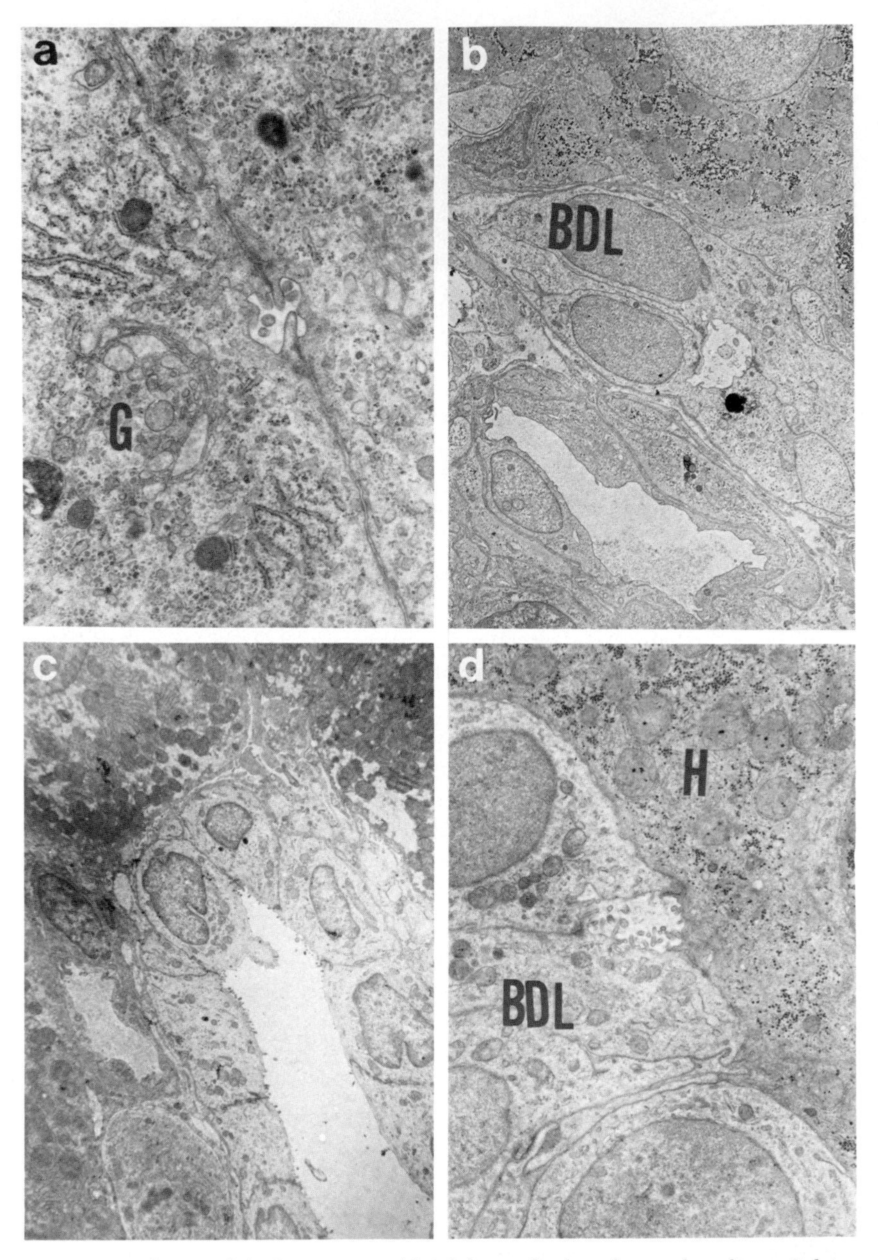


**Fig 2.** Intrahepatic bile duct system. (a) A bile canaliculus of normal rat liver. Golgi apparatus, G. ×8000. (b) A bile ductule (BDL) of normal human liver. ×3000. (c) An interlobular bile duct of normal rat liver. ×2000. (d) A canalicular-ductular junction of normal human liver. Hepatocyte, H; ductular epithelial cell, BDL. ×5000. (From Tanikawa, 1979.)

glycocalix is unknown. However, it may be closely related with canalicular enzymes, such as ATPase and 5′-nucleotidase which appear to be involved in the excretion of bile. Recently it was made possible to separate the bile canalicular membrane from the canaliculus-free plasma membrane for biochemical examinations, and electron microscopic observations and studies on the compositions of lipids and proteins and activities of enzymes in the bile canalicular membrane show that they are different from those of the canaliculus-free plasma membrane, suggesting the different physiological functions between the two membranes (Fisher *et al.*, 1975; Wisher and Evans, 1975; Toda *et al.*, 1975).

The pericanalicular ectoplasm, a narrow layer of homogeneous, condensed, and organelle-free zone, is noted around the bile canaliculus. A recent study using ruthenium red demonstrated a well-developed microfilamentous web around the bile canaliculus, which may regulate the caliber of the canalicular lumen in addition to providing structural stability (Oda *et al.*, 1974). Under the electron microscope, microfilaments appear to be inserted in the microvillous membranes and also in the zonulae adhaerens of the junctional complexes (Fig. 3), and are decorated by heavy

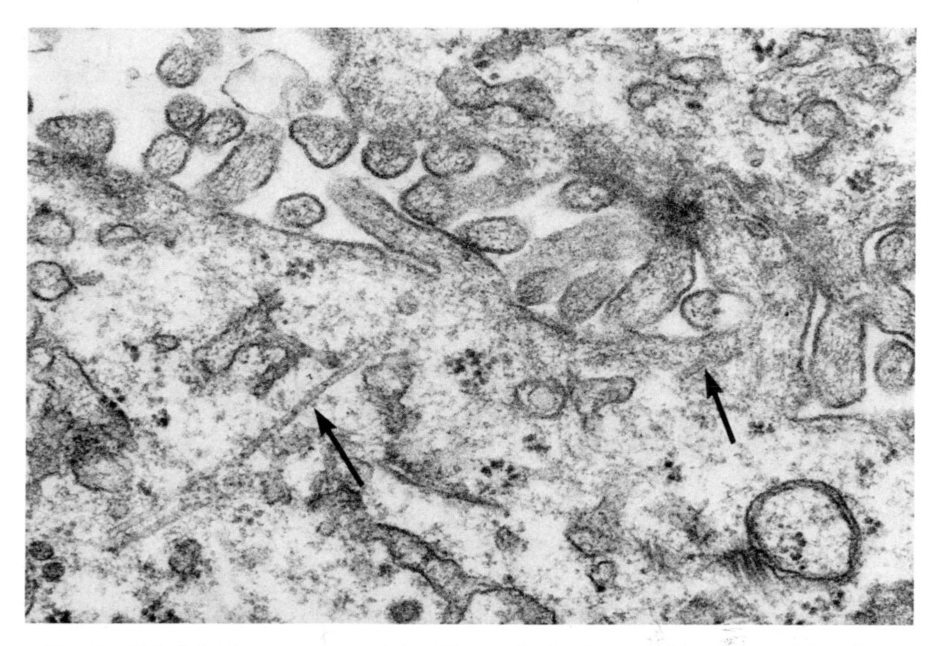

**Fig. 3.** Cytoskeletal system around the bile canaliculus in normal hepatocyte. Microfilaments are observed around the bile canaliculus. They appear to insert in the canalicular microvilli and also in the zone adhaerens. Several microtubules (at arrows) are also seen in the vicinity of the bile canaliculus. ×42,000.

meromyosin (French and Davies, 1975; Fig. 4). The role of the microfilaments in bile excretion had been focused on because their destruction causes cholestasis (discussed later, IIIA). Microtubules are also observed around the bile canaliculus (Fig. 3). Although their exact role in bile excretion is not clear, biliary lipid excretion and bile flow are inhibited by the microtubular blocking agents, vinblastine and colchicine. Microtubules may participate in the translocation of lipids from the endoplasmic reticulum to the canalicular membrane (Gregory *et al.*, 1978).

At present, however, the mode of bile excretion into the bile canaliculus cannot be identified morphologically.

The bile canaliculi which form grooves in the one-cell thick hepatocyte plates communicate with each other to form a continuous canalicular mesh and drain into larger bile ductules near the portal tract and then to form a bile duct in the portal area.

The bile ductule or interlobular bile duct is lined by biliary epithelial cells, which differ morphologically from the hepatocyte by having fewer organelles. The biliary epithelial cell, cuboidal or columnar in shape, appears less dense than the hepatocyte and contains smaller mitochondria, less well-developed endoplasmic reticulum, and a relatively large nucleus. However,

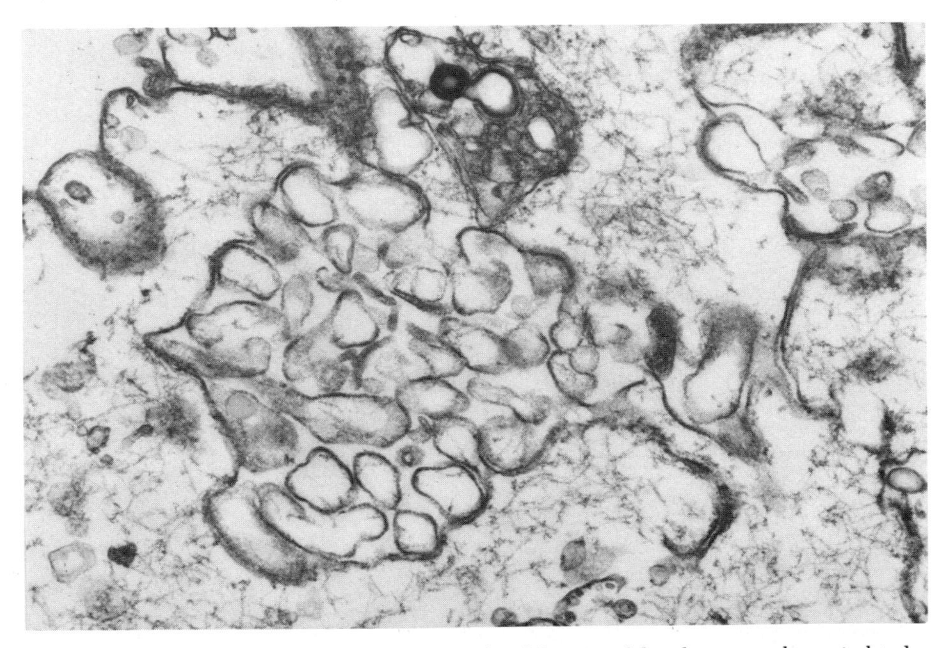

**Fig. 4.** Microfilaments in the canaliculus-rich subfraction of the plasma membrane isolated from normal liver. Microfilaments are well decorated with heavy meromyosin. ×24,000.

in the Golgi apparatus, usually located between the nucleus and the apical pole, well developed and free ribosomes are abundant, and fine fibrils are present in groups (Fig. 2b and c). Numerous microvilli, projecting from the apical pole of these cell into the lumen, appear short in length and occasionally edematous in shape. The cohesion among the biliary epithelial cells is achieved by tight junctions generally located in the apical zone of the cell, and the membranes of adjacent cells appear to be well interdigitated.

The bile ductule or bile duct is surrounded by a basement membrane of varying thickness. The cell membrane facing the basal space shows pinocytic activity under the electron microscope, suggesting some transport of materials between the biliary epithelial cell or biliary lumen and surrounding tissue space. In other words, biliary epithelial cells in the area have active excretory and adsorptive functions. A bicarbonate-rich fluid and mucoprotein seem to be actively excreted into the lumen.

The connection between the bile canaliculus and the smallest perilobular bile ductule has long been under discussion. Most observers, however, recognize the transitional part, which is composed of hepatocytes and ductular epithelial cells. This has been called the ampulla, canal of Hering, intermediate piece, and recently canalicular–ductular junction by Schaffner and Popper (1961). In this part, the lumen is lined by both biliary epithelial cells and hepatocytes (Fig. 2d). The epithelial cells are only separated by the adjacent tissue by a basement membrane. Junctional complexes are observed between the cell membranes of the biliary epithelial cell and of the hepatocyte. No particular dilation of the lumen in this part is noted. Thus, the term *ampulla* seems unsuitable.

The bile ductule most frequently located in the perilobular or portal area is lined by a rosette of three or four ductular epithelial cells and is not usually accompanied by tributaries of the hepatic artery or portal vein, in contrast to the bile duct in the portal tract.

## II. Constitutional Hyperbilirubinemia

At present, four clinical entities of diseases manifested by jaundice and based on genetic disorders of bilirubin metabolism have been identified. They are Dubin–Johnson syndrome, Rotor's hyperbilirubinemia, Gilbert's disease and Crigler–Najjar syndrome. The first two are characterized by conjugated hyperbilirubinemia and the last two by unconjugated hyperbilirubinemia. Crigler-Najjar syndrome is clinically seen only in the infant.

For many years congenital hyperbilirubinemia has received much attention from many investigators interested in the mechanism of jaundice. Among the many approaches to the understanding of their mechanism, electron microscopic studies have been increasingly important and the fine structural alterations they have shown seem to be well correlated with biochemical, hemodynamic or clinical changes seen in these disorders. Recently, morphometric analysis of ultrastructural changes has been well established and our recent investigations (Tanikawa *et al.*, 1971; Miyakoda, 1975) depended mainly on such analysis.

### A. *Dubin-Johnson Syndrome*

Dubin-Johnson syndrome is a chronic, benign, intermittent jaundice with conjugated hyperbilirubinemia. Under the light microscope, the liver biopsy specimens appear normal except for excess pigment deposition in the hepatocyte.

In this disease, in addition to bilirubin diglucuronide, test dyes such as bromosulphalein (BSP) and gallbladder contrast media are delayed in excretion to the bile canaliculus. The $T_m$ for BSP is greatly reduced but its storage and conjugation are intact. From such clinical observations this disease is considered to be defective excretion of organic anions other than bile salts into the bile canaliculus.

Under the electron microscope (Fig. 5), numerous electron-dense granules, round or oval in shape and measuring from 0.5 to 2.5 $\mu$m in diameter, are distributed in the cytoplasm of the hepatocyte and more frequently around the bile canaliculus. These granules are delimited by a single membrane and their matrix contains a great many ferritinlike fine grains with larger dense particles (Ichida and Funahashi, 1964; Tanikawa, 1965). They resemble lipofuscin bodies in structure, but are morphologically distinguishable from them by their prominent lipid component (Toker and Trevino, 1965; Tanikawa *et al.*, 1971). They have been considered to be closely related with lysosomes because of their high acid phosphatase activity (Essner and Novikoff, 1960; Baba and Ruppert, 1972). These granules have been shown to be melanin by histochemical or infrared spectrometric studies (Wegmann *et al.*, 1960; Johnson, 1970). However, the relationship between a defect of bilirubin metabolism seen in this disease and melanin deposition is not well understood. Similar granules are also observed in the Kupffer cells and portal macrophages (Toker and Trevino, 1965). These granules may have resulted from the ingestion by these cells of material liberated from the hepatocytes. In cases with relatively high serum bilirubin level, electron-dense laminated granules in vacuoles are noted around the bile canaliculus (Tanikawa, 1965; Barone, 1966). They may be nonspecific.

The mitochondria in the hepatocyte generally appear fairly normal, however, the SER is somewhat increased in the form of small vesicles and the rough endoplasmic reticulum (RER) appears to be somewhat reduced in

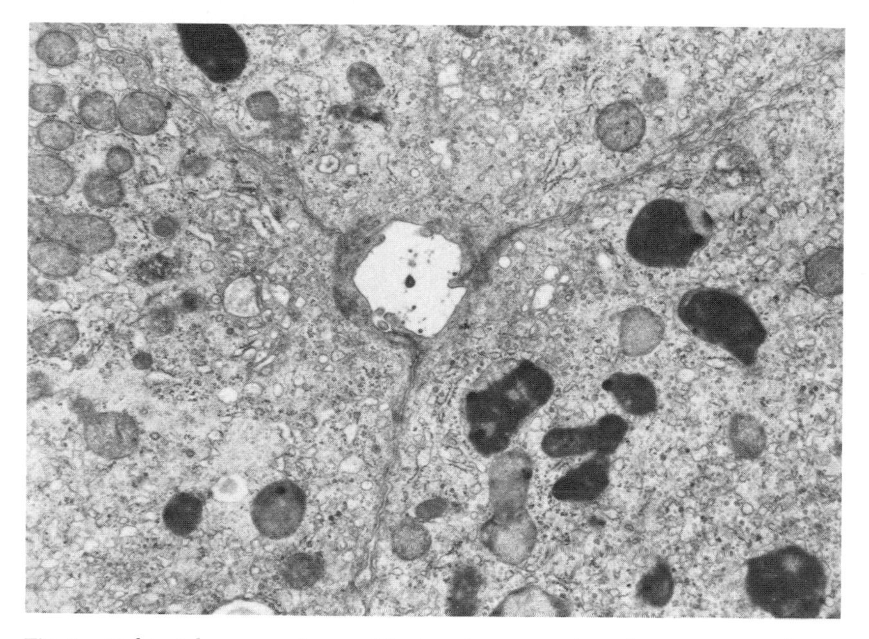

**Fig. 5.** Dubin–Johnson syndrome. Electron-dense granules, characteristic to this syndrome, are observed around the bile canaliculus, which is moderately dilated with loss of its microvilli and contains electron-dense granules in the lumen ×9000. (From Tanikawa, 1979.)

number. The cell membrane and the Golgi apparatus are well preserved. Inferrera and Motta (1965) reported that centrioles were frequently noted in this disease.

According to our observations, characteristic alterations were seen in the canaliculus. Most of the bile canaliculi are more or less dilated with stunt or loss of their microvilli (Tanikawa, 1965; Pages and Baldet, 1969; Kobayashi, 1971). Morphometric study (Tanikawa *et al.*, 1971; Miyakoda, 1975) indicates definite dilatation of the canalicular lumen with fewer and shortened microvilli compared to normal ones or those in Gilbert's disease. In addition, electron-dense bodies, usually oval or round, are often observed in the lumen (Tanikawa *et al.*, 1971). These bodies are seen in about half of the bile canaliculi and sometimes in the intercellular space; their origin is not clear. However, they may be evacuated dense granules from the hepatocyte to the canalicular lumen. Further study on the bile of this syndrome would be of interest because of frequent appearance of these bodies in the canaliculus.

The alterations of the bile canaliculus are important in etiological consideration. Though it is not clear whether these canalicular changes are due to congenital abnormalities or secondary ones, such alterations are highly suggestive of impaired excretion of certain bile component such as bilirubin. Toker and Trevino (1965), however, having found no abnormalities in the bile canaliculi or ductular cell, postulate that the secretory process within the hepatocyte is interrupted at a stage subsequent to conjugation.

Similar fine structural alterations have been also observed in mutant Corridale sheep with Dubin–Johnson syndrome (Cornelius *et al.*, 1968).

### B. *Gilbert's Disease*

The term *Gilbert's disease* is, at present, applied to a heterogeneous group of benign disorders, which are characterized by relatively low-grade chronic unconjugated hyperbilirubinemia not due to over hemolysis.

From a pathogenic standpoint, this disease could be divided into at least two types. One variety of Gilbert's disease is considered to result from a disturbance of bilirubin uptake on the cell membrane facing the space of Disse. Another one is due to a partial defect of bilirubin conjugation in which glucuronyltransferase is deficient. However, these two types have not been clearly distinguished on a clinical basis. In addition, fine structural differences in the liver between the two types have not been well studied.

Under the light microscope, the hepatocyte appears normal. However, under the electron microscope (Fig. 6), obvious alterations are observed in the cell membrane facing the space of Disse and the endoplasmic reticulum in the hepatocyte. The hepatocytic cell membrane facing the space of Disse appears to be fragile (Simon and Varonier, 1963) and frequently flattened with loss or stunting of its microvilli associated with increased accumulation of the reticulum fibers in the space (Tanikawa and Emura, 1965; Kobayashi, 1971; Akeda *et al.*, 1973). Decrease in number of the microvilli along the cell membrane facing the space of Disse has been also shown by morphometric analysis (Miyakoda, 1975).

The cell membrane facing the space of Disse normally has numerous microvilli, which effectively increase the uptake surface. The changes in this membrane imply a defect in hepatocytic uptake of some materials from the plasma in this disease.

Giant mitochondria with paracrystalline inclusions are seen relatively often in the hepatocyte especially near the cell membrane facing the space of Disse (Minio and Gauter., 1966; Schaff *et al.*, 1969; Tanikawa *et al.*, 1971). This could suggest some defect of mitochondrial energy supply to the cell membrane.

The SER is markedly increased in number in the hepatocyte (Sasaki and Ichida, 1961; Tanikawa and Emura, 1965), and its cytoplasm appears to be mostly occupied by them. Under high-power magnification, these smooth surfaced profiles of the endoplasmic reticulum appear mostly to be vesicular or dilated in shape, and few tubular forms are observed. The RER, on the

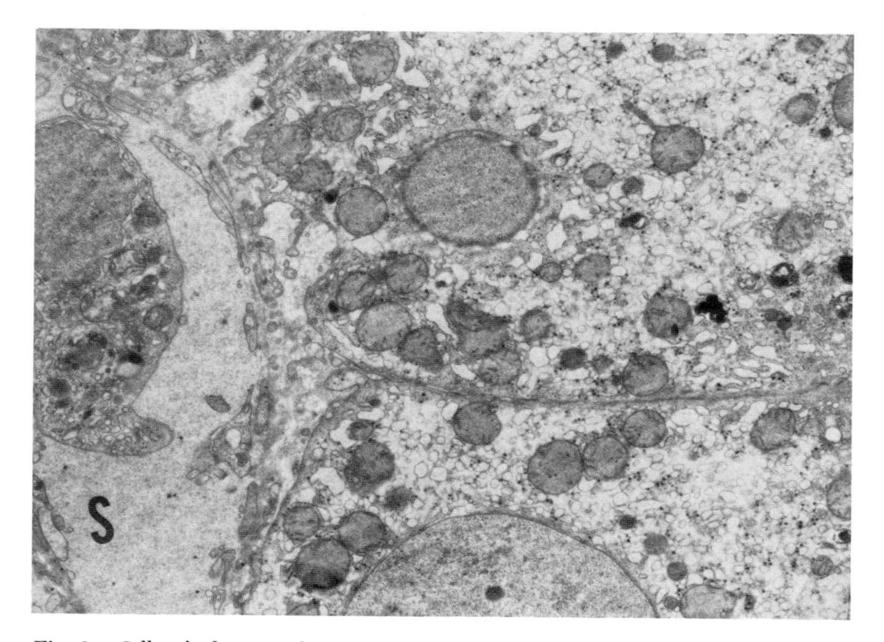


**Fig. 6.** Gilbert's disease. The cytoplasm of hepatocyte is mostly occupied by the smooth endoplasmic reticulum (SER), which appear to be vesicular and dilated. The plasma membrane of the hepatocyte facing the space of Disse is flattened with loss of microvilli. Sinusoid, S×6000. (From Tanikawa, 1979.)

other hand, are reduced in number and seen only among the mitochondria, but are normal in appearance. Similar changes have been observed in Gunn rats by Novikoff and Essner (1960). Though it is not clear whether such changes of the SER could be a manifestation of hypoactive hypertrophy or a result of cytoplasmic response to increase the site of bilirubin conjugation, such alterations of the SER could be closely related with some defect of bilirubin conjugation in one variety of this disease.

The excess accumulation of lipofuscin granules are often observed in the cytoplasm of the hepatocyte (Herman *et al.*, 1964; Minio and Gauter, 1966; Scaff *et al.*, 1969; Barth *et al.*, 1971; Akeda *et al.*, 1973), however, its significance is unknown. Brown and Smuckler (1970) noted peroxisomes with electron-dense crystalloids in siblings affected with alkaptonuria and Gilbert's disease.

### C. *Crigler–Najjar Syndrome*

The Crigler–Najjar syndrome is an extremely rare form of familiar, nonhemolytic jaundice with unconjugated hyperbilirubinemia in infants and

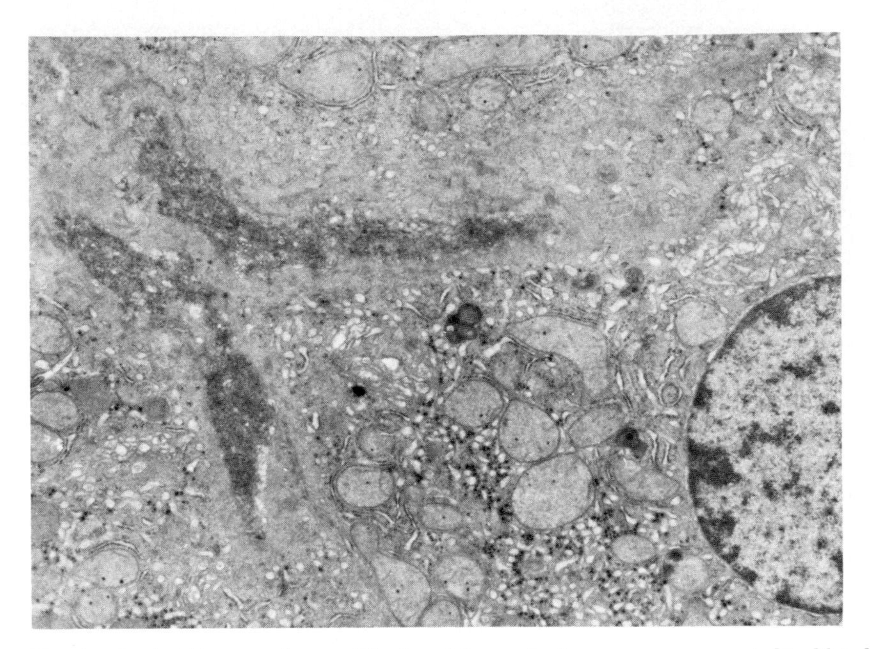

**Fig. 7.** Crigler–Najjar syndrome. Bile thrombi, granular in appearance, are noted in dilated bile canaliculi. ×8000. (By courtesy of Dr. K. Yamaguchi.)

young children, and deficiency of glucuronyltransferase can be demonstrated in the liver. Liver biopsy specimens reveal no remarkable changes except for intracanalicular bile thrombi under the light microscope.

There are only a few fine structural studies on the liver of this disease. Minio-Paluello and associates (1969) reported an enlargement of intercellular space which may be an expression of immaturity.

Bile canalicular changes and bile thrombi in the lumen, which have been examined under the electron microscope by Yamaguchi *et al.* (1975) (Fig. 7), are essentially the same as the ones in other types of cholestasis. The mechanism of cholestasis in this disease is not known. This could be related with immaturity of bile acid metabolism in the infant liver. Examinations of bile acids in infant would be of great interest.

### *D. Rotor Syndrome*

The Rotor syndrome is a chronic, familiar, conjugated hyperbilirubinemia. Although this syndrome is similar to Dubin–Johnson syndrome, under the light microscope, liver biopsy specimens appear normal with absence of pigment deposition in the hepatocyte. The removal of both unconjugated and conjugated bilirubin, infused intravenously, is considerably delayed as compared with removal in a normal subject. The $T_m$ and storage for BSP are both reduced in this syndrome. These data suggest a diminished uptake of bilirubin by the hepatocyte as well as a defect in excretion of conjugated bilirubin into the bile canaliculus.

Under the electron microscope (Fig. 8), several interesting findings are observed in spite of normal light microscopic appearances. Organelles in the hepatocyte are generally well preserved, but megamitochondria with numerous cristae and myelinlike structure are often noted (Molbert and Marx, 1966; De Brito *et al.*, 1966; Tanikawa *et al.*, 1971; Kobayashi, 1971; Czarnecki *et al.*, 1973), and these mitochondria occasionally reach the size of the nucleus.

The bile canaliculus is frequently altered with mild dilation of the lumen and loss or stunt of its microvilli (Tanikawa *et al.*, 1971; Kobayashi, 1971; Akeda *et al.*, 1973). Such canalicular alterations, though less prominent than those in Dubin–Johnson syndrome, may indicate an excretory disturbance of some bile components in this disease.

Intercellular space is occasionally widened with intact junctional complexes around the bile canaliculus (Molbert and Marx, 1966; Tanikawa *et al.*, 1971). Formation of microvilli on the cell membrane facing widened intercellular space is also noted. The cell membrane facing the space of Disse shows some loss and stunt of its microvilli associated with some increased accumulation of reticulum fibers in the space (Tanikawa *et al.*, 1971; Akeda *et al.*, 1973). However, these changes are not remarkable as compared with Gilbert's disease.

Those changes seen in this disease definitely seem different from those in Dubin–Johnson syndrome. Thus, Rotor syndrome and Dubin–Johnson syndrome may each have a different congenital basis, although some deny any pathogenic distinction between two diseases (Patrassi *et al.*, 1965).

From fine structural observations, it could be concluded that Rotor syndrome has some morphological abnormalities of both Dubin–Johnson syndrome and Gilbert's disease. In other words, this syndrome seems to be situated between the two.

The fine structural changes of congenital hyperbilirubinemia are summarized in Table I.

Recent techniques like subfractionation of the plasma membrane, scanning electron microscopy, etc. would provide us with more informations on these constitutional hyperbilirubinemia and mechanism of jaundice in these disorders would be more clarified. Especially, the cytoskeletal system of the hepatocyte is one of the most interesting subjects to investigate in these congenital diseases.

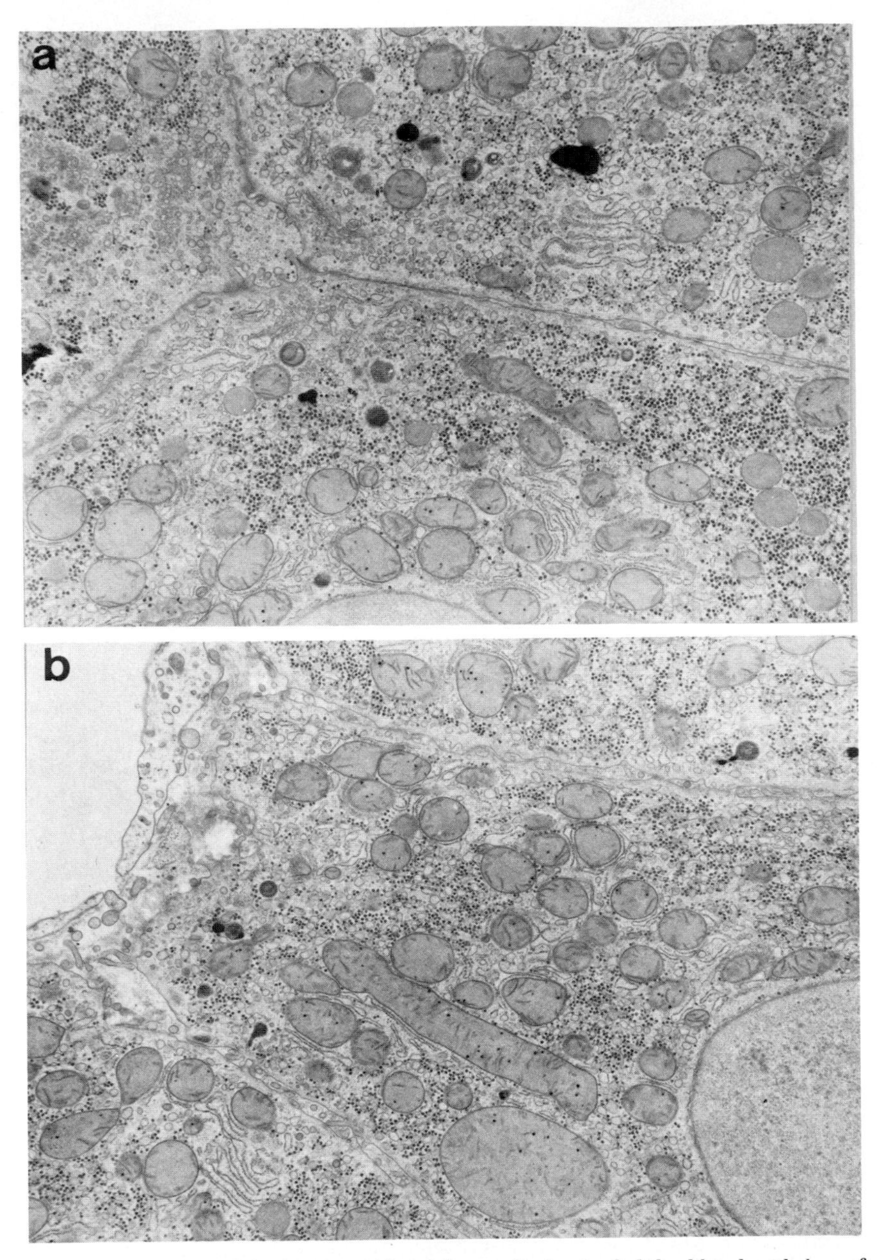


**Fig. 8.** Rotor's hyperbilirubinemia. (a) A bile canaliculus is slightly dilated with loss of microvilli. ×8000. (b) The hepatocytic plasma membrane facing the space of Disse is flattened with moderate loss of microvilli. The intercellular space is slightly dilated with development of microvilli, which extend from the lateral surface of the hepatocyte. Giant mitochondria are noted in the cytoplasm of the hepatocyte. ×8000. (From Tanikawa, 1979.)

TABLE I

*Characteristic Fine Structure of Liver in Constitutional Hyperbilirubinemia*[a]

| | *Gilbert* | *D-J* | *Rotor* |
|---|---|---|---|
| Hepatocyte | | | |
| Mitochondria | Giant mitochondria (not frequent) | Normal | Giant motochondria (frequent) |
| Smooth surfaced ER | Markedly increased vesicular | Slightly increased | Slightly increased |
| Glycogen particles | Scattered | Normal | Normal |
| Lysosome | Slightly increased number of lipofuscin granules | Characteristic granules | Slightly increased number of lipofuscin granules |
| Plasma membrane | Flattened cell membrane facing space of Disse | Normal | Flattened cell membrane facing space of Disse |
| Bile canaliculus | Normal | Decreased number of microvilli<br>Slightly dilated lumen<br>Ring-formed dense granules in lumen | Decreased number of microvilli |
| Space of Disse | Increased fiber | Normal | Increased fiber |
| Intercellular space | Normal | Normal | Slightly dilated |
| Kupffer cell | Normal | Characteristic granules | Normal |

[a] Gilbert, Gilbert's disease, D-J, Dubin-Johnson syndrome; Rotor, Rotor's hyperbilirubinemia.

## III. Cholestasis

Cholestasis has been defined as a stagnation of bile in the liver with an accumulation in the blood of all constituents normally excreted in the bile. From the etiological and pathogenic points of view, cholestasis is divided into two groups; extrahepatic and intrahepatic cholestasis. However, under the electron microscope no remarkable differences between the two have so far been noted at the hepatocytic or bile canalicular level.

### A. *Fine Structure of Cholestasis and Proposed Mechanisms of Intrahepatic Cholestasis*

In cholestasis, the bile canaliculus appears to be mostly dilated with partial or complete disappearance of its microvilli. The dilation is variable in degrees ranging from 1 to 10 $\mu$m in diameter (Fig. 9a). Sometimes, edematous microvillar bleb appears to occlude the canalicular lumen (Steiner and

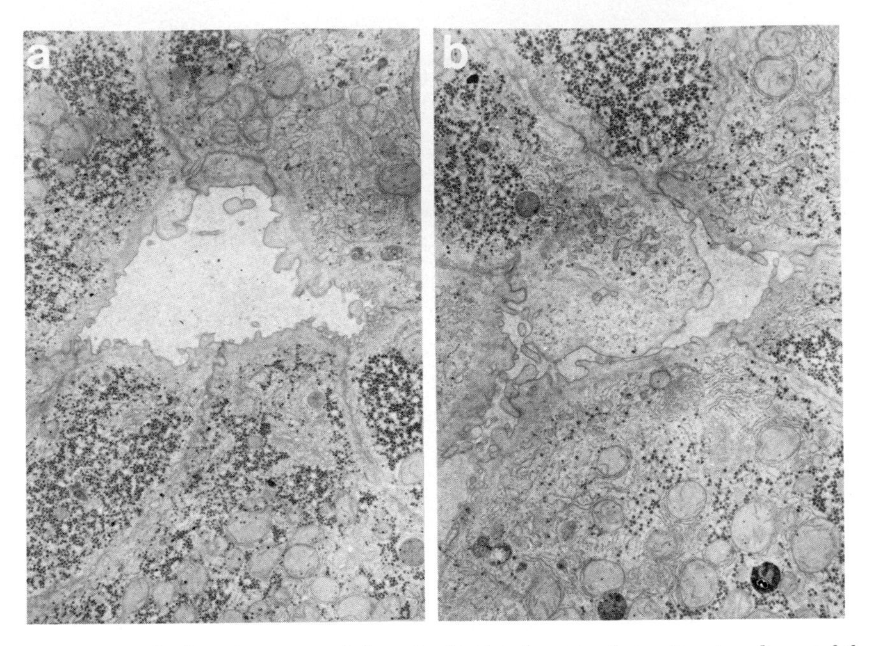


**Fig. 9.** Canalicular changes in cholestasis. (a) Extrahepatic obstructive jaundice. A bile canaliculus is markedly dilated with loss of microvilli. The pericanalicular ectoplasm appears to be thickened. ×4500. (b) Extrahepatic obstructive jaundice. A large edematous microvillus, projecting into the lumen, occupies the most of the canalicular lumen. ×4500. (From Tanikawa, 1979.)

Carruthers, 1961; Fig. 9b). The microvilli derived from the marginal ridge, however appear to be intact in cholestasis (Vial *et al.*, 1976). An increased number of canalicular sections has been noted (Steiner *et al.*, 1963). This could be caused by increased canalicular tortuosity in cholestasis. Focal dilation or evagination of canaliculi often noted in cholestasis has been described as canalicular diverticulosis (Steiner *et al.*, 1963).

Dilated canalicular lumen may be empty or filled with bile thrombi. Canalicular bile thrombi show many different appearances; homogeneous, granular, crystalline, lipidlike, lamellar, or whorled (Fig. 10). Bile thrombi seem to contain bile materials and altered canalicular membranes. However. at present, how each component in the bile thrombus appears respectively in fine structure is not known. The pericanalicular ectoplasm usually appears thickened and frequently shows focal swelling and disruption suggesting that bile plugs arise at least in part from the leakage of ectoplasmic material in the canalicular lumen (Biava, 1964b). Bile thrombi are occasionally found in the dilated intercellular space and the space of Disse (Orlandi, 1962; Zaki, 1966). An increase in the number of Kupffer cells, which contain electron-dense

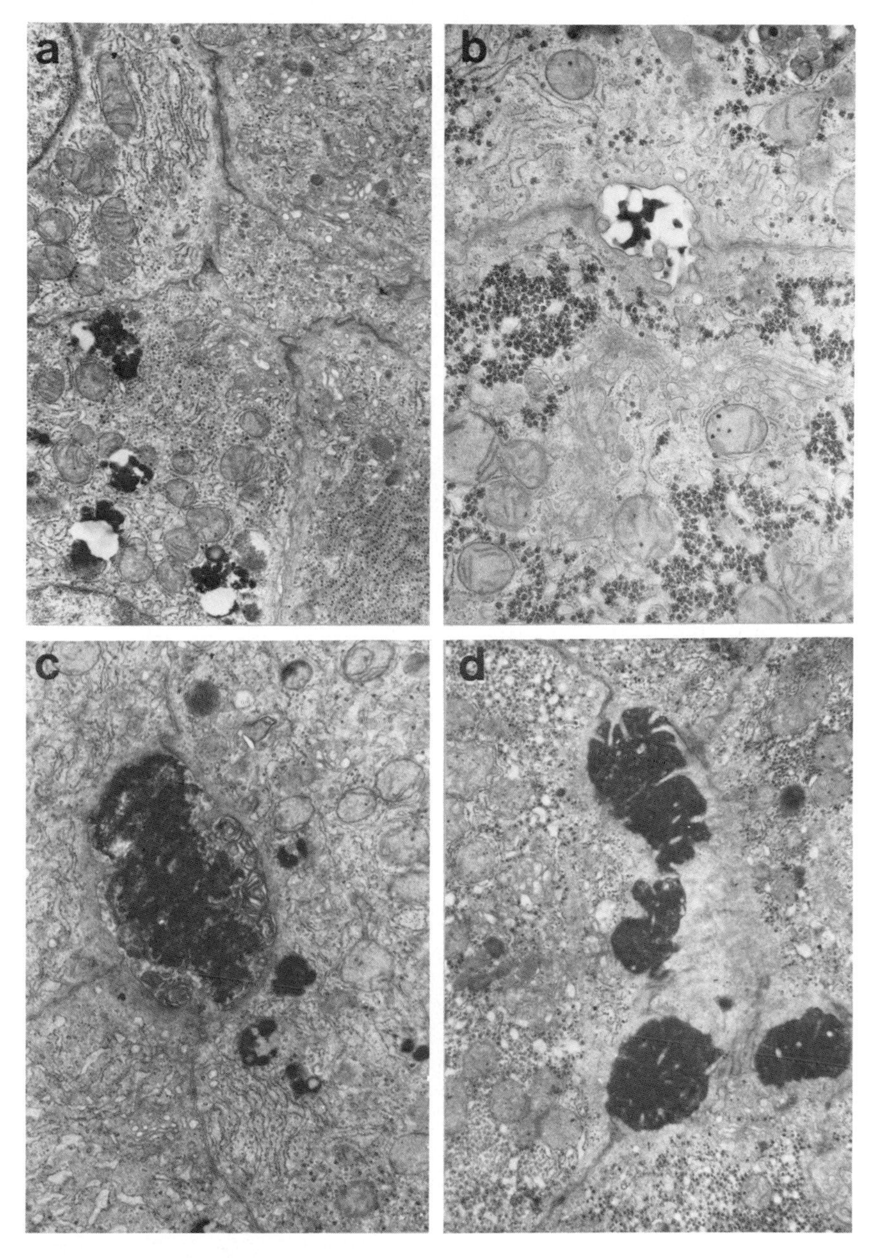


**Fig. 10.** Various appearances of bile thrombi. (a) Cholestatic type of acute viral hepatitis. A bile thrombus appears to be granular. ×6000. (b) Intrahepatic cholestasis of pregnancy. A bile thrombus appears to be electron dense and irregularly shaped. ×10,000. (c) Drug-induced cholestasis. A bile thrombus appears to be partly lamellar and the canalicular membrane seems to be continuous with lamellar part of the bile thrombus. ×8000. (d) Recurrent intrahepatic cholestasis. Bile canaliculi are completely filled with bile thrombi. However, the microvilli appear to be fairly well preserved. ×8000. (From Tanikawa, 1979.)

material, probably bile, is also noted in cholestasis (Tanikawa, and Ikejiri, 1977; Djaldetti *et al.*, 1978).

The Golgi apparatus in the hepatocyte usually appears to be hypertrophied and accompanied by small granules, vesicles, or stringy materials in its lumens. The changes of the SER such as dilatation and hypertrophy of their profiles has received much attention because these alterations in intrahepatic cholestasis is considered to be the starting phenomenon, leading to changes in bile salt metabolism favoring the production of monohydroxy bile salts and consequent disturbance in micelle formation of bile acids (Popper and Schaffner, 1970).

Recent studies show that disturbed micelle formation of bile salts is one of the important causative factors in formation of bile thrombi or canalicular cholestasis. Cholestasis can be induced by sodium taurolithocholate (Schaffner and Javitt, 1966; Javitt and Emerman, 1968) or other monohydroxy bile acid in rats and hamsters. Figure 11 shows canalicular bile thrombi in taurolithocholate-induced cholestasis in rat.

Monohydroxy bile acids such as lithocholate tend to form large, unstable micelles in comparison with dihydroxy or trihydroxy bile acid. Thus, in-

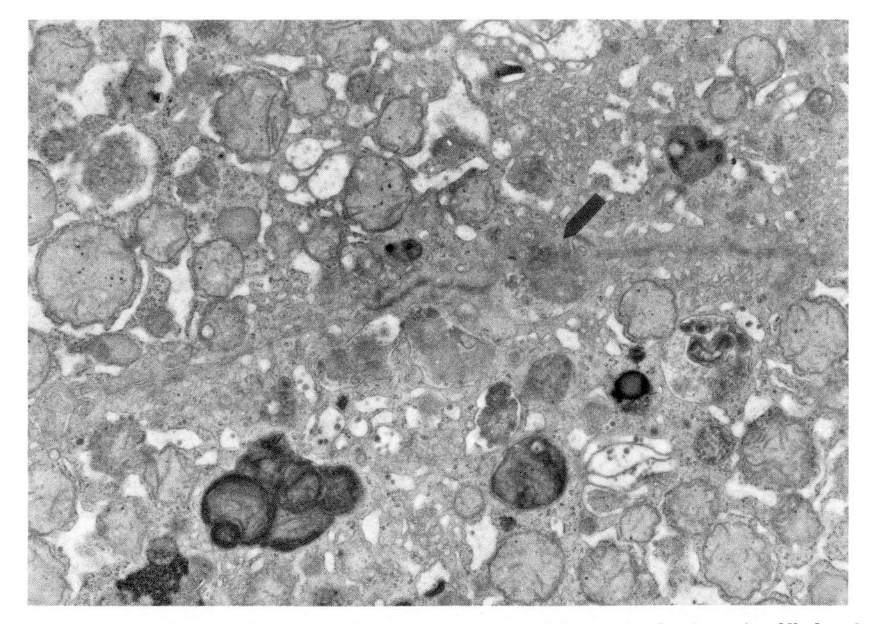

**Fig. 11.** Taurolithocholate-induced cholestasis in rat. A bile canaliculus (arrow) is filled with fine granular bile materials. The Golgi apparatus is markedly hypertropied with small granules in the cisternae. ×10,000.

creased concentration or ratio of monohydroxy bile acids in bile would result in inhibition of canalicular bile flow and subsequent increase of bile viscosity.

Ring hydroxylation of cholesterol occurs in the SER. However, when the cholesterol side chain is first completely oxidized to C-24 in the mitochondria, ring hydroxylation is inhibited and subsequently formation of monohydroxy bile acids increases. In most cases of intrahepatic cholestasis, the SER seems to be the primary site of alterations, caused by either disease or injury from drugs (Popper and Schaffner, 1970). Drugs or steroid hormones, many of which are metabolized in the SER, may influence bile acid metabolism, which leads to inhibition of the ring hydroxylation by competition or direct injuries. Our data revealed an increased concentration of lithocholate in the bile of estrogen-treated rats, suggesting that administration of estrogen affects bile acid metabolism, favoring the production of monohydroxy bile salts. Such disturbed function of the SER in turn interferes with micelle formation, resulting in cholestasis. In cholestasis of both intrahepatic and extrahepatic origins, ultrastructural changes appear similar, especially in long-standing situations. In fact, cholestasis with primary lesion in the hepatocyte may be complicated by secondary bile ductular obstruction (Desmet *et al.*, 1968), whereas extrahepatic obstruction may lead to secondary hepatocellular changes.

However, fine structural differences could be found especially in early stage of cholestasis between intrahepatic and extrahepatic origins, or among various types of intrahepatic cholestasis with different etiologies. At present, however, no sufficient experiences have been gained at this point probably because of the difficulty in observing such instances. Indeed, as shown in Fig. 10, various ultrastructural appearances of bile thrombi have been noted. Although their fine structural differences do not seem specifically related to their own etiological causes, such heterogeneous ultrastructural appearances indicate that many factors are involved in their morphogenesis.

Whether the canalicular changes such as dilation or changes of microvilli are the cause or the consequence of cholestasis remain uncertain. Canalicular dilation in extrahepatic cholestasis is easily understood as the result of increased hydrostatic pressure in the intrahepatic bile duct system. However, similar dilation is also observed in the intrahepatic cholestasis. As pericanalicular microfilaments seem to provide structural stability of the canaliculus, alteration of these structures may be expected in canalicular dilation and cholestasis.

Recently, the role of the pericanalicular microfilaments on bile excretion has obtained much attention. The bile excretion is disturbed and cholestasis occurs when the microfilaments are destroyed by cytochalasin B (Phillips *et al.*, 1975) (Fig. 12), and detachment of the microfilaments from the canalicular membrane is considered as an important factor in the mechanism of the

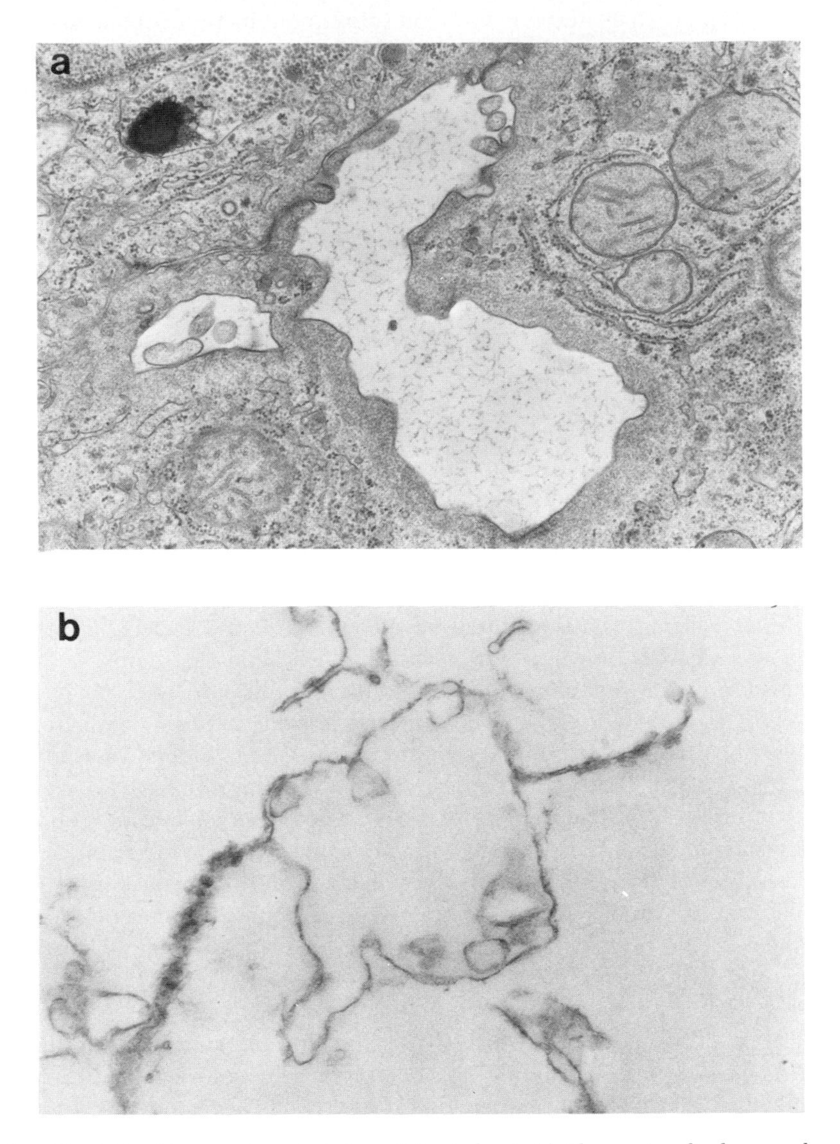


**Fig. 12.** Microfilaments in cytochalasin B-treated rat. (a) The pericanalicular ectoplasm appears thickened and granular. No filaments can be observed around the bile canaliculus. ×16,000. (b) Absolutely no microfilaments are seen around the bile canaliculus in the canaliculus-rich subfraction of the plasma membrane. ×22,000.

cholestasis (Oda and Phillips, 1977). In addition, it has been noted that administration of phalloidin induces biliary excretory failure by irreversible polymerization of the microfilaments (Gabbiani *et al.*, 1975) (Fig. 13), and a remarkable decrease in the bile acid independent bile flow is noted in phalloidin-induced cholestasis (Dubin *et al.*, 1978). Thus, it seems likely that the pericanalicular microfilaments have some role in bile excretion. In human cholestasis a widening of the pericanalicular ectoplasm with changes of the microfilaments are generally noted. However, it is still not clear that these microfilamentous changes are primary lesions in cholestasis or secondary responses to cholestasis.

These canalicular alterations would be simply explained by bile stagnation in the canalicular lumen. In fact, in some of intrahepatic cholestasis (Fig. 10d), canalicular microvilli appear to be well preserved in spite of extreme accumulation of bile in the lumen. However, in most instances, microvillar changes are associated with canalicular dilation and such alteration of microvilli would be secondary changes by the stagnated bile in the lumen.

On the other hand, it would also be possible that canalicular stagnation of bile occurs from the changes in the canalicular membrane which alters bile flow or bile concentration.

Many proposals and hypotheses have been made about the mechanism of intrahepatic cholestasis, and recent development of experimentally induced intrahepatic cholestasis in animals by monohydroxy bile acids is one of major advances in this field. However, we know very little about the factors inducing such cholestasis in selective human individuals by virus infection or drug administration including alcohol.

Cholestasis is usually predominant in the centrolobular zone in almost any form of cholestasis and, in severe cases of longer duration, it may be also found in the periphery and portal tract. Such preferential distrubution of cholestasis within the lobule may be related, at least, partially with greater bile flow and canalicular flushing in the lobular periphery and with emptying of periportal canaliculi by formation of ductular-hepatocellular recirculation in complete biliary obstruction (Popper, 1968). In addition, monohydroxy bile acids would be formed in greater amount in the centrolobular hepatocytes which have the greatest amount of the SER (Loud, 1968), the site of bile acid metabolism.

### *B. Bile Regurgitation into the Bloodstream in Cholestasis*

Cholestasis is clinically manifested by an accumulation in the blood of all constituents normally excreted in the bile. Thus, the routes or mechanism of bile regurgitation into the bloodstream have been long discussed. Electron microscopy with tracers and the assistance of histochemical studies have

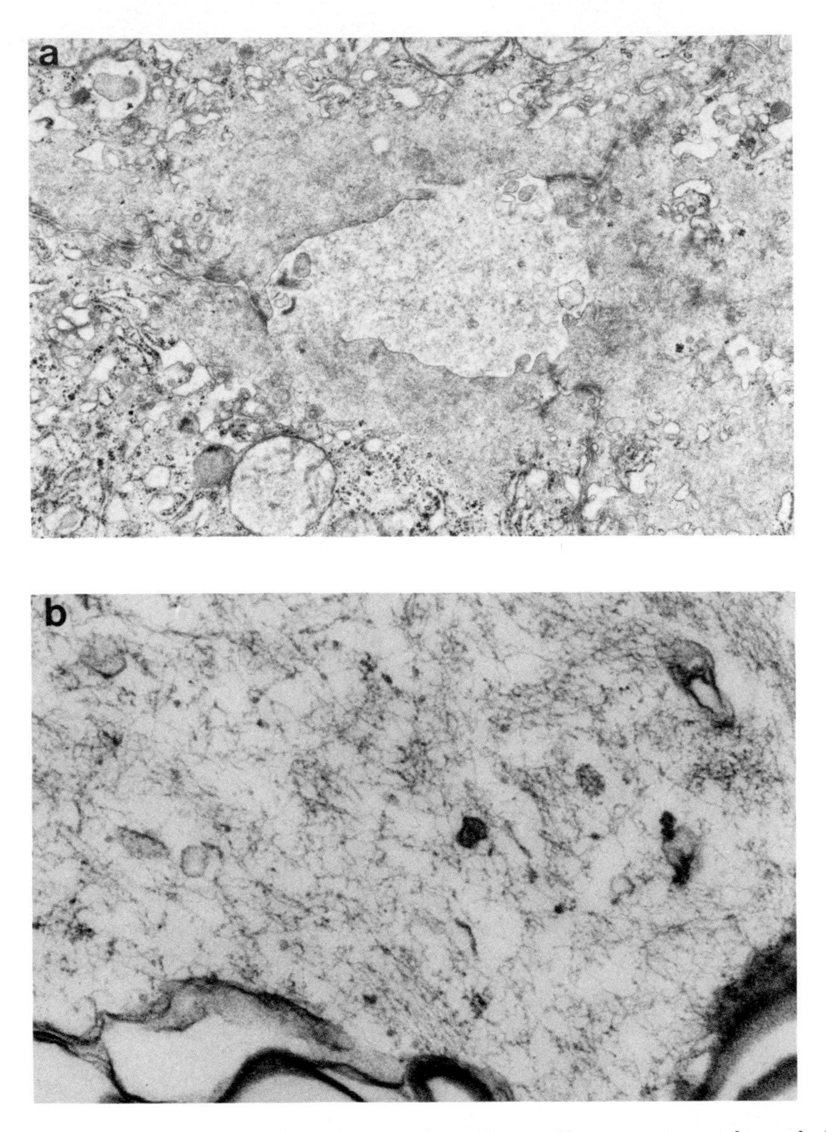


**Fig. 13.** Microfilaments in phalloidin-treated rat. (a) Microfilaments appear to be markedly increased around a dilated bile canaliculus. ×12,750. (b) Abundant microfilaments are noted around the bile canaliculus in the canaliculus-rich subfraction of the plasma membrane. ×36,000.

contributed much to the solution of this problem. At the present, at least three main routes have been considered as the sites of bile regurgitation into the bloodstream; transhepatocytic, through the communication of bile canaliculus with the space of Disse, and through the ductule (Tanikawa, 1968).

Evidence of bile stasis in the hepatocyte demonstrated under light and electron microscopes suggests the transhepatocytic regurgitation due to impaired excretion of bile into the bile canaliculus. Bilirubin precipitations of both conjugated and unconjugated types can be demonstrated in tissue sections by histochemical staining technique (Rais, 1965; Desmet *et al.*, 1967). Small bile granules in the hepatocyte usually react as conjugated bilirubin. However, larger precipitations are often stained as unconjugated ones. Under the electron microscope (Fig. 14) such a bile deposition appears as an accumulation of finely granular and fibrillar materials without any surrounding membrane. On the other hand, a larger accumulation corresponding to unconjugated bilirubin deposits, is surrounded by a membrane (Biava, 1964b; Hubner, 1968). This may be interpreted as sequestration of conjugated bilirubin in secondary lysosomes where conjugated bilirubin is deconjugated by their $\beta$-glucuronidase (Billing *et al.*, 1968; Desmet *et al.*, 1970).

ATPase is localized mainly on the membrane of canalicular microvilli in normal hepatocyte (Wachstein and Meisel, 1957; Sandstroom, 1971) and, in cholestasis, canalicular ATPase disappears (Schatzki, 1962; Breitfellner *et al.*, 1966; Wills and Epstein, 1966; Desmet *et al.*, 1968; Krstulovic *et al.*, 1968; Garay *et al.*, 1969; Chou and Gikson, 1971) and is related to canalicular changes such as loss or stunting of microvilli, or dilation. This disappearance is quicker in intrahepatic cholestasis than in extrahepatic obstruction (Holzner, 1960). The disappearance of ATPase is also associated with increased staining of this enzyme on the sinusoidal and lateral cell membranes (Krstulovic *et al.*, 1968; Desmet *et al.*, 1968; Wills and Epstein, 1966). Moreover, in cholestasis, widening of intercellular space is noted with development of microvilli along the lateral cell membrane. Such findings may indicate reversed polarity of the hepatocyte and reorientation of bile excretion toward the intercellualr space communicating with the sinusoid (Desmet, 1972). In addition, the mitochondria and the RER are significantly grouped along the sinusoidal and lateral border of the hepatocyte in the centro- and mediolobular areas (Orlandi, 1962), suggesting some specific energy supply to those membranes in such cholestasis. Intrahepatocytic bile regurgitation in cholestasis is best explained by reversed polarity of the hepatocyte (Desmet, 1972).

In addition, retrogradely injected tracer materials, such as mercuric sulfatide, Thorotrast (Hampton, 1958), or horseradish peroxidase (Matter *et al.*, 1969) are shown to traverse the cytoplasm of the hepatocyte in vacuoles to

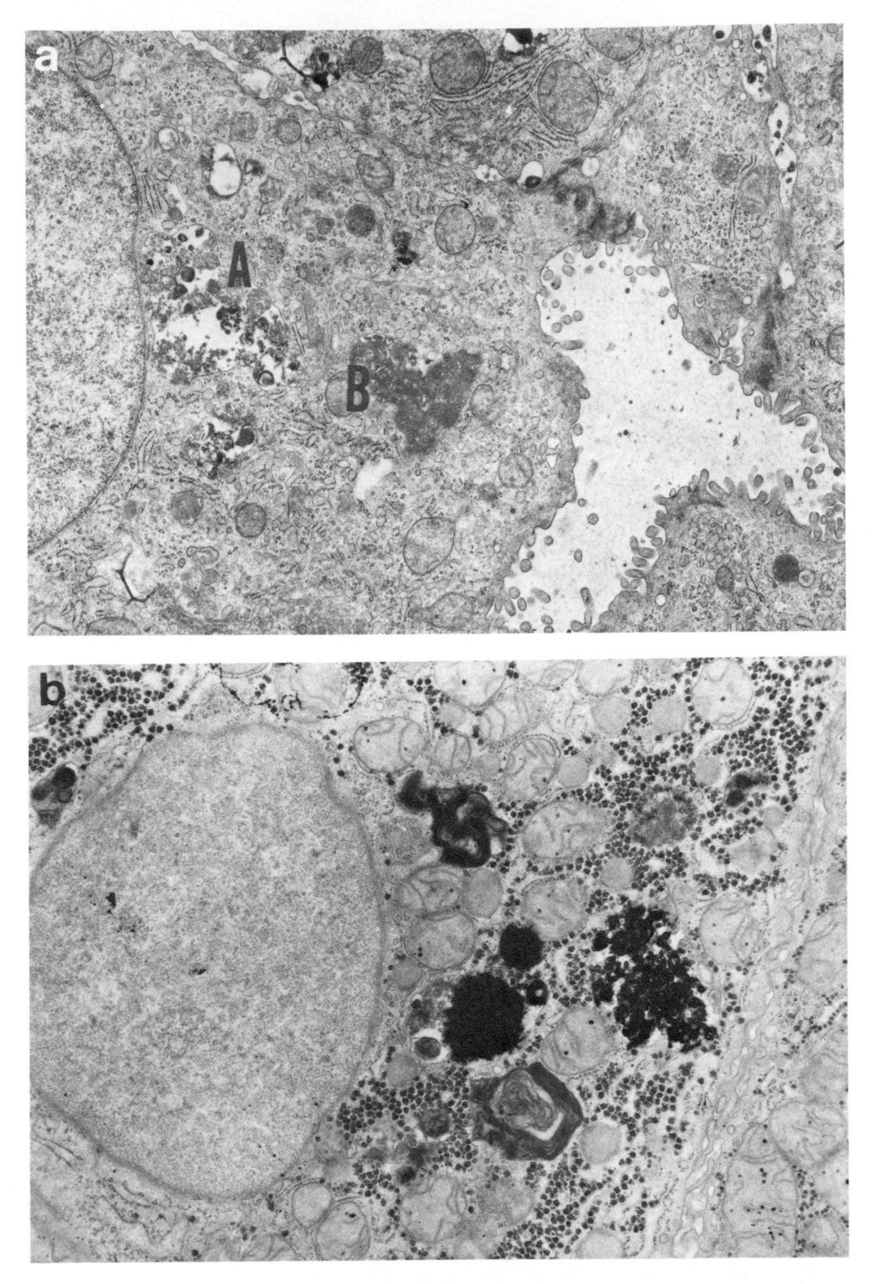


**Fig. 14.** Deposits of bile materials in the cytoplasm of the hepatocyte. (a) Congenital atresia of the bile duct. A bile canaliculus is markedly dilated. Electron-dense materials are noted in the cytoplasm of the hepatocyte. Some of them appear to be coarsely granular and surrounded by a membrane (AB). ×6000. (b) Extrahepatic obstructive jaundice. Several different types of electron dense deposits are noted in the cytoplasm of the hepatocyte. ×9500. (From Tanikawa, 1979.)

the sinusoidal cell membrane and to be discharged into the intercellular space.

Those findings indicate that bile constituents may be transferred from the hepatocyte directly to the tissue space or they may be returned from the canaliculus to the hepatocyte before they are discharged to the tissue space. At present, this transhepatocytic regurgitation is considered to be the main route in cholestasis.

The route of bile regurgitation through the communication between the bile canaliculus and the space of Disse has been also long discussed. Such open communication has been described even in normal liver (Rouiller, 1956). Most subsequent studies, however, denied its existence and recent study on serial sections failed to find it (Matter *et al.*, 1969). In cholestasis, some bile canaliculi appear to be 5 to 10 $\mu$m in diameter (Figs. 9a and 14a) and such extreme dilation should result in subsequent rupture or dissociation of the junctional complexes around the canaliculi. In fact, such communication is occasionally observed with dissociation or destruction of the junctional complexes (Tanikawa and Okuda, 1970) as shown in Fig. 15. Thus, such communication seems to occur more frequently than has generally

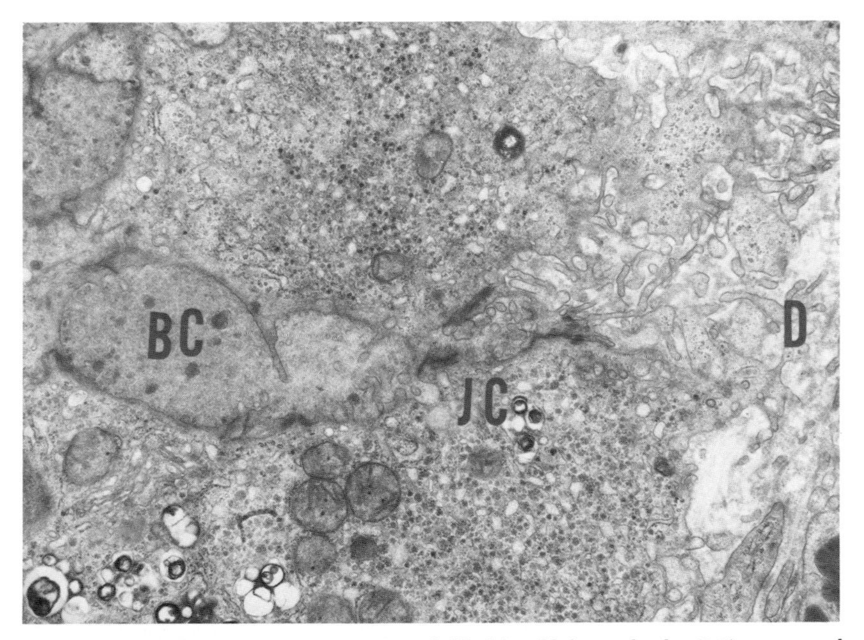


**Fig. 15.** Drug-induced cholestasis. A remarkably dilated bile canaliculus (BC) appears to be communicated with the space of Disse (D). Junctional complexes (JC) seem to be destructed. ×8000. (From Tanikawa, 1979.)

been thought. However, generally speaking, even in case of extreme canalicular dilation, the junctional complexes remain intact.

Another possibility for bile regurgitation at this side lies in an eventual permeability of junctional complexes without destruction or dissociation. Studies with tracers such as lanthanum (Schatzki, 1969, 1971; Tanikawa and Okuda, 1970) (Fig. 16), lead-gum solution (Yodaiken, 1966), and peroxidase (Tanikawa and Okuda, 1970) demonstrated a passage of these substances through the junctional complexes without destruction or dissociation, although in another study peroxidase was impermeable (Matter *et al.*, 1969). The problem remains to what extent studies using nonphysiologic tracers are reliable indicators of the eventual transjunctional passage of physiological bile constituents. However, if bile constituents of smaller molecules, such as water or electrolyte, could be passed through the junctional complexes under increased intraluminal pressure in the canaliculus, this may result in an increased viscosity of bile which leads to canalicular cholestasis.

Bile regurgitation through the bile ductile has also been considered.

The bile ductule has been noted to have excretory and absorptive functions (Steiner and Carruthers, 1961). In cholestasis, absorptive function

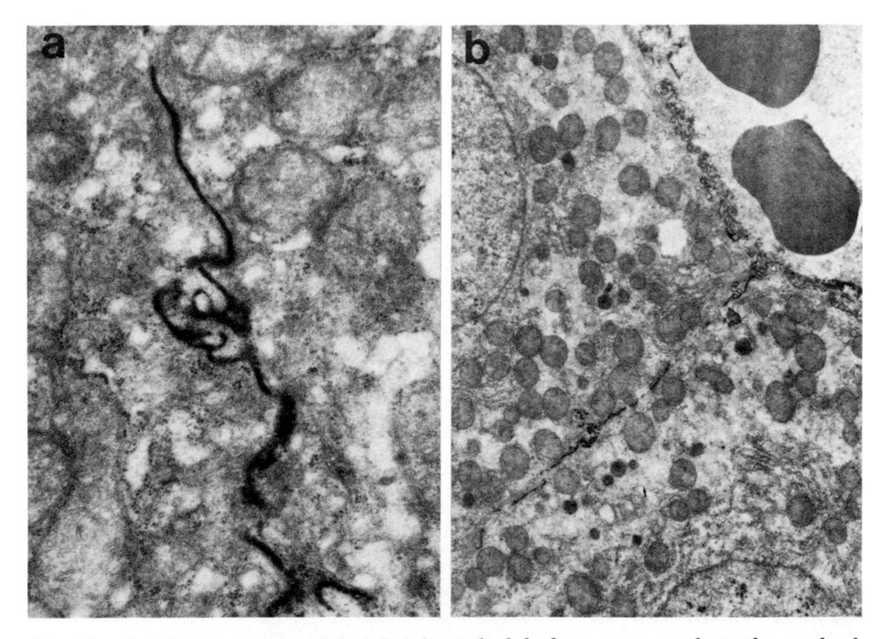


**Fig. 16.** Lanthanum, retrogradely infused into the bile duct, appears to be in the canalicular lumen and passes through the junctional complexes (a) to the space of Disse (b). a ×18,000; b ×3000. (From Tanikawa, 1979.)

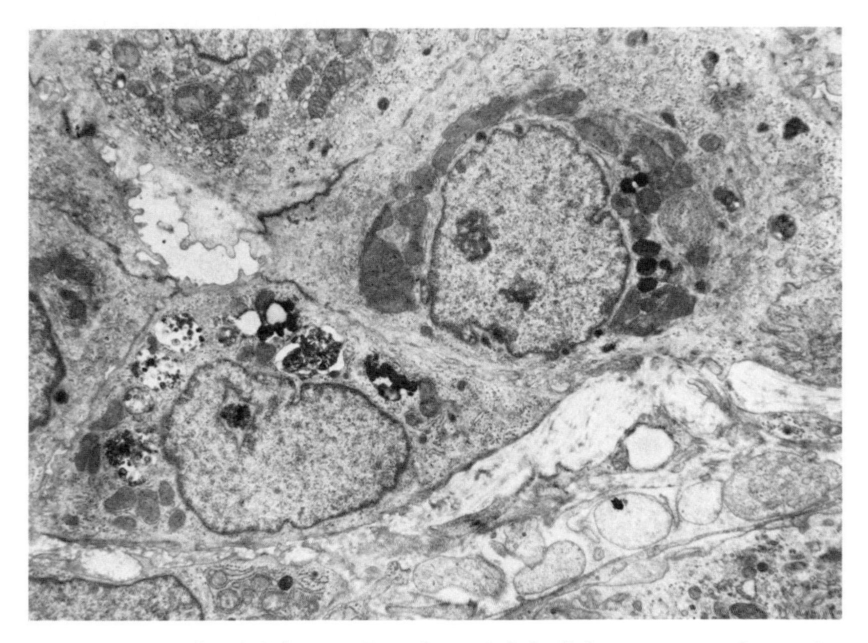

**Fig. 17.** Drug-induced cholestasis. Ductular epithelial cells have numerous electron-dense materials in their cytoplasms. ×3500. (From Tanikawa, 1979.)

is presumably enhanced, especially in extrahepatic obstruction by increased intraluminal pressure. In the epithelial cells of bile ductules in cholestasis, numerous pinocytic vesicles, myelinlike or granular materials (Fig. 17), interpreted as precipitates of bile constituents, are noted associated with widening of the intercellular space at the basal portion of the cells and with focal duplication of the surrounding basement membrane (Schaffner and Sasaki, 1965; Schaffner, 1965; Hollander and Schaffner, 1968; Sasaki *et al.*, 1967). These findings indicate an evidence of increased reabsorption of biliary constituents to the biliary epithelial cell and their discharge to the portal tissue space.

Experimentally gold colloids, retrogradely infused into the common bile duct of dogs under a slightly higher pressure than in normal excretion, can be traced from the ductular lumen through the biliary epithelial cell into the surrounding tissue space by electron microscope (Okuda and Tanikawa, 1967; Tanikawa, 1968)(Fig. 18). Thus, biliary constituents such as bilirubin and bile acids are probably regurgitated through the ductular epithelium into the portal tissue space. Although such bile regurgitation through the ductule seems to be confirmed, its quantitative importance remains to be elucidated.

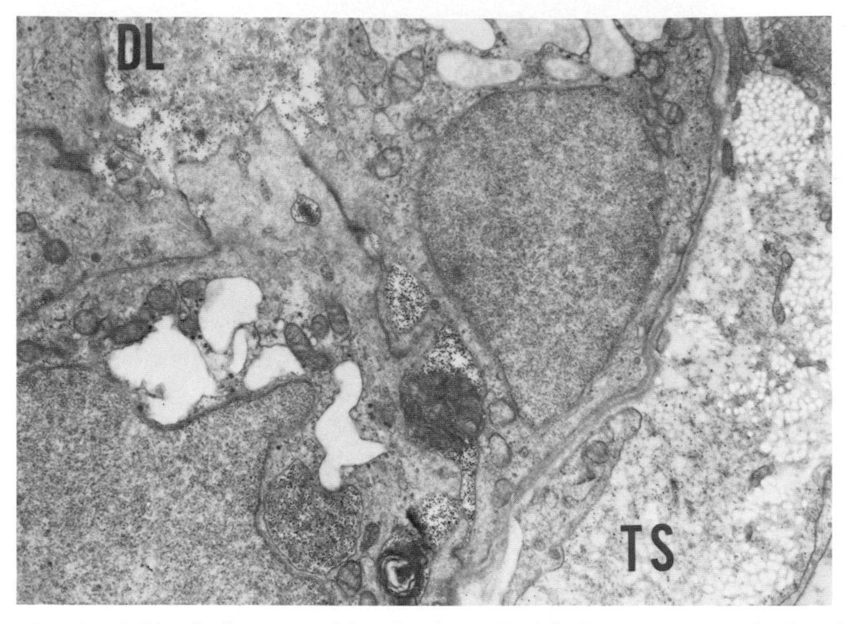


**Fig. 18.** Gold colloids, retrogradely infused into the bile duct, are seen in the ductular lumen (DL) and appear to be taken up to the epithelial cells and to be discharged to the surrounding tissue space (TS). ×10,000. (From Tanikawa, 1979.)

In addition, under the electron microscope, dissociation or destruction of junctional complexes between adjacent biliary ductular epithelial cells are rarely noted in extrahepatic obstructive jaundice (Yamaguchi, 1975) as shown in Fig. 19. Such changes could be explained by light microscopic evidence of bile lake, often seen in the periportal area of the lobule in extrahepatic biliary obstruction.

Bile constituents leaked out into the portal area, either through the biliary epithelial cells of ductules or through the open communication between the ductular lumen and the surrounding tissue space, are drained, either into the lymph vessels which open at the portal area, or directly to the sinusoidal bloodstream. The former route has been demonstrated by a study using gold colloids (Tanikawa, 1968). In fact, following ligation of the common bile duct in experimental animals, bilirubin appears in hilar hepatic lymph much earlier and in higher level than in the peripheral blood (Tanikawa, 1968). Such ductular regurgitation of bile constituents forms ductular-hepatocytic circulation (Schaffner, 1965).

The routes of bile regurgitation discussed above are shown schematically in Fig. 20.

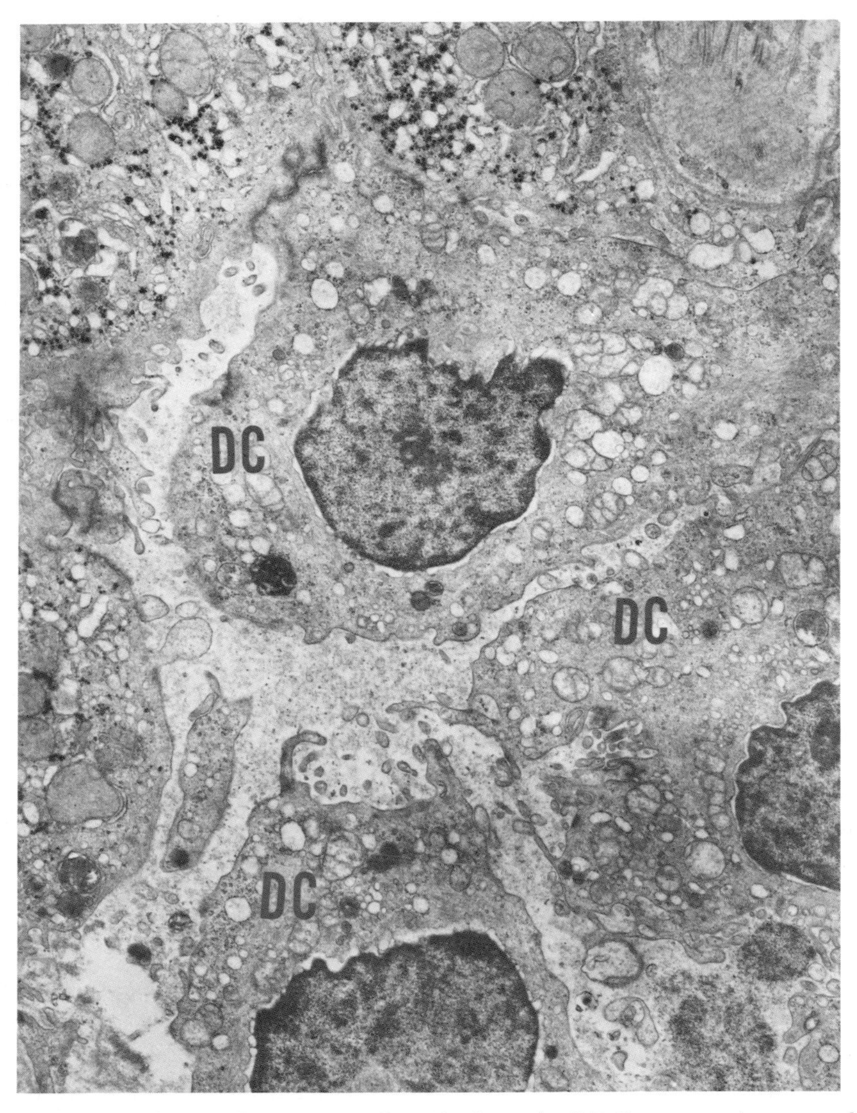


**Fig. 19.** Extrahepatic obstructive jaundice. The ductural cell (DC) appears to be separated at the tight junctions and the communication between the ductular lumen and surrounding tissue space seems to be occurred. ×8000. (From Tanikawa, 1979.)

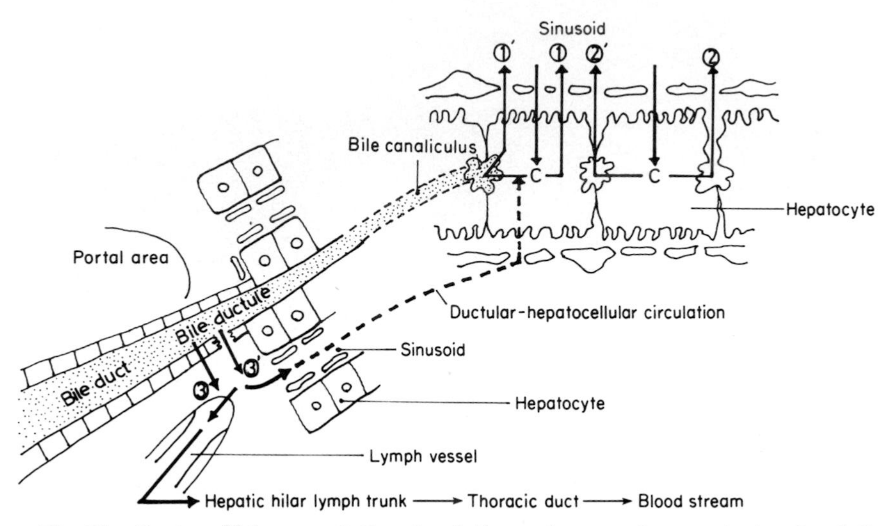


**Fig. 20.** Routes of bile regurgitation. 1 and 1′, transhepatocytic regurgitation; 2 and 2′, regurgitation through the communication between the bile canaliculus and the space of Disse; 3 and 3′, regurgitation through the bile ductule; C, site of bilirubin conjugation.

### *C. Secondary Effects of Cholestasis on Hepatocytes*

Bile constituents retained in the hepatocyte lead to further cellular injuries. Bile acids and bilirubin, in concentration within the range found in cholestatic liver tissue, have been shown to inhibit and uncouple oxidative phosphorylation and to stimulate ATPase activity of isolated rat liver mitochondria (Pressman and Lardy, 1956; Lee and Whitehouse, 1965). Similar results on the mitochondrial function of liver biopsy specimens from patients with obstructive jaundice have been reported (Schersten *et al.*, 1966; 1970). So far as bilirubin toxicity is concerned, unconjugated bilirubin, deconjugated by β-glucuronidase in lysosomes which increase in cholestasis, would play a role because of its ability to penetrate through the mitochondrial membrane. Recently, precipitated unconjugated bilirubin has been demonstrated in the hepatocyte and Kupffer cell of cholestatic liver by histochemical technique (Desmet *et al.*, 1970).

Morphologically one of the most frequently and characteristically involved organelles in cholestasis is the mitochondria, which show enlargement, increase in number, increased intramitochondrial granules, curling of cristae, and myelinlike inclusions in the matrix (Schaffner, 1965; Tanikawa, 1968; Schaffner and Popper, 1969; Perez *et al.*, 1969). Morphometric analysis has confirmed these changes (Yamaguchi *et al.*, 1975). Such morphological alterations may be correlated with the above-mentioned functional disturbances of the mitochondria.

In any type of cholestasis, the SER is increased. However, impaired microsomal drug metabolism has been noted in cholestasis (McLuen and Founts, 1961). Thus, the hypertrophy of the SER may be hypoactive. Which of bile substances retained in the hepatocyte causes such impairment is not established. But, the detergent action of bile acids seems to be the most important. The retained bile acids may interfere with transformation of cholesterol to bile salts by reduced hydroxylation, which leads to enhanced formation of monohydroxy bile acids. This in turn disturbs bile salt micelle formation. Such a cycle may result in perpetuation of cholestasis (Popper and Schaffner, 1970).

Increased amounts of peroxisomes have been noted in cholestasis (Schaffner and Kniffen, 1963), but its significance remains uncertain.

From light and electron microscopic observations, the size and the number of lysosomes are increased in cholestasis. Increased activity of $\beta$-glucuronidase has been also demonstrated (Sussi and Rubaltelli, 1968). Such an increase in number and size associated with increased activity of lysosomal enzymes would be a response to bile materials accumulated in the hepatocyte.

In cholestatic human liver, especially of a long-standing jaundice, lytic necrosis of the hepatocyte are fairly frequently observed with bile retention. Under the electron microscope (Fig. 21), these cells contain numerous electron-dense materials in their cytoplasm (Tanikawa, 1968), although it is difficult to obtain clear evidence as to whether the lysosome is significant during the onset and development of the necrosis. Addition *in vitro* of bile acids or bilirubin to lysosomal suspensions of liver homogenates from human liver causes a release of lysosomal enzymes (Bjorkerud *et al.*, 1967).

Fortunately a safety mechanism seems to be operating in cholestatic liver to prevent further injuries. In bile duct-ligated rat liver, ursodeoxycholic and $\beta$-muricholic acids, both having much less detergent action, seem to be formed in great amounts from chenodeoxycholic acid, which is a strong detergent (Greim *et al.*, 1972). Enhancement of such a pathway, which normally is minor, can therefore be considered a safety mechanism to prevent the destruction of organelles by high concentrations of strong detergent bile acids in the hepatocyte (Hutterer *et al.*, 1972).

## D. *Portal and Periportal Reactions in Cholestasis*

The bile ductule, usually located in the periportal or portal area, has the peculiar property of proliferating in many liver diseases. Recently, the proliferating bile ductules surrounded by a longitudinal array of collagen fibers

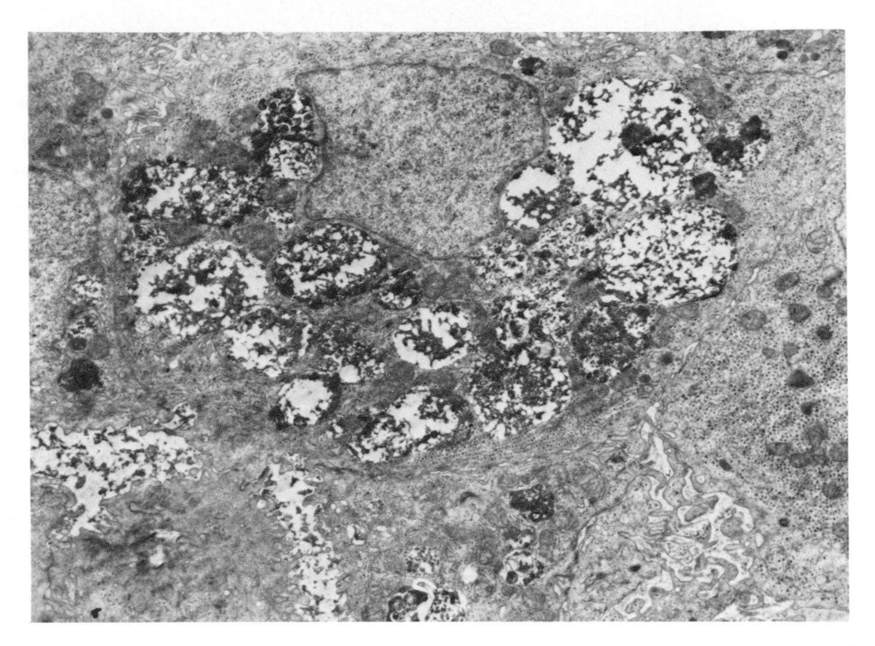

**Fig. 21.** Extrahepatic obstructive jaundice. The cytoplasm of the hepatocyte is occupied by electron-dense bile materials. This may represent a state of bile necrosis. ×2000.

in the ligation of the common bile duct was observed by scanning electron microscopy (Brooks *et al.*, 1975). In cholestasis of both intrahepatic and extrahepatic origins, such ductular proliferation occurs frequently. In extrahepatic biliary obstruction, elevated intralobular biliary pressure could stimulate the proliferation. However, in intrahepatic cholestasis, elevation of intralobular biliary pressure in the early stage does not occur. Recently, bile acids have been suspected as main stimulants for such ductular proliferation. In cholestasis and even in any type of human liver diseases, bile acid metabolism is more or less disturbed in the hepatocyte and consequently altered constituents or concentration of bile acids could be excreted from the hepatocyte. These changes of bile acids may be responsible for such proliferation. In fact, experimentally lithocholate can produce such ductular proliferation in animals (Leveille *et al.*, 1964; Hunt *et al.*, 1964; Schaffner and Javitt, 1966).

In addition, such ductular proliferation in cholestasis may become a secondary cause of intrahepatic obstruction because of their complex pathway and accompanying periductular inflammation and fibrosis.

In cholestasis, especially of extrahepatic origin, ductules or interlobular bile ducts are dilated with a decrease in number and size of microvilli, formation of edametous microvillous blebs and occurrence of luminal diverticula.

In the ductular epithelial cells, electron-dense materials, interpreted as bile precipitates, are numerously noted (Fig. 17). Moreover, electron microscopic study (Tanikawa, 1968) indicates that regurgitated colloid particles into the bile duct can pass through the ductular cell into the surrounding tissue space (Fig. 18). Thus, periductular inflammation and fibrosis associated with ductular proliferation are considered to be caused by leakage of bile constituents, especially bile acids, into the surrounding portal tissue space, either through the biliary epithelial cells by absorption or the communication between the biliary lumen and portal tissue space by rupture at the tight junction of the biliary epitherial cells.

Progressive inflammation and fibrosis associated with ductular proliferation may further cause a secondary mechanical obstruction of the bile flow.

In primary biliary cirrhosis distinctive fine structural features have been observed in the small and medium-sized bile ducts which are considered as a primary site of the lesion in this disease (Chedid *et al.*, 1974).

### *E. Cholestasis and Associated Plasma Changes*

In cholestasis, clinical liver function test is characterized by elevated cholesterol level and increased activity of alkaline phosphatase in serum besides conjugated hyperbilirubinemia. Those plasma changes in cholestasis could be correlated with fine structural alterations of the hepatocyte.

In normal human liver, alkaline phosphatase is localized mainly on the canalicular membrane of the hepatocyte, but also positive on the sinusoidal side of the plasma membrane. In cholestasis, great increase of this enzyme activity is noted not only on the canalicular, but also on the lateral and sinusoidal hepatocellular membranes. Though the role of alkaline phosphatase is not well known, this is generally considered to be related to transport of materials through the cell membrane. In cholestasis, such increase in membrane-localized alkaline phosphatase may be related to increased local concentrations of bile acids as shown by studies on biliary nephrosis in bile duct-ligated rats (De Vos *et al.*, 1972). It has been known that one of serum alkaline phosphatase fractions elevated in cholestasis has a higher molecular weight and membrane bound (Hattori *et al.*, 1969). From fine structural observations, the peripheral part of bile thrombus in the bile canaliculus appears occasionally membranous and often seems to be continuous with the canalicular membrane in which alkaline phosphate exists. If the membranous part of bile thrombi originates from desquamated canalicular membrane, fragments of these membrane might explain the nature of the serum alkaline phosphatase in cholestasis. Some part of the alkaline phosphatase elevated in cholestasis may be released from the sinusoidal and lateral plasma membranes where this enzyme activity increased.

Plasma lipid changes in cholestasis are characterized by increased concentrations of nonesterified cholesterol and of phospholipid. The increase of the serum cholesterol concentration may be ascribed partly to an increased synthesis of cholesterol with loss of the negative feedback control mechanism for cholesterol (Katterman and Creutzfeldt, 1970) and partly to the diminished excretion into bile. In electron microscopy free cholesterol can be demonstrated in spicular formation in digitonin-treated tissue, and in cholestasis such spicular formation is greatly increased in the hepatocyte, and dilated bile canaliculus, mainly in the centrolobular zone (Scharnbeck and Schaffner, 1970). The lipoprotein pattern in serum is also changed in cholestasis. An abnormal lipoprotein, called lipoprotein X, is demonstrated in the low density lipoprotein fraction. They appear as unique disk-shaped particles measuring about 400 to 700 Å in diameter in negatively stained preparations of the serum (Seidel *et al.*, 1972; Hamilton *et al.*, 1971). These particles are also observed in ultra-thin sections in bile duct-ligated mice (Stein *et al.*, 1973) and in humans with obstructive jaundice as shown in Fig. 22. A recent study suggests that lipoprotein X is formed from bile lipids regurgitated into

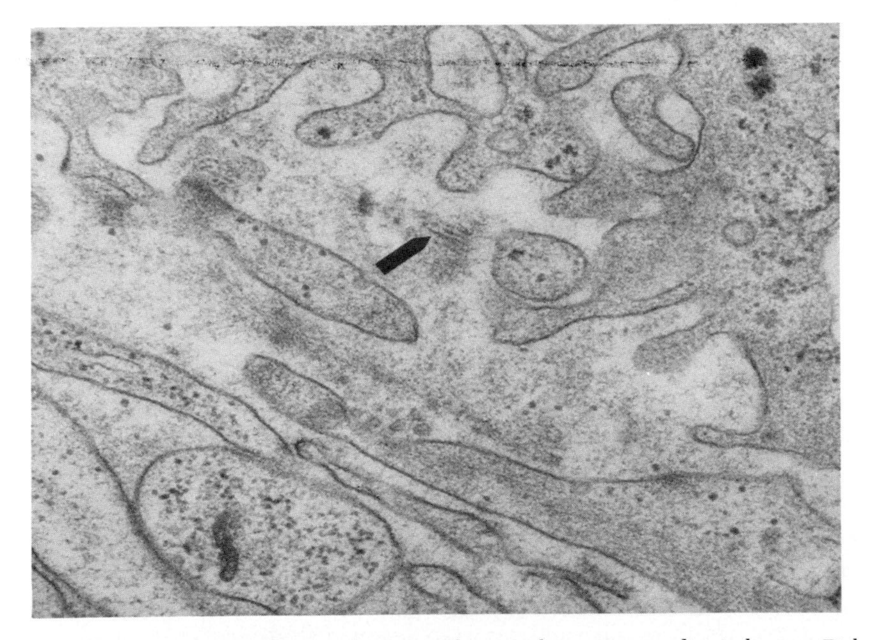

**Fig. 22.** Lipoprotein X like particles in extrahepatic obstructive jaundice in human. Disk-shaped particles of bilayer structure (arrow) measuring about 500–1000 Å in diameter are noted in the space of Disse. ×40,000.

the bloodstream because the lipid composition of lipoprotein X is almost identical to the lipids found in bile and also the lipoprotein-like material is formed *in vitro* by adding albumin or serum to native bile (Manzano *et al.*, 1976).

About twenty years have passed since we began ultrastructural studies of jaundice. During that time, in addition to significant advances in the field of bilirubin and bile acid metabolism, development of experimentally induced intrahepatic cholestasis is specially noteworthy.

From the morphological point of view, scanning electron microscopy and numerous other new techniques in tracer studies, autoradiography, and histochemical procedures have been introduced, and those morphological approaches at the ultrastructural level also have to be evaluated in this field. Figures 2b, 2c, 2d, 5, 6, 7, 8a, 8b, 9b, 10b, 10c, 14a, 14b, 15, 16a, 16b, 18, and 19 are from Tanikawa (1979), Ultrastructural Aspects of the Liver and Its Disorders, 2nd ed. Igaku-shoin Ltd., Tokyo and New York.

Recent excellent reviews by Popper and Schaffner (1970), Schaffner and Popper (1974), and Desmet (1972, 1977) on cholestasis, and by Billing (1973), Schmid (1974, 1978), and Arias (1974) on bilirubin metabolism, have contributed much to the understanding of jaundice and cholestasis, and they have stimulated us for further studies on the subject. In spite of extensive studies, however, we still have numerous problems to solve regarding the old and new subject "jaundice."

## References

Akeda, S., Ninomiya, T., Shima, K., Miyaji, K., and Takeda, S. (1973). *Acta Hepatologica Japopica* **14**, 266.

Akeda, S., Takeda, S., Tameda, Y., Kondo, I., Tagawa, S., Kosaka, Y., Takezawa, H., and Takasaki, H. (1973). *Mie Med. J.* **17**, 139.

Arias, I. M. (1974). In "The Liver and Its Diseases" (F. Schaffner, S. Sherlock, and C. M. Leevy, eds.), pp. 97–104. Intercontinental Medical Corp., New York.

Arias, I. M., Gartner, L. M., Cohen, M., Ben Ezzer, J., and Levi, A. J. (1969). *Am. J. Med.* **47**, 395.

Baba, N., and Ruppert, RmN. (1972). *Am. J. Clin. Pathol.* **57**, 306.

Barone, P. (1966). *Virchows Arch. Pathol. Anat. Physiol.* **341**, 43.

Barth, R. F., Crimliy, P. M., Beck, P. D. Bloomer, J. R., and Howe, R. B. (1971). *Arch. Pathol.* **91**, 41.

Bianchi, L., Otto, R., and Rohr, H. (1971). *Acta Hepatosplen.* **18**, 305.

Biava, C. (1964a). *Lab. Invest.* **13**, 840.

Biava, C. (1964b). *Lab. Invest.* **13**, 1099.

Billing, B. H. (1973). *In* "The Liver" (E. A. Gall and F. K. Mostofi, eds.), pp. 1–11. William Wilkins, Baltimore, Maryland.

Billing, B., Raia, S., and Armas-Merino, R. (1968). *In* "Ikterus" (K. Beck ed.), pp. 49–52. Schattauer, Stuttgart.

Bjorkerud, S., Bjorntorp, P., and Scherstern, T. (1967). *J. Clin. Lab. Invest.* **20**, 224.

Blanckaert, N., Heirwegh, K. P. M., and Zaman, Z. (1977). *Biochem. J.* **164**, 229.

Bonnett, R., and Davis, J. E. (1976). *Nature (London)* **263**, 326.
Breitfellener, G., Holzner, J., Schumacher, A., and Stefenelli, N. (1966). *Beitr. Pathol. Anat.* **134**, 267.
Brooks, S. E. H., Reynolds, P., Audretsch, J. J., and Haggis, G. (1975). *Lab. Invest.* **33**, 311.
Brown, N. K., and Smucker, E. A. (1970). *Am. J. Med.* **48**, 759.
Carruthers, J. and Steiner, J. (1962). *Gastroenterology* **42**, 419.
Chedid, A., Spellberg, M. A., and DeBeer, R. A. (1974). *Gastroenterology* **67**, 858.
Chou, S. and Gibson, J. (1971). *J. Pathol.* **103**, 163.
Colleran, E. and O'Carra, P. (1970). *Biochem. J.* **119**, 16.
Cornelius, C. E., Osburn, B. I., Gronwall, R. R., and Cardinet III, G. H. (1968). *Am. J. Dig. Dis.* **13**, 1072.
Czarnecki, J., Nowakowski, A., and Pawlowska-Tochman, A. (1973). *Pat. Pol* **24**, 139.
De Brito, T., Borges, M. A., and daSilva, L. C. (1966). *Gastroenterologia* **106**, 325.
Desmet, V. J. (1972). *In* "Progress in Liver Desease" (H. Popper and F. Schaffner, eds'.) Vol IV, pp. 97–132. Grune & Stratton, New York.
Desmet, V. J. (1977). *In* "Liver and Bile" (L. Bianchi, W. Gerok, and K. Sickinger, eds). pp. 3–31. MTP Press, Lancaster.
Desmet, V., Bullens, A., and Heriwegh, K. (1967). *In* "Bilirubin Metabolism" (I. Bouchier and B. Billing, eds.), pp. 281–284. Blackwell, Oxford.
Desmet, V., Krstulovic B., and Van Damme, B. (1968). *Am. J. Pathol.* **52**, 401.
Desmet, V., Bullens, A., and De Groote, J. (1970). *Gut* **11**, 516.
De Vos, R., De Wolf-Peeters, C., and Desmet, V. (1972). *Beitr Pathol.* **145**, 315.
Djaldetti, M., Zahav, L. H., Gafter, U., and Mandel, E. M. (1978). *Acta Pathol. Lab. Med.* **102**, 49.
Dubin, M., Maurice, M., Feldmann, G., and Erlinger, S. (1978). *Gastroenterology* **75**, 450.
Erlinger, S., and Dhumeaux, D. (1974). *Gastroenterology* **66**, 281.
Erlinger, S., Dhumeaux, D., Berthelot, P., and Dumont, M. (1970). *Am. J. Physiol.* **219**, 416.
Essner, E., and Novikoff, A. B. (1960). *J. Ultrastruct. Res.* 347.
Feverly, J., Van Damme, B., and Michiels, R. (1972). *J. Clin. Invest.* **51**, 2482.
Fisher, M. M., Bloxam, D. L., Oda, M., and Phillips. (1975). *Proc. Soc. Exp. Biol. Med.* **150**, 177.
Fog, J., and Jellum, E. (1963) *Nature (London)* **198**, 88.
French, S. W., and Davies, P. L. (1975). *Gastoenterology* **68**, 765.
Gabbiani, G., Montesano, R., Tuchweber, B., Salas, M., and Orci, L. (1975). *Lab. Invest.* **33**, 562.
Garay, E., Piccaluga, A., Perez, V., and Royer, M. (1969). *Acta Hepatosplenol.* **16**, 221.
Gordon, E. R., Tak-Hang, C., and Perlin, A. S. (1976). *Biochem. J.* **155**, 477.
Goresky, C., and Back, G. G. (1970). *Ann. N.Y. Acad. Sci.* **170**, 18.
Gregory, D. H., Vlahcevic, Z. R., Prugh, M. F., and Swell, L. (1978). *Gastroenterology* **74**, 93.
Greim, H., Trulzsch, K., Czygan, P., Dressler, K., and Hutterer, F. (1972). *Fed. Proc Fed. Am. Soc. Exp. Biol.* **31**, 612.
Hamilton, R. L., Havel, R. J., Kane, J. P., Blaurock, A. E., and Sata, T. (1971). *Science* **172**, 475.
Hampton, J. (1958). *Acta Anat.* **32**, 262.
Hattori, N., Murayama, S., Mitsui, H., and Arima, M. (1969). *Acta Hepat. Jap.* **10**, 40.
Herman, J., Cooper, E. B., Takeuchi, A., and Sprinz, H. (1964). *Am. J. Dig. Dis.* **9**, 160.
Hollander, M., and Schaffner, F. (1968). *Am. J. Dis. Child.* **116**, 49.
Holzner, J. (1960). *Verh Deutsc Ges. Pathol.* **44**, 233.
Hubner, G. (1968). *In* "Ikterus" (K. Beck. ed.), pp. 115–123. Schattauer, Stuttgart.
Hunt, R., Veiveille, G., and Sauberlich, H. (1964). *Proc. Soc. Exp. Biol. Med.* **115**, 277.

Hutterer, F., Greim, H., Trulzsch, D., Czygan, P., and Schenkman, J. B. (1972). *In* "Progress in Liver Disease" (H. Popper and F. Schaffner, eds.), Vol. IV, pp. 151–171. Grune & Stratton, New York.
Ichida, F., and Funahasi, H. (1964). *Acta Hepatosplenol.* **11**, 332.
Inferrera, C., and Motta, P. (1965). *Virchows Arch. pathol. Anat.* **339**, 327.
Jacobson, J. A. (1969). *Acta Chem. Scand.* **23**, 3023.
Jansen, P. L. M., Chowdhury, J. R. and Fishgerg, E. G. (1977). *J. Biol. Chem.* **252**, 2710.
Javitt, N. B., and Emerman, S. (1968). *J. Clin. Invest.* **47**, 1002.
Johnson, F. B. (1970). *J. Histochem. Cytochem.* **18**, 674.
Kattermann, R., and Creutzfeldt, W. (1970). *Scand. J. Gastroent.* **5**, 337.
Kobayashe, T. (1971). *Acta Hepat. Jap.* **12**, 157.
Krstulovic, D., Van Damme, B., and Desmet, V. (1968). *Am. J. Pathol.* **52**, 423.
Lee, M. J., and Whitehouse, M. W. (1965). *Biochem. Biophys. Acta* **100**, 317.
Leveille, G., Hunt, R., and Sauberlich, H. (1964). *Proc. Soc. Exp. Biol. Med.* **115**, 573.
Levi, A. J., Gatmaitan, Z., and Arias, I. M. (1969). *J. Clin. Invest.* **48**, 2156.
Loud, A. U. (1968). *J. Cell Biol.* **37**, 27.
Manzano, E., Rellin, R., Baggio, G., Walch, S., Neubeck, W., and Seidel, D. (1976). *J. Clin. Invest.* **57**, 1248.
Mc Luen, E. F., and Fouts, J. R. (1961). *J. Pharmacol Exp. Ther.* **131**, 7.
Matter, A., Orchi, L., and Rouiller, C. (1969). *J. Ultrastruct. Suppl.* **11**, 1.
Minio, F., and Gauter, A., (1966). *Z. Zellforsch. Mikrosk. Anat.* **72**, 168.
Minio-Pauello, F., Gautier, A., and Magnenat, P. (1969). *Acta Hepatosplenol.* **15**, 65.
Miyakoda, U. (1975). *Acta Hepat. Jap.* **16**, 121.
Miyai, K., Mayr, M., Richardson, A., and Fisher, M. M. (1974). *In* "Jaundice" (C. A. Goresky and M. M. Fisher, eds.), pp. 383–400. Plenum, New York.
Molbert, E., and Marx, R. (1966). *Acta Hepatosplenol.* **13**, 160.
Noir, B. A., dewalz, A. T., and Rodriguez Garay, E. Z. (1970). *Biochem. Biophys. Acta* **222**, 15.
Novikoff, A. B., and Essner, E. (1960). *Am. J. Med.* **29**, 102.
Oda, M., and Phillips, M. J. (1977). *Lab. Invest.* **37**, 350.
Oda, M., Prince, V. M., Fisher, M. M., and Phillips, M. J. (1974). *Lab. Invest.* **31**, 314.
Okuda, K., and Tanikawa, K. (1967). *In* "Liver Research" (J. Vandenbroucke, J. De Groote, and L. O. Standaert, eds.), pp. 459–468. Tifdschrift voor Gastroenterologie Autwerpen.
Orlandi, F. (1962). *Acta Hepatosplenol.* **9**, 155.
Ostrow, J. D., and Murphy, C. L. (1970). *Biochem. J.* **120**, 311.
Pages, A., and Baldet, P. (1969). *Ann. Anat. Pathol.* **14**, 77.
Patrassi, G., Sandre, G., and Leonardi, P. (1965). *Rev. Int. Hepat.* **15**, 481.
Perez, V., Gorodisch, S., De Martive, J., Nicholson, R., and Di Paola, G. (1969). *Science* **165**, 805.
Phillips, M. J., Oda, M., Max, E., Fishee, M. M. and Jeejeebhoy, K. N. (1975). *Gastroenterology* **69**, 48.
Popper, H. (1968). *Annu. Rev. Med.* **19**, 39.
Popper, H., and Schaffner, F. (1970). *Huma. Pathol.* **1**, 1.
Pressman, B. C. and Lardy, H. A. (1956). *Biochem. Biophys. Acta* **21**, 458.
Raia, S. (1965). *Nature (London).* **205**, 304.
Remmer, H., and Merker, H. J. (1963). *Klin, Wschr.* **41**, 276.
Rouiller, C. (1956). *Acta Anat.* **26**, 94.
Sandstroom, B. (1971). *Histochemie* **25**, 9.
Sasaki, H., and Ichida, F. (1961). *Report Inst. Virus Res. Kyoto Univ., 1961* **4**, 172.
Sasaki, H., Schaffner, F., and Popper, H. (1967). *Lab. Invest.* **16**, 84.
Shaff, Z., Lapis, K., and Safrany, L. (1969). *Beitr. Pathol. Anat.* **140**, 54.

Schaffner, F. (1965). *Am. J. Dig. Dis.* **10,** 99.
Schaffner, F., and Javitt, N. B. (1966). *Lab. Invest.* **15,** 1783.
Schaffner, F., and Kniffen, J. (1963). *Ann. N.Y. Acad. Sci.* **104,** 847.
Schaffner, F., and Popper, H. (1961). *Am. J. Pathol.* **38,** 393.
Schaffner, F., and Popper, H. (1974). *In* "Jaundice" (C. A. Goresky and M. M. Fisher, eds.). pp. 329–349. Plenum, New York.
Schaffner, F., and Sasaki, H. (1965). *Rev. Int. Hepat.* **15,** 461.
Schaffner, F., Bacchin, P., Hutterer, F., Schanbeck, H., Sarkozi, L., Denk, H., and Popper, H. (1971). *Gastroenterology* **60,** 888.
Scharnbeck, H. and Schaffner, F. (1970). *Am. J. Pathol.* **61,** 479.
Schatzki, P. (1962). *J. Pathol.* **73,** 511.
Schatzki, P. (1969). *Lab. Invest.* **20,** 87.
Schatzki, P. (1971). *Z. Zellforsch. Mikrosk. Anat.* **119,** 451.
Schersten, T., Bjorkerud, B., Jakoi, L., and Bjorntorp, P. (1966). *Scan. J. Gastroent.* **1,** 284.
Schersten, T., Bjorntorp, P., Bjorkerud, B., Smeds, S., and Ekhol, R. (1970). *Acta Hepatosplenol.* **17,** 375.
Schmid, R. (1974). *In* "The Liver and Its Deseases". (F. Schaffner, S. Sherloch, and C. M. Leevy, eds.). pp. 85–96. Intercontinental Medical Corp., New York.
Schmid, R. (1978). *Gastroenterology* **74,** 1307.
Seidel, D., Agostini, B., and Muller, P. (1972). *Biochem. Biophys. Acta* **260,** 146.
Simon, G., and Varonier, H. P. (1963). *Schweiz. Med. Wochschr.* **93,** 459.
Stein, O., Alkan, M., and Stein, Y. (1973). *Lab Invest.* **29,** 166.
Steiner, J. and Baglio, C. (1963). *Lab. Invest.* **12,** 765.
Steiner, J., and Carruthers, J. S. (1961). *Am. J. Pathol.* **39,** 41.
Steiner, J., Phillips, M., and Boglio, C. (1963). *Am. J. Pathol.* **43,** 677.
Sussi, P. L., and Rubaltelli, F. F. (1968). *Lancet* **II,** 1396.
Tanikawa, K., (1965). *Kurume Med. J.* **12,** 86.
Tanikawa, K. (1968). "Ultrastructural Aspects of the Liver and Its Disorders." Springer-Verlag, Berlin and New York.
Tanikawa, K., and Eruma, T. (1965). *Kurume Med. J.* **12,** 27.
Tanikawa, K., and Ikejiri, N. (1977). *In* "Kupffer cells and Other Liver Sinusoidal Cells" (E. Wisse and D. L. Knook, eds., pp. 153–162, Elsevier, Amsterdam.
Tanikawa, K., and Okuda, K. (1970). *Proc. 4th World Cong. Gastroent., Copenhagen 1970,* 126.
Tanikawa, K., Abe, H., Miyakoda, U., and Okuda, K. (1971). *Acta Hepat. Jap.* **12,** 160.
Toda, G., Oka, H., Oda, T., and Ikeda, Y. (1975). *Biochem. Biophys. Acta* **413,** 52.
Toker, C., and Trevino, N. (1965). *Arch. Pathol.* **80,** 453.
Vial, J. D., Simon, F. R., and Mackinnon, A. M. (1976). *Gastroenterology* **70,** 85.
Wachstein, M., and Heisel, E. (1957). *Am. J. Clin. Pathol.* **27,** 13.
Wegmann, R., Caroli, J., Eteve, J., Rangier, M., and Charbonrier, A. (1960). *Ann.Histochem.* **5,** 71.
Weibel, E., Staubli, W., Gnagi, H., and Hess, F. (1969). *J. Cell Biol.* **42,** 68.
Wills, E., and Epstein, N. (1966). *Am. J. Pathol.* **49,** 605.
Wisher, M. H., and Evans, W. H. (1975). *Biochem. J.* **146,** 375.
Yamaguchi, H., Koyama, K., Matsuo, Y., Kashimura, S., Takagi, Y., Muto, I., Otoda, Y., Otowa, T., Ouchi, K., Anezaki, T., and Itoh, K. (1975). *Jap. J. Gastroent.* **72,** 392.
Yodaiken, R. (1966). *Lab. Invest.* **15,** 403.
Zaki, F. (1966). *Medicine* **45,** 537.

# EDITORS' SUMMARY TO CHAPTER IX

In this chapter Dr. Tanikawa reviews the currently available information concerning the cellular mechanisms involved in jaundice and cholestasis. The ultrastructural aspects of these problems are reviewed and correlated with the altered pathophysiology. Many questions remain unanswered in this particular field, however, there is considerable evidence that this problem relates closely to cell membrane function both in intracellular organelles and at the cell surface. Recently, it has been suggested that intrahepatic cholestasis is related to a failure of microfilament-membrane interaction. It is suggested that normally contraction of these filaments propels bile along the canaliculi. In a recent paper (Trump *et al.*, 1976) (Fig. 23) we explored the possible mechanisms of cholestasis in viral hepatitis and in another paper the relationship of hemorrhagic and septic shock in human patient to jaundice (Champion *et al.*, 1976). Both may relate to the original suggestion of Phillips and Steiner (1965) that filament contraction within the hepatocyte is an important force involved in the propulsion of bile along the canaliculi and that failure of such contraction may represent a fundamental mechanism in jaundice. Certainly, filament condensations have been seen along the cell membrane in hepatocytes and one could argue that following cell injury, for example, after complement activation which may occur in viral hepatitis cell membrane damage with calcium influx could lead to impaired filament contraction. Some of the cell membrane changes described here including formation of bulbous dilated villi, blebs, and circular profiles could also be related to this filament failure. The role of membrane ATPases in filament contraction needs further explanation, however, ATP utilization is probably clearly involved in the contraction of such filaments and secondarily in bile secretion.

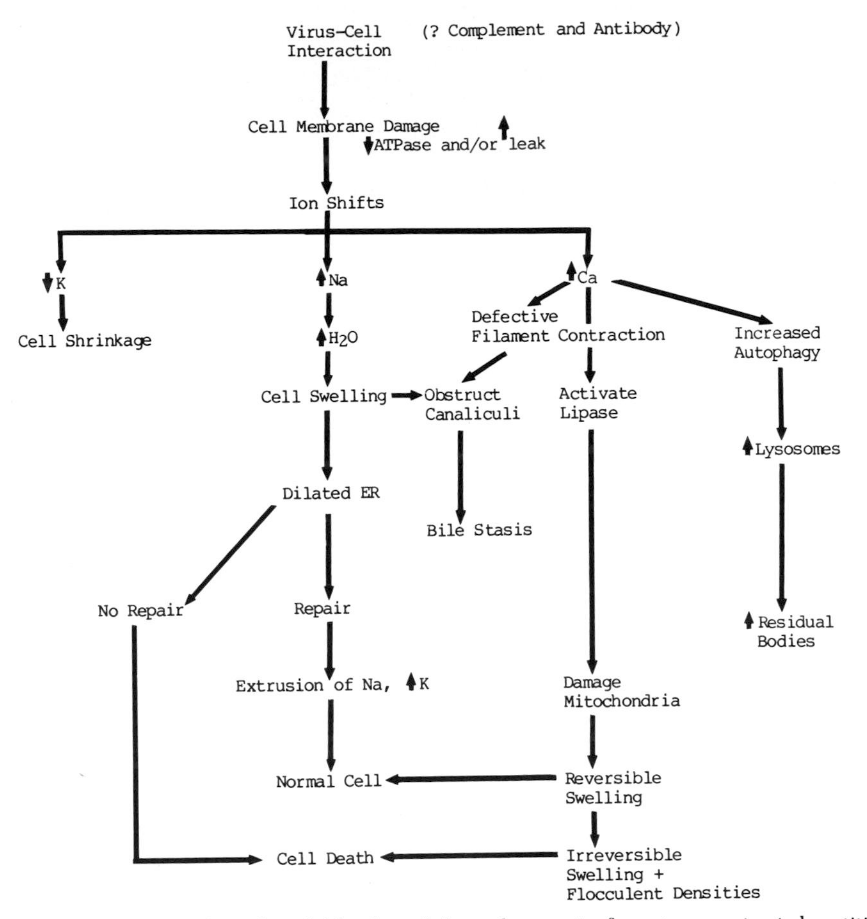


**Fig. 23.** A hypothetical model for the cellular pathogenesis of events occurring in hepatitis. The initial event involves virus–cell interaction with hepatic parenchymal cells possibly involving antibody and complement. This damages the plasma membrane, increasing the passive permeability to ions and possibly also involves inactivation of the transport pumps. This soon leads to the production of ionic and water shifts. The increased sodium content soon leads to swelling of various compartments such as the cell sap and endoplasmic reticulum. This swelling obstructs the bile canaliculi and probably contributes to bile stasis. A serious event leading ultimately to irreversible cell damage is redistribution of calcium inside the mitochondria and in the cell sap, which activates some very damaging mitochondrial enzymes, so-called mitochondrial phosphalipases.

## References

Champion, H. R., Jones, R. T., Trump, B. F., Decker, R., Wilson, S., Miginski, M., and Gill, W. (1976). *Surg. Gynecol. Obstet.* **142**, 657.

Phillips, M. J., and Steiner, J. W. (1965). *Am. J. Pathol.* **46**, 985.

Trump, B. F., Kim, K. M., and Iseri, O. A. (1976). *Am. J. Clin. Pathol.* **65**, 828.

# CHAPTER X

# PATHOLOGY OF THE ENDOTHELIUM

Reginald G. Mason and John U. Balis



## I. Introduction

The endothelium has been studied morphologically for many years. Indeed, until recently the majority of studies of endothelium were of a



PATHOBIOLOGY OF CELL MEMBRANES, VOL. II

ISBN 0-12-701502-7

morphological nature. Earlier studies of endothelium with light microscopy indicated that this cell type responded to injury in a limited number of ways. Injured endothelial cells were known to put forth surface projections, to swell, and to desquamate. Aside from these relatively gross changes, little was known of the reactivity of endothelium to injurious agents until recently.

The advent of the electron microscope permitted a more detailed investigation of the normal structure of endothelium as well as the reactivity of endothelium to injury. Endothelial cells from most parts of the vascular system were found to contain a surprisingly limited number of cell organelles (Majno, 1965). The most striking feature of the cytoplasm of endothelial cells was the presence of numerous pinocytotic vesicles. In endothelium from many areas, these vesicles are the most numerous structures in the cell's cytoplasm. Additional cytoplasmic structures include centrioles, mitochondria, endoplasmic reticulum, Golgi, occasional lipid droplets, granules that may be lysosomes or storage structures, cytoplasmic microfibrils, numerous small vesicles, and peculiar tubular structures termed Weibel–Palade bodies (Fig. 1) (Weibel and Palade, 1964). The later are thought to be found only in endothelial cells. The paucity of endoplasmic reticulum and storage granules suggests that endothelial cells normally synthesize few products that will be released into the pericellular environment.

A striking feature of endothelial cells is their ability to spread themselves over large surface areas of basal lamina (Zweifach, 1973). This produces only a thin coating of endothelium in many areas and thus an amazingly attenuated lining for the vascular system. This thin cellular lining in some parts of the vascular system apparently controls permeability, while in other areas such as the glomerulus, it plays only a limited role.

The question of control of vascular permeability has been pursued for many years (Luft, 1973). The presence of viable endothelial cells appears necessary for control of vascular permeability in most cases, but even now there is debate as to the exact role of the endothelium in this phenomenon (Renkin, 1978). In some areas, particularly in capillaries, endothelium is fenestrated, and those diaphragmed openings are thought to control permeability (Majno, 1965; Zweifach, 1973; Simionescu *et al.*, 1975). In organs such as liver, spleen, and bone marrow capillary endothelium generally is discontinuous and in rat liver it is fenestrated (Wisse, 1970). In many capillary beds and most larger blood vessels, endothelium is continuous; the cells overlap and are connected by a number of different types of junctions. Electron microscopic tracers in permeability experiments have been useful for the identification of pathways involved in the transport of relatively large molecules. However, electron microscopy has been of limited value in determining the relative importance of the various pathways. At present, most

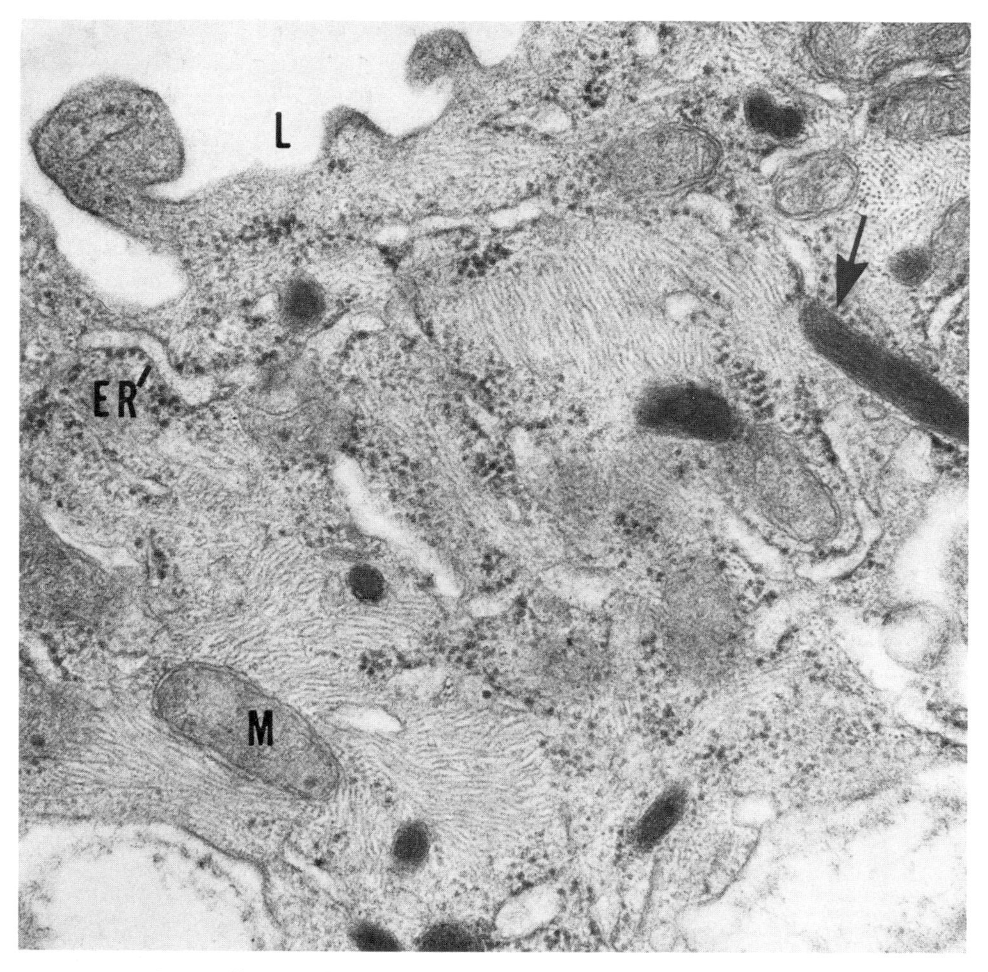


**Fig. 1.** Postcapillary venule in connective tissue stroma of a mediastinal paraganglioma invading the spinal epidural space. The cytoplasm contains numerous microfilaments and frequent Weibel-Palade bodies (arrow). Vascular lumen, L; rough endoplasmic reticulum, ER; mitochondria, M. ×42,800. (Courtesy of Dr. Raoul Fresco.)

investigators agree that the major site of egress of larger molecular weight substances and cells from the vessel lumen to perivascular tissues is through the intercellular junctions of endothelium. These junctions normally are closed and open only in response to specific physiological stimuli or to injurious agents. Small molecules appear to pass normally through the pinocytotic or transendothelial channel systems that exclude molecules of a diameter greater than 20 Å. This subject will not be pursued further at this point, but several review and other articles pertinent to the transport of molecules with a diameter greater than 20 Å are available (Zweifach, 1973; Renkin, 1978; Gabbiani and Guido, 1976; Simionescu *et al.*, 1978).

Endothelial cells with apparent phagocytic properties are seen on occasion in vessels of various organs (Fig. 2). It also has been reported that under certain conditions, such as loading of the reticuloendothelial system by carbon, the endothelium of multiple vascular beds manifests enhanced phagocytic activity (Cotran, 1965). However, this potential function of the endothelium has not been studied adequately.

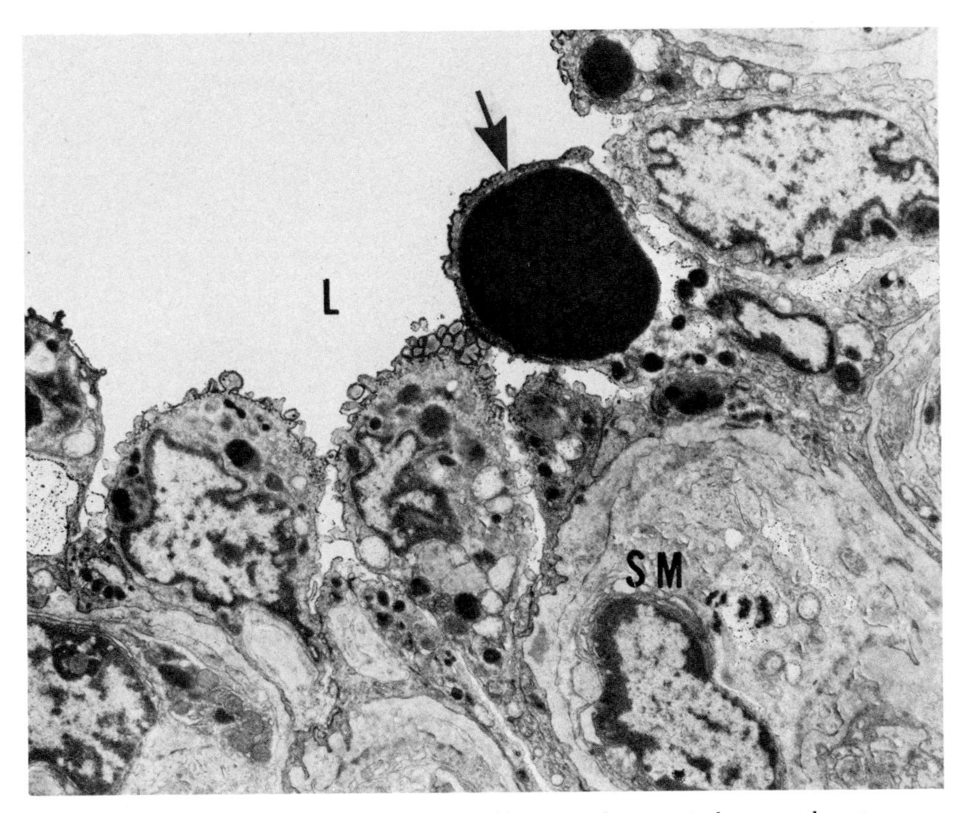


**Fig. 2.** Splenic arteriole from an 18-year-old man with systemic lupus erythematosus, thrombocytopenia, and Coombs' positive hemolytic anemia. An endothelial cell with an engulfed erythrocyte is noted at the arrow. Other endothelial cells contain numerous dense bodies that apparently represent fragments of erythrocytes in various stages of degradation. Vascular lumen, L, smooth muscle cell, SM. ×6100. (Courtesy of Dr. Raoul Fresco.)

## II. Morphology of Normal Endothelium

The morphology of endothelium varies somewhat in different parts of the vascular system of the human and between vascular systems of certain species (Majno, 1965; Wisse, 1972). This must be appreciated, if one is to discern minor changes in the normal morphology of endothelium. There are sometimes marked morphological differences between the endothelium of the human and those of lower species (Rhodin, 1968). Hence, particular attention must be given to the species in many of the published reports of endothelial research. Obviously, most studies of experimentally induced alterations in endothelium are done with nonhuman endothelial cells. Species differences in endothelial morphology as well as in the reactivity of blood components to various agents can be marked.

Recently, there has been considerable attention paid to artifacts that may be induced in endothelium by the fixation and processing of tissues for ultrastructural studies (Gertz *et al.*, 1975a; Davies and Bowyer, 1975; Clark and Glagov, 1976). Of particular importance are the types of fixative, the intraluminal pressure maintained during fixation of blood vessels, osmotic properties of fixatives, and shrinkage and contraction of tissue during fixation and processing. The possibility of producing artifacts in endothelium has been appreciated for many years but first became critical with transmission electron microscopic studies. Of particular importance were artifacts that produced shrinkage of cells in studies where emphasis was placed on evaluation of vascular permeability (Majno, 1965). More recently, considerable attention has been paid to artifacts created during fixation and processing of specimens for scanning electron microscopy (Gertz *et al.*, 1975a; Davies and Bowyer, 1975; Clark and Glagov, 1976). Here the possible surface distortion of endothelium has been well documented during fixation and processing procedures. The importance of maintaining intraluminal vascular pressures at near physiological levels during vessel fixation has been stressed both in scanning and transmission electron microscopic studies. Nevertheless, when careful attention is paid to details of intraluminal pressure, osmolarity of fixatives, avoidance of mechanical injury, and proper dehydration procedures, it is possible to obtain excellent transmission and scanning electron microscopic views of endothelium.

Technical advances in recent years have made possible studies of some of the functions of endothelium (Mason *et al.*, 1977a,b). Endothelial function will not be considered here except as it relates to morphological changes. Several reviews on endothelial function are now available, and the reader is referred to them (Majno, 1965, Mason *et al.*, 1977a,b).

## III. Endothelial Responses to Injury

The ability of endothelium to respond to injurious stimuli appears to be relatively limited so far as the extent of morphological change is concerned. Early changes may include widening of interendothelial cell junctions (Figs. 3–6), adherence of leukocytes (Fig. 7) or platelets (Fig. 8), formation of cytoplasmic blebs or vesicles (Figs. 9 and 10), subendothelial edema (Figs. 4, 8, and 11), separation of adjacent endothelial cells with exposure of subendothelium (Fig. 12), and finally, necrosis of endothelium (Figs. 13 and 14). On occasion, desquamated endothelial cells are found free in the vascular lumen (Fig. 15). Many different chemical and physical stimuli elicit some or all of these changes in endothelium as will be discussed in detail below. In at least some instances, the degree of endothelial cell change appears to be related directly to the strength of the injurious stimulus.

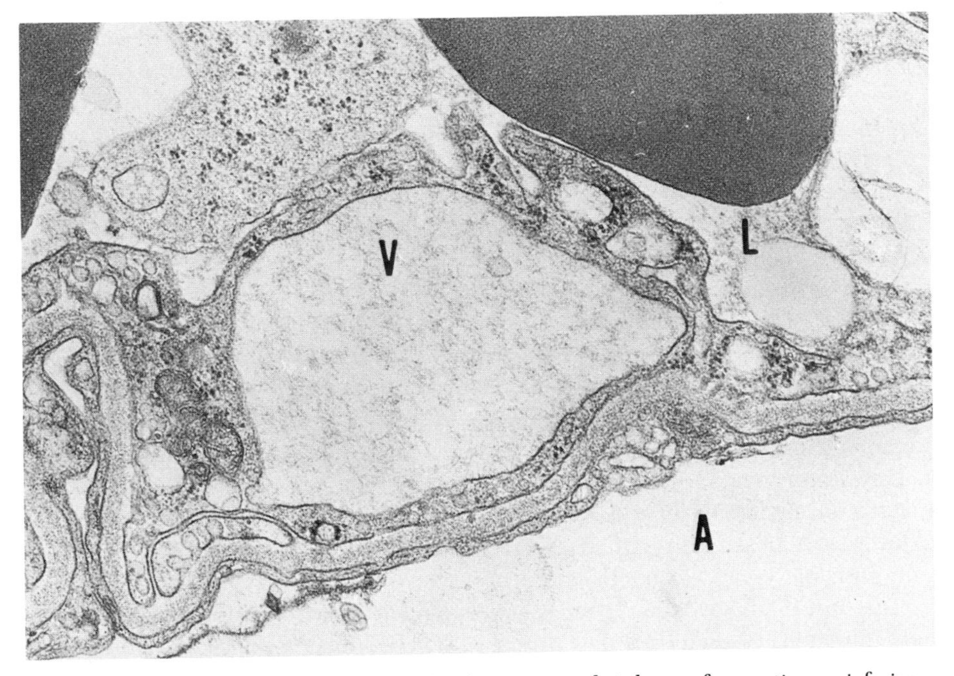


**Fig. 3.** Blood–air barrier from a monkey lung examined six hours after continuous infusion with *E. coli* endotoxin at a rate of 10 mg/kg $hr^{-1}$. The endothelial lining contains a markedly dilated interendothelial junction (V). Alveolar space, A; vascular lumen, L. ×17,400.

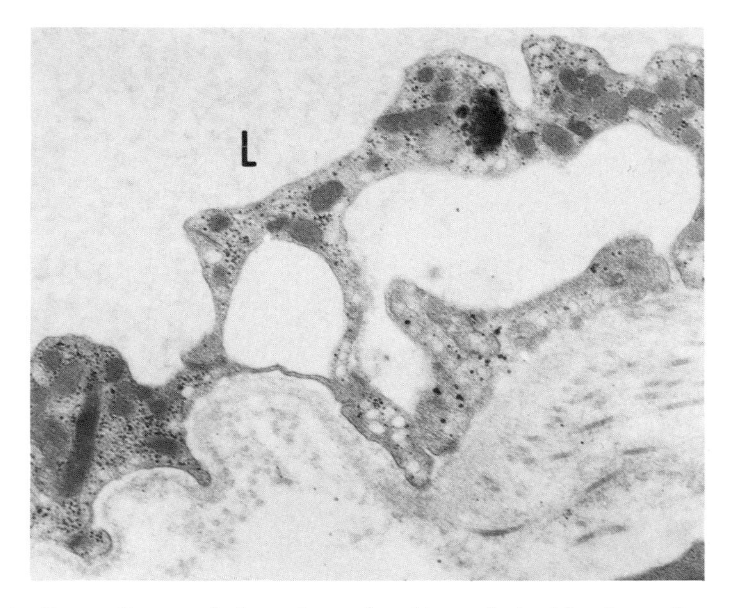


**Fig. 4.** Postcapillary venule from a human lung biopsy obtained four hours after cardiopulmonary bypass. The endothelium shows dilated interendothelial junctions and subendothelial edema. Vascular lumen, L. ×21,300.

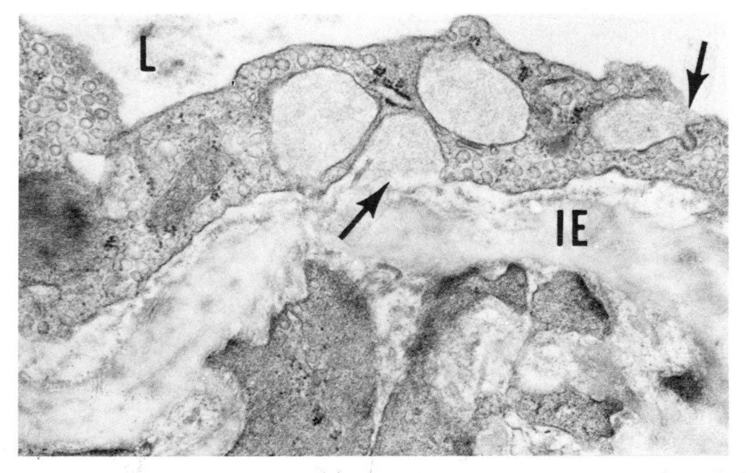


**Fig. 5.** Small pulmonary artery from a mouse with radiation pneumonitis. The endothelium contains dilated interendothelial junctions that mimic cytoplasmic vacuoles opening both in the vascular lumen and subendothelial space (arrows). Vascular lumen, L; internal elastic lamina, IE. ×12,600.

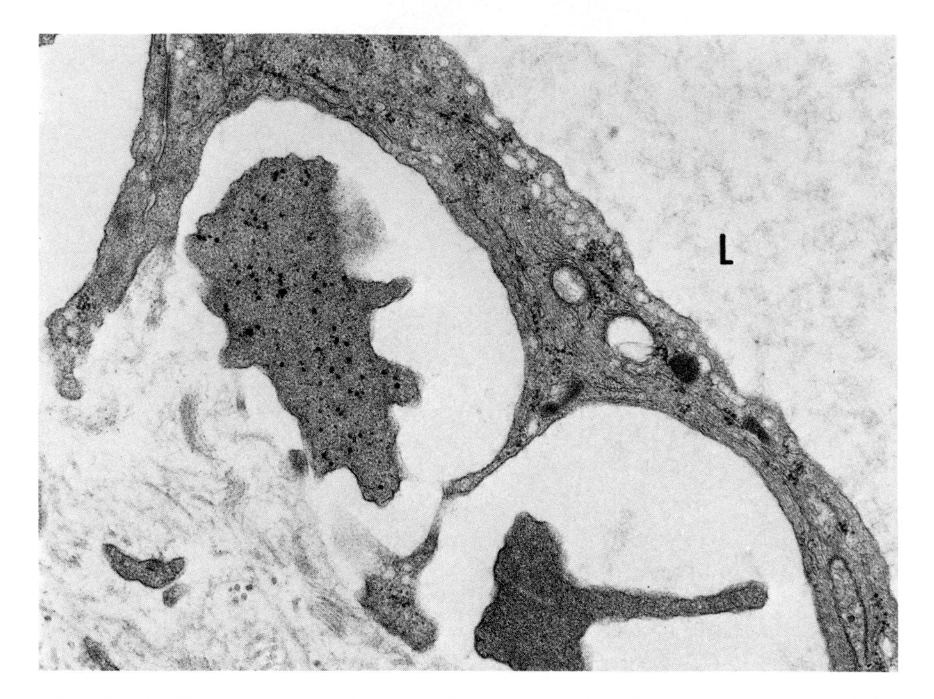


**Fig. 6.** Small pulmonary vein from a monkey examined three hours after infusion of a bolus of endotoxin, 10 mg/kg body weight. The endothelial lining shows subendothelial blisters that represent markedly dilated junctions opening into the subendothelial space. The cytoplasm of the endothelium contains numerous microfilaments. Vascular lumen, L. ×19,500.

## IV. Morphological Changes in Endothelium Produced by Endogenous or Exogenous Agents

### A. *Acute Inflammation*

There are numerous studies of the enhanced adhesion of leukocytes to endothelium in acute inflammation. The mechanisms responsible for this enhanced adhesiveness of leukocytes are not yet clear (Movat and Fernando, 1963; Lackie and deBono, 1977; Jones, 1970; Craddock *et al.*, 1977; Sacks *et al.*, 1978). Recent evidence suggests that in certain instances the complement system may be involved with a specific role played by $C_5$ (Craddock *et al.*, 1977; Sacks *et al.*, 1978). In many studies of acute inflammation in the vascular system, particulate materials such as carbon or ferritin have been used in order to document increased vascular permeability (Renkin, 1978; Simionescu *et al.*, 1978; Cotran, 1965; Robertson and Khirallah, 1973; Bell *et al.*, 1974). It is of interest that enhanced vascular permeability can be demonstrated in many cases before there are detectable morphological changes in endothelial cells. Similarly, enhanced adhesion of leukocytes to endothelium can be demonstrated where no morphological alteration in endothelium is detected, and such adhesion may well be reversible.

### *B. Anoxic Injury*

One of the most frequently used mechanisms for producing experimental injury to endothelium is anoxia. This is particularly pertinent, since anoxic changes in endothelium occur with some frequency in a number of medical conditions as well as during various surgical procedures. Studies of anoxic injury to endothelium have been carried out in a number of different species including humans. Numerous reports indicate that anoxia produces a pre-

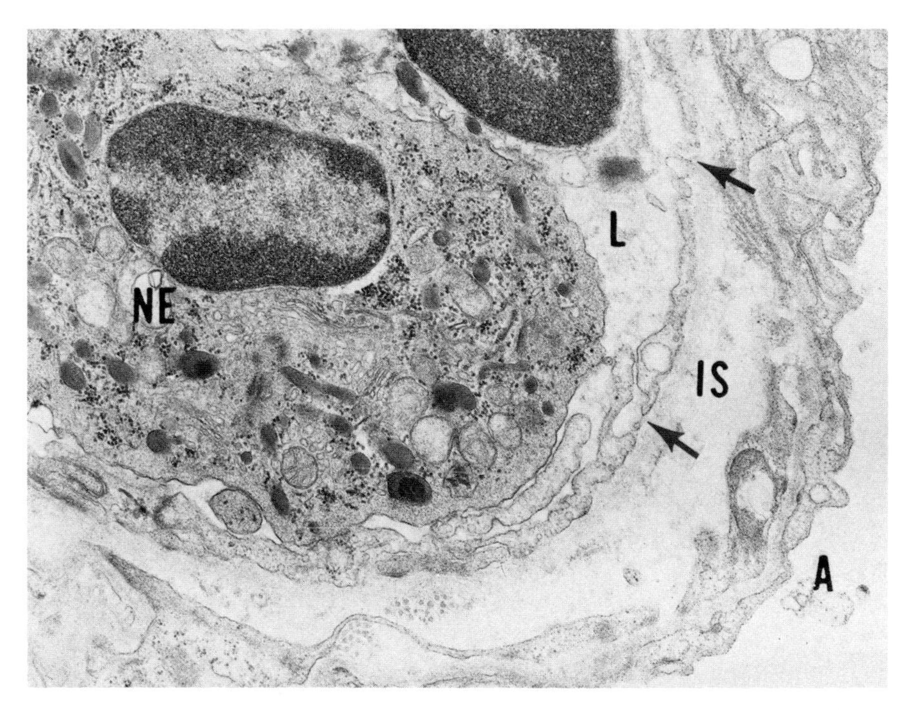


**Fig. 7.** Blood–air barrier from a monkey lung 10 hours after continuous infusion with endotoxin (10 mg/kg -hr$^{-1}$). The endothelial lining of the septal capillary shows loose junctions or interendothelial gaps (arrows) in association with marginated and partially degranulated neutrophils (NE). The interstitial space (IS) is dilated reflecting the presence of interstitial edema. Vascular lumen, L; alveolar space, A. ×14,200.

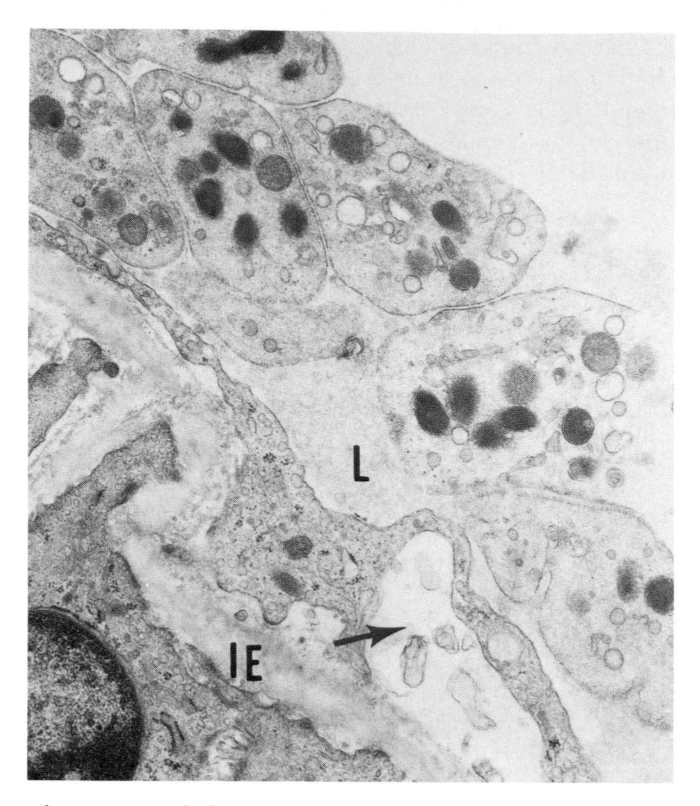


**Fig. 8.** Pulmonary arteriole from a mouse with radiation pneumonitis. Aggregated platelets with partial degranulation and disruption are adherent to the endothelial lining. A large subendothelial blister is noted at the arrow. Vascular lumen, L; internal elastic lamina, IE. ×11,100.

dictable series of morphological changes in endothelium in various parts of the vascular tree (Buck, 1961; Ashton and Pedler, 1962; Tedder and Shorey, 1965; Willms-Kretschmer and Majno, 1969; Meyrick *et al.*, 1972; Little *et al.*, 1973; Kawamura *et al.*, 1974a; Stewart, *et al.*, 1974; Gertz *et al.*, 1975; Kjeldsen and Thomsen, 1975; Nelson *et al.*, 1975; Barnhart and Chen, 1976; Fonkalsrud *et al.*, 1976; Gertz *et al.*, 1976a, b; Nelson *et al.*, 1976; Nelson, 1976; Bhawan *et al.*, 1977; Fonkalsrud *et al.*, 1977a; Johnston and Latta, 1977). These changes begin within minutes of the onset of anoxia and consist initially of the formation of irregular cytoplasmic protrusions sometimes termed pseudopods, filipods, or cytoplasmic veils. This is followed by the formation of generally empty cytoplasmic structures that have been described as blebs, vesicles, or balloons that later appear to burst leaving craters on the surface of the endothelial cell (Willms-Kretschmer and

Majno, 1969; Meyrick *et al.*, 1972; Kawamura *et al.*, 1974a; Stewart, *et al.*, 1974; Barnhart and Chen, 1976; Fonkalsrud *et al.*, 1976; Gertz *et al.*, 1976a, b; Bhawan *et al.*, 1977).

While these changes are in progress, there appears to be a swelling of the entire endothlial cell as is demonstrated by a general decrease in the staining of cell cytoplasm. In both large blood vessels and in capillaries, endothelial damage is usually associated with subendothelial edema, widening of intercellular junctions, and formation of subendothelial blisters (Fig. 4). In some cases, prominent subendothelial edema develops without appreciable structural damage to the endothelium (Figs. 4 and 11). In longer standing anoxia, endothelial cells are reported to become rounded and to have one or more protruding surface blebs. Subendothelial edema becomes prominent in some cases. At least in certain species, leukocytes begin to penetrate between altered endothelial cells and invade the subendothelial space (Stewart

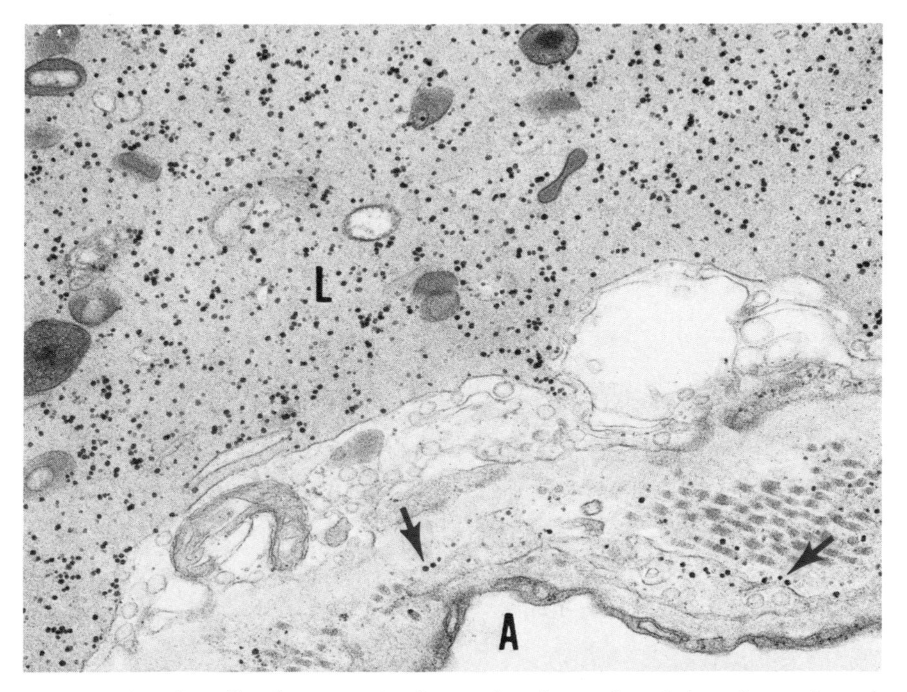


**Fig. 9.** Septal capillary from a monkey lung at three hours after a bolus infusion of *E. coli* endotoxin, 10 mg/kg. The capillary lumen (L) contains cytoplasmic granules and debris from fragmented PMN-leukocytes as well as abundant glycogen particles. Similar glycogen particles are seen in the extracellular space (arrows) indicating increased vascular permeability. The endothelial cytoplasm shows swelling and, in places, bleb formation. Alveolar space, A. ×27,200.

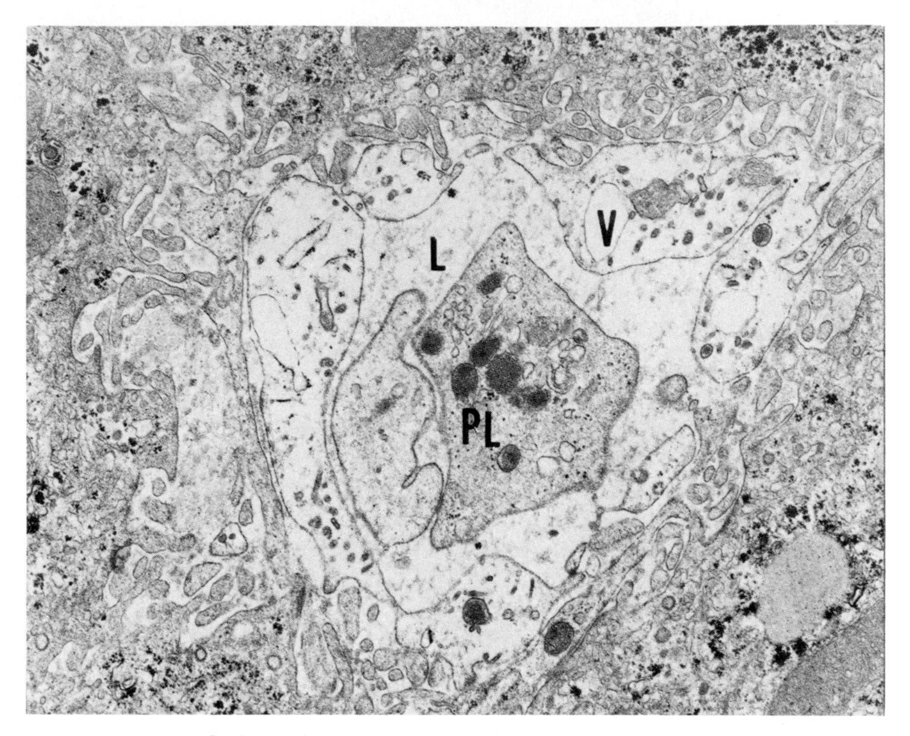


**Fig. 10.** Sinusoid of a rat liver examined 1 hour after a bolus infusion of live *E. coli* organisms ($2.6 \times 10^9$). The endothelium appears swollen with frequent membrane-bound vacuoles (V). Sinusoidal lumen, L; platelet, PL. ×12,600.

*et al.*, 1974). In addition, some reports indicate that there is sticking of leukocytes and platelets to the endothelial surface, particularly at points of damage (Tedder and Shorey, 1965; Willms-Kretschmer and Majno, 1969; Meyrick *et al.*, 1972; Kawamura *et al.*, 1974; Stewart *et al.*, 1974; Gertz *et al.*, 1975b; Kjeldsen and Thomsen, 1975; Nelson *et al.*, 1975; Fonkalsrud *et al.*, 1976; Gertz *et al.*, 1976a; Nelson *et al.*, 1976; Nelson, 1976; Bhawan *et al.*, 1977; Johnston and Latta, 1977).

There is controversy concerning this latter statement, since some investigators have reported no sticking or adhesion of platelets to apparently damaged endothelium (Warren *et al.*, 1973; Fishman *et al.*, 1975). Other workers have reported in anoxic endothelium a decrease in the thickness of the layer of concanavalin A stainable material normally found on the outer surface of the endothelial cell plasma membrane (Baumann *et al.*, 1976). The significance of this latter observation is not known at present.

With the progression of anoxia, there is increasing damage to the endothelial cell with increased surface vesicle and crater formation, increased swelling of the cell, and increased association with leukocytes. If anoxia continues past a certain point, the endothelial cell usually detaches from the basal lamina (Nelson, 1976; Fonkalsrud *et al.*, 1977a). In this detachment, the endothelial cell may be assisted by the numerous leukocytes that have invaded the subendothelial space (Stewart *et al.*, 1974). In other instances, investigators have reported damaged endothelial cells remaining attached to the basal lamina; in these cases only a few particles of cytoplasm and fragments of cell membrane may remain. Such variance may reflect species differences, differences in the function of endothelium in various parts of the vascular system, or differences in the response to specific injurious stimuli. Nevertheless, the end result of severe anoxic damage to endothelium appears to be cell death with the formation of a small mural thrombus composed predominantly of platelets.

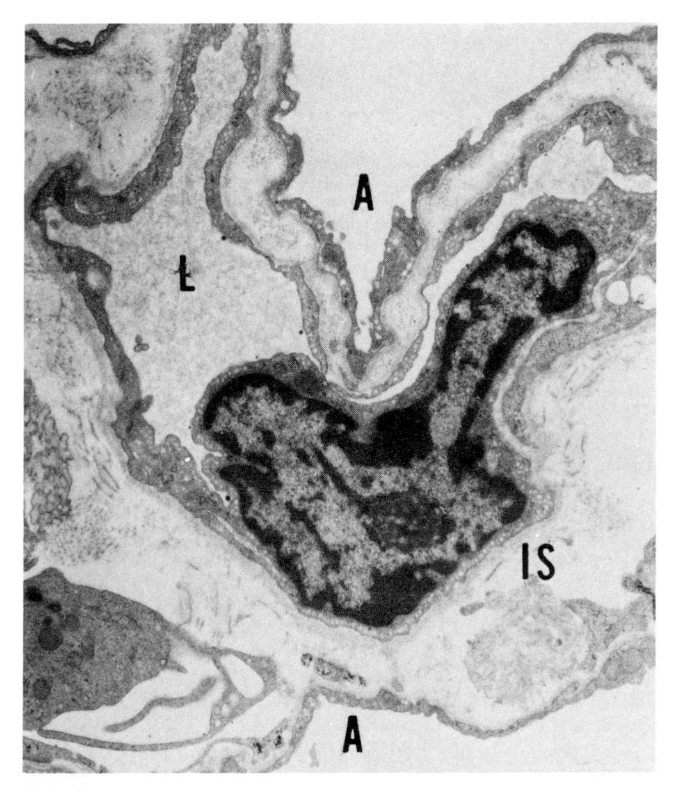


**Fig. 11.** Alveolar septum from the same lung biopsy as Fig. 4. The endothelium of a capillary appears intact but shows numerous pinocytic vesicles. The interstitial spaces (IS) of the septum are markedly dilated and edematous. Capillary lumen, L; aveolar space, A. ×7900.

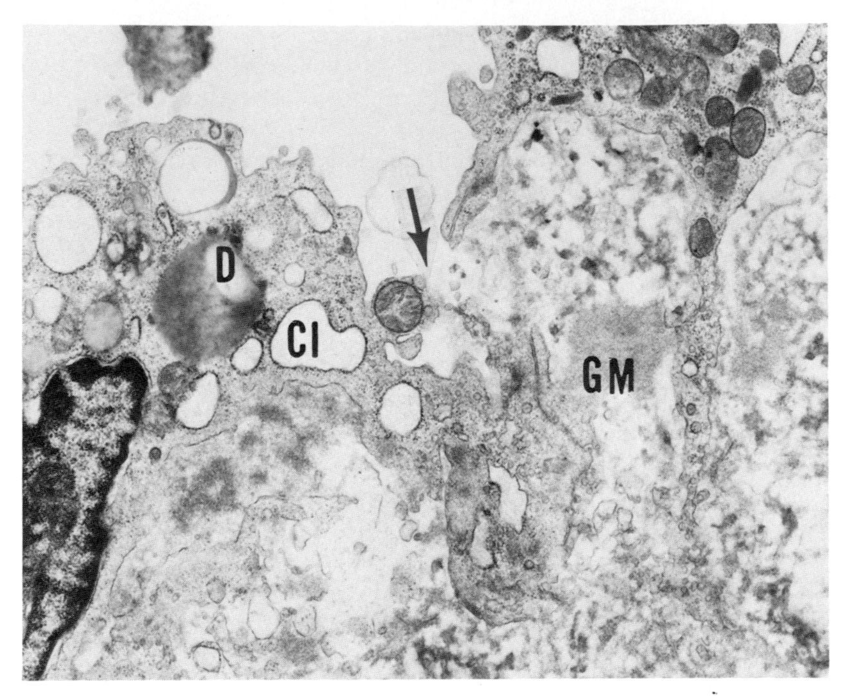


**Fig. 12.** Aorta of a rat that was on a hyperlipemic diet for 10 weeks. The endothelial lining shows a gap at the arrow. Masses of homogenous granular material (GM), probably plasma proteins, are noted in the subendothelial spaces. The endothelium contains partially vacuolated lipid droplets (D) and cisternae of the endoplasmic reticulum (CI). ×24,600. (Reproduced with the permission of the publisher from Figure 14, *Exp. Mol. Pathol.* 8, 90, 1968.)

## *C. Damage Produced by Gases*

Several different gasses have been reported to injure endothelium. Air bubbles have been shown to embolize within the vascular system and produce injury to endothelium of the microcirculation (Warren *et al.*, 1973). The presence of air bubbles within the vascular system appears to damage endothelium primarily through pressure effects, since herniation of endothelial cells through basal lamina has been observed. On the other hand, exposure of endothelium to air by permitting air to flow through a vessel or by opening the vessel results in marked injury to this cell type, and this has been used as a model for endothelial injury and regeneration (Fishman *et al.*, 1975). An increase in the carbon monoxide (Kjeldsen *et al.*, 1972) content of inspired air was found to cause a severe edematous reaction in rabbit endothelial cells with formation of cytoplasmic blebs. Elevated levels of oxygen in inspired air produce widespread pulmonary damage with injury to endothelium and thrombus formation (Kistler *et al.*, 1967; Teplitz, 1976).

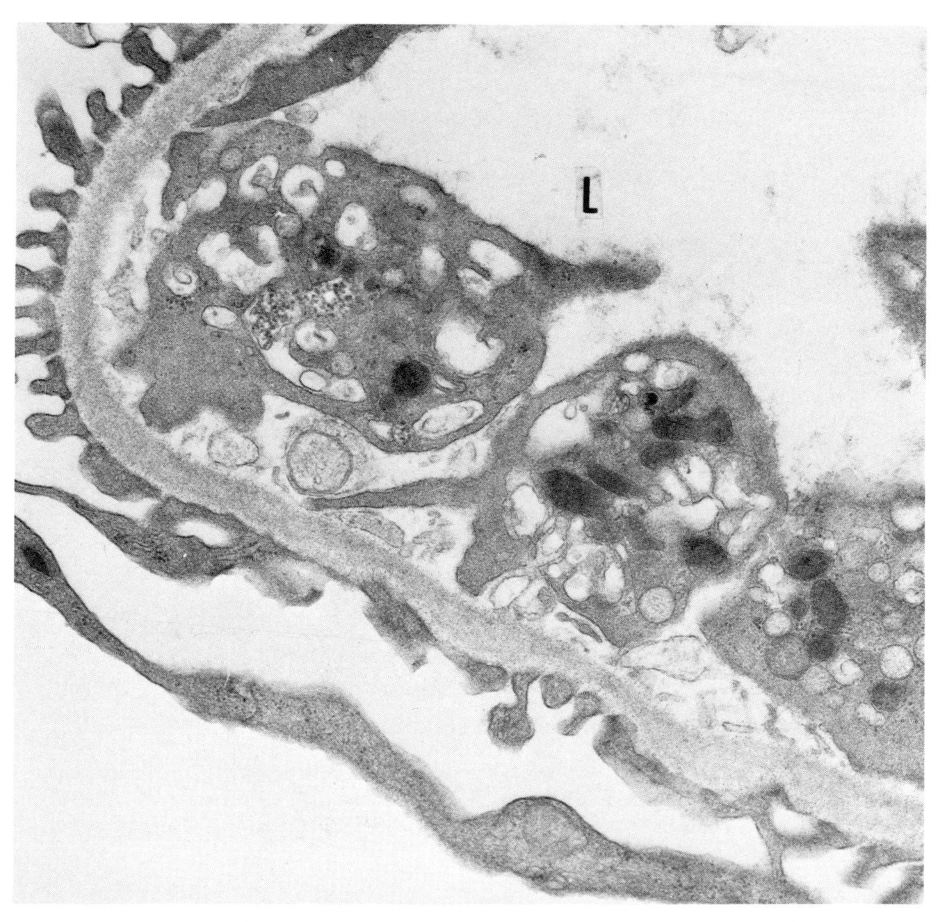


**Fig. 13.** Canine glomerulus following hyperacute rejection. The capillary endothelium is replaced by cell debris and aggregated platelets that are adherent to the underlying basement membrane. Capillary lumen, L. ×21,600. (Courtesy of Dr. Raoul Fresco.)

### *D. Vasoactive Amines and Kinins*

A number of studies of effects of epinephrine (Strum and Junod, 1972; Burri and Weibel, 1968; Sunaga *et al.*, 1969; Wang *et al.*, 1971; Griffiths and Irving, 1976; Constantinides and Robinson, 1969b; Bevan and Duckles, 1975), norepinephrine (Bevan and Duckles, 1975; Iwasawa *et al.*, 1973), histamine (Majno *et al.*, 1969), serotonin (Iwasawa *et al.*, 1973), bradykinin (Mason *et al.*, 1977a), prostaglandins and endoperoxides (Mason *et al.*, 1977a), and angiotensin II (Mason *et al.*, 1977a) indicate that these agents can influence the morphology of endothelium. The agents listed above cause

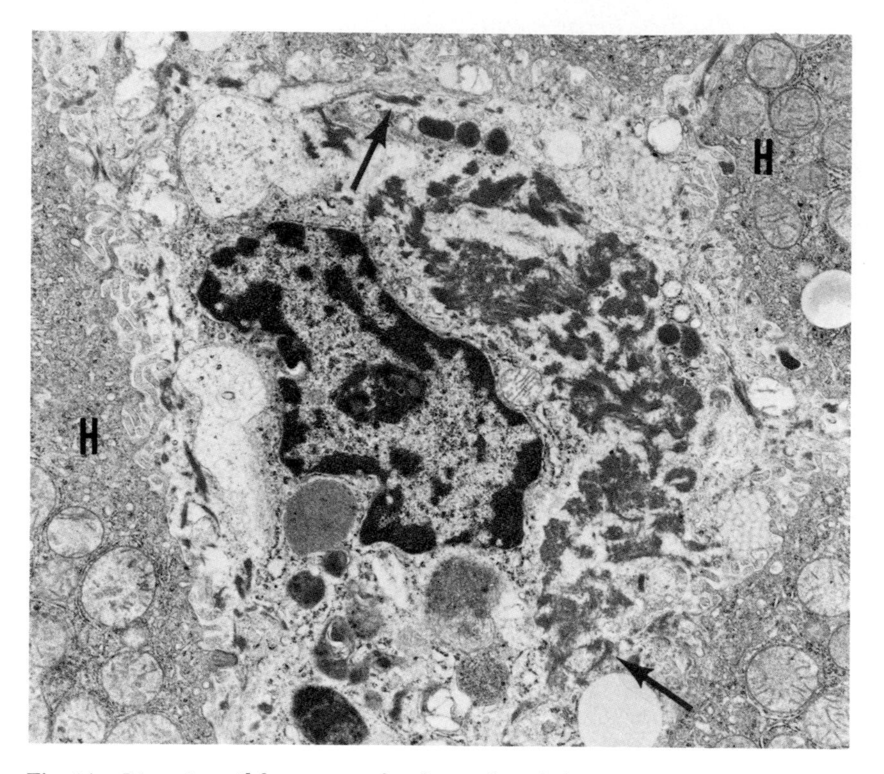


**Fig. 14.** Liver sinusoid from a rat at four hours after a bolus injection of *E. coli* organisms. The sinusoidal lumen is occluded by fibrinous deposits and debris. The sinusoidal endothelium is disrupted and strands of fibrin are noted in the spaces of Disse (arrows). Hepatocyte, H. ×9200.

focal separation of endothelial junctions, primarily of the venular endothelium, leading to the appearance of large interendothelial gaps. Recent studies have indicated that the intercellular junctions of the venular endothelium are characterized by a loose organization that may, at least in part, be responsible for the susceptibility of these junctions to local mediators of change in vascular permeability (Simionescu *et al.*, 1978). In cases where high concentrations of these vasoactive substances were used experimentally or were present in patients, there was apparent direct damage to endothelial cells with rupture of cell plasma membranes (Constantinides and Robinson, 1969b). In general, when experimental studies are conducted with lower concentrations of these agents, the most frequently noted change is contraction of endothelial cells with wrinkling of the nucleus (Majno *et al.*, 1969). The vasoactive amines produce increased vascular permeability, but aside

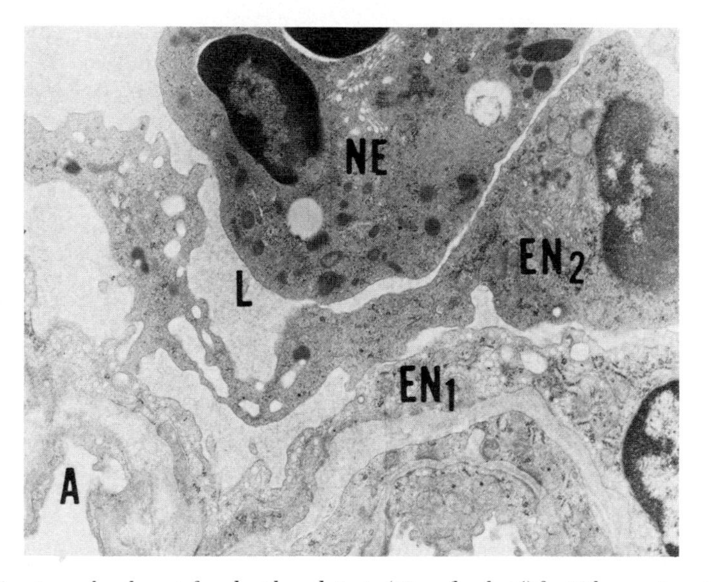


**Fig. 15.** A monkey lung infused with endotoxin (10 mg/kg $hr^{-1}$) for 10 hours. A portion of a septal capillary with an intact endothelial lining ($EN_1$) is present. The capillary lumen contains a partially degranulated neutrophil (NE) and a completely desquamated endothelial cell ($EN_2$) that can be recognized by its long cytoplasmic process. Capillary lumen, L; alveolar space, A. ×11,400.

from cell contraction these agents did not appear to damage endothelial cells unless present at exceedingly high concentrations. Several investigators have reported swelling of endothelial cells and formation of projections from endothelial cell cytoplasm in response to physiologically attainable but still high levels of vasoactive amines or kinins (Strum and Junod, 1972; Constantinides and Robinson, 1969b).

### *E. Atherosclerosis*

Much has been written concerning the possible role of endothelium in atherosclerosis (Constantinides, 1976a, b). Particular attention has been paid of late to alterations in vascular permeability induced by agents that are thought to initiate atherosclerosis (Constantinides, 1968; Shimamoto *et al.*, 1971; Weber *et al.*, 1974; Maca and Hoak, 1974; Jeppsson and Schoefl, 1974; Silkworth *et al.*, 1975; Shimamoto, 1975; Björkerud and Bondjers, 1976; Shimamoto *et al.*, 1976; Goode *et al.*, 1977; Lewis and Kottke, 1977; Bondjers *et al.*, 1977; Bylock *et al.*, 1977a). On the one hand, endothelium has been injured deliberately through mechanical or other means, and the generation of atherosclerosis within the underlying segment of vascular wall studied (Constantinides, 1976a, b; Björkerud and Bondjers, 1971, 1972). On the other hand, attempts have been made to alter endothelial cell function by less traumatic means (Fishman *et al.*, 1975). In other studies (*vice infra*), when injurious agents were used at high concentration, there frequently was dramatic swelling of endothelial cells usually followed by desquamation. When injurious agents were used at more physiological concentrations, the morphological changes in endothelium usually were predictable. Increases in the level of plasma cholesterol (Constantinides, 1968; Shimamoto *et al.*, 1971; Weber *et al.*, 1974; Maca and Hoak, 1974; Jeppsson and Schoefl, 1974; Silkworth *et al.*, 1975; Shimamoto, 1975; Björkerud and Bondjers, 1976; Shimamoto *et al.*, 1976; Goode *et al.*, 1977; Lewis and Kottke, 1977; Bondjers *et al.*, 1977; Bylock *et al.*, 1977a) and other lipids have been reported to produce generalized swelling of endothelial cells with formation of cytoplasmic blebs or pseudopods. In addition, cholesterol and other agents have been reported to produce increased vascular permeability presumably through induction of endothelial cell contraction.

Considerable attention has been given to the effects of hypercholesterolemia on endothelial cell function (Constantinides, 1968; Shimamoto *et al.*, 1971; Weber *et al.*, 1974; Maca and Hoak, 1974; Jeppsson and Schoefl, 1974; Silkworth *et al.*, 1975; Shimamoto, 1975; Björkerud and Bondjers, 1976; Shimamoto *et al.*, 1976; Goode *et al.*, 1977; Lewis and Kootke, 1977; Bondjers *et al.*, 1977; Bylock *et al.*, 1977a).

While numerous studies of effects of hypercholesterolemia have been reported, there are few reports of specific injury to endothelium. Animals that were rendered hypercholesterolemic have been reported to have endothelium with increased numbers of stomata, and the presence of multinucleated endothelial cells as well as alterations in the silver staining properties of endothelial cell junctions have been noted as have alterations in permeability to Evans Blue. While there appears to be definite damage to endothelium in hypercholesterolemic animals, the mechanisms of this damage have not yet been elucidated. Morphological changes of endothelial cells in hypercholesterolemia are of a nonspecific type. Of interest is the fact that enhanced adhesion of platelets to altered endothelial cells has been found in hypercholesterolemic animals.

The importance of changes in endothelium in the genesis of atherosclerosis is obvious. Numerous studies of atherosclerosis have been conducted at a late stage of the disease, when endothelial cell damage already is marked, and, indeed, endothelium may well be missing in areas overlying atherosclerotic plaques (Constantinides, 1966; Sunaga *et al.*, 1970; Davies *et al.*, 1976 Minick *et al.*, 1977a). It has not been difficult to produce marked injury in endothelium when high concentrations of injurious agents are infused. On the other hand, the more subtle changes that have been reported in endothelium in long-standing hypercholesterolemia suggest that this state is indeed injurious to endothelium. Several investigators report an apparent decrease in endothelial cell glycocalyx in atheroscelerosis (Weber *et al.*, 1973; Balint *et al.*, 1974). More refined techniques are needed, if we are to study the earlier changes in endothelium that may preceed the accumulation of lipid deposits within blood vessel walls.

### *F. Bacteria*

Numerous interesting studies have documented effects of injury to endothelium apparently produced by bacteria or bacterial products. Marked injury to endothelium with intracellular swelling and formation of cytoplasmic blebs with subsequent thrombus formation was reported to occur following anthrax septicemia (Dalldorf and Beall, 1967; Dalldorf *et al.*, 1969). Cholera toxin has been reported to produce extensive ultrastructural alteration in endothelial cell cytoplasm without disruption of intercellular junctions (Hashimoto *et al.*, 1974). Cholera toxin produced increased numbers of cytoplasmic multivesicular bodies and systems of caveolae and vesicles. These changes were thought to be responsible for the perivascular edema that follows intradermal injection of cholera vibrio toxin. Certain bacterial exotoxins appear to injure endothelium (Carne, 1978). Endothelial injury induced by gram-negative bacteria (Figs. 10, 14) may lead to thrombosis in certain vascular beds, and this is thought to be essentially similar to that observed in response to endotoxin (*vide infra*). However, this notion is not based on solid morphological evidence. For example, at sites of microembolization with *Escherichia coli* organisms, the endothelial lining was found to be completely necrotic in the absence of fibrinous deposits, platelet aggregates, or marginated leukocytes (Fig. 16, unpublished). Similar changes have not been described follwing endotoxin injections. Mycoplasma (Manuelidis and Thomas, 1973) infection has been shown to induce occlusion of capillaries due to swelling of endothelial cells. Finally, it is known that a number of organisms including rickettsia can live within endothelial cells (DeBrito *et al.*, 1973). The rickettsia apparently multiply within endothelial cell cytoplasm without immediate, pronounced injury to their host. On the other hand, the immunological response to rickettsia in the form of antibody formation and liberation of lysosomal enzymes from immunologically attracted neutrophils damages endothelial cells, so that they desquamate initiating the formation of thrombi. Proliferation of rickettsia intracellularly increases endothelial cell size, and this can disturb flow in the microcirculation. Damage to endothelium in meningococcal septicemia (Dalldorf and Jennette, 1977) appears to be due more to thrombus formation (Dalldorf *et al.*, 1968) in the

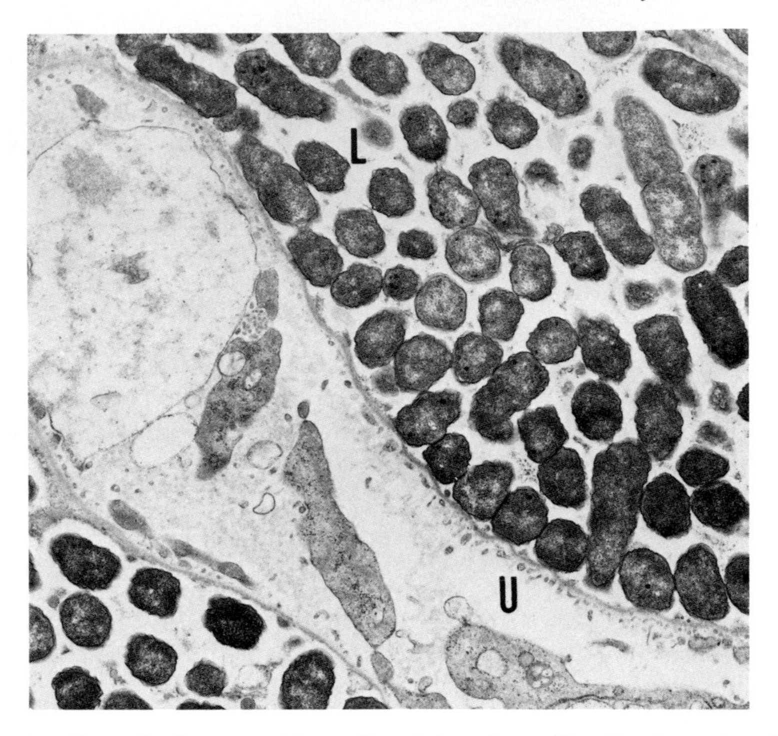


**Fig. 16.** Glomerulus from a rat 10 hours after a bolus infusion of live *E. coli* organisms. The endothelial lining is in contact with clumps of bacterial microemboli, and it is almost completely fragmented or desquamated. Similar fragmentation and desquamation involves the epithelial cells. Capillary lumen, L; urinary space, U. ×12,600.

microcirculation than to injurious effects of the organism on the endothelial cell itself.

## G. *Endotoxin*

Endotoxin has been reported by a number of different investigators to produce severe injury to endothelium (McGrath and Stewart, 1968; Stehbens, 1965; Gaynor, 1971, 1972; Wright, 1973; Evensen and Shepro, 1974; Evensen *et al.*, 1975; Gerrity *et al.*, 1976; Reidy and Bowyer, 1977a; Balis *et al.*, 1974, 1978). Unfortunately many studies of this type have been conducted only at the light microscopic level. Nevertheless, several electron microscopic studies of endothelium exposed to endotoxin have been carried out, and these indicate marked swelling of the cells, formation of cytoplasmic projections, and adhesion of leukocytes (Fig. 7). Additional changes include dilatation of interendothelial junctions and formation of subendothelial blisters. These changes often are seen in capillaries (Fig. 3) but are especially prominent in arterioles and venules (Fig. 6). Patchy endothelial desquamation is often prominent in endotoxemia and fragmented or intact (Fig. 15) endothelial cells may be found free in the vascular lumen. Once endothelial cells have desquamated, thrombus formation can be marked in some but not necessarily all parts of the vascular system. Moreover, there are considerable species differences in the distribution of vascular lesions including thrombosis.

The action of endotoxin in the vascular system is complex, since endotoxin frequently produces a state of disseminated intravascular coagulation (Balis *et al.*, 1978). Thus, it is difficult to say that infusion of endotoxin in an animal has a direct effect upon endothelium itself, because endotoxin activates the blood coagulation, complement, fibrinolytic, and kinin systems and alters platelets and leukocytes. The role of leukocyte margination in initiating endothelial injury has been emphasized especially, because this process represents the earliest event described in both endotoxemia and gram-negative bacteremia. Endotoxin-induced damage to endothelium could be an important precursor of more chronic vascular injury including the production of atherosclerosis. More studies of effects of endotoxin on endothelium are needed.

### *H. Hypertension*

The predominant change in endothelium in hypertension appears to be contraction of the endothelial cell with enhanced vascular permeability (Suzuki *et al.*, 1971a, b; Wiener and Giacomelli, 1973; Robertson and Khairallah, 1972; Still and Dennison, 1974; Jones, 1974; Schwartz and Benditt, 1977; Gabbiani *et al.*, 1975).

Several workers have reported an increase in the number of cytoplasmic microfilaments in endothelial cells of hypertensive subjects (Suzuki *et al.*, 1971b; Gabbiani *et al.*, 1975; Huttner and Rona, 1975). Other workers have documented the increase in vascular permeability in hypertensive animals (Suzuki *et al.*, 1971a, b; Robertson and Khairallah, 1972). Still others have reported an increase in the cylindrical cytoplasmic surface projections demonstrated by both transmission and scanning electron microscopy with an increase in adhesion of leukocytes to endothelium (Still and Dennison, 1974). Whether the latter changes are secondary to contraction of the muscularis or due to direct effects of toxic agents on endothelium, or both, remains unanswered. Since endothelium catabolizes a number of vasoactive kinins, amines, prostaglandins, and angiotensin II (Mason *et al.*, 1977a), hypertension may reflect an alteration in this catabolic function not demonstrable by presently employed ultrastructural techniques.

### *I. Biochemical and Chemical Injury*

Certain disease states produce biochemically mediated injury to endothelial cells. A lack of vitamin C (Gore, *et al.*, 1965; McKinney, 1976) in scurvy has been reported to produce endothelial cell swelling and distortion as well as reduced stainability of intercellular cement lines. Elevated levels of plasma homocystine (Harker *et al.*, 1976) can induce injury to endothelium, and if these levels are maintained for a sufficient period of time in a hyperlipemic animal, the recipient may develop atherosclerosis. Homocystine-induced atherosclerosis is an interesting animal model for study of this disease, since patients with elevated levels of homocystine experience an increased incidence of thrombosis. Obstruction of the bile ducts with consequently elevated levels of palsma bile (Gustein and Park, 1973) has pro duced injury to endothelial cells. This bile-associated injury consisted of the formation of cytoplasmic vesicles or blebs, widening intercellular junctions, and subendothelial swelling.

A wide variety of exogenous or nonbiological agents as well as alteration in certain physical states have been found to injure endothelium. Endothelial cells exposed to fresh water (Nopanitaya *et al.*, 1974) in drowning victims show cytoplasmic swelling with microvesicle formation of the luminal surface and detachment of cells from the basal membrane; these changes are most marked in pulmonary vessels. Endothelial cells lining human umbilical arteries of mothers who smoked (Asmussen and Kjeldsen, 1975) were found to show cellular swelling and cytoplasmic bleb formation to a much greater degree than did cells from cords of mothers who did not smoke. The administration of fulvine (Wagenvoort *et al.*, 1974) to rats produced swelling of endothelial cells that was severe enough to compromise the lumens of smaller blood vessels. Oral contraceptives (Irey *et al.*, 1970; Almen *et al.*, 1975; Widemann and Fahimi, 1976) have been reported to enhance the mitotic rate of endothelium, but ultrastructural studies revealed no significant morphological changes in this cell type. The low pH of some intravenous glucose (Fonkalsrud *et al.*, 1968a, b) preparations has been found to damage endothelium and to produce thrombosis. Perfusion of hypertonic solutions of glucose produced cytolysis of rabbit endothelium and even saline infusion alters endothelial morphology locally (O'Connel *et al.*, 1974).

Other studies have shown that extremes of pH, osmolarity, and temperature induce endothelial alterations (Constantinides and Robinson, 1969a; Bowers *et al.*, 1973a). Infusion of DMSO (Josnson *et al.*, 1967) and certain radioopaque media has been shown to injury endothelium and to produce thrombosis. Ultrastructural changes consisting of cytoplasmic swelling with formation of cytoplasmic blebs and irregular cell surface projections have been seen in animals treated with ammonium sulfate (Hayes and Shiga, 1970), amylase (Constantinides and Robinson, 1969c), neuraminidase (Constantinides and Robinson, 1969c), Russell's viper venom, sodium deoxycholate, turpentine (Ham and Hurley, 1965), lysolecithin, certain snake and scorpion venoms (Rossi *et al.*, 1974; Suzuki and Ohashi, 1975), elevated levels of cadmium (Gabbiani *et al.*, 1974), or a vitamin-E-deficient diet (Nafstad, 1974) associated with high levels of polyunsaturated fatty acids. Warfarin therapy (Kahn *et al.*, 1971) alters the untrastructural appearance of endothelium by producing a loss of cytoplasmic ground substance and a decrease in organelle content. Of interest is the fact that tantalum particles (Grant *et al.*, 1974) were reproted to adhere to damaged rather than to normal endothelium.

### *J. Agents that Protect Endothelium*

Several agents appear to protect endothelium from certain types of injury. Pyridinocarbamate (Shimamoto *et al.*, 1974; Shimamoto and Sunaga, 1976) has been reported to protect against the endothelial cell contracting effects of certain injurious agents. Treatment with heparin (Gregorius and Rand, 1976) has been reported to decrease the alterations in endothelium produced by ischemic injury. Finally, endothelial cells in culture have been shown to respond to several pharmaceutical agents (Maynard *et al.*, 1978) in a manner different from that of fibroblasts or smooth muscle cells with respect to alterations in the availability in tissue factor on the surfaces of these various cell types; morphological studies showed little change.

### *K. Mechanical Injury*

Mechanical injury (Björkerud, 1974) in experimental settings frequently has taken the form of direct mechanical destruction of the endothelial layer by use of clamps (Bhawan *et al.*, 1977), needles, scalpels, or catheters (Hirsch and Robertson, 1977; Christensen *et al.*, 1977; Chemnitz *et al.*, 1977). On the other hand, a more recently favored means for damaging endothelium has been the use of a balloon catheter (Stemerman, 1974) that can be inflated within the vascular lumen and pulled through a segment of vessel to remove the endothelial layer by abrasion. Criticism of this latter technique is that it may produce injury not only to endothelium but also to subendothelial structures including the internal elastic lamina. The presence of an intraluminal catheter itself can damage endothelium by direct mechanical injury and produce thrombosis. An interesting concept is that the thickening of basal lamina may be due to repeated episodes of loss of or damage to endothelium with each new endothelial layer producing a new basal lamina (Vracko and Benditt, 1970).

In general the reactions of endothelium to low levels of mechanical injury are similar to those changes produced in response to other types of injury. Endothelial cells will show cytoplasmic swelling with production of cytoplasmic blebs and other cytoplasmic projections. Most often the injurious stimulus is of such magnitude that the endothelium is removed completely from the involved area, and a thrombus forms upon the exposed basal lamina.

### *L. Radiation*

Several investigations have indicated that endothelial cells are sensitive to effects of radiation (Reinhold and Buisman, 1973; DeGowin *et al.*, 1974; Reinhold and Buisman, 1975; DeGowin *et al.*, 1976; Fonkalsrud *et al.*, 1977b; Sholley *et al.*, 1977a). In one study (Sholley *et al.*, 1977a) radiation injury to endothelial cells in culture resulted in abolishment of tritiated thymidine uptake but did not inhibit the migration of endothelial cells into areas of injury. Other studies have reported that endothelial cells are approximately as sensitive to radiation effects as are connective tissue cells and epithelial cells (Reinhold and Buisman, 1973). Bleeding following radiation injury has been attributed to thrombocytopenia more than to major endothelial alteration. However, endothelial injury and associated microthrombosis (Figs. 5, 8) have been reported to play an initiating role in radiation pneumonitis.

### *M. Rheologic Effects*

There is now little doubt that endothelial cells are changed in areas of certain types of altered rheology (Fry, 1968, 1969a; Gessner, 1973). Numerous reports indicate that endothelial cells near the origin of branches from major arteries (Fry, 1969b; Gutstein *et al.*, 1973; Fallon and Stehbens, 1972; Stehbens and Ludatscher, 1973; Robertson and Rosen, 1976; Flaherty *et al.*, 1972; Imparato *et al.*, 1974a; Bylock *et al.*, 1977b; Reidy and Bowyer, 1977b; Hertzer *et al.*, 1977; Svendsen and Jorgensen, 1978), in areas of partial vascular occlusion, in aneurysms (Fallon and Stehbens, 1973), and in certain other locations show both ultrastructural and functional changes. These ultrastructural changes include an alteration of endothelial cell polarity accompanied by a decrease in staining of the intercellular cement by ruthenium red, increased permeability to Evans Blue, cell swelling with formation of cytoplasmic blebs and thin cylindrical cytoplasmic projections, and even cell desquamation. Various hypotheses have been put forth to explain how increased shear stress can damage endothelial cells in specific areas of the vascular tree. Of particular interest have been alterations in vascular permeability in areas of altered rheology where morphological changes in endothelium can be shown. These alterations of vascular permeability most frequently have been demonstrated by use of Evans Blue dye. In addition, it has been shown that bovine endothelial cells contain an increased level of histidine decarboxylase (Rosen *et al.*, 1974) when subjected to increased shear stresses *in vitro*. Alterations in endothelium in areas of increased shear stress in the vascular system have been demonstrated in numerous animal species and in many instances have been correlated with the appearance of the early stages of atherosclerosis (Fonkalsrud *et al.*, 1977b; Sholley *et al.*, 1977a; Fry, 1968, 1969a, b; Gessner, 1973; Gustein *et al.*, 1973; Fallon and Stehbens, 1972). These findings are of particular importance, since they may indicate mechanisms by which atherosclerotic changes begin in the vascular system. On the other hand, these rheologic changes alone cannot be the explanation for production of atherosclerosis, since these same changes are present in populations that have a low rate of this disease.

### *N. Thermal Injury*

Extremes of both heat (Leak and Kato, 1972; Gabbiani and Badonnel, 1975; Sholley *et al.*, 1977b) and cold (Belzer *et al.*, 1972; Bowers *et al.*, 1973b; Malczak and Buck, 1977a) have been shown to alter endothelium. Indeed, damage to endothelium produced by increases in temperature has been used as a model for vascular injury by a number of investigators (Leak and Kato, 1972). Heat can be delivered directly or by use of lasers or infusion of fluorescent dye followed by ultraviolet irradiation. Elevations in temperature can be shown to increase vascular permeability by damage to endothelium with production of actual openings between adjacent endothelial cells as well as endothelial cell swelling, dilatation of rough endoplasmic reticulum, and an irregularity of the surface plasma membrane (Leak and Kato, 1972; Gabbiani and Badonnel, 1975; Sholley *et al.*, 1977b). Such injury was accompanied by an increased mitotic rate of surrounding endothelial cells. Enhanced vascular permeability can be demonstrated by use of ferritin and other particulate material. On the other hand, endothelial cells are sensitive also to low temperatures. Freezing injury produces desquamation of endothelium that is later accompanied by regeneration from nearby viable cells (Belzer *et al.*, 1972; Bowers *et al.*, 1973b; Malczak and Buck, 1977a). Less drastic lowering of temperatures produces ultrastructural changes of swelling and cytoplasmic bleb formation.

### O. *Thrombosis*

It is generally agreed that normal or even moderately damaged endothelium is not attractive to platelets (Mason *et al.*, 1977a; Ashford and Freiman, 1967). Some investigators have reported that severely damaged endothelium attracts platelets and induces formation of mural platelet thrombi (Fig. 13) (Branemark and Ekholm, 1968; Spaet and Gaynor, 1970; Haudenschild and Studer, 1971; Tazawa *et al.*, 1971; Gavin *et al.*, 1973; Sumiyoshi and Tanaka, 1976). However, endothelium covered by a thrombus that originated at a point downstream can be damaged due to interference with exchange of gases, nutrients, and wastes (Lough and Moore, 1975) as well as by action of lysosomal enzymes released from platelets and leukocytes. Endothelial cells in many parts of the vascular system contain an activator of fibrinolysin, and thus the endothelium can protect itself to a certain extent from damage due to an encroaching thrombus (Warren and Brock, 1964; Almer and Janzon, 1975; Loskutoff and Edgington, 1977).

There is an interesting difference between the regenerative powers of human and nonhuman mammalian endothelial cells. On the luminal surface of vascular grafts of various types, human endothelium seldom regenerates to a significant degree (Ross *et al.*, 1970; Sauvage *et al.*, 1972, 1973). On the other hand, these same grafts, when placed in the vascular system of nonhuman mammals, will become covered by endothelium within a period of weeks to months (Ghidoni *et al.*, 1968; O'Neal *et al.*, 1964; Stemerman *et al.*, 1977). The difference in this mitotic activity between human endothelium and nonhuman endothelium is important, since most vascular prostheses generally are evaluated in nonhumans before being used in humans. In at least certain nonhuman mammals, such as the dog or cow, endothelium will grow to cover an exposed vascular surface and eventually assume a normal pattern of cell orientation, while vascular grafts in humans are lined by thrombus or organized thrombus. In addition to species differences, the speed of endothelial regeneration may be influenced by local factors (Folkman and Cotran, 1976a; Schwartz and Benditt, 1976; Eisenstein *et al.*, 1978), relating to the integrity of the subendothelial structures (Fishman *et al.*, 1975) and to factors that enhance or inhibit mitosis (Mason *et al.*, 1977a; Saba and Mason, 1975; D'Amore and Shepro, 1977; Gospodarowicz *et al.*, 1976, 1977).

### P. *Thrombocytopenia*

There is ample evidence that markedly decreased levels of circulating platelets produce an increase in vascular permeability as well as morphological changes in endothelial cells (Johnson, 1971; Gimbrone *et al.*, 1969; Aursnes, 1974; Kitchen and Weiss, 1975; Dale and Hurley, 1977). The mechanism for these changes is not known. Endothelial cells from thrombocytopenic animals have been reported to show alterations of cytoplasmic projections and folds as well as cellular thinning (Kitchen and Weiss, 1975; Dale and Hurley, 1977). Restoration of circulating platelet levels to normal resulted in a prompt restoration of endothelial cell morphology and a return of normal or near normal vascular permeability. Their relationship between platelets and endothelium is poorly understood and has been termed platelet support of endothelium. This platelet support role has been found to be of critical importance in some studies of organ preservation (Gimbrone *et al.*, 1969). In the past, it has been shown that intact platelets are not needed for maintenance of normally functioning endothelium *in vivo,* since platelets disrupted by freezing and thawing can be infused in place of intact platelets to restore normal function of endothelium (Klein *et al.*, 1956). An unanswered question is whether this effect of platelet particles might be due to metabolic products of arachidonic acid.

## V. Immunological Injury

### A. *Antigen–Antibody Complexes*

Several investigators have reported that circulating antigen–antibody complexes damage endothelium (Wright and Giacometti, 1972; Yamaguchi *et al.*, 1973; Friedman *et al.*, 1975). Endothelial cell swelling in response to antigen–antibody complexes with formation of cytoplasmic blebs and irregular cytoplasmic protrusions has been reported as has the detachment of endothelial cells from basal lamina. A most common finding in these studies was the formation of cytoplasmic vacuoles that appeared within minutes of the time of infusion of antigen–antibody complexes (Yamaguchi *et al.*, 1973). Indeed, the infusion of human serum with antibodies against rabbit leukocytes has induced alterations in rabbit endothelium (Friedman *et al.*, 1975). It has been reported that in anaphylactic shock there is considerable loss of endothelium from basal lamina and an increase in the mitotic rate of remaining endothelium (Wright and Giacometti, 1972). All of these findings suggest that circulating antigen–antibody complexes are toxic to endothelial cells. These findings may have important implications in the pathogenesis of vascular disease, although exact mechanisms whereby antigen–antibody complexes alter endothelium are not yet known.

### B. *Vascular and Organ Transplants*

Numerous attempts have been made to transplant segments of blood vessel (Wyatt and Taylor, 1966; Imparato *et al.*, 1974b; O'Connell and

Mowbray, 1973; Williams and Harr, 1975; Reidy and Levesque, 1977; Ramos *et al.*, 1976; Shibata *et al.*, 1978; Reidy and Bowyer, 1978; Barboriak *et al.*, 1978; Hobbs, 1973), organs containing an intact vascular system (Hobbs, 1973), and even intact hearts (Biever *et al.*, 1970; Alonso *et al.*, 1977) from one animal to another, from animals to humans, or from one human to another. Evidence has accumulated slowly that endothelial cells must remain intact in vascular grafts, if edema and thrombosis are to be avoided. In organ transplantation survival of graft endothelium appears highly desirable, if not essential (Kitchen and Weiss, 1975; Dale and Hurley, 1977; Klein *et al.*, 1956; Wright and Giacometti, 1972; Yamaguchi *et al.*, 1973; Friedman *et al.*, 1975; Wyatt and Taylor, 1966; Imparato *et al.*, 1974b; O'Connell and Mowbray, 1973; Williams *et al.*, 1975; Shibata *et al.*, 1978). In addition, it is now appreciated that endothelial cells contain specific antigens that will elicit antibody formation (Reidy and Bowyer, 1978; Barboriak *et al.*, 1978; Hobbs, 1973; Biever *et al.*, 1970; Alonso *et al.*, 1977). If there is a mismatch in the antigenic makeup of endothelial cells in the graft relative to those in the recipient, antibodies against the grafted endothelial cells will be formed and these cells will be killed. In such mismatches, there is adhesion of platelets to grafted endothelium (Hobbs, 1973) and considerable invasion of the endothelial cell layer by leukocytes (Reidy and Bowyer, 1978), but changes in endothelium sometimes can be kept to a minimum by use of azathioprine and prednisolone (Reidy and Bowyer, 1978).

The alloantigens of human (lindqvist and Osterland, 1971; Hirschberg *et al.*, 1975; Moraes and Stastny, 1976; Moraes and Stastny, 1977) and certain other mammalian (Vetto and Burger, 1971; DeBono, 1974) endothelial cells have been characterized. It is now known that there are at least eight human endothelial cell alloantigens (Moraes and Stastny, 1976). Endothelial cells damaged by reaction with antibodies become attractive to platelets, later become rounded in shape, and eventually detach from the basal lamina. Other endothelial cells may remain attached to the basal lamina but extrude their cytoplasmic contents. These findings of specific endothelial cell alloantigens have important consequences in any transplantation procedure (Moraes and Stastny, 1976, 1977). Damage to endothelium has been seen in transplants of human kidneys (Shibata *et al.*, 1978; Hobbs, 1973), blood vessels (O'Connell and Mowbray, 1973), hearts (Biever *et al.*, 1970), and in porcine livers (Belzer *et al.*, 1970).

## VI. Endothelial Cell Inclusions

There are numerous reports of cytoplasmic tubular inclusions, tubuloreticular structures, in endothelial cells (Grausz *et al.*, 1970; Dreyer *et al.*, 1973; Aizawa *et al.*, 1973; Datsis, 1973; Bariéty *et al.*, 1973; Mims and

Murphy, 1973; Herrlinger *et al.*, 1974; Kawamura *et al.*, 1974b; Macadam *et al.*, 1975). These inclusions appear to be different from Weibel–Palade bodies (Weibel and Palade, 1964) (Fig. 1) that are normal constituents of endothelial cells. Cytoplasmic inclusions have been reported in endothelial cells of patients with anaphylactoid purpura (Robertson and Khirallah, 1973), in a number of different skin diseases (Macadam *et al.*, 1975), in lupus erythematosus (Fig. 17) (Grausz *et al.*, 1970; Dreyer *et al.*, 1973), in a

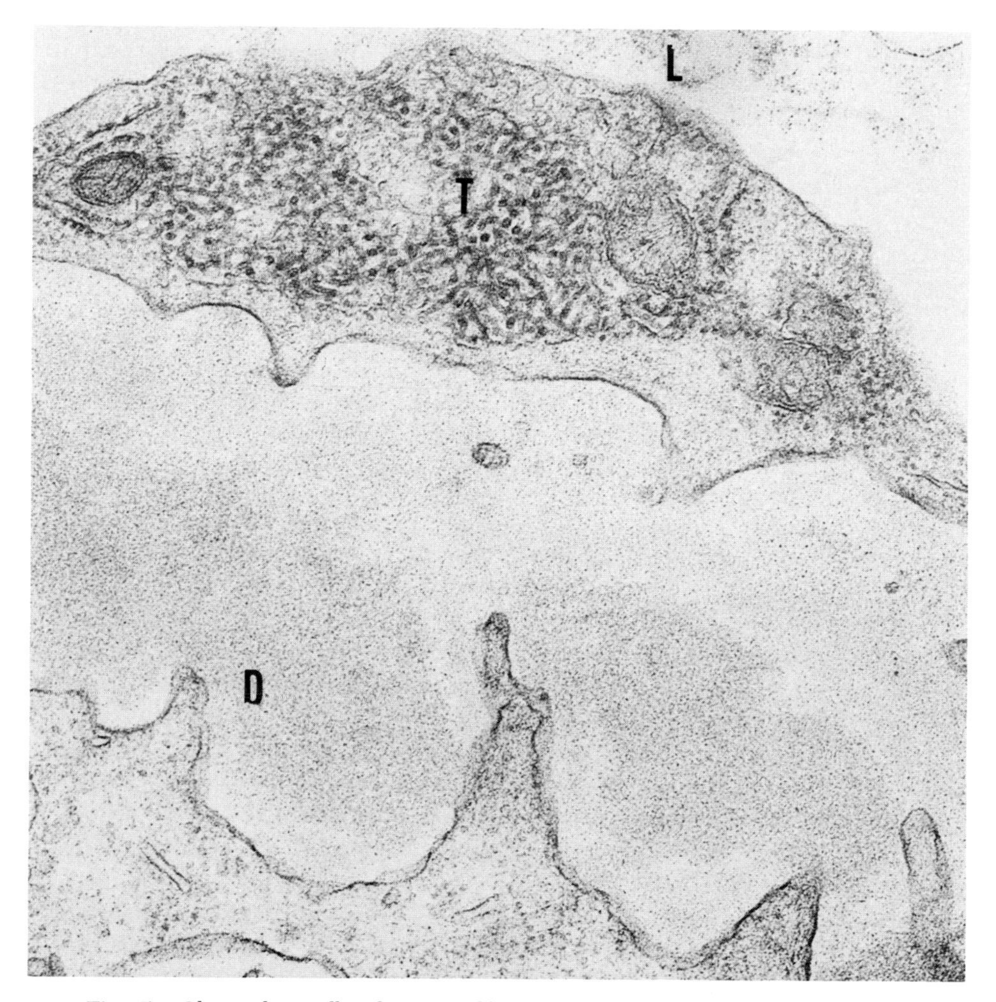


**Fig. 17.** Glomerular capillary from a renal biopsy of an 18-year-old man with systemic lupus erythematosus and membranous nephropathy. Characteristic tubulo-reticular structures (T) are noted in the endothelial cytoplasm. The basement membrane is irregular and contains subepithelial deposits (D). Capillary lumen, L. ×42,500. (Courtesy of Dr. Raoul Fresco.)

cerebellar neoplasm (Kawamura *et al.*, 1974b), in congenital infantile nephrosis (Datsis, 1973), and in a number of supposedly normal blood vessels (Bariéty *et al.*, 1973; Herrlinger *et al.*, 1974). The nature and significance of these structures is unknown, but their widespread occurrence suggests that they are a nonspecific finding. Indeed, the function and significance of the Weibel–Palade body is unknown to date. Intralysosomal collections of glycogen (Fig. 18) and other polymers and lipids may occur in a large number of heritable "storage" diseases.

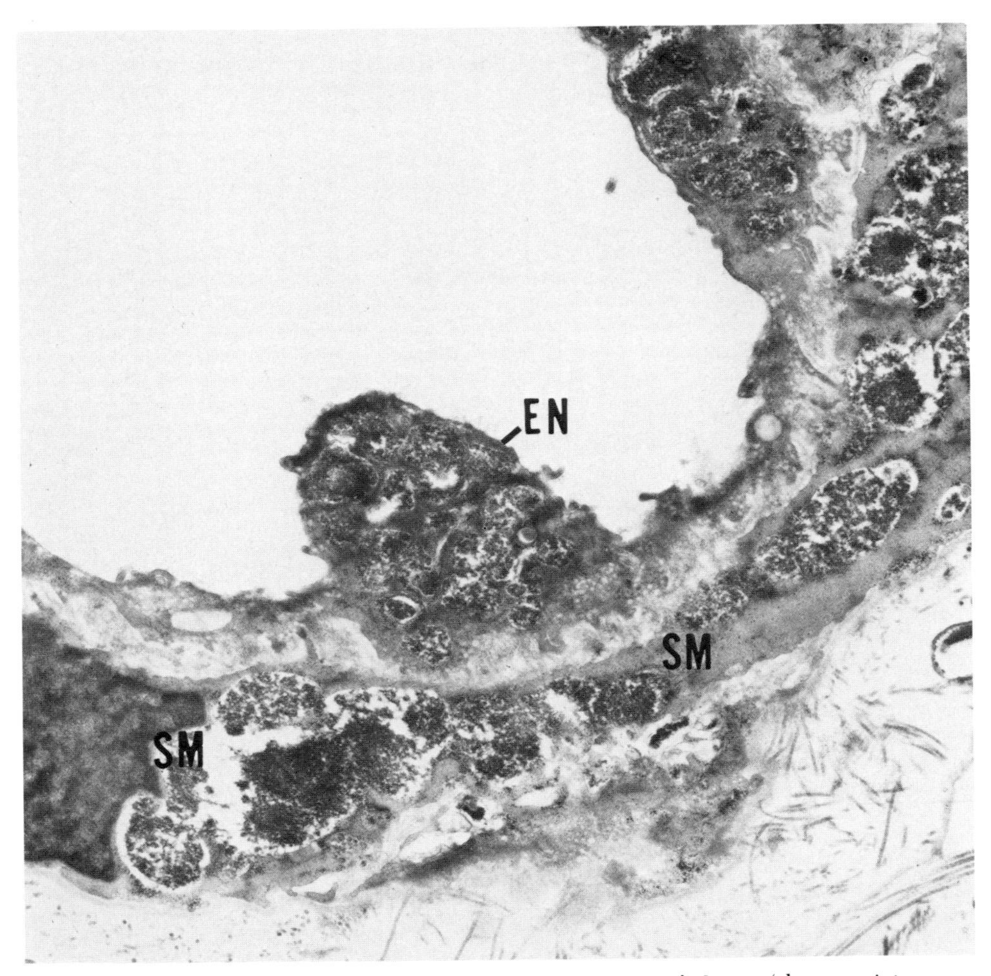


**Fig. 18.** Arteriole in a muscle biopsy from a patient with Pompe's disease (glycogenosis type II). Intralysosomal accumulations of glycogen are noted in endothelial cells (EN) and underlying smooth muscle cells (SM). ×10,400. (Courtsey of Dr. Raoul Fresco.)

## VII. Morphogenesis and Reactivity of the Regenerating Endothelium

It is generally appreciated that endothelial damage and desquamation due to physical, mechanical, or chemical injury is followed by increased mitotic activity near the edges of the intimal defects. The regenerating cells cease to proliferate after reestablishment of endothelial continuity but their permeability and orientation may remain altered for some time (Webster *et al.*, 1974). At present, the factors that control mitotic behavior as well as the movement and permeability of the endothelial cells covering intimal defects are not well understood. However, studies of angiogenesis induced by tumor or wounding have provided considerable insight with respect to the morphogenetic and functional interrelationships of the regenerating endothelium (Ausprunk and Folkman, 1977; Folkman and Cotran, 1976b; Schoefl, 1963; Yamagami, 1970). Using these models, it has been shown clearly that as a consequence of chemotactic attraction endothelial cells of existing capillaries actively migrate in the direction of the angiogenic stimulus. This movement appears to result in loosening of cell-to-cell junctions and formation of gaps between existing endothelial cells that are behind the advancing tip (Ausprunk and Folkman, 1977). The formation of interendothelial gaps is associated with increased permeability, and it is followed by mitosis that characteristically occurs several cells proximal to the migrating endothelium (Yamagami, 1970; Schwartz *et al.*, 1978).

In large vessels, endothelial regeneration over intimal defects proceeds in a fashion similar to that occurring in microvessels (Fishman *et al.*, 1975; Hirsch and Robertson, 1977; Schwartz and Benditt, 1976; Webster *et al.*, 1974; Wright, 1972; Malczak and Buck, 1977b; Schwartz *et al.*, 1975; Gerrity *et al.*, 1977; Lowenstein, 1975). Endothelial proliferation near the edges of the defects occurs after the onset of cell migration. In this situation, however, the process of the cell migration occurs preferentially along the axis of the vessel (Schwartz *et al.*, 1978) and it is apparently due to lack of cell contact rather than to chemotactic attraction. An analogous but not identical (Schwartz *et al.*, 1978) sequence of events has been described in cultured endothelium following focal injury to the monolayer: DNA synthesis occurs after random migration of cells into the defects (Sholley *et al.*, 1977c).

There is evidence that regenerating endothelium over intimal defects is characterized by increased permeability in association with loose or incompletely formed junctions with frequent interendothelial gaps (Webster *et al.*, 1974; Gerrity *et al.*, 1977; Shimamoto, 1974). It is also of interest that similar alterations as well as foci of replicating endothelial cells have been described in "normal" aortas at sites of increased hemodynamic stress (Schwartz and Benditt, 1976; Wright, 1972; Gerrity *et al.*, 1977). The similarity of regeneration processes observed in large vessels with those found using neovascularization models have led to the concept that formation of interendothelial gaps not only results in increased permeability but also provides the stimulus for endothelial proliferation (Ausprunk and Folkman, 1977). It has been suggested that the actual signal for cell mitosis is electrical uncoupling caused by loss of gap junctions, and that this uncoupling persists until regeneration is complete (Webster *et al.*, 1974; Ausprunk and Folkman, 1977).

From published reports it appears that various mechanisms operate, singly or in combination, in the development of interendothelial gaps. These include endothelial contraction in response to various types of nonspecific injury, unfolding of junctions as a consequence of cell migration (Ausprunk and Folkman, 1977), incomplete development of junctions in regenerating endothelium (Gerrity *et al.*, 1977), and possibly direct damage of loose junctions by certain vasoactive agents (Simionescu *et al.*, 1978). Therefore gap formation is a common demoninator in both endothelial injury and repair and apparently plays a central role in the reactivity of preformed and regenerating endothelium. In addition, the topological association of interendothelial gaps with enhanced permeability and mitotic activity provides a structural and functional basis for the development of localized lesions in

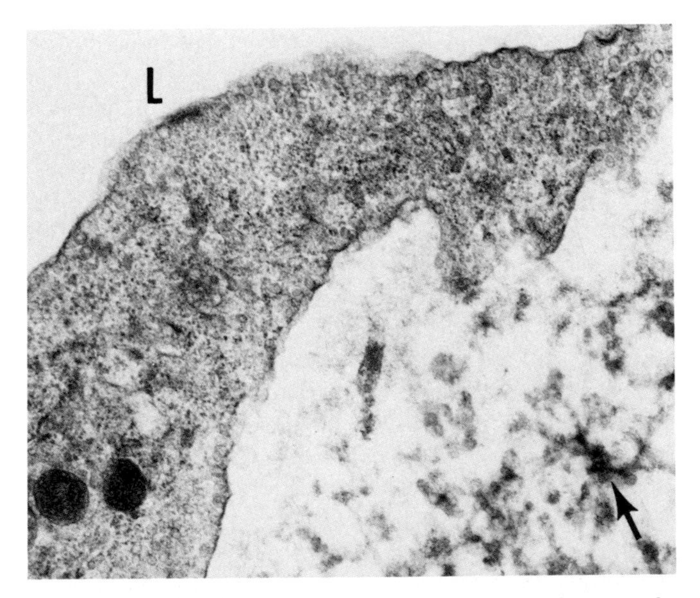


**Fig. 19.** Aorta from a rat fed a high fat diet for 12 weeks. A portion of a seemingly regenerating endothelial cell containing predominantly ribosome particles is present. Such endothelial cells usually lack an underlying basement membrane. Numerous lipid granules (arrow) are noted in the edematous subendothelial space. Vascular lumen, L. ×28,500. (Reproduced with the permission of the publisher from Figure 13, *Exp. Mol. Pathol.* 8, 90, 1968.)

response to prolonged, intermittent, or chronic effects of various systemic factors such as hyperlipemia. Other localizing factors probably relate to the random (Webster *et al.*, 1974) or slow rate of circumferential (Schwartz *et al.*, 1978) migration of the regenerating endothelium over intimal defects prior to remodeling. In addition, there is recent evidence that the speed of endothelial regeneration is markedly accelerated when damage to the subendothelial structures is avoided (Hirsch and Robertson, 1977). Conversely,

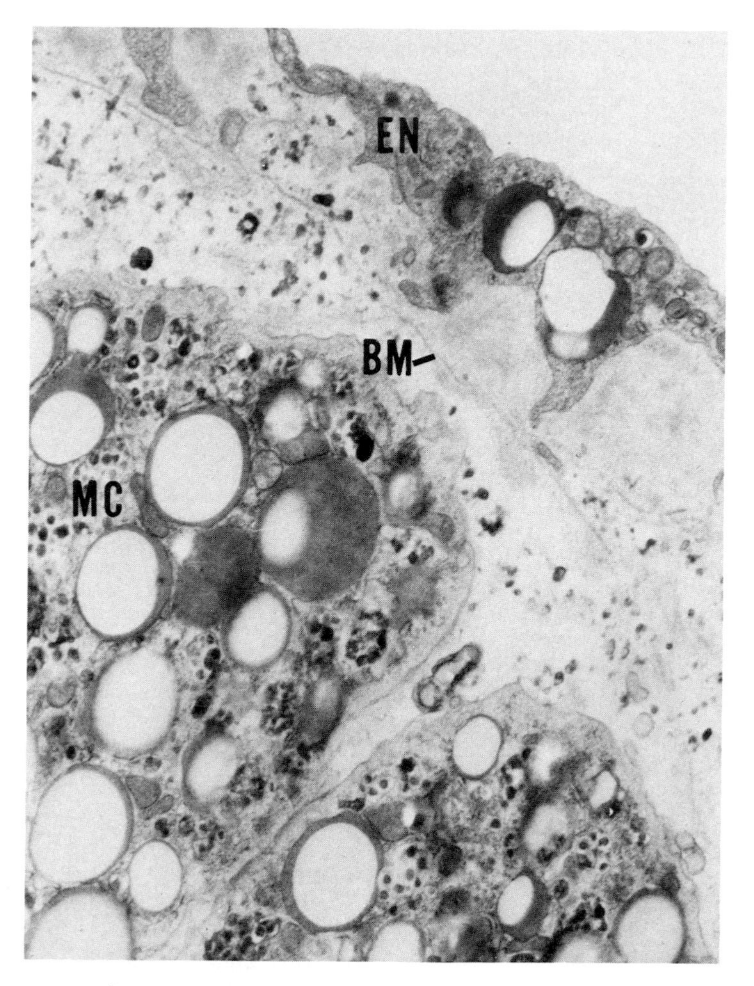


**Fig. 20.** Aorta from a rat on a high fat diet for 12 weeks. The endothelium (EN) shows partially vacuolated cytoplasmic fat droplets and several subendothelial blisters containing moderately dense material and lipid granules. In addition, the underlying basement membrane (BM) is displaced. Macrophage with numerous membrane-bound vacuoles filled with partially vacuolated fat droplets and lipid granules, MC. ×18,700.

in the presence of intimal thickening and medial damage, reendothelialization is delayed and interendothelial gaps persist (Webster *et al.*, 1974; Schwartz *et al.*, 1975).

Various lines of evidence support the concept that the presence of an actively regenerating endothelium is a major localizing factor for the accumulation of lipids in the arterial wall. Thus, in animals receiving cholesterol after partial reendothelialization of aortic segments, lipid accumulation was preferentially found to be localized in areas of intimal thickening covered by regenerating endothelial cells rather than in areas devoid of endothelium (Minick *et al.*, 1977b). In addition, a hyperlipemic diet fed to rats with organizing medial scars of the aorta resulted in formation of accelerated atheromatous lesions overlying the medial scars (Balis *et al.*, 1968). In these locations, the endothelial lining demonstrated a spectrum of changes including lipid accumulation and interendothelial gap formation in association with

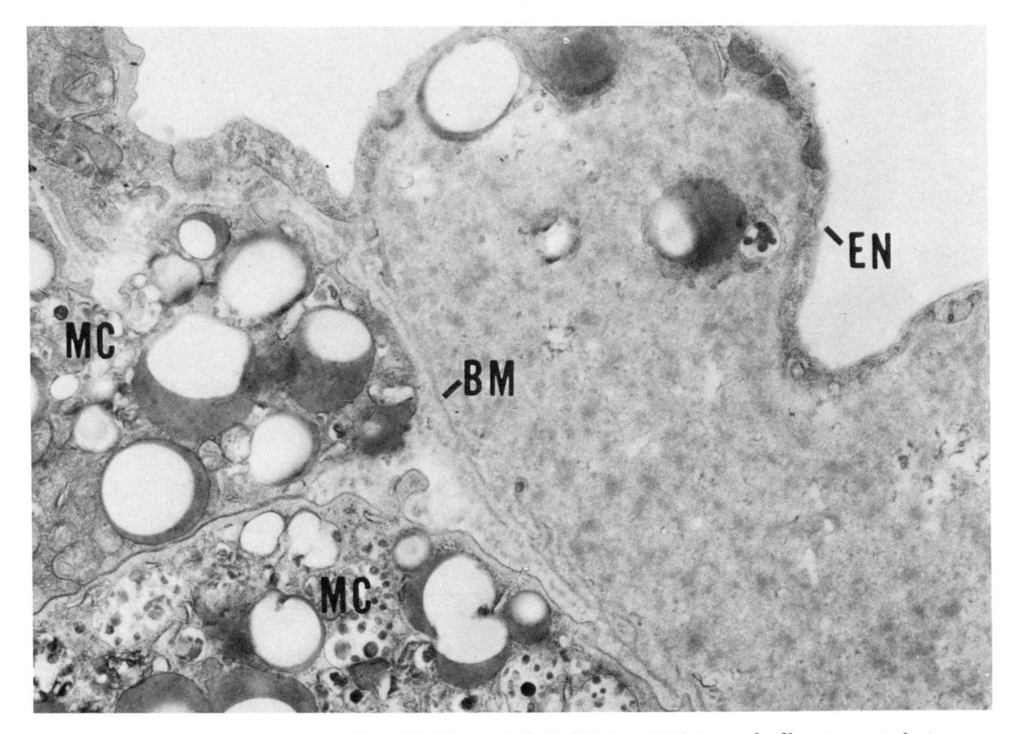


**Fig. 21.** Same aorta as in Fig. 17. The endothelial lining (EN) is markedly attenuated. A large subendothelial compartment containing homogenous granular material is present between endothelium and its basement membrane (BM). Macrophage filled with lipids, MC. ×18,700. (Reproduced with the permission of the publisher from Figure 7, *Exp. Mol. Pathol.* **8**, 90, 1968).

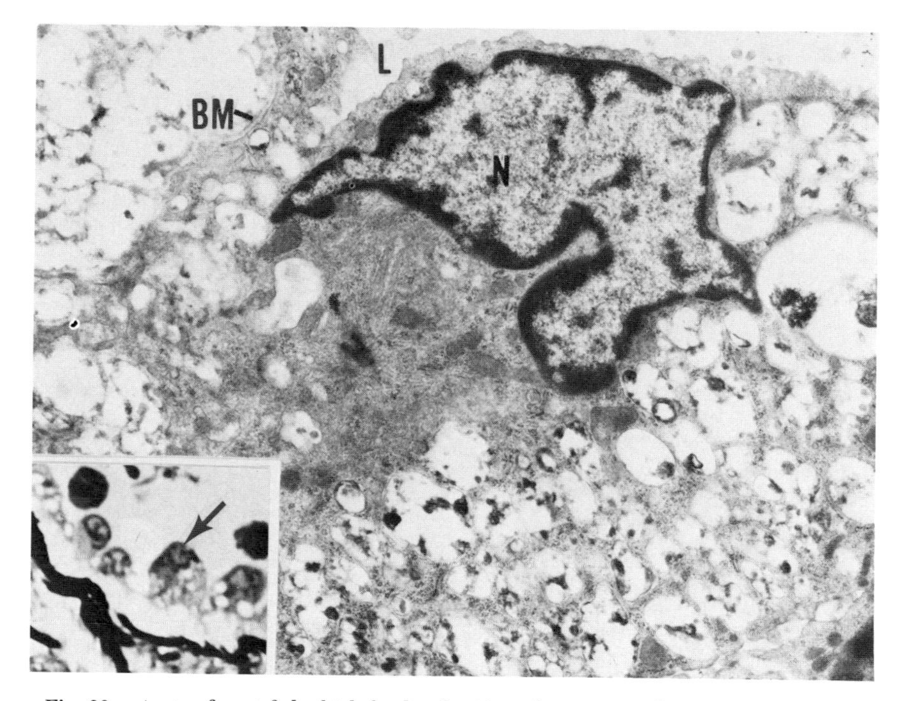


**Fig. 22.** Aorta of a rat fed a high fat diet for 16 weeks. Portion of an *endothelial foam cell* with cytoplasm replaced for the most part by membrane-bound vacuoles that contain lipid granules similar to those found in macrophages. Nucleus of the endothelial foam cell, N; basement membrane of the endothelium, BM; vascular lumen, L. ×17,300. In an adjacent, toluidine blue stained section (insert, ×800), the same cell (arrow) is shown by light microscopy. (Reproduced with the permission of the publisher from Figure 15, *Exp. Mol. Pathol.* 8, 90, 1968).

increased permeability and, in places, changes consistent with endothelial regeneration (Figs. 19–21). In such atheromatous lesions foamy macrophages seemingly originating from endothelial cells were observed also (Fig. 22), supporting the hypothesis that responses of the regenerated endothelium to hyperlipemia includes significant phagocytic activity in addition to enhanced lipid transport. These considerations serve to emphasize the need for comprehensive studies primarily focusing on the structural and functional properties of the regenerating endothelium and its reactivity to hyperlipemia and other systemic factors with or without coexistent subendothelial damage.

## Acknowledgments

Supported in part by Grant HL-20679, NIH, PHS and Grant 939A from the Council for Tobacco Research-U.S.A., Inc.

## References

Aizawa, S., Hamaguchi, K., Ogoshi, E., Ishikawa, E., and Adachi, T. (1973). Virus-like microtubular inclusions in the glomerular endothelium of patients with anaphylactoid purpura. *Acta Pathol. Jap.* **23**, 27.

Almen, T., Hartsel, M., Nylander, G., and Olivecrona, H. (1975). The effect of estrogen on the vascular endothelium and its possible relation to thrombosis. *Surg Gyn Obstet.* **140**, 938.

Almer, L. O., and Janzon, L.(1975). Low vascular fibrinolytic activity in obesity. *Thromb Res* **6**, 171.

Alonso, D. R., Starek, P. K., Minick C. R. (1977). Studies on the pathogenesis of athero-arteriosclerosis induced in rabbit cardiac allografts by the synergy of graft rejection and hypercholesterolemia. *Am J Path* **87**, 415.

Ashford, T. P., and Freiman, D. G. (1967). The role of the endothelium in the initial phases of thrombosis. *Am J. Pathol.* **50**, 257.

Ashton, N., and Pedler, C. (1962). Studies on developing retinal vessels. IX. Reaction of endothelial cells to oxygen. *Br J Ophthalmol.* **46**, 257.

Asmussen, I., and Kjeldsen, K. (1975). Intimal ultrastructure of human umbilical arteries. *Circ. Res.* **36**, 579.

Aursnes, I. (1974). Increased permeability of capillaries to protein during thrombocytopenia. An experimental study in the rabbit. *Microvasc Res* **7**, 283.

Ausprunk, D. H., and Folkman, J. (1977). Migration and proliferation of endothelial cells in performed and newly formed blood vessels during tumor angiogenesis. *Microvasc Res* **14**, 53.

Balint, A., Veress, B., and Jellinek, H. (1974). Modifications of surface coat of aortic endothelial cells in hyperlipemic rats. *Pathol. Eur.* **9**, 105.

Balis, J. U., Chan, A. S., and Corien, E. (1968). The effect of postlathyritic medial scar in the progression of experimental atherosclerosis in rats. *Exp Mol. Pathol.* **8**, 90.

Balis, J. U., Gerber, L. I., Rappaport, S., and Neville, W. E. (1974). Mechanisms of blood vascular reactions of the primate lung to acute endotoxemia. *Exp Mol. Pathol.* **21**, 123.

Balis, J. U., Rappaport, E. S., Gerger, L., Fareed, J., Buddingh, F., and Messmore, H. L. (1978). A primate model for prolonged endotoxin shock. *Lab. Invest.* **38**, 511.

Barboriak, J. J., Pintar, K., Van Horn, D. L., Batayias, G. E., and Korns, M. E. (1978). Pathologic findings in the aortocoronary vein grafts. A scanning electron microscope study. *Atherosclerosis* **29**, 69.

Bariéty, J., Richer, D., Appay, M. D., Grossetete, J., and Callard, P. (1973). Frequency of intra-endothelial 'virus-like' particles: An electron microscopy study of 376 human renal biopsies. *J. Clin. Pathol.* **26**, 21.

Barnhart, M. I., and Chen, S. (1976). Platelet-vessel wall dynamics, in Gastpar, Kuhn, Marx and Eibl (eds): Collagen, platelet, plasma protein interaction. *1st Munich Symposium on Biology of Connective Tissue, Munich, 1976*, 1.

Baumann, F. G., Imparato, A. M., and Kin, G. E. (1976). The evolution of early fibromuscular lesions hemodynamically induced in the dog renal artery. *Circ. Res.* **39**, 809.

Bell, F. P., Adamson, I. L., and Schwartz, C. J. (1974). Aortic endothelial permeability to albumin: focal and regional patterns of uptake and transmural distribution of $^{131}$I-albumin in the young pig. *Exp. Mol. Pathol.* **20**, 57.

Belzer, F. O., May, R., Berry, M. N., and Lee, J. C. (1970). Short term preservation of procine livers. *J. Surg. Res.* **10**, 55.

Belzer, F. O., Hoffman, R., Huang, J., and Downes, G. (1972). Endothelial damage in perfused dog kidney and cold sensitivity of vascular Na-K-ATPase. *Cryo* **9**, 457.

Bevan, J. A., and Duckles, S. P. (1975). Evidence for α-adrenergic receptors on intimal endothelium. *Blood Vessels* **12**, 307.

Bhawan, J., Joris, I., DeGirolami, U., and Majno, G. (1977). Effect of occlusion on large vessels. *Am. J. Pathol.* **88**, 355.

Biever, C. P., Stinson, E. B., Shumway, N. E., Payne, R., and Kosek, J. (1970). Cardiac transplantation in man. VII. Cardiac allograft pathology. *Circ.* **61**, 753.

Björkerud, S. (1974). Injury and repair in arterial tissue. Experimental models: types and relevance to human vascular diseases— a survey. *Angiology* **25**, 636.

Björkerud, S., and Bondjers, G. (1971). Arterial repair and atherosclerosis after mechanical injury. *Atherosclerosis* **13**, 355.

Björkerud, S., and Bondjers, G. (1972). Endothelial integrity and viability in the aorta of the normal rabbit and rat as evaluated with dye exclusion tests and interference contrast microscopy. *Atherosclerosis* **15**, 285.

Björkerud, S., and Bondjers, G. (1976). Repair responses and tissue lipid after experimental injury to the artery. *Ann. N.Y. Acad. Sci.* **275**, 180.

Bondjers, G., Brattsand, R., Bylock, A., Hansson, G. K., and Björkerud, S. (1977). Endothelial integrity and atherogenesis in rabbits with moderate hypercholesterolemia. *Artery* **3**, 395.

Bowers, W. D., Jr., Hubbard, R. W., Daum, R. C., Ashbaugh, P., and Nilson, E. (1973a) Ultrastructural studies of muscle cells and vascular endothelium immediately after freeze—thaw injury. *Cryo* **10**, 9.

Bowers, W. D., Jr., Hubbard, R. W., Daum, R. C., Ashbaugh, P., and Nilson, E. (1973b). Ultrastructural studies of muscle cells and vascular endothelium immediately after freeze—thaw injury. *Cryo* **10**, 9.

Branemark, P. I., and Ekholm, R. (1968). Adherence of blood cells to vascular endothelium. *Blut* **16**, 274.

Buck, R. C. (1961). Intimal thickening after ligature of arteries. *Circ. Res.* **9**, 418.

Burri, P. H., and Weibel, E. R. (1968). Effect of epinephrine on a specific organelle of endothelial cells of blood vessels. *Z. Zellforschung. Mikrosk. Anat.* **88**, 426.

Bylock, A., Björkerud, S., Brattsand, R., Hansson, G. K., Hansson, H., and Bondjers, G. (1977a). Endothelial structure in rabbits with moderate hypercholesterolaemia. *Acta Pathol. Microbiol. Scand.* **85**, 671.

Bylock, A., Björkerud, S., Brattsand, R., Hansson, G. K., Hansson, H. A., and Bondjers, G. (1977b). Endothelial structure in rabbits with moderate hypercholesterolaemia. *Acta Pathol. Microbiol. Scand.* **85**, 671.

Carne, H. R. (1978). Action of corynebacterium ovis exotoxin on endothelial cells of blood vessels. *Nature (London)* **271**, 100.

Chemnitz, J., Collatz, B., and Tkocz, I. (1977). Enface organ cultures of rabbit aortic segments after a single diatation trauma in vivo. *Virchows Arch. A* **375**, 257.

Christensen, B. C., Chemnitz, J., Tkocz, I., and Blaabjerk, O. (1977). Repair in arterial tissue. *Acta Pathol. Microbiol. Scand.* **85**, 297.

Clark, J. M., and Glagov, S. (1976). Luminal surface of distended arteries by scanning electron microscopy: eliminating configuration and technical artefacts. *Br. J. Exp. Pathol.* **57**, 129.

Constantinides, P. (1966). Plaque fissures in human coronary thrombosis. *J. Atheroscl. Res.* **6**, 1.

Constantinides, P. (1968). Lipid deposition in injured arteries. *Arch. Pathol.* **85**, 280.

Constantinides, P. (1976a). The important role of endothelial changes in atherogenesis. *Triangle* **15**, 51.

Constantinides, P. (1976b). Importance of the endothelium and blood platelets in the pathogenesis of atherosclerosis. *Triangle* **15**, 53.

Constantinides, P., and Robinson, M. (1969a). Ultrastructural injury of arterial endothelium. I. Effects of pH, osmolarity, anoxia, and temperature. *Arch. Pathol.* **88**, 99.

Constantinides, P. and Robinson, M. (1969b). Ultrastructural injury of arterial endothelium. II. Effects of vaso-active amines. *Arch. Pathol.* **88**, 106.

Constantinides, P., and Robinson, M. (1969c). Ultrastructural injury of arterial endothelium. III. Effects of enzymes and surfactants. *Arch. Pathol.* **88**, 113.

Cotran, R. S. (1965). Endothelial phagocytosis: an electron-microscopic study. *Exp. Mol. Pathol.* **4**, 217.

Craddock, P. R., Hammerschmidt, D., White, J. G., Dalmasso, A. P., and Jacob, H. S. (1977). Complement (C5a)-induced granulocyte aggregation in vitro. *J. Clin. Invest.* **60**, 260.

Dale, C., and Hurley, J. V. (1977). An electron-microscope study of the mechanism of bleeding in experimental thrombocytopenia. *J. Pathol.* **121**, 193.

Dalldorf, F. G., and Beall, F. A. (1967). Capillary thrombosis as a cause of death in experimental anthrax. *Arch Pathol.* **83**, 154.

Dalldorf, F. G., and Jennette, J. C. (1977). Fatal meningococcal septicemia. *Arch. Pathol. Lab. Med.* **101**, 6.

Dalldorf, F. G., Carney, C. N., Rackley, C. E., and Raney, R. B. (1968). Pulmonary capillary thrombosis in septicemia due to gram-positive bacteria. *J. Am. Med. Assoc.* **206**, 583.

Dalldorf, F. G., Beall, F. A., Krigman, M. R., Goyer, R. A., and Livingston, H. L. (1969). Transcellular permeability and thrombosis of capillaries in anthrax toxemia. *Lab. Invest.* **21**, 42.

D'Amore, P., and Shepro, D. (1977). Stimulation of growth and calcium influx in cultured bovine, aortic endothelial cells by platelets and vasoactive substances. *J. Cell Biol.* **92**, 177.

Datsis, A. G. (1973). Endothelial inclusions in congenital infantile nephrosis. *Virchows Arch. A* **359**, 105.

Davies, P. F., and Bowyer, D. E. (1975). Scanning electron microscopy: arterial endothelial integrity after fixation at physiological pressure. *Atherosclerosis* **21**, 463.

Davies, P. F., Reidy, M. A., Goode, T. B., Bowyer, D. E. (1976). Scanning electron microscopy in the evaluation of endothelial integrity of the fatty lesion in atherosclerosis. *Atherosclerosis* **25**, 125.

DeBono, D. (1974). Effects of cytotoxic sera on endothelium *in vitro*. *Nature (London)* **252**, 83.

DeBrito, T., Hoshino-Shimizu, S., Pereira, M. O., and Rigolon, N. (1973). The pathogenesis of the vascular lesions in experimental rickettsial disease of the guinea pig (Rocky Mountain spotted fever group). *Virchows Arch. A* **358**, 205.

DeGowin, R. L., Lewis, L. J., Haok, J. C., Mueller, A. L., and Gibson, D. P. (1974). Radiosensitivity of human endothelial cells in culture. *J. Lab. Clin. Med.* **84**, 42.

DeGowin, R. L., Lewis, Mason, R. E., Borke, M. K., and Hoak, J. C. (1976). Radiation-induced inhibition of human endothelial cells replicating in culture. *Radiat. Res.* **68**, 244.

Dreyer, D. O., Muldiyarov, P. Y., Nassonova, V. A., and Alekberova, Z. S. (1973). Endothelial inclusions and 'nuclear bodies' in systemic lupus erythematosus. *Ann. Rheum. Dis.* **32**, 444.

Eisenstein, R., Schumacher, B., Meineke, C., Branislav, M., and Kuettner, K. E. (1978). Growth regulators in connective tissue. *Am. J. Pathol.* **91**, 1.

Evensen, S. A., and Shepro, D. (1974). DNA synthesis in rat aortic endothelium: effect of bacterial endotoxin and trauma. *Microvasc. Res.* **8**, 90.

Evensen, S. A., Pickering, R. J., Batbouta, J., and Shepro, D. (1975). Endothelial injury induced by bacterial endotoxin: effect of complement depletion. *Euro. J. Clin. Invest.* **5**, 463.

Fallon, J. T., and Stehbens, W. E. (1972). Venous endothelium of experimental arteriovenous fistuals in rabbits. *Circ. Res.* **31**, 546.

Fallon, J. T., and Stehbens, W. E. (1973). The endothelium of experimental saccular aneurysms of the abdominal aorta in rabbits. *Br. J. Exp. Pathol.* **54**, 13.

Fishman, J. A., Ryan, G. B., and Karnovsky, M. J. (1975). Endothelial regeneration in the rat carotid artery and the significance of endothelial denudation in the pathogenesis of myointimal thickening. *Lab. Invest.* **32**, 339.

Flaherty, J. T., Pierce, J. E., Ferrans, V. J., Patel, D. J., Tucker, W. K., and Fry, D. L. (1972). Endothelial nuclear patterns in the canine arterial tree with particular reference to hemodynamic events. *Circ. Res.* **30**, 23.

Folkman, J., and Cotran, R. (1976a). I. Endothelial turnover and capillary regeneration. *Intern. Rev. Exp. Pathol.* **16**, 208.

Folkman, J., and Cotran, R. (1976). Relation of vascular proliferation to tumor growth. *In* "International Review of Experimental Pathology" (G. W. Richter, ed.), pp. 208–245. Academic Press, New York.

Fonkalsrud, E. W., Murphy, J., and Smith, F. G. (1968a). Effect of pH in glucose infusions on development of thrombophlebitis. *J. Surg. Res.* **8**, 539.

Fonkalsrud, E. W., Pederson, B. M., Murphy, J., and Beckerman, J. H. (1968b). Reduction of infusion thrombophlebitis with buffered glucose solutions. *Surg.* **63**, 280.

Fonkalsrud, E. W., Sanchez, M., Zerubavel, R., Lassaletta, L., Smeesters, C., Mahoney, A. (1976). Arterial endothelial changes after ischemia and perfusion. *Surg. Gyn. Obstet.* **42**, 715.

Fonkalsrud, E. W., Sanchez, M., Zerubavel, R., and Mahoney, A. (1977a). Serial changes in arterial endothelium following ischemia and perfusion. *Surg.* **81**, 527.

Fonkalsrud, E. W., Sanchez, M., Zerubavel, R., and Mahoney, A. (1977b). Serial changes in arterial structure following radiation therapy. *Surg. Gyn. Obstet.* **145**, 395.

Friedman, R. J., Moore, S., and Singal, D. P. (1975). Repeated endothelial injury and induction of antherosclerosis in normolipemic rabbits by human serum. *Lab. Invest.* **30**, 404.

Fry, D. L. (1968). Acute vascular endothelial changes associated with increased blood velocity gradients. *Circ. Res.* **22**, 165.

Fry, D. L. (1969a). Responses of the arterial wall to certain physical factors. *Circ. Res.* **24**, 93.

Fry, D. L. (1969b). Certain histological and chemical responses of the vascular interface to acutely induced mechanical stress in the aorta of the dog. *Circ. Res.* **24**, 93.

Gabbiani, G., and Badonnel, M. C. (1975). Early changes of endothelial clefts after thermal injury. *Microvasc. Res.* **10**, 65.

Gabbiani, G., and Guido, M. (1976). Fine structure of endothelium. "Microcirculation" (G. Kaley and G. M. Altura, eds.), Vol. 1, pp. 133–144. Univ. Park Press, Baltimore, Maryland.

Gabbiani, G., Badonnel, M. C., Mathewson, S. M., and Ryan, G. B. (1974). Acute cadmium intoxication. Early selective lesions of endothelial clefts. *Lab. Invest.* **30**, 686.

Gabbiani, G., Badonnel, M. C., and Rona, G. (1975). Cytoplasmic contractile apparatus in aortic endothelial cells of hypertensive rats. *Lab. Invest.* **32**, 227.

Gavin, J. B., Wheeler, E. E., and Herdson, P. B. (1973). Scanning electron microscopy of the endocardial endothelium overlying early myocardial infarcts. *Pathology* **5**, 145.

Gaynor, E. (1971). Increased mitotic activity in rabbit endothelium after endotoxin. An autoradiographic study. *Lab. Invest.* **24**, 318.

Gaynor, E. (1972). Vascular lesions in endotoxemia. The fundamental mechanisms of shock. *Adv. Exp. Med. Biol.* **23**, 337.

Gerrity, R. G., Richardson, M., Caplan, B. A., Cade, J. F., Hirsh, J., and Schwartz, C. J. (1976). Endotoxin-induced endothelial injury and repair. *Exp. Mol. Pathol.* **24**, 59.

Gerrity, R. G., Richardson, M., Somer, J. B., Bell, F. P., and Schwartz, C. J. (1977). Endothelial cell morphology in areas of in vivo Evans blue uptake in the aorta of young pigs. II. Ultrastructure of the intima in areas of different permeability to proteins. *Am. J. Pathol.* **89**, 313.

Gertz, S. D., Rennels, M. L., Forbes, M. S., and Nelson, E. (1975a). Preparation of vascular endothelium for scanning electron microscopy: a comparison of the effects of perfusion and immersion fixation. *J. Microsc.* **105,** 309.

Gertz, S. D., Rennels, M. L., and Nelson, E. (1975b). Endothelial cell ischemic injury: protective effect of heparin or aspirin assessed by scanning electron microscopy. *Stroke* **6,** 357.

Gertz, S. D., Rennels, M. L., Forbes, M. S., Kawamura, J., Sunaga, T., and Nelson, E. (1976a). Endothelial cell damage by temporary arterial occlusion with surgical clips. *J Neurosurg.* **45,** 514.

Gertz, S. D., Forbes, M. S., Sunaga, T., Kawamura, J., Rennels, M. L., Shimamoto, T., and Nelson, E. (1976b). Ischemic carotid endothelium. *Arch. Pathol. Lab. Med.* **100,** 522.

Gessner, F. B. (1973). Hemodynamic theories of atherogenesis. *Circ. Res.* **33,** 259.

Ghidoni, J. J., Liotta, D., Hall, C. W., Adams, J. G., Lechter, A., Barrionueva, M., O'Neal, R. M., and DeBakey, M. E. (1968). Healing of pseudointimas in velour-lined, impermeable arterial prostheses. *Am. J. Pathol.* **53,** 375.

Gimbrone, M. A., Jr., Aster, R. H., Cotran, R. S., Corkery, J., Jandl, J. H., and Folkman, J. (1969). Preservation of vascular integrity in organs perfused in vitro with a platelet-rich medium. *Nature (London)* **222,** 33.

Goode, T. B., Davies, P. F., Reidy, M. A., and Bowyer, D. E. (1977). Aortic endothelial cell morphology observed in situ by scanning electron microscopy during atherogenesis in the rabbit. *Atherosclerosis* **27,** 235.

Gore, I., Fujinami, T., and Shirahama, T. (1965). Endothelial changes produced by ascorbic acid deficiency in guinea pigs. *Arch. Pathol.* **80,** 371.

Gospodarowicz, D., Moran, J., Braun, D., and Birdwell, C. (1976). Clonal growth of bovine vascular endothelial cells: fibroblast growth factor as a survival agent. *Proc. Nat. Acad. Sci. USA* **73,** 4120.

Gospodarowicz, D., Moran, J. S., and Braun, D. L. (1977). Control of proliferation of bovine vascular endothelial cells. *J. Cell Physiol.* **91,** 377.

Grant, L., Martelli, A. B., and Dumont, A. E. (1974). Selective deposition of tantalum particles on injured vascular endothelium. *Microvasc. Res.* **7,** 376.

Grausz, H., Earley, L. E., Stephens, B. G., Lee, J. C., and Hopper, J. (1970). Diagnostic import of virus-like particles in the glomerular endothelium of patients with systemic lupus erythematosus. *New. Eng. J. Med.* **283,** 506.

Gregorius, F. K., and Rand, R. W. (1976). Scanning electron microscopic observations of the common carotid artery of the rat. III. Heparin effect on platelets. *Surg.* **79,** 584.

Griffiths, N. J., and Irving, M. H. (1976). Changes in the fine structure of venous endothelium in the rabbit following adrenalin infusion. *Cell Tissue Res.* **169,** 123.

Gutstein, W. H., and Park, F. (1973). Ultrastructural changes in coronary artery endothelium associated with biliary obstruction in the rat. *Am. J. Pathol.* **71,** 49.

Gutstein, W. H., Farrell, G. A., and Armellini, C. (1973). Blood flow disturbance and endothe lial cell injury in pre-atherosclerotic swine. *Lab. Invest.* **29,** 134.

Ham, K. N., and Hurley, J. V. (1965). Acute inflammation: an electron-microscope study of turpentine-induced pleurisy in the rat. *J. Pathol. Bacterol.* **90,** 365.

Harker, L. A., Ross, R., Slichter, S. J., and Scott, C. R. (1976). Homocystine-induced arteriosclerosis. *J. Clin. Invest.* **58,** 731.

Hashimoto, P. H., Takaesu, S., Chazono, M., and Amano, T. (1974). Vascular leakage through intra-endothelial channels induced by cholera toxin in the skin of guinea pigs. *Am. J. Pathol.* **75,** 171.

Haudenschild, C., and Studer, A. (1971). Early interactions between blood cells and severely damaged rabbit aorta. *Eur. J. Clin. Invest.* **2,** 1.

Hayes, J. A., and Shiga, A. (1970). Ultrastructural changes in pulmonary oedema produced experimentally with ammonium sulphate. *J. Pathol.* **100,** 281.

Herrlinger, H., Anzil, A. P., Blinzinger, K., and Kronski, D. (1974). Endothelial microtubular bodies in human brain capillaries and venules. *J. Anat.* **118,** 205.

Hertzer, N. R., Beven, E. G., and Benjamin, S. P. (1977). Ultramicroscopic ulcerations and thrombi of the carotid bifurcation. *Arch. Surg.* **112,** 1394.

Hirsch, E. Z., and Robertson, L., Jr. (1977). Selective acute arterial endothelial injury and repair. I. Methodology and surface characteristics. *Atherosclerosis* **28,** 271.

Hirschberg, H., Thorsby, E., and Rolstad, B. (1975). Antibody-induced cell-mediated damage to human endothelium cells in vitro. *Nature (London)* **255,** 62.

Hobbs, J. B. (1973). Platelets and the renal vascular endothelium. *Perspect Nephrol Hypertension* **1,** 907.

Huttner, I., and Rona, G. (1975). Contractile apparatus in aortic endothelium of hypertensive rat. *Rec. Adv. Stud. Carb. Struct. Metab.* **10,** 591.

Imparato, A. M., Baumann, F. G., Pearson, J., Kim, G. E., Davidson, T., Ibrahim, I., and Nathan, I. (1974a). Electron microscopic studies of experimentally produced fibromuscular arterial lesions. *Surg. Gyn. Obstet.* **139,** 1.

Imparato, A. M., Kim, G. E., Madayag, M., and Habeson, S. P. (1974b). The results of tibial artery reconstruction procedures. *Surg Gyn Obstet.* **138,** 33.

Irey, N. S., Manion, W. C., and Taylor, H. B. (1970). Vascular lesions in women taking oral contraceptives. *Arch. Pathol.* **89,** 1.

Iwasawa, Y., Gillis, C. N., and Aghajanian, G. (1973). Hypothermic inhibition of 5-hydroxytryptamine and norepinephrine uptake by lung: cellular location of amines after uptake. *J Pharmacol. Exp. Ther.* **186,** 498.

Jeppsson, R., and Schoefl, G. I. (1974). Electron microscopic observations on cerebral and pulmonary blood vessels after the intravenous injection of artificial lipid emulsions containing barbituric acids. *AJEBAK* **52,** 703.

Johnson, J. H., Baker, R. R., Wood, S., Jr. (1967). Effects of DMSO on blood and vascular endothelium. *Bibli. Anat.* **9,** 214.

Johnson, S. A. (1971). Endothelial supporting function of platelets. *In* "The Circulating Platelet," pp. 283–299. Academic Press, New York.

Johnston, W. H., and Latta, H. (1977). Glomerular mesangial and endothelial cell swelling following temporary renal ischemia and its role in the no-reflow phenomenon. *Am. J. Pathol.* **89,** 153.

Jones, D. B. (1970). The morphology of acid mucosubstances in leukocytic sticking to endothelium in acute inflammation. *Lab. Invest.* **23,** 606.

Jones, D. B. (1974). Arterial and glomerular lesions associated with severe hypertension. *Lab. Invest.* **31,** 303.

Kahn, R. A., Johnson, S. A., and DeGraff, A. F. (1971). Effects of sodium warfarin on capillary ultrastructure. *Am. J. Pathol.* **65,** 149.

Kawamura, J., Gertz, S. D., and Sunaga, T. (1974a). Scanning electron microscopic observations of the luminal surface of the rabbit common carotid artery subjected to ischemia by arterial occlusion. *Stroke* **5,** 765.

Kawamura, J., Kamijyo, Y., Sunaga, T., and Nelson, E. (1974b). Tubular bodies in vascular endothelium of a cerebellar neoplasm. *Lab. Invest.* **30,** 358.

Kistler, G. S., Caldwell, P. R. B., Weibel, E. R. (1967). Development of fine structural damage to alveolar and capillary lining cells in oxygen-poisoned rat lungs. *J. Cell Biol.* **32,** 605.

Kitchen, C. S., and Weiss, L. (1975). Ultrastructural changes of endothelium associated with thrombocytopenia. *Blood* **46,** 567.

Kjeldsen, K., and Thomsen, H. K. (1975). The effect of hypoxia on the fine structure of the aortic intima in rabbits. *Lab. Invest.* **33,** 533.

Kjeldsen, K., Astrup, P., and Wanstrup, J. (1972). Ultrastructure intimal changes in the rabbit aorta after a moderate carbon monoxide exposure. *Atherosclerosis* **16,** 67.

Klein, E., Farber, S., Djerassi, I., Tock, R., Freeman, G., and Arnold, P. (1956). The preparation and clinical administration of lyophilized platelet material to children with acute leukemia and aplastic anemia. *J. Pediatr.* **49**, 517.

Lackie, J. M., and deBono, D. (1977). Interactions of neutrophil granulocytes (PMNs) and endothelium in vitro. *Microvasc. Res.* **3**, 107.

Leak, L. V., and Kato, F. (1972). Electron microscopic studies of lymphatic capillaries during early inflammation. I. Mild and severe thermal injuries. *Lab. Invest.* **26**, 572.

Lewis, J. C., and Kottke, B. A. (1977). Endothelial damage and thrombocyte adhesion in pidgeon atherosclerosis. *Science* **196**, 1007.

Lindqvist, K. J., and Osterland, C. K. (1971). Human antibodies to vascular endothelium. *Clin. Exp. Immunol.* **9**, 753.

Little, J. H., Cooper, P., Sarwat, A., Waisman, J., and Fonkalsrud, E. W. (1973). Factors influencing endothelial injury and vascular thrombosis after perfusion. *J. Surg. Res.* **14**, 221.

Loskutoff, D. J., and Edgington, T. S. (1977). Synthesis of a fibrinolytic activator and inhibitor by endothelial cells. *Proc. Nat. Acad. Sci. USA* **74**, 3903.

Lough, J., and Moore, S. (1975). Endothelial injury induced by thrombin or thrombi. *Lab. Invest.* **33**, 130.

Lowenstein, W. R. (1975). Permeable junctions. *Cold Spring Harbor Symp. Quant. Biol.* **4**, 49.

Luft, J. H. (1973). Capillary permeability. I. Structural Considerations. *In* "The Inflammatory Process" (B. W. Zweifach, L. Grant, R. T. McCluskey, eds.), pp. 47–93. New York, Academic Press.

Maca, R. D., and Hoak, J. C. (1974). Endothelial injury and platelet aggregation associated with acute lipid mobilization. *Lab. Invest.* **30**, 589.

Macadam, R. F., Vetters, J. M., and Saikia, N. K. (1975). A search for microtubular inclusions in endothelial cells in a variety of skin diseases. *Br. J. Dermatol.* **92**, 175.

McKinney, R. V., Jr. (1976). The structure of scorbutic regenerating capillaries in skeletal muscle wounds. *Microvasc. Res.* **11**, 361.

McGrath, J. M., and Stewart, G. J. (1968). The effects of endotoxin on vascular endothelium. *J. Exp. Med.* **129**, 833.

Majno, G. (1965). Ultrastructure of the vascular membrane. *In* "Handbook of Physiology," Vol. 3, Section 2, pp. 2293–2375. Williams & Wilkins, Baltimore, Maryland.

Majno, G., Shea, S. M., and Leventhal, M. (1969). Endothelial contraction induced by histamine-type mediators. *J. Cell Biol.* **42**, 647.

Malczak, H. T., and Buck, R. C. (1977b). Regeneration of endothelium in rat aorta after local freezing. *Am. J. Pathol.* **86**, 133.

Manuelidis, E. E., and Thomas, L. (1973). Occlusion of brain capillaries by endothelial swelling in mycoplasma infections. *Proc. Nat. Acad. Sci. USA* **70**, 706.

Mason, R. G., Chuang, H. Y. K., Mohammad, S. F., and Sharp, D. E. (1977a). Endothelium: newly discovered functions and methods of study. *J. Bioenerg.* **1**, 3.

Mason, R. G., Sharp, D. E., Chuang, H. Y. K., and Mohammad, S. F. (1977b). Endothelial functions in prevention of thrombosis and the participation of damaged endothelium in thrombosis and hemostasis. *Arch. Pathol. Lab. Med.* **101**, 64.

Maynard, J. R., Burkholder, D. E., and Pizzuti, D. J. (1978). Comparative pharmacologic effects on tissue factor activity in normal cells and an established cell line. *Lab. Invest.* **38**, 14.

Meyrick, B., Miller, J., and Reid, L. (1972). Pulmonary oedema induced by antu, or by high or low oxygen concentrations in rat—an electron microscopic study. *Br. J. Exp. Pathol.* **53**, 347.

Mims, C. A., and Murphy, F. A. (1973). Para-influenza virus sendai infection in macrophages, ependyma, choroid plexus, vascular endothelium and respiratory tract of mice. *Am. J. Pathol.* **70**, 315.

Minick, C. R., Stemerman, M. B., and Insull, W., Jr. (1977a). Effect of regenerated endothelium on lipid accumulation in the arterial wall. *Proc. Nat. Acad. Sci. USA* **74**, 1724.

Minick, R. C., Stemerman, M. B., and Insull, W. (1977b). Effect of regenerating endothelium on lipid accumulation in the arterial wall. *Proc. Nat. Acad. Sci. USA* **74**, 1724.

Moraes, J. R., and Stastny, P. (1976). Eight groups of human endothelial cell allo-antigens. *Tissue Antigens* **8**, 273.

Moraes, J. R., and Stastny, P. (1977). A new antigen system expressed in human endothelial cells. *J. Clin. Invest.* **60**, 449.

Movat, H. Z., and Fernando, N. V. P. (1963). Acute inflammation. The earliest fine structural changes at the blood-tissue barrier. *Lab. Invest.* **12**, 895.

Nafstad, I. (1974). Endothelial damage and platelet thrombosis associated with pufa-rich, vitamin E deficient diet fed to pig. *Thromb. Res.* **5**, 251.

Nelson, E. (1976). Endothelial ischemia as studied by correlated scanning and transmission electron microscopy and by fluorescent antibody staining. *Triangle* **15**, 66.

Nelson, E., Sunaga, T., Shimamoto, T., Lawamura, J., Rennels, M. L., and Hebel, R. (1975). Ischemic carotid endothelium. *Arch. Pathol.* **99**, 125.

Nelson, E., Gertz, S. D., Rennels, M. L., Forbes, M. S., and Kawamura, J. (1976). Scanning and transmission electron microscopic studies of arterial endothelium following experimental vascular occlusion. *In* "The Cerebral Vessel Wall" (J. Cervos-Navarro *et al.*, eds.), pp. 33–39. Raven, New York.

Nopanitaya, W., Gambill, T. G., and Brinkhous, K. M. (1974). Fresh water drowning. *Arch. Pathol.* **98**, 361.

O'Connell, T. X., and Mowbray, J. F. (1973). Effects of humoral transplatation antibody on the arterial intima of rabbits. *Surgery* **74**, 145.

O'Connell, T. X., Sanchez, M., Mowbray, J. F., and Fonkalsrud, E. W. (1974). Effects on arterial intima of saline infusions. *J. Surg. Res.* **16**, 197.

O'Neal, R. M., Jordan, G. L., Rabin, E. R., DeBakey, M. E., and Halpert, B. (1964). Cells grown on isolated intravascular dacron hub. An electron microscopic study. *Exp. Mol. Pathol.* **3**, 403.

Ramos, J. R., Berger, K., Mansfield, P. B., and Sauvage, L. R. (1976). Histologic fate and endothelial changes of distended and nondistended vein grafts. *Ann. Surg.* **183**, 205.

Reidy, M. A., and Bowyer, D. E. (1977a). Scanning electron microscopy: morphology of aortic endothelium following injury by endotoxin and during subsequent repair. *Atherosclerosis* **26**, 319.

Reidy, M. A., and Bowyer, D. E. (1977b). Scanning electron microscopy of arteries. The morphology of aortic endothelium in haemodynamically stressed areas associated with branches. *Atherosclerosis* **26**, 181.

Reidy, M. A., and Bowyer, D. E. (1978). Scanning electron-microscope studies of the endothelium of aortic allografts in the rabit. Effect of azathioprine, prednisolone, and promethazine on early cellular invasion. *J. Pathol.* **124**, 1.

Reidy, M. A., and Levesque, M. J. (1977). A scanning electron microscopic study of arterial endothelial cells using vascular casts. *Atherosclerosis* **28**, 463.

Reinhold, H. S., and Buisman, G. H. (1973). Radiosensitivity of capillary endothelium. *Br. J. Radiol.* **46**, 54.

Reinhold, H. S., and Buisman, G. H. (1975). Repair of radiation damage to capillary endothelium. *Br. J. Radiol.* **48**, 727.

Renkin, E. M. (1978). Transport pathways through capillary endothelium. *Microvasc. Res.* **15**, 123.

Rhodin, J. A. G. (1968). Ultrastructure of mammalian venous capillaries, venules, and small collecting veins. *J. Ultrastruct. Res.* **25**, 452.

Robertson, A. L. and Khirallah, P. A. (1972). Effects of angiotensin II and some analogues on vascular permeability in the rabbit. *Circ. Res.* **41**, 923.

Robertson, A. L., Jr., and Khirallah, P. A. (1973). Arterial endothelial permeability and vascular disease. The "trap door" effect. *Exp. Mol. Pathol.* **18**, 241.

Robertson, A. L., Jr., and Rosen, L. A. (1976). The arterial endothelium: characteristics and function of the endothelial lining of larger arteries. *In* "Microcirculation" (G. Kaley and G. M. Altura, eds.), Vol. 1. pp. 145–165. Univ. Park Press, Baltimore, Maryland.

Rosen, L. A., Hollis, T. M., and Sharma, M. G. (1974). Alterations and molecular histidine decarboxylase activity following exposure to shearing stresses. *Exp. Mol. Pathol.* **20**, 329.

Ross, R., Everett, N. B., and Tyler, R. (1970). Wound healing and collagen formation. VI. The origin of the wound fibroblast studies in parabiosis. *J. Cell Biol.* **44**, 645.

Rossi, M. A., Ferreira, A. L., and Paiva, S. M. (1974). Fine structures of pulmonary changes induced by Brazilian scorpion venom. *Arch. Pathol.* **97**, 284.

Saba, S. R., and Mason, R. G. (1975). Effects of platelets and certain platelet components on growth of cultured human endothelial cells. *Thromb. Res.* **7**, 807.

Sacks, T., Moldow, C. F., Craddock, P. R., Bowers, T. K., and Jacob, H. S. (1978). Oxygen radicals mediate endothelial cell damage by complement-stimulated granulocytes. *J. Clin. Invest.* **61**, 1161.

Sauvage, L. R., Viggers, R. F., Berger, K., Robel, S. B., Sawyer, P. N., and Wood, S. J. (1972). "Prosthetic Replacement of the Aortic Valve" (L. R. Sauvage, ed.), pp. 81–119. Thomas, Springfield, Illinois.

Sauvage, L. R., Yates, S. G. II, Berger, K., Nakagawa, Y., and Wood, S. J. (1973). Prosthetic arteries and valves: thrombogenicity, healing and design. *In* "Coagulation", (G. Schmer and P. E. Strandjord, eds.), pp. 189–200. Academic Press, New York.

Schoefl, G. I. (1963). Studies on inflammation. III. Growing capillaries: Their structure and permeability. *Virchows Arch. A* **337**, 97.

Schwartz, S. M., and Benditt, E. P. (1973). Cell Replication in the aortic endothelium: A new method for study of the problem. *Lab. Invest.* **28**, 699.

Schwartz, S. M., and Benditt, E. P. (1976). Clustering of replicating cells in aortic endothelium. *Proc. Nat. Acad. Sci. USA* **73**, 651.

Schwartz, S. M., and Benditt, E. P. (1977). Aortic endothelial cell replication. I. Effects of age and hypertension in the rat. *Circ. Res.* **41**, 248.

Schwartz, S. M., Stemerman, M. B., and Benditt, E. P. (1975). The aortic intima. *Am. J. Pathol.* **81**, 15.

Schwartz, S. M., Haudenschild, C. C., and Eddy, E. M. (1978). Endothelial regeneration. I. Quantitative analysis of initial stages of endothelial regeneration in rat aortic intima. *Lab. Invest.* **38**, 568.

Shibata, S., Sakaguchi, H., and Nagasawa, T. (1978). Exfoliation of endothelial cytoplasm in nephrotoxic serum nephritis. *Lab. Invest.* **38**, 201.

Shimamoto, T. (1974). Injury and repair in arterial tissue. *Angiology* **25**, 682.

Shimamoto, T. (1975). Drugs and foods on contraction of endothelial cells as a key mechanism in atherogenesis and treatment of atherosclerosis with endothelial cell relaxants (cyclic AMP phosphodiesterase inhibitors). *Adv. Exp. Med. Biol.* **60**, 77.

Shimamoto, T., and Sunaga, T. (1976). The contraction and blebbing of endothelial cells accompanied by acute infiltration of plasma substances into the vessel wall and their prevention. *Triangle* **15**, 3.

Shimamoto, T., Yamashita, Y., Numano, F., and Sunaga, T. (1971). Scanning and transmission electron microscopic observation of endothelial cells in the normal condition and in initial stages of atherosclerosis. *Acta Pathol. Jap.* **21**, 93.

Shimamoto, T., Sagara, A., and Numano, F. (1974). Treatment of arterial thrombosis with endothelial cell relaxant. *Thromb. Diath. Haemorrh. Suppl.* **60**, 517.

Shimamoto, T., Hidaka, H., Moriya, K., Kobayashi, M., Takahashi, T., and Numano, F. (1976). Hyper-reactive arterial endothelial cells: A clue for the treatment of atherosclerosis. *Ann. N.Y. Acad. Sci.* **275**, 266.

Sholley, M. M., and Cotran, R. S. (1978). Endothelial proliferation in inflammation. II. Autoradiographic studies in X-irradiated leukopenic rats after thermal injury to the skin. *Am. J. Pathol.* **91**, 229.

Sholley, M. M., Gimbrone, M. A., Jr., and Cotran, R. S. (1977a). Cellular migration and replication in endothelial regeneration. A study using irradiated endothelial cultures. *Lab. Invest.* **36**, 18.

Sholley, M. M., Cavallo, T., and Cotran, R. S. (1977b). Endothelial proliferation in inflammation. *Am. J. Pathol.* **89**, 277.

Sholley, M. M., Gimbrone, M. A., and Cotran, R. S. (1977c). Cellular migration and replication in endothelial regeneration. *Lab. Invest.* **36**, 18.

Silkworth, J. B., McLean, B., and Stehbens, W. E. (1975). The effect of hypercholesterolemia on aortic endothelium studied EN FACE. *Atherosclerosis* **22**, 335.

Simionescu, M., Simionescu, N., and Palade, G. E. (1975a). Segmental differentiations of cell junctions in the vascular endothelium. *J. Cell Biol.* **67**, 863.

Simionescu, N., Simionescu, M., and Palade, G. E. (1975b). Permeability of muscle capillaries to small heme-peptides. *J. Cell Biol.* **64**, 586.

Simionescu, N., Simionescu, M., and Palade, G. E. (1978). Structural basis of permeability in sequential segments of the microvasculature of the diaphragm. *Microvasc. Res.* **15**, 17.

Spaet, T. H., and Gaynor, E. (1970). Vascular endothelial damage and thrombosis. *Adv. Cardiol.* **4**, 47.

Stehbens, W. E. (1965). Reaction of venous endothelium to injury. *Lab. Invest.* **14**, 449.

Stehbens, W. E., and Ludatscher, R. M. (1973). Ultrastructure of the renal arterial bifurcation of rabbits. *Exp. Mol. Pathol.* **18**, 50.

Stemerman, M. B. (1974). Platelet interaction with intimal connective tissue. *In* "Platelets: Production Function, Transfusion and Storage" (M. G. Baldini and S. Ebbe, eds.), pp. 157–170. Grune & Stratton, New York.

Stemerman, M. B., Spaet, T. H., Pitlick, F., Cintron, J., Lejnieks, I., and Tiell, M. L. (1977). Intimal healing. The pattern of re-endothelialization and intimal thickening. *Am. J. Pathol.* **87**, 125.

Stewart, G. J., Ritchie, W. G. M., and Lynch, P. R. (1974). Venous endothelial damage produced by massive sticking and emigration of leukocytes. *Am. J. Pathol.* **74**, 507.

Still, W. J. S., and Dennison, S. (1974). The arterial endothelium of the hypertensive rat. *Arch. Pathol.* **97**, 337.

Strum, J. M., and Junod, A. F. (1972). Radioautographic demonstration of 5-hydroxytryptamine-$^3$H uptake by pulmonary endothelial cells. *J. Cell Biol.* **54**, 456.

Sumiyoshi, A., and Tanaka, K. (1976). Endothelial damage and thrombosis: a scanning and transmission electron microscopic study. *Thromb. Res.* **8**, 277.

Sunaga, T., Yamashita, Y., and Shimamoto, T. (1969). Epinephrine effect on arterial endothelial cells observed by scanning electron microscope. *Proc. Jap. Acad.* **45**, 808.

Sunaga, T., Yamashita, Y., Numano, F., and Shimamoto, T. (1970). Luminal surface of normal and atherosclerotic arteries observed by scanning electron microscope. *Third Annu. Scanning Electron Microsc. Symp., Chicago, Ill. Inst. Tech., 1970*, 243–248.

Suzuki, A. O., and Ohashi, M. (1975). The spurting of erythrocytes through junctions of the vascular endothelium treated with snake venom. *Microvasc. Res.* **10**, 208.
Suzuki, K., Ookawara, S., and Ooneda, G. (1971a). Increased permeability of the arteries in hypertensive rats: an electron microscopic study. *Exp. Mol. Pathol.* **15**, 198.
Svendsen, E., and Jorgensen, L. (1978). Focal "spontaneous" alterations and loss of endothelial cells in rabbit aorta. *Acta Pathol. Microbiol. Scand.* **86**, 1.
Tazawa, Y., Mariscal, I., Moffat, C., Huebner, B., Seaman, A. J. (1971). The endothelial role in thrombosis. *Invest. Ophthal.* **10**, 481.
Tedder, E., and Shorey, C. D. (1965). Intimal changes in venous stasis. *Lab. Invest.* **14**, 208.
Teplitz, C. (1976). The core pathobiology and integrated medical science of adult acute respiratory insufficiency. *Surg. Clin. North Am.* **56**, 1091.
Vetto, R. M., and Burger, D.R. (1971). The identification and comparison of transplantation antigens on canine vascular endothelium and lymphocytes. *Transplantation* **11**, 374.
Vracko, R., and Benditt, E. P. (1970). Capillary basal lamina thickening. Its relationship to endothelial cell death and replacement. *J. Cell Biol.* **47**, 281.
Wagenvoort, C. A., Dingemans, K. P., and Lotgering, G. G. (1974). Electron microscopy of pulmonary vasculature after application of fulvine. *Thorax* **29**, 551.
Wang, N. S., Huang, S. N., Sheldon, H., and Thurlbeck, W. M. (1971). Ultrastructural changes of clara and type II alveolar cells in adrenalin-induced pulmonary edema in mice. *Am. J. Pathol.* **62**, 237.
Warren, B. A., and Brock, L. G. (1964). The electron microscopic features and fibrinolytic properties of "neo-intima". *Br. J. Exp. Pathol.* **65**, 612.
Warren, B. A., Philip, R. B., and Inwood, M. J. (1973). The ultrastructural morphology of air embolism: platelet adhesion to the interface and endothelial damage. *Br. J. Exp. Pathol.* **54**, 163.
Weber, G., Fabbrini, P., and Resi, L. (1973). On the presence of a concanavalin-A reactive coat over the endothelial aortic surface and its modifications during early experimental cholesterol atherogenesis in rabbits. *Virchows Arch. A* **359**, 299.
Weber, G., Fabbrini, P., and Resi, L. (1974). Scanning and transmission electron microscopy observations on the surface lining of aortic intimal plaques in rabbits on a hypercholesterolic diet. *Virchows Arch. A* **364**, 325.
Webster, W. S., Bishop, S. P., and Geer, J. C. (1974). Experimental aortic intimal thickening. II. Endothelialization and permeability. *Am. J. Pathol.* **76**, 265.
Weibel, E. R., and Palade, G. E. (1964). New cytoplasmic components in arterial endothelia. *J. Cell Biol.* **23**, 101.
Widemann, J., and Fahimi, H. D. (1976). Proliferation of endothelial cells in estrogen-stimulated rat liver. *Lab. Invest.* **34**, 141.
Wiener, J., and Giacomelli, F. (1973). The cellular pathology of experimental hypertension. *Am. J. Pathol.* **72**, 221.
Williams, G. M., and Harr, A. (1975). Krajewski, C., Parks, L. C., Roth J. Rejection and repair of endothelium in major vessel transplants. *Surgery* **78**, 694.
Willms-Kretschmer, D., Majno, G. (1969). Ischemia of the skin. *Am. J. Pathol.* **54**, 327.
Wisse, E. (1970). An electron microscopic study of the fenestrated endothelial lining of rat liver sinusoids. *J Ultrastruct. Res.* **31**, 125.
Wisse, E. (1972). An ultrastructural characterization of the endothelial cell in the rat liver under normal and various experimental conditions, as a contribution to the distinction between endothelial and kupffer cells. *J. Ultrastruct. Res.* **38**, 528.
Wright, H. P. (1972). Mitosis patterns in aortic endothelium. *Atherosclerosis* **15**, 93.
Wright, H. P. (1973). Endothelial injury and repair. *Bibl. Anat.* **12**, 87.

Wright, H. P., and Giacometti, N. J. (1972). Circulating endothelial cells and arterial endothelial mitosis in anaphylactic shock. *Br. J. Exp. Pathol.* **53**, 1.
Wyatt, A. P., and Taylor, G. W. (1966). Vein grafts: Changes in the endothelium of autogenous free vein grafts used as arterial replacements. *Br. J. Surg.* **53**, 943.
Yamagami, I. (1970). Electron microscopic study on the cornea. I. The mechanism of experimental new vessel formation. *Jap. J. Ophthal.* **14**, 41.
Yamaguchi, H., Torikata, C., Takeuchi, H., and Kageyama, K. (1973). Morphological changes of the endothelium in lung with administration of soluble immune complexes. *Acta Pathol. Jap.* **23**, 51.
Zweifach, B. W. (1973). Microvascular aspects of tissue injury, *In* "The Inflammatory Process" (B. W. Zweifach, L. Grant and R. T. McCluskey, eds.), pp. 3–46. Academic Press, New York.

# EDITORS' SUMMARY TO CHAPTER X

The vascular endothelium representing the initial barrier between the blood and the interstitium potentially represents the most important epithelial membrane barrier in the body. In this chapter the authors review a wide variety of pathological changes involving this important epithelial barrier. Many pathological conditions may relate to this barrier including inflammation, shock, anoxia, total ischemia, atherosclerosis, hypertention, and a variety of injuries including mechanical, thermal, and radiation, as well as chemical agents. The integrity of the endothelial barrier depends both on the function of individual cells themselves as well as on their interrelationships as exemplified by cell junctions. Agents within the vascular system can often modulate either the cell function or the function of the cell junctions and therefore modify endothelial barrier permeability. At the present time it is difficult to estimate the qualitative or quantitative significance of these, however, as discussed in the chapter by Ratliff, the initial damage to the endothelium appears to be important in the shock lung. Damage induced by leukocytes, and/or platelets may modulate the function of the endothelium in the atherosclerosis in ischemia, and transport of particles across the endothelial wall through junctions through basal lamina and through vesicular transport has been frequently studied. It is interesting that Leaf and coworkers in discussing the problem of reflow and cell swelling following parenchymal cell damage may reflect damage to the endothelium which may modify the reinstitution of flow following ischemic injury. This may be of particular importance in certain organ systems including the brain, the heart, and the kidney. Damage to the endothelium of the glomerulus may play a key role in acute glomerulonephritis and damage to the endothelium in the pulmonary circulation may play a key role in the pathophysiology of the shock lung. Increasingly, it appears that endotoxin-induced alterations in endothelial permeability in other functions may be vital in the pathophysiology of septic and endotoxic shock.

# INDEX

**R**